Student Solutions Manual for *Mathematical M* *and Engineering*, third editic

Mathematical Methods for Physics and Engineering, third edition, is a highly acclaimed undergraduate textbook that teaches all the mathematics needed for an undergraduate course in any of the physical sciences. As well as lucid descriptions of the topics and many worked examples, it contains over 800 exercises. New stand-alone chapters give a systematic account of the 'special functions' of physical science, cover an extended range of practical applications of complex variables, and give an introduction to quantum operators.

This solutions manual accompanies the third edition of *Mathematical Methods for Physics and Engineering*. It contains complete worked solutions to over 400 exercises in the main textbook, the odd-numbered exercises that are provided with hints and answers. The even-numbered exercises have no hints, answers or worked solutions and are intended for unaided homework problems; full solutions are available to instructors on a password-protected website, www.cambridge.org/9780521679718.

KEN RILEY read mathematics at the University of Cambridge and proceeded to a Ph.D. there in theoretical and experimental nuclear physics. He became a research associate in elementary particle physics at Brookhaven, and then, having taken up a lectureship at the Cavendish Laboratory, Cambridge, continued this research at the Rutherford Laboratory and Stanford; in particular he was involved in the experimental discovery of a number of the early baryonic resonances. As well as having been Senior Tutor at Clare College, where he has taught physics and mathematics for over 40 years, he has served on many committees concerned with the teaching and examining of these subjects at all levels of tertiary and undergraduate education. He is also one of the authors of *200 Puzzling Physics Problems*.

MICHAEL HOBSON read natural sciences at the University of Cambridge, specialising in theoretical physics, and remained at the Cavendish Laboratory to complete a Ph.D. in the physics of star-formation. As a research fellow at Trinity Hall, Cambridge and subsequently an advanced fellow of the Particle Physics and Astronomy Research Council, he developed an interest in cosmology, and in particular in the study of fluctuations in the cosmic microwave background. He was involved in the first detection of these fluctuations using a ground-based interferometer. He is currently a University Reader at the Cavendish Laboratory, his research interests include both theoretical and observational aspects of cosmology, and he is the principal author of *General Relativity: An Introduction for Physicists*. He is also a Director of Studies in Natural Sciences at Trinity Hall and enjoys an active role in the teaching of undergraduate physics and mathematics.

Student Solutions Manual for

Mathematical Methods for Physics and Engineering

Third Edition

K. F. RILEY and M. P. HOBSON

CAMBRIDGE
UNIVERSITY PRESS

CAMBRIDGE UNIVERSITY PRESS
Cambridge, New York, Melbourne, Madrid, Cape Town, Singapore, São Paulo

Cambridge University Press
The Edinburgh Building, Cambridge CB2 2RU, UK

Published in the United States of America by Cambridge University Press, New York

www.cambridge.org
Information on this title: www.cambridge.org/9780521861533

First published 2006

Printed in the United Kingdom at the University Press, Cambridge

A catalogue record for this publication is available from the British Library

Library of Congress Cataloguing in Publication data

ISBN-13   978-0-521-67973-2 paperback
ISBN-10   0-521-67973-7 paperback

Contents

Preface

The second edition of *Mathematical Methods for Physics and Engineering* carried more than twice as many exercises, based on its various chapters, as did the first. In the Preface we discussed the general question of how such exercises should be treated but, in the end, decided to provide hints and outline answers to all problems, as in the first edition. This decision was an uneasy one as, on the one hand, it did not allow the exercises to be set as totally unaided homework that could be used for assessment purposes, but, on the other, it did not give a full explanation of how to tackle a problem when a student needed explicit guidance or a model answer.

In order to allow both of these educationally desirable goals to be achieved, we have, in the third edition, completely changed the way this matter is handled. All of the exercises from the second edition, plus a number of additional ones testing the newly added material, have been included in penultimate subsections of the appropriate, sometimes reorganised, chapters. Hints and outline answers are given, as previously, in the final subsections, *but only to the odd-numbered exercises.* This leaves all even-numbered exercises free to be set as unaided homework, as described below.

For the four hundred plus *odd-numbered* exercises, complete solutions are available, to both students and their teachers, in the form of *this* manual; these are in addition to the hints and outline answers given in the main text. For each exercise, the original question is reproduced and then followed by a fully worked solution. For those original exercises that make internal reference to the text or to other (even-numbered) exercises not included in this solutions manual, the questions have been reworded, usually by including additional information, so that the questions can stand alone. Some further minor rewording has been included to improve the page layout.

In many cases the solution given is even fuller than one that might be expected of a good student who has understood the material. This is because we have aimed to make the solutions instructional as well as utilitarian. To this end, we have included comments that are intended to show how the plan for the solution is formulated and have provided the justifications for particular intermediate steps (something not always done, even by the best of students). We have also tried to write each individual substituted formula in the form that best indicates how it was obtained, before simplifying it at the next or a subsequent stage. Where several lines of algebraic manipulation or calculus are needed to obtain a final result, they are normally included in full; this should enable the student to determine whether an incorrect answer is due to a misunderstanding of principles or to a technical error.

The remaining four hundred or so *even-numbered* exercises have no hints or answers (outlined or detailed) available for general access. They can therefore be used by instructors as a basis for setting unaided homework. Full solutions to these exercises, in the same general format as those appearing in this manual (though they may contain references to the main text or to other exercises), are available without charge to accredited teachers as downloadable pdf files on the password-protected website http://www.cambridge.org/9780521679718. Teachers wishing to have access to the website should contact solutions@cambridge.org for registration details.

As noted above, the original questions are reproduced in full, or in a suitably modified stand-alone form, at the start of each exercise. Reference to the main text is not needed provided that standard formulae are known (and a set of tables is available for a few of the statistical and numerical exercises). This means that, although it is not its prime purpose, this manual could be used as a test or quiz book by a student who has learned, or thinks that he or she has learned, the material covered in the main text.

In all new publications, errors and typographical mistakes are virtually unavoidable, and we would be grateful to any reader who brings instances to our attention. Finally, we are extremely grateful to Dave Green for his considerable and continuing advice concerning typesetting in LaTeX.

Ken Riley, Michael Hobson,
Cambridge, 2006

Preliminary algebra

Polynomial equations

1.1 *It can be shown that the polynomial*

$$g(x) = 4x^3 + 3x^2 - 6x - 1$$

has turning points at $x = -1$ *and* $x = \frac{1}{2}$ *and three real roots altogether. Continue an investigation of its properties as follows.*

(a) *Make a table of values of* $g(x)$ *for integer values of* x *between* -2 *and* 2. *Use it and the information given above to draw a graph and so determine the roots of* $g(x) = 0$ *as accurately as possible.*

(b) *Find one accurate root of* $g(x) = 0$ *by inspection and hence determine precise values for the other two roots.*

(c) *Show that* $f(x) = 4x^3 + 3x^2 - 6x - k = 0$ *has only one real root unless* $-5 \le k \le \frac{7}{4}$.

(a) Straightforward evaluation of $g(x)$ at integer values of x gives the following table:

x	-2	-1	0	1	2
$g(x)$	-9	4	-1	0	31

(b) It is apparent from the table alone that $x = 1$ is an exact root of $g(x) = 0$ and so $g(x)$ can be factorised as $g(x) = (x-1)h(x) = (x-1)(b_2 x^2 + b_1 x + b_0)$. Equating the coefficients of x^3, x^2, x and the constant term gives $4 = b_2$, $b_1 - b_2 = 3$, $b_0 - b_1 = -6$ and $-b_0 = -1$, respectively, which are consistent if $b_1 = 7$. To find the two remaining roots we set $h(x) = 0$:

$$4x^2 + 7x + 1 = 0.$$

The roots of this quadratic equation are given by the standard formula as

$$\alpha_{1,2} = \frac{-7 \pm \sqrt{49 - 16}}{8}.$$

(c) When $k = 1$ (i.e. the original equation) the values of $g(x)$ at its turning points, $x = -1$ and $x = \frac{1}{2}$, are 4 and $-\frac{11}{4}$, respectively. Thus $g(x)$ can have up to 4 subtracted from it or up to $\frac{11}{4}$ added to it and still satisfy the condition for three (or, at the limit, two) distinct roots of $g(x) = 0$. It follows that for k outside the range $-5 \le k \le \frac{7}{4}$, $f(x)$ [$= g(x) + 1 - k$] has only one real root.

1.3 *Investigate the properties of the polynomial equation*

$$f(x) = x^7 + 5x^6 + x^4 - x^3 + x^2 - 2 = 0,$$

by proceeding as follows.

(a) *By writing the fifth-degree polynomial appearing in the expression for $f'(x)$ in the form $7x^5 + 30x^4 + a(x - b)^2 + c$, show that there is in fact only one positive root of $f(x) = 0$.*

(b) *By evaluating $f(1)$, $f(0)$ and $f(-1)$, and by inspecting the form of $f(x)$ for negative values of x, determine what you can about the positions of the real roots of $f(x) = 0$.*

(a) We start by finding the derivative of $f(x)$ and note that, because f contains no linear term, f' can be written as the product of x and a fifth-degree polynomial:

$$f(x) = x^7 + 5x^6 + x^4 - x^3 + x^2 - 2 = 0,$$
$$f'(x) = x(7x^5 + 30x^4 + 4x^2 - 3x + 2)$$
$$= x[7x^5 + 30x^4 + 4(x - \tfrac{3}{8})^2 - 4(\tfrac{3}{8})^2 + 2]$$
$$= x[7x^5 + 30x^4 + 4(x - \tfrac{3}{8})^2 + \tfrac{23}{16}].$$

Since, for positive x, every term in this last expression is necessarily positive, it follows that $f'(x)$ can have no zeros in the range $0 < x < \infty$. Consequently, $f(x)$ can have no turning points in that range and $f(x) = 0$ can have at most one root in the same range. However, $f(+\infty) = +\infty$ and $f(0) = -2 < 0$ and so $f(x) = 0$ has at least one root in $0 < x < \infty$. Consequently it has exactly one root in the range.

(b) $f(1) = 5$, $f(0) = -2$ and $f(-1) = 5$, and so there is at least one root in each of the ranges $0 < x < 1$ and $-1 < x < 0$.

There is no simple systematic way to examine the form of a general polynomial function for the purpose of determining where its zeros lie, but it is sometimes helpful to group terms in the polynomial and determine how the sign of each group depends upon the range in which x lies. Here grouping successive pairs of terms yields some information as follows:

$$x^7 + 5x^6 \text{ is positive for } x > -5,$$
$$x^4 - x^3 \text{ is positive for } x > 1 \quad \text{and} \quad x < 0,$$
$$x^2 - 2 \text{ is positive for } x > \sqrt{2} \quad \text{and} \quad x < -\sqrt{2}.$$

Thus, all three terms are positive in the range(s) common to these, namely $-5 < x < -\sqrt{2}$ and $x > 1$. It follows that $f(x)$ is positive definite in these ranges and there can be no roots of $f(x) = 0$ within them. However, since $f(x)$ is negative for large negative x, there must be at least one root α with $\alpha < -5$.

1.5 *Construct the quadratic equations that have the following pairs of roots:*
(a) $-6, -3$; (b) $0, 4$; (c) $2, 2$; (d) $3 + 2i, 3 - 2i$, where $i^2 = -1$.

Starting in each case from the 'product of factors' form of the quadratic equation, $(x - \alpha_1)(x - \alpha_2) = 0$, we obtain:

(a) $\qquad (x + 6)(x + 3) = x^2 + 9x + 18 = 0;$

(b) $\qquad (x - 0)(x - 4) = x^2 - 4x = 0;$

(c) $\qquad (x - 2)(x - 2) = x^2 - 4x + 4 = 0;$

(d) $\quad (x - 3 - 2i)(x - 3 + 2i) = x^2 + x(-3 - 2i - 3 + 2i)$
$$+ (9 - 6i + 6i - 4i^2)$$
$$= x^2 - 6x + 13 = 0.$$

Trigonometric identities

1.7 *Prove that*

$$\cos \frac{\pi}{12} = \frac{\sqrt{3} + 1}{2\sqrt{2}}$$

by considering

(a) *the sum of the sines of $\pi/3$ and $\pi/6$,*
(b) *the sine of the sum of $\pi/3$ and $\pi/4$.*

(a) Using

$$\sin A + \sin B = 2 \sin \left(\frac{A + B}{2} \right) \cos \left(\frac{A - B}{2} \right),$$

we have

$$\sin\frac{\pi}{3} + \sin\frac{\pi}{6} = 2\sin\frac{\pi}{4}\cos\frac{\pi}{12},$$

$$\frac{\sqrt{3}}{2} + \frac{1}{2} = 2\frac{1}{\sqrt{2}}\cos\frac{\pi}{12},$$

$$\cos\frac{\pi}{12} = \frac{\sqrt{3}+1}{2\sqrt{2}}.$$

(b) Using, successively, the identities

$$\sin(A+B) = \sin A\cos B + \cos A\sin B,$$
$$\sin(\pi - \theta) = \sin\theta$$
$$\text{and } \cos(\tfrac{1}{2}\pi - \theta) = \sin\theta,$$

we obtain

$$\sin\left(\frac{\pi}{3} + \frac{\pi}{4}\right) = \sin\frac{\pi}{3}\cos\frac{\pi}{4} + \cos\frac{\pi}{3}\sin\frac{\pi}{4},$$

$$\sin\frac{7\pi}{12} = \frac{\sqrt{3}}{2}\frac{1}{\sqrt{2}} + \frac{1}{2}\frac{1}{\sqrt{2}},$$

$$\sin\frac{5\pi}{12} = \frac{\sqrt{3}+1}{2\sqrt{2}},$$

$$\cos\frac{\pi}{12} = \frac{\sqrt{3}+1}{2\sqrt{2}}.$$

1.9 *Find the real solutions of*

(a) $3\sin\theta - 4\cos\theta = 2$,
(b) $4\sin\theta + 3\cos\theta = 6$,
(c) $12\sin\theta - 5\cos\theta = -6$.

We use the result that if

$$a\sin\theta + b\cos\theta = k$$

then

$$\theta = \sin^{-1}\left(\frac{k}{K}\right) - \phi,$$

where

$$K^2 = a^2 + b^2 \quad \text{and} \quad \phi = \tan^{-1}\frac{b}{a}.$$

4

Recalling that the inverse sine yields two values and that the individual signs of a and b have to be taken into account, we have

(a) $k = 2$, $K = \sqrt{3^2 + 4^2} = 5$, $\phi = \tan^{-1}(-4/3)$ and so

$$\theta = \sin^{-1} \tfrac{2}{5} - \tan^{-1} \tfrac{-4}{3} = 1.339 \text{ or } -2.626.$$

(b) $k = 6$, $K = \sqrt{4^2 + 3^2} = 5$. Since $k > K$ there is no solution for a real angle θ.

(c) $k = -6$, $K = \sqrt{12^2 + 5^2} = 13$, $\phi = \tan^{-1}(-5/12)$ and so

$$\theta = \sin^{-1} \tfrac{-6}{13} - \tan^{-1} \tfrac{-5}{12} = -0.0849 \text{ or } -2.267.$$

1.11 *Find all the solutions of*

$$\sin \theta + \sin 4\theta = \sin 2\theta + \sin 3\theta$$

that lie in the range $-\pi < \theta \leq \pi$. *What is the multiplicity of the solution* $\theta = 0$?

Using

$$\sin(A + B) = \sin A \cos B + \cos A \sin B,$$

$$\text{and} \quad \cos A - \cos B = -2 \sin \left(\frac{A + B}{2} \right) \sin \left(\frac{A - B}{2} \right),$$

and recalling that $\cos(-\phi) = \cos(\phi)$, the equation can be written successively as

$$2 \sin \frac{5\theta}{2} \cos \left(-\frac{3\theta}{2} \right) = 2 \sin \frac{5\theta}{2} \cos \left(-\frac{\theta}{2} \right),$$

$$\sin \frac{5\theta}{2} \left(\cos \frac{3\theta}{2} - \cos \frac{\theta}{2} \right) = 0,$$

$$-2 \sin \frac{5\theta}{2} \sin \theta \sin \frac{\theta}{2} = 0.$$

The first factor gives solutions for θ of $-4\pi/5$, $-2\pi/5$, 0, $2\pi/5$ and $4\pi/5$. The second factor gives rise to solutions 0 and π, whilst the only value making the third factor zero is $\theta = 0$. The solution $\theta = 0$ appears in each of the above sets and so has multiplicity 3.

Coordinate geometry

1.13 *Determine the forms of the conic sections described by the following equations:*

(a) $x^2 + y^2 + 6x + 8y = 0$;
(b) $9x^2 - 4y^2 - 54x - 16y + 29 = 0$;
(c) $2x^2 + 2y^2 + 5xy - 4x + y - 6 = 0$;
(d) $x^2 + y^2 + 2xy - 8x + 8y = 0$.

(a) $x^2 + y^2 + 6x + 8y = 0$. The coefficients of x^2 and y^2 are equal and there is no xy term; it follows that this must represent a circle. Rewriting the equation in standard circle form by 'completing the squares' in the terms that involve x and y, each variable treated separately, we obtain

$$(x + 3)^2 + (y + 4)^2 - (3^2 + 4^2) = 0.$$

The equation is therefore that of a circle of radius $\sqrt{3^2 + 4^2} = 5$ centred on $(-3, -4)$.

(b) $9x^2 - 4y^2 - 54x - 16y + 29 = 0$. This equation contains no xy term and so the centre of the curve will be at $(54/(2 \times 9), 16/[2 \times (-4)]) = (3, -2)$, and in standardised form the equation is

$$9(x - 3)^2 - 4(y + 2)^2 + 29 - 81 + 16 = 0,$$

or

$$\frac{(x - 3)^2}{4} - \frac{(y + 2)^2}{9} = 1.$$

The minus sign between the terms on the LHS implies that this conic section is a hyperbola with asymptotes (the form for large x and y and obtained by ignoring the constant on the RHS) given by $3(x - 3) = \pm 2(y + 2)$, i.e. lines of slope $\pm\frac{3}{2}$ passing through its 'centre' at $(3, -2)$.

(c) $2x^2 + 2y^2 + 5xy - 4x + y - 6 = 0$. As an xy term is present the equation cannot represent an ellipse or hyperbola in standard form. Whether it represents two straight lines can be most easily investigated by taking the lines in the form $a_i x + b_i y + 1 = 0$, $(i = 1, 2)$ and comparing the product $(a_1 x + b_1 y + 1)(a_2 x + b_2 y + 1)$ with $-\frac{1}{6}(2x^2 + 2y^2 + 5xy - 4x + y - 6)$. The comparison produces five equations which the four constants a_i, b_i, $(i = 1, 2)$ must satisfy:

$$a_1 a_2 = \frac{2}{-6}, \quad b_1 b_2 = \frac{2}{-6}, \quad a_1 + a_2 = \frac{-4}{-6}, \quad b_1 + b_2 = \frac{1}{-6}$$

and

$$a_1 b_2 + b_1 a_2 = \frac{5}{-6}.$$

Combining the first and third equations gives $3a_1^2 - 2a_1 - 1 = 0$ leading to a_1 and a_2 having the values 1 and $-\frac{1}{3}$, in either order. Similarly, combining the second and fourth equations gives $6b_1^2 + b_1 - 2 = 0$ leading to b_1 and b_2 having the values $\frac{1}{2}$ and $-\frac{2}{3}$, again in either order.

Either of the two combinations $(a_1 = -\frac{1}{3}, b_1 = -\frac{2}{3}, a_2 = 1, b_2 = \frac{1}{2})$ and $(a_1 = 1, b_1 = \frac{1}{2}, a_2 = -\frac{1}{3}, b_2 = -\frac{2}{3})$ also satisfies the fifth equation [note that the two alternative pairings do not do so]. That a consistent set can be found shows that the equation does indeed represent a pair of straight lines, $x + 2y - 3 = 0$ and $2x + y + 2 = 0$.

(d) $x^2 + y^2 + 2xy - 8x + 8y = 0$. We note that the first three terms can be written as a perfect square and so the equation can be rewritten as

$$(x + y)^2 = 8(x - y).$$

The two lines given by $x + y = 0$ and $x - y = 0$ are orthogonal and so the equation is of the form $u^2 = 4av$, which, for Cartesian coordinates u, v, represents a parabola passing through the origin, symmetric about the v-axis ($u = 0$) and defined for $v \geq 0$. Thus the original equation is that of a parabola, symmetric about the line $x + y = 0$, passing through the origin and defined in the region $x \geq y$.

Partial fractions

1.15 *Resolve*

$$\text{(a)} \ \frac{2x + 1}{x^2 + 3x - 10}, \qquad \text{(b)} \ \frac{4}{x^2 - 3x}$$

into partial fractions using each of the following three methods:

   (i) *Expressing the supposed expansion in a form in which all terms have the same denominator and then equating coefficients of the various powers of x.*

   (ii) *Substituting specific numerical values for x and solving the resulting simultaneous equations.*

   (iii) *Evaluation of the fraction at each of the roots of its denominator, imagining a factored denominator with the factor corresponding to the root omitted – often known as the 'cover-up' method.*

Verify that the decomposition obtained is independent of the method used.

(a) As the denominator factorises as $(x + 5)(x - 2)$, the partial fraction expansion must have the form

$$\frac{2x + 1}{x^2 + 3x - 10} = \frac{A}{x + 5} + \frac{B}{x - 2}.$$

(i).

$$\frac{A}{x+5} + \frac{B}{x-2} = \frac{x(A+B)+(5B-2A)}{(x+5)(x-2)}.$$

Solving $A + B = 2$ and $-2A + 5B = 1$ gives $A = \frac{9}{7}$ and $B = \frac{5}{7}$.

(ii) Setting x equal to 0 and 1, say, gives the pair of equations

$$\frac{1}{-10} = \frac{A}{5} + \frac{B}{-2}; \quad \frac{3}{-6} = \frac{A}{6} + \frac{B}{-1},$$

$$-1 = 2A - 5B; \quad -3 = A - 6B,$$

with solution $A = \frac{9}{7}$ and $B = \frac{5}{7}$.

(iii)

$$A = \frac{2(-5)+1}{-5-2} = \frac{9}{7}; \quad B = \frac{2(2)+1}{2+5} = \frac{5}{7}.$$

All three methods give the same decomposition.

(b) Here the factorisation of the denominator is simply $x(x-3)$ or, more formally, $(x-0)(x-3)$, and the expansion takes the form

$$\frac{4}{x^2-3x} = \frac{A}{x} + \frac{B}{x-3}.$$

(i)

$$\frac{A}{x} + \frac{B}{x-3} = \frac{x(A+B)-3A}{(x-0)(x-3)}.$$

Solving $A + B = 0$ and $-3A = 4$ gives $A = -\frac{4}{3}$ and $B = \frac{4}{3}$.

(ii) Setting x equal to 1 and 2, say, gives the pair of equations

$$\frac{4}{-2} = \frac{A}{1} + \frac{B}{-2}; \quad \frac{4}{-2} = \frac{A}{2} + \frac{B}{-1},$$

$$-4 = 2A - B; \quad -4 = A - 2B,$$

with solution $A = -\frac{4}{3}$ and $B = \frac{4}{3}$.

(iii)

$$A = \frac{4}{0-3} = -\frac{4}{3}; \quad B = \frac{4}{3-0} = \frac{4}{3}.$$

Again, all three methods give the same decomposition.

8

1.17 *Rearrange the following functions in partial fraction form:*

(a) $\dfrac{x-6}{x^3-x^2+4x-4}$, (b) $\dfrac{x^3+3x^2+x+19}{x^4+10x^2+9}$.

(a) For the function

$$f(x) = \frac{x-6}{x^3-x^2+4x-4} = \frac{g(x)}{h(x)}$$

the first task is to factorise the denominator. By inspection, $h(1) = 0$ and so $x-1$ is a factor of the denominator.

Write

$$x^3 - x^2 + 4x - 4 = (x-1)(x^2 + b_1 x + b_0).$$

Equating coefficients: $-1 = b_1 - 1$, $4 = -b_1 + b_0$ and $-4 = -b_0$, giving $b_1 = 0$ and $b_0 = 4$. Thus,

$$f(x) = \frac{x-6}{(x-1)(x^2+4)}.$$

The factor $x^2 + 4$ cannot be factorised further without using complex numbers and so we include a term with this factor as the denominator, but 'at the price of' having a linear term, and not just a number, in the numerator.

$$f(x) = \frac{A}{x-1} + \frac{Bx+C}{x^2+4}$$
$$= \frac{Ax^2 + 4A + Bx^2 + Cx - Bx - C}{(x-1)(x^2+4)}.$$

Comparing the coefficients of the various powers of x in this numerator with those in the numerator of the original expression gives $A + B = 0$, $C - B = 1$ and $4A - C = -6$, which in turn yield $A = -1$, $B = 1$ and $C = 2$. Thus,

$$f(x) = -\frac{1}{x-1} + \frac{x+2}{x^2+4}.$$

(b) By inspection, the denominator of

$$\frac{x^3+3x^2+x+19}{x^4+10x^2+9}$$

factorises simply into $(x^2+9)(x^2+1)$, but neither factor can be broken down further. Thus, as in (a), we write

$$f(x) = \frac{Ax+B}{x^2+9} + \frac{Cx+D}{x^2+1}$$
$$= \frac{(A+C)x^3 + (B+D)x^2 + (A+9C)x + (B+9D)}{(x^2+9)(x^2+1)}.$$

Equating coefficients gives

$$A + C = 1,$$
$$B + D = 3,$$
$$A + 9C = 1,$$
$$B + 9D = 19.$$

From the first and third equations, $A = 1$ and $C = 0$. The second and fourth yield $B = 1$ and $D = 2$. Thus

$$f(x) = \frac{x+1}{x^2+9} + \frac{2}{x^2+1}.$$

Binomial expansion

1.19 *Evaluate those of the following that are defined:* (a) $^5C_3$, (b) $^3C_5$, (c) $^{-5}C_3$, (d) $^{-3}C_5$.

(a) $^5C_3 = \frac{5!}{3!\,2!} = 10$.

(b) $^3C_5$. This is not defined as $5 > 3 > 0$.

For (c) and (d) we will need to use the identity

$$^{-m}C_k = (-1)^k \frac{m(m+1)\cdots(m+k-1)}{k!} = (-1)^k \,^{m+k-1}C_k.$$

(c) $^{-5}C_3 = (-1)^3 \,^{5+3-1}C_3 = -\frac{7!}{3!\,4!} = -35$.

(d) $^{-3}C_5 = (-1)^5 \,^{5+3-1}C_5 = -\frac{7!}{5!\,2!} = -21$.

Proof by induction and contradiction

1.21 *Prove by induction that*

$$\sum_{r=1}^{n} r = \tfrac{1}{2}n(n+1) \qquad \text{and} \qquad \sum_{r=1}^{n} r^3 = \tfrac{1}{4}n^2(n+1)^2.$$

To prove that

$$\sum_{r=1}^{n} r = \tfrac{1}{2}n(n+1),$$

assume that the result is valid for $n = N$ and consider

$$\sum_{r=1}^{N+1} r = \sum_{r=1}^{N} r + (N+1)$$
$$= \tfrac{1}{2}N(N+1) + (N+1), \quad \text{using the assumption,}$$
$$= (N+1)(\tfrac{1}{2}N+1)$$
$$= \tfrac{1}{2}(N+1)(N+2).$$

This is the same form as in the assumption except that N has been replaced by $N+1$; this shows that the result is valid for $n = N+1$ if it is valid for $n = N$. But the assumed result is trivially valid for $n = 1$ and is therefore valid for all n.

To prove that

$$\sum_{r=1}^{n} r^3 = \tfrac{1}{4}n^2(n+1)^2,$$

assume that the result is valid for $n = N$ and consider

$$\sum_{r=1}^{N+1} r^3 = \sum_{r=1}^{N} r^3 + (N+1)^3$$
$$= \tfrac{1}{4}N^2(N+1)^2 + (N+1)^3, \quad \text{using the assumption,}$$
$$= \tfrac{1}{4}(N+1)^2[N^2 + 4(N+1)]$$
$$= \tfrac{1}{4}(N+1)^2(N+2)^2.$$

This is the same form as in the assumption except that N has been replaced by $N+1$ and shows that the result is valid for $n = N+1$ if it is valid for $n = N$. But the assumed result is trivially valid for $n = 1$ and is therefore valid for all n.

1.23 *Prove that $3^{2n} + 7$, where n is a non-negative integer, is divisible by 8.*

As usual, we assume that the result is valid for $n = N$ and consider the expression with N replaced by $N + 1$:

$$3^{2(N+1)} + 7 = 3^{2N+2} + 7 + 3^{2N} - 3^{2N}$$
$$= (3^{2N} + 7) + 3^{2N}(9 - 1).$$

By the assumption, the first term on the RHS is divisible by 8; the second is clearly so. Thus $3^{2(N+1)} + 7$ is divisible by 8. This shows that the result is valid for $n = N+1$ if it is valid for $n = N$. But the assumed result is trivially valid for $n = 0$ and is therefore valid for all n.

1.25 *Prove by induction that*

$$\sum_{r=1}^{n} \frac{1}{2^r} \tan\left(\frac{\theta}{2^r}\right) = \frac{1}{2^n} \cot\left(\frac{\theta}{2^n}\right) - \cot\theta. \qquad (*)$$

Assume that the result is valid for $n = N$ and consider

$$\sum_{r=1}^{N+1} \frac{1}{2^r} \tan\left(\frac{\theta}{2^r}\right) = \frac{1}{2^N} \cot\left(\frac{\theta}{2^N}\right) - \cot\theta + \frac{1}{2^{N+1}} \tan\left(\frac{\theta}{2^{N+1}}\right).$$

Using the half-angle formula

$$\tan\phi = \frac{2r}{1-r^2}, \qquad \text{where } r = \tan\tfrac{1}{2}\phi,$$

to write $\cot(\theta/2^N)$ in terms of $t = \tan(\theta/2^{N+1})$, we have that the RHS is

$$\frac{1}{2^N}\left(\frac{1-t^2}{2t}\right) - \cot\theta + \frac{1}{2^{N+1}}t = \frac{1}{2^{N+1}}\left(\frac{1-t^2+t^2}{t}\right) - \cot\theta$$

$$= \frac{1}{2^{N+1}}\cot\left(\frac{\theta}{2^{N+1}}\right) - \cot\theta.$$

This is the same form as in the assumption except that N has been replaced by $N + 1$ and shows that the result is valid for $n = N + 1$ if it is valid for $n = N$.

But, for $n = 1$, the LHS of $(*)$ is $\frac{1}{2}\tan(\theta/2)$. The RHS can be written in terms of $s = \tan(\theta/2)$:

$$\frac{1}{2}\cot\left(\frac{\theta}{2}\right) - \cot\theta = \frac{1}{2s} - \frac{1-s^2}{2s} = \frac{s}{2},$$

i.e. the same as the LHS. Thus the result is valid for $n = 1$ and hence for all n.

1.27 *Establish the values of k for which the binomial coefficient ^pC_k is divisible by p when p is a prime number. Use your result and the method of induction to prove that $n^p - n$ is divisible by p for all integers n and all prime numbers p. Deduce that $n^5 - n$ is divisible by 30 for any integer n.*

Since

$$^pC_k = \frac{p!}{k!(p-k)!},$$

its numerator will always contain a factor p. Therefore, the fraction will be divisible by p unless the denominator happens to contain a (cancelling) factor of p. Since p is prime, this latter factor cannot arise from the product of two or more terms in the denominator; nor can p have any factor that cancels with a term in the denominator. Thus, for cancellation to occur, either $k!$ or $(p - k)!$ must contain a term p; this can only happen for $k = p$ or $k = 0$; for all other values of k, ^pC_k will be divisible by p.

Assume that $n^p - n$ is divisible by prime number p for $n = N$. Clearly this is true for $N = 1$ and any p. Now, using the binomial expansion of $(N + 1)^p$, consider

$$(N + 1)^p - (N + 1) = \sum_{k=0}^{p} {}^pC_k N^k - (N + 1)$$

$$= 1 + \sum_{k=1}^{p-1} {}^pC_k N^k + N^p - N - 1.$$

But, as shown above, ^pC_k is divisible by p for all k in the range $1 \le k \le p - 1$, and $N^p - N$ is divisible by p, by assumption. Thus $(N + 1)^p - (N + 1)$ is divisible by p if it is true that $N^p - N$ is divisible by p. Taking $N = 1$, for which, as noted above, the assumption is valid by inspection for any p, the result follows for all positive integers n and all primes p.

Now consider $f(n) = n^5 - n$. By the result just proved $f(n)$ is divisible by (prime number) 5. Further, $f(n) = n(n^4 - 1) = n(n^2 - 1)(n^2 + 1) = n(n - 1)(n + 1)(n^2 + 1)$. Thus the factorisation of $f(n)$ contains three consecutive integers; one of them must be divisible by 3 and at least one must be even and hence divisible by 2. Thus, $f(n)$ has the prime numbers 2, 3 and 5 as its divisors and must therefore be divisible by 30.

1.29 *Prove, by the method of contradiction, that the equation*

$$x^n + a_{n-1}x^{n-1} + \cdots + a_1 x + a_0 = 0,$$

in which all the coefficients a_i are integers, cannot have a rational root, unless that root is an integer. Deduce that any integral root must be a divisor of a_0 and hence find all rational roots of

(a) $x^4 + 6x^3 + 4x^2 + 5x + 4 = 0$,
(b) $x^4 + 5x^3 + 2x^2 - 10x + 6 = 0$.

Suppose that the equation has a rational root $x = p/q$, where integers p and q have no common factor and q is neither 0 nor 1. Then substituting the root and multiplying the resulting equation by q^{n-1} gives

$$\frac{p^n}{q} + a_{n-1}p^{n-1} + \cdots + a_1 pq^{n-2} + a_0 q^{n-1} = 0.$$

But the first term of this equation is not an integer (since p and q have no factor in common) whilst each of the remaining terms is a product of integers and is therefore an integer. Thus we have an integer equal to (minus) a non-integer. This is a contradiction and shows that it was wrong to suppose that the original equation has a rational non-integer root.

From the general properties of polynomial equations we have that the product of the roots of the equation $\sum_{i=0}^{n} b_i x^i = 0$ is $(-1)^n b_0/b_n$. For our original equation, $b_n = 1$ and $b_0 = a_0$. Consequently, the product of its roots is equal to the integral value $(-1)^n a_0$. Since there are no non-integral rational roots it follows that any integral root must be a divisor of a_0.

(a) $x^4 + 6x^3 + 4x^2 + 5x + 4 = 0$. This equation has integer coefficients and a leading coefficient equal to unity. We can thus apply the above result, which shows that its only possible rational roots are the six integers ± 1, ± 2 and ± 4. Of these, all positive values are impossible (since then every term would be positive) and trial and error will show that none of the negative values is a root either.

(b) $x^4 + 5x^3 + 2x^2 - 10x + 6 = 0$. In the same way as above, we deduce that for this equation the only possible rational roots are the eight values ± 1, ± 2, ± 3 and ± 6. Substituting each in turn shows that only $x = -3$ satisfies the equation.

Necessary and sufficient conditions

1.31 *For the real variable x, show that a sufficient, but not necessary, condition for $f(x) = x(x + 1)(2x + 1)$ to be divisible by 6 is that x is an integer.*

First suppose that x is an integer and consider $f(x)$ expressed as

$$f(x) = x(x + 1)(2x + 1) = x(x + 1)(x + 2) + x(x + 1)(x - 1).$$

Each term on the RHS consists of the product of three consecutive integers. In such a product one of the integers must divide by 3 and at least one of the other integers must be even. Thus each product separately divides by both 3 and 2, and hence by 6, and therefore so does their sum $f(x)$. Thus x being an integer *is a sufficient* condition for $f(x)$ to be divisible by 6.

That it is *not* a necessary condition can be shown by considering an equation of the form

$$f(x) = x(x + 1)(2x + 1) = 2x^3 + 3x^2 + x = 6m,$$

where m is an integer. As a specific counter-example consider the case $m = 4$. We note that $f(1) = 6$ whilst $f(2) = 30$. Thus there must be a root of the equation that lies strictly between the values 1 and 2, i.e a non-integer value of x that makes $f(x)$ equal to 24 and hence divisible by 6. This establishes the result that x being an integer is *not a necessary* condition for $f(x)$ to be divisible by 6.

> **1.33** The coefficients a_i in the polynomial $Q(x) = a_4x^4 + a_3x^3 + a_2x^2 + a_1x$ are all integers. Show that $Q(n)$ is divisible by 24 for all integers $n \geq 0$ if and only if all of the following conditions are satisfied:
>
>   (i) $2a_4 + a_3$ is divisible by 4;
>   (ii) $a_4 + a_2$ is divisible by 12;
> (iii) $a_4 + a_3 + a_2 + a_1$ is divisible by 24.

This problem involves both proof by induction and proof of the 'if and only if' variety. Firstly, assume that the three conditions are satisfied:

$$2a_4 + a_3 = 4\alpha,$$

$$a_4 + a_2 = 12\beta,$$

$$a_4 + a_3 + a_2 + a_1 = 24\gamma,$$

where α, β and γ are integers. We now have to prove that $Q(n) = a_4n^4 + a_3n^3 + a_2n^2 + a_1n$ is divisible by 24 for all integers $n \geq 0$. It is clearly true for $n = 0$, and we assume that it is true for $n = N$ and that $Q(N) = 24m$ for some integer m. Now consider $Q(N + 1)$:

$$
\begin{aligned}
Q(N + 1) &= a_4(N + 1)^4 + a_3(N + 1)^3 + a_2(N + 1)^2 + a_1(N + 1) \\
&= a_4N^4 + a_3N^3 + a_2N^2 + a_1N + 4a_4N^3 + (6a_4 + 3a_3)N^2 \\
&\quad + (4a_4 + 3a_3 + 2a_2)N + (a_4 + a_3 + a_2 + a_1) \\
&= 24m + 4a_4N^3 + 3(4\alpha)N^2 \\
&\quad + [4a_4 + (12\alpha - 6a_4) + (24\beta - 2a_4)]N + 24\gamma \\
&= 24(m + \gamma + \beta N) + 12\alpha N(N + 1) + 4a_4(N - 1)N(N + 1).
\end{aligned}
$$

Now $N(N + 1)$ is the product of two consecutive integers and so one must be even and contain a factor of 2; likewise $(N - 1)N(N + 1)$, being the product of three consecutive integers, must contain both 2 and 3 as factors. Thus every term in the expression for $Q(N + 1)$ divides by 24 and so, therefore, does $Q(N + 1)$. Thus the proposal is true for $n = N + 1$ if it is true for $n = N$, and this, together with our observation for $n = 0$, completes the 'if' part of the proof.

Now suppose that $Q(n) = a_4n^4 + a_3n^3 + a_2n^2 + a_1n$ is divisible by 24 for all integers $n \geq 0$. Setting n equal to 1, 2 and 3 in turn, we have

$$a_4 + a_3 + a_2 + a_1 = 24p,$$

$$16a_4 + 8a_3 + 4a_2 + 2a_1 = 24q,$$

$$81a_4 + 27a_3 + 9a_2 + 3a_1 = 24r,$$

for some integers p, q and r. The first of these equations is condition (iii). The other conditions are established by combining the above equations as follows:

$$14a_4 + 6a_3 + 2a_2 = 24(q - 2p),$$
$$78a_4 + 24a_3 + 6a_2 = 24(r - 3p),$$
$$36a_4 + 6a_3 = 24(r - 3p - 3q + 6p),$$
$$22a_4 - 2a_2 = 24(r - 3p - 4q + 8p).$$

The two final equations show that $6a_4 + a_3$ is divisible by 4 and that $11a_4 - a_2$ is divisible by 12. But, if $6a_4 + a_3$ is divisible by 4 then so is $(6 - 4)a_4 + a_3$, i.e. $2a_4 + a_3$. Similarly, $11a_4 - a_2$ being divisible by 12 implies that $12a_4 - (11a_4 - a_2)$, i.e. $a_4 + a_2$, is also divisible by 12. Thus, conditions (i) and (ii) are established and the 'only if' part of the proof is complete.

2

Preliminary calculus

(a) From the definition of the derivative as a limit, we have

$$f'(x) = \lim_{\Delta x \to 0} \frac{[3(x + \Delta x) + 4] - (3x + 4)}{\Delta x} = \lim_{\Delta x \to 0} \frac{3\Delta x}{\Delta x} = 3.$$

(b) These are calculated similarly, but using each calculated derivative as the input function for finding the next higher derivative.

$$f'(x) = \lim_{\Delta x \to 0} \frac{[(x + \Delta x)^2 + (x + \Delta x)] - (x^2 + x)}{\Delta x}$$

$$= \lim_{\Delta x \to 0} \frac{[(x^2 + 2x\Delta x + (\Delta x)^2) + (x + \Delta x)] - (x^2 + x)}{\Delta x}$$

$$= \lim_{\Delta x \to 0} \frac{[(2x\Delta x + (\Delta x)^2) + \Delta x]}{\Delta x}$$

$$= 2x + 1;$$

$$f''(x) = \lim_{\Delta x \to 0} \frac{[2(x + \Delta x) + 1] - (2x + 1)}{\Delta x} = \lim_{\Delta x \to 0} \frac{2\Delta x}{\Delta x} = 2;$$

$$f'''(x) = \lim_{\Delta x \to 0} \frac{2 - 2}{\Delta x} = 0.$$

(c) We use the expansion formula for $\sin(A + B)$ and then the series definitions of the sine and cosine functions to write $\cos \Delta x$ and $\sin \Delta x$ as series involving

17

increasing powers of Δx.

$$
\begin{aligned}
f'(x) &= \lim_{\Delta x \to 0} \frac{\sin(x + \Delta x) - \sin x}{\Delta x} \\
&= \lim_{\Delta x \to 0} \frac{(\sin x \cos \Delta x + \cos x \sin \Delta x) - \sin x}{\Delta x} \\
&= \lim_{\Delta x \to 0} \frac{\sin x (1 - \frac{(\Delta x)^2}{2!} + \cdots) + \cos x (\Delta x - \frac{(\Delta x)^3}{3!} + \cdots) - \sin x}{\Delta x} \\
&= \lim_{\Delta x \to 0} -\tfrac{1}{2}\Delta x \sin x + \cos x - \tfrac{1}{6}(\Delta x)^2 \cos x + \cdots \\
&= \cos x.
\end{aligned}
$$

2.3 *Find the first derivatives of*

(a) $x^2 \exp x$, (b) $2 \sin x \cos x$, (c) $\sin 2x$, (d) $x \sin ax$,
(e) $(e^{ax})(\sin ax) \tan^{-1} ax$, (f) $\ln(x^a + x^{-a})$,
(g) $\ln(a^x + a^{-x})$, (h) x^x.

(a) $x^2 \exp x$ is the product of two functions, both of which can be differentiated simply. We therefore apply the product rule and obtain:

$$
f'(x) = x^2 \frac{d(\exp x)}{dx} + \exp x \frac{d(x^2)}{dx} = x^2 \exp x + (2x) \exp x = (x^2 + 2x) \exp x.
$$

(b) Again, the product rule is appropriate:

$$
\begin{aligned}
f'(x) &= 2 \sin x \frac{d(\cos x)}{dx} + 2 \cos x \frac{d(\sin x)}{dx} \\
&= 2 \sin x(- \sin x) + 2 \cos x(\cos x) \\
&= 2(- \sin^2 x + \cos^2 x) = 2 \cos 2x.
\end{aligned}
$$

(c) Rewriting the function as $f(x) = \sin u$, where $u(x) = 2x$, and using the chain rule:

$$
f'(x) = \cos u \times \frac{du}{dx} = \cos u \times 2 = 2 \cos(2x).
$$

We note that this is the same result as in part (b); this is not surprising as the two functions to be differentiated are identical, i.e. $2 \sin x \cos x \equiv \sin 2x$.

(d) Once again, the product rule can be applied:

$$
f'(x) = x \frac{d(\sin ax)}{dx} + \sin ax \frac{d(x)}{dx} = xa \cos ax + \sin ax \times 1 = \sin ax + ax \cos ax.
$$

(e) This requires the product rule for three factors:

$$f'(x) = (e^{ax})(\sin ax)\frac{d(\tan^{-1} ax)}{dx} + (e^{ax})(\tan^{-1} ax)\frac{d(\sin ax)}{dx}$$

$$+ (\sin ax)(\tan^{-1} ax)\frac{d(e^{ax})}{dx}$$

$$= (e^{ax})(\sin ax)\left(\frac{a}{1 + a^2 x^2}\right) + (e^{ax})(\tan^{-1} ax)(a \cos ax)$$

$$+ (\sin ax)(\tan^{-1} ax)(ae^{ax})$$

$$= ae^{ax}\left[\frac{\sin ax}{1 + a^2 x^2} + (\tan^{-1} ax)(\cos ax + \sin ax)\right].$$

(f) Rewriting the function as $f(x) = \ln u$, where $u(x) = x^a + x^{-a}$, and using the chain rule:

$$f'(x) = \frac{1}{u} \times \frac{du}{dx} = \frac{1}{x^a + x^{-a}} \times (ax^{a-1} - ax^{-a-1}) = \frac{a(x^a - x^{-a})}{x(x^a + x^{-a})}.$$

(g) Using logarithmic differentiation and the chain rule as in (f):

$$f'(x) = \frac{1}{a^x + a^{-x}} \times (\ln a\, a^x - \ln a\, a^{-x}) = \frac{\ln a(a^x - a^{-x})}{a^x + a^{-x}}.$$

(h) In order to remove the independent variable x from the exponent in $y = x^x$, we first take logarithms and then differentiate implicitly:

$$y = x^x,$$

$$\ln y = x \ln x,$$

$$\frac{1}{y}\frac{dy}{dx} = \ln x + \frac{x}{x}, \quad \text{using the product rule,}$$

$$\frac{dy}{dx} = (1 + \ln x)x^x.$$

2.5 *Use the result that $d[v(x)^{-1}]/dx = -v^{-2}dv/dx$ to find the first derivatives of*
(a) $(2x + 3)^{-3}$, (b) $\sec^2 x$, (c) $\operatorname{cosech}^3 3x$, (d) $1/\ln x$, (e) $1/[\sin^{-1}(x/a)]$.

(a) Writing $(2x + 3)^3$ as $v(x)$ and using the chain rule, we have

$$f'(x) = -\frac{1}{v^2}\frac{dv}{dx} = -\frac{1}{(2x + 3)^6}[3(2x + 3)^2 (2)] = -\frac{6}{(2x + 3)^4}.$$

(b) Writing $\cos^2 x$ as $v(x)$, we have

$$f'(x) = -\frac{1}{v^2}\frac{dv}{dx} = -\frac{1}{\cos^4 x}[2 \cos x(-\sin x)] = 2\sec^2 x \tan x.$$

(c) Writing $\sinh^3 3x$ as $v(x)$, we have

$$f'(x) = -\frac{1}{v^2}\frac{dv}{dx} = -\frac{1}{\sinh^6 3x}\,[\,3\sinh^2 3x(\cosh 3x)(3)\,]$$

$$= -9\,\text{cosech}^3 3x\,\coth 3x.$$

(d) Writing $\ln x$ as $v(x)$, we have

$$f'(x) = -\frac{1}{v^2}\frac{dv}{dx} = -\frac{1}{(\ln x)^2}\frac{1}{x} = -\frac{1}{x\ln^2 x}.$$

(e) Writing $\sin^{-1}(x/a)$ as $v(x)$, we have

$$f'(x) = -\frac{1}{v^2}\frac{dv}{dx} = -\frac{1}{[\,\sin^{-1}(x/a)\,]^2}\frac{1}{\sqrt{a^2-x^2}}.$$

2.7 Find dy/dx if $x = (t-2)/(t+2)$ and $y = 2t/(t+1)$ for $-\infty < t < \infty$. Show that it is always non-negative, and make use of this result in sketching the curve of y as a function of x.

We calculate dy/dx as $dy/dt \div dx/dt$:

$$\frac{dy}{dt} = \frac{(t+1)2 - 2t(1)}{(t+1)^2} = \frac{2}{(t+1)^2},$$

$$\frac{dx}{dt} = \frac{(t+2)(1) - (t-2)(1)}{(t+2)^2} = \frac{4}{(t+2)^2},$$

$$\Rightarrow \quad \frac{dy}{dx} = \frac{2}{(t+1)^2} \div \frac{4}{(t+2)^2} = \frac{(t+2)^2}{2(t+1)^2},$$

which is clearly positive for all t.

By evaluating x and y for a range of values of t and recalling that its slope is always positive, the curve can be plotted as in figure 2.1. Alternatively, we may eliminate t using

$$t = \frac{2x+2}{1-x} \quad \text{and} \quad t = \frac{y}{2-y},$$

to obtain the equation of the curve in x-y coordinates as

$$2(x+1)(2-y) = y(1-x),$$
$$xy - 4x + 3y - 4 = 0,$$
$$(x+3)(y-4) = 4 - 12 = -8.$$

20

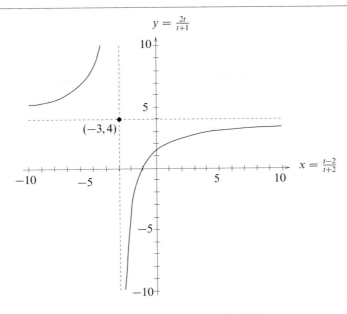

Figure 2.1   The solution to exercise 2.7.

This shows that the curve is a rectangular hyperbola in the second and fourth quadrants with asymptotes, parallel to the x- and y-axes, passing through $(-3, 4)$.

2.9 *Find the second derivative of $y(x) = \cos[(\pi/2) - ax]$. Now set $a = 1$ and verify that the result is the same as that obtained by first setting $a = 1$ and simplifying $y(x)$ before differentiating.*

We use the chain rule at each stage and, either finally or initially, the equality of $\cos(\frac{1}{2}\pi - \theta)$ and $\sin\theta$:

$$y(x) = \cos\left(\frac{\pi}{2} - ax\right),$$

$$y'(x) = a\sin\left(\frac{\pi}{2} - ax\right),$$

$$y''(x) = -a^2\cos\left(\frac{\pi}{2} - ax\right).$$

For $a = 1$, $y''(x) = -\cos\left(\frac{\pi}{2} - x\right) = -\sin x.$

Setting $a = 1$ initially, gives $y = \cos(\frac{1}{2}\pi - x) = \sin x$. Hence $y' = \cos x$ and $y'' = -\sin x$, yielding the same result as before.

> **2.11** *Show by differentiation and substitution that the differential equation*
> $$4x^2\frac{d^2y}{dx^2} - 4x\frac{dy}{dx} + (4x^2 + 3)y = 0$$
> *has a solution of the form* $y(x) = x^n \sin x$, *and find the value of* n.

The solution plan is to calculate the derivatives as functions of n and x and then, after substitution, require that the equation is identically satisfied for all x. This will impose conditions on n.

We have, by successive differentiation or by the use of Leibnitz' theorem, that

$$y(x) = x^n \sin x,$$
$$y'(x) = nx^{n-1} \sin x + x^n \cos x,$$
$$y''(x) = n(n-1)x^{n-2} \sin x + 2nx^{n-1} \cos x - x^n \sin x.$$

Substituting these into

$$4x^2\frac{d^2y}{dx^2} - 4x\frac{dy}{dx} + (4x^2 + 3)y = 0$$

gives

$$(4n^2 - 4n - 4n + 3)x^n \sin x + (-4 + 4)x^{n+2} \sin x + (8n - 4)x^{n+1} \cos x = 0.$$

For this to be true for all x, both $4n^2 - 8n + 3 = (2n - 3)(2n - 1) = 0$ and $8n - 4 = 0$ have to be satisfied. If $n = \frac{1}{2}$, they *are* both satisfied, thus establishing $y(x) = x^{1/2} \sin x$ as a solution of the given equation.

> **2.13** *Show that the lowest value taken by the function* $3x^4 + 4x^3 - 12x^2 + 6$ *is* -26.

We need to calculate the first and second derivatives of the function in order to establish the positions and natures of its turning points:

$$y(x) = 3x^4 + 4x^3 - 12x^2 + 6,$$
$$y'(x) = 12x^3 + 12x^2 - 24x,$$
$$y''(x) = 36x^2 + 24x - 24.$$

Setting $y'(x) = 0$ gives $x(x + 2)(x - 1) = 0$ with roots 0, 1 and -2. The corresponding values of $y''(x)$ are -24, 36 and 72.

Since $y(\pm\infty) = \infty$, the lowest value of y is that corresponding to the lowest minimum, which can only be at $x = 1$ or $x = -2$, as y'' must be positive at a minimum. The values of $y(x)$ at these two points are $y(1) = 1$ and $y(-2) = -26$, and so the lowest value taken is -26.

> **2.15** *Show that* $y(x) = xa^{2x} \exp x^2$ *has no stationary points other than* $x = 0$, *if*
> $$\exp(-\sqrt{2}) < a < \exp(\sqrt{2}).$$

Since the logarithm of a variable varies monotonically with the variable, the stationary points of the logarithm of a function of x occur at the same values of x as the stationary points of the function. As x appears as an exponent in the given function, we take logarithms before differentiating and obtain:

$$\ln y = \ln x + 2x \ln a + x^2,$$
$$\frac{1}{y}\frac{dy}{dx} = \frac{1}{x} + 2\ln a + 2x.$$

For a stationary point $dy/dx = 0$. Except at $x = 0$ (where y is also 0), this equation reduces to

$$2x^2 + 2x \ln a + 1 = 0.$$

This quadratic equation has no real roots for x if $4(\ln a)^2 < 4 \times 2 \times 1$, i.e. $|\ln a| < \sqrt{2}$; a result that can also be written as $\exp(-\sqrt{2}) < a < \exp(\sqrt{2})$.

> **2.17** *The parametric equations for the motion of a charged particle released from rest in electric and magnetic fields at right angles to each other take the forms*
> $$x = a(\theta - \sin\theta), \qquad y = a(1 - \cos\theta).$$
> *Show that the tangent to the curve has slope* $\cot(\theta/2)$. *Use this result at a few calculated values of* x *and* y *to sketch the form of the particle's trajectory.*

With the given parameterisation,

$$\frac{dx}{d\theta} = a - a\cos\theta,$$
$$\frac{dy}{d\theta} = a\sin\theta,$$
$$\Rightarrow \quad \frac{dy}{dx} = \frac{dy}{d\theta}\frac{d\theta}{dx} = \frac{\sin\theta}{1-\cos\theta} = \frac{2\sin\frac{1}{2}\theta\cos\frac{1}{2}\theta}{2\sin^2\frac{1}{2}\theta} = \cot\tfrac{1}{2}\theta.$$

Clearly, $y = 0$ whenever $\theta = 2n\pi$ with n an integer; dy/dx becomes infinite at the same points. The slope is zero whenever $\theta = (2n+1)\pi$ and the value of y is then $2a$. These results are plotted in figure 2.2.

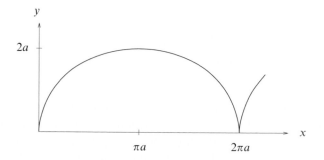

Figure 2.2   The solution to exercise 2.17.

2.19 *The curve whose equation is $x^{2/3} + y^{2/3} = a^{2/3}$ for positive x and y and which is completed by its symmetric reflections in both axes is known as an astroid. Sketch it and show that its radius of curvature in the first quadrant is $3(axy)^{1/3}$.*

For the asteroid curve (see figure 2.3) and its first derivative in the first quadrant, where all fractional roots are positive, we have

$$x^{2/3} + y^{2/3} = a^{2/3},$$

$$\frac{2}{3x^{1/3}} + \frac{2}{3y^{1/3}}\frac{dy}{dx} = 0,$$

$$\Rightarrow \quad \frac{dy}{dx} = -\left(\frac{y}{x}\right)^{1/3}.$$

Differentiating again,

$$\frac{d^2y}{dx^2} = -\frac{1}{3}\left(\frac{y}{x}\right)^{-2/3}\left[\frac{-x(\frac{y}{x})^{1/3} - y}{x^2}\right]$$

$$= \frac{1}{3}(x^{-2/3}y^{-1/3} + x^{-4/3}y^{1/3})$$

$$= \frac{1}{3}y^{-1/3}x^{-4/3}(x^{2/3} + y^{2/3})$$

$$= \frac{1}{3}y^{-1/3}x^{-4/3}a^{2/3}.$$

Hence, the radius of curvature is

$$\rho = \frac{\left[1 + \left(\dfrac{dy}{dx}\right)^2\right]^{3/2}}{\dfrac{d^2y}{dx^2}} = \frac{\left[1 + \left(\dfrac{y}{x}\right)^{2/3}\right]^{3/2}}{\frac{1}{3}y^{-1/3}x^{-4/3}a^{2/3}}$$

$$= 3(x^{2/3} + y^{2/3})^{3/2}x^{1/3}y^{1/3}a^{-2/3} = 3a^{1/3}x^{1/3}y^{1/3},$$

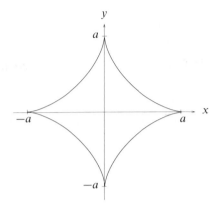

Figure 2.3   The astroid discussed in exercise 2.19.

as stated in the question.

2.21 *Use Leibnitz' theorem to find*

(a) *the second derivative of* $\cos x \sin 2x$,
(b) *the third derivative of* $\sin x \ln x$,
(c) *the fourth derivative of* $(2x^3 + 3x^2 + x + 2)e^{2x}$.

Leibnitz' theorem states that if $y(x) = u(x)v(x)$ and the rth derivative of a function $f(x)$ is denoted by $f^{(r)}$ then

$$y^{(n)} = \sum_{k=0}^{n} {}^{n}C_k \, u^{(k)} \, v^{(n-k)}.$$

So,

(a) $\dfrac{d^2(\cos x \sin 2x)}{dx^2} = (-\cos x)(\sin 2x) + 2(-\sin x)(2\cos 2x)$

$$+ (\cos x)(-4\sin 2x)$$

$$= -5\cos x \sin 2x - 4\sin x \cos 2x$$

$$= 2\sin x[-5\cos^2 x - 2(2\cos^2 x - 1)]$$

$$= 2\sin x(2 - 9\cos^2 x).$$

(b) $\dfrac{d^3(\sin x \ln x)}{dx^3} = (-\cos x)(\ln x) + 3(-\sin x)(x^{-1})$

$$+3(\cos x)(-x^{-2}) + (\sin x)(2x^{-3})$$

$$= (2x^{-3} - 3x^{-1})\sin x - (3x^{-2} + \ln x)\cos x.$$

25

(c) We note that the nth derivative of e^{2x} is $2^n e^{2x}$ and that the 4th derivative of a cubic polynomial is zero. And so,

$$\frac{d^4[(2x^3 + 3x^2 + x + 2)e^{2x}]}{dx^4}$$
$$= (0)(e^{2x}) + 4(12)(2e^{2x}) + 6(12x + 6)(4e^{2x})$$
$$\quad + 4(6x^2 + 6x + 1)(8e^{2x}) + (2x^3 + 3x^2 + x + 2)(16e^{2x})$$
$$= 16(2x^3 + 15x^2 + 31x + 19)e^{2x}.$$

2.23 *Use the properties of functions at their turning points to do the following.*

> (a) *By considering its properties near $x = 1$, show that $f(x) = 5x^4 - 11x^3 + 26x^2 - 44x + 24$ takes negative values for some range of x.*
> (b) *Show that $f(x) = \tan x - x$ cannot be negative for $0 \le x < \pi/2$, and deduce that $g(x) = x^{-1} \sin x$ decreases monotonically in the same range.*

(a) We begin by evaluating $f(1)$ and find that $f(1) = 5 - 11 + 26 - 44 + 24 = 0$. This suggests that $f(x)$ will be positive on one side of $x = 1$ and negative on the other. However, to be sure of this we need to establish that $x = 1$ is *not* a turning point of $f(x)$. To do this we calculate its derivative there:

$$f(x) = 5x^4 - 11x^3 + 26x^2 - 44x + 24,$$
$$f'(x) = 20x^3 - 33x^2 + 52x - 44,$$
$$f'(1) = 20 - 33 + 52 - 44 = -5 \ne 0.$$

So, $f'(1)$ is negative and f is decreasing at this point, where its value is 0. Therefore $f(x)$ must be negative in the range $1 < x < \alpha$ for some $\alpha > 1$.

(b) The function $f(x) = \tan x - x$ is differentiable in the range $0 \le x < \pi/2$, and $f'(x) = \sec^2 x - 1 = \tan^2 x$ which is > 0 for all x in the range; taken together with $f(0) = 0$, this establishes the result.

For $g(x) = (\sin x)/x$, the rule for differentiating quotients gives

$$g'(x) = \frac{x \cos x - \sin x}{x^2} = -\frac{\cos x(\tan x - x)}{x^2}.$$

The term in parenthesis cannot be negative in the range $0 \le x < \pi/2$, and in the same range $\cos x > 0$. Thus $g'(x)$ is never positive in the range and $g(x)$ decreases monotonically [from its value of $g(0) = 1$].

2.25 *By applying Rolle's theorem to $x^n \sin nx$, where n is an arbitrary positive integer, show that $\tan nx + x = 0$ has a solution α_1 with $0 < \alpha_1 < \pi/n$. Apply the theorem a second time to obtain the nonsensical result that there is a real α_2 in $0 < \alpha_2 < \pi/n$, such that $\cos^2(n\alpha_2) = -n$. Explain why this incorrect result arises.*

Clearly, the function $f(x) = x^n \sin nx$ has zeroes at $x = 0$ and $x = \pi/n$. Therefore, by Rolle's theorem, its derivative,

$$f'(x) = nx^{n-1} \sin nx + nx^n \cos nx,$$

must have a zero in the range $0 < x < \pi/n$. But, since $x \neq 0$ and $n \neq 0$, this is equivalent to a root α_1 of $\tan nx + x = 0$ in the same range. To obtain this result we have divided $f'(x) = 0$ through by $\cos nx$; this is allowed, since $x = \pi/(2n)$, the value that makes $\cos nx = 0$, is not a solution of $f'(x) = 0$.

We now note that $g(x) = \tan nx + x$ has zeroes at $x = 0$ and $x = \alpha_1$. Applying Rolle's theorem again (blindly) then shows that $g'(x) = n \sec^2 nx + 1$ has a zero α_2 in the range $0 < \alpha_2 < \alpha_1 < \pi/n$, with $\cos^2(n\alpha_2) = -n$.

The false result arises because $\tan nx$ is not differentiable at $x = \pi/(2n)$, which lies in the range $0 < x < \pi/n$, and so the conditions for applying Rolle's theorem are not satisfied.

2.27 *For the function $y(x) = x^2 \exp(-x)$ obtain a simple relationship between y and dy/dx and then, by applying Leibnitz' theorem, prove that*

$$xy^{(n+1)} + (n + x - 2)y^{(n)} + ny^{(n-1)} = 0.$$

The required function and its first derivative are

$$y(x) = x^2 e^{-x},$$
$$y'(x) = 2xe^{-x} - x^2 e^{-x}$$
$$= 2xe^{-x} - y.$$

Multiplying through by a factor x will enable us to express the first term on the RHS in terms of y and obtain

$$xy' = 2y - xy.$$

Now we apply Leibnitz' theorem to obtain the nth derivatives of both sides of this last equation, noting that the only non-zero derivative of x is the first derivative. We obtain

$$xy^{(n+1)} + n(1)y^{(n)} = 2y^{(n)} - [xy^{(n)} + n(1)y^{(n-1)}],$$

which can be rearranged as

$$xy^{(n+1)} + (n + x - 2)y^{(n)} + ny^{(n-1)} = 0,$$

thus completing the proof.

2.29 *Show that the curve $x^3 + y^3 - 12x - 8y - 16 = 0$ touches the x-axis.*

We first find an expression for the slope of the curve as a function of x and y. From

$$x^3 + y^3 - 12x - 8y - 16 = 0$$

we obtain, by implicit differentiation, that

$$3x^2 + 3y^2 y' - 12 - 8y' = 0 \quad \Rightarrow \quad y' = \frac{3x^2 - 12}{8 - 3y^2}.$$

Clearly $y' = 0$ at $x = \pm 2$. At $x = 2$,

$$8 + y^3 - 24 - 8y - 16 = 0 \quad \Rightarrow \quad y \neq 0.$$

However, at $x = -2$,

$$-8 + y^3 + 24 - 8y - 16 = 0, \quad \text{with one solution } y = 0.$$

Thus the point $(-2, 0)$ lies on the curve and $y' = 0$ there. It follows that the curve touches the x-axis at that point.

2.31 *Find the indefinite integrals J of the following ratios of polynomials:*

(a) $(x + 3)/(x^2 + x - 2)$;
(b) $(x^3 + 5x^2 + 8x + 12)/(2x^2 + 10x + 12)$;
(c) $(3x^2 + 20x + 28)/(x^2 + 6x + 9)$;
(d) $x^3/(a^8 + x^8)$.

(a) We first need to express the ratio in partial fractions:

$$\frac{x + 3}{x^2 + x - 2} = \frac{x + 3}{(x + 2)(x - 1)} = \frac{A}{x + 2} + \frac{B}{x - 1}.$$

Using any of the methods employed in exercise 1.15, we obtain the unknown

28

coefficients as $A = -\frac{1}{3}$ and $B = \frac{4}{3}$. Thus,

$$\int \frac{x+3}{x^2+x-2}\,dx = \int \frac{-1}{3(x+2)}\,dx + \int \frac{4}{3(x-1)}\,dx$$

$$= -\frac{1}{3}\ln(x+2) + \frac{4}{3}\ln(x-1) + c$$

$$= \frac{1}{3}\ln\frac{(x-1)^4}{x+2} + c.$$

(b) As the numerator is of higher degree than the denominator, we need to divide the numerator by the denominator and express the remainder in partial fractions before starting any integration:

$$x^3 + 5x^2 + 8x + 12 = (\tfrac{1}{2}x + a_0)(2x^2 + 10x + 12) + (b_1 x + b_0)$$

$$= x^3 + (2a_0 + 5)x^2 + (10a_0 + 6 + b_1)x$$

$$+ (12a_0 + b_0),$$

yielding $a_0 = 0$, $b_1 = 2$ and $b_0 = 12$. Now, expressed as partial fractions,

$$\frac{2x+12}{2x^2+10x+12} = \frac{x+6}{(x+2)(x+3)} = \frac{4}{x+2} + \frac{-3}{x+3},$$

where, again, we have used one of the three methods available for determining coefficients in partial fraction expansions. Thus,

$$\int \frac{x^3+5x^2+8x+12}{2x^2+10x+12}\,dx = \int \left(\frac{1}{2}x + \frac{4}{x+2} - \frac{3}{x+3}\right)\,dx$$

$$= \tfrac{1}{4}x^2 + 4\ln(x+2) - 3\ln(x+3) + c.$$

(c) By inspection,

$$3x^2 + 20x + 28 = 3(x^2 + 6x + 9) + 2x + 1.$$

Expressing the remainder after dividing through by x^2+6x+9 in partial fractions, and noting that the denominator has a double factor, we obtain

$$\frac{2x+1}{x^2+6x+9} = \frac{A}{(x+3)^2} + \frac{B}{x+3},$$

where $B(x+3) + A = 2x+1$. This requires that $B = 2$ and $A = -5$. Thus,

$$\int \frac{3x^2+20x+28}{x^2+6x+9}\,dx = \int \left[3 + \frac{2}{x+3} - \frac{5}{(x+3)^2}\right]\,dx$$

$$= 3x + 2\ln(x+3) + \frac{5}{x+3} + c.$$

(d) Noting the form of the numerator, we set $x^4 = u$ with $4x^3\,dx = du$. Then,

$$\int \frac{x^3}{a^8 + x^8}\,dx = \int \frac{1}{4(a^8 + u^2)}\,du$$

$$= \frac{1}{4a^4} \tan^{-1} \frac{u}{a^4} + c = \frac{1}{4a^4} \tan^{-1} \left(\frac{x^4}{a^4}\right) + c.$$

2.33 *Find the integral J of $(ax^2 + bx + c)^{-1}$, with $a \neq 0$, distinguishing between the cases* (i) $b^2 > 4ac$, (ii) $b^2 < 4ac$ *and* (iii) $b^2 = 4ac$.

In each case, we first 'complete the square' in the denominator, i.e. write it in such a form that x appears *only* in a term that is the square of a linear function of x. We then examine the overall sign of the terms that do not contain x; this determines the form of the integral. In case (iii) there is no such term. We write $b^2 - 4ac$ as $\Delta^2 > 0$, or $4ac - b^2$ as $\Delta'^2 > 0$, as needed.

(i) For $\Delta^2 = b^2 - 4ac > 0$,

$$J = \int \frac{dx}{a\left[\left(x + \frac{b}{2a}\right)^2 - \left(\frac{b^2}{4a^2} - \frac{c}{a}\right)\right]}$$

$$= \frac{1}{a} \int \frac{dx}{\left(x + \frac{b}{2a}\right)^2 - \frac{\Delta^2}{4a^2}}$$

$$= \frac{1}{a} \frac{a}{\Delta} \ln \frac{x + \frac{b}{2a} - \frac{\Delta}{2a}}{x + \frac{b}{2a} + \frac{\Delta}{2a}} + k$$

$$= \frac{1}{\Delta} \ln \frac{2ax + b - \Delta}{2ax + b + \Delta} + k.$$

(ii) For $-\Delta'^2 = b^2 - 4ac < 0$,

$$J = \int \frac{dx}{a\left[\left(x + \frac{b}{2a}\right)^2 - \left(\frac{b^2}{4a^2} - \frac{c}{a}\right)\right]}$$

$$= \frac{1}{a} \int \frac{dx}{\left(x + \frac{b}{2a}\right)^2 + \frac{\Delta'^2}{4a^2}}$$

$$= \frac{1}{a} \frac{2a}{\Delta'} \tan^{-1} \left(\frac{x + \frac{b}{2a}}{\frac{\Delta'}{2a}}\right) + k$$

$$= \frac{2}{\Delta'} \tan^{-1} \left(\frac{2ax + b}{\Delta'}\right) + k.$$

(iii) For $b^2 - 4ac = 0$,

$$J = \int \frac{dx}{ax^2 + bx + \frac{b^2}{4a}}$$

$$= \frac{1}{a} \int \frac{dx}{\left(x + \frac{b}{2a}\right)^2}$$

$$= \frac{-1}{a\left(x + \frac{b}{2a}\right)} + k$$

$$= -\frac{2}{2ax + b} + k.$$

2.35 *Find the derivative of $f(x) = (1 + \sin x)/\cos x$ and hence determine the indefinite integral J of $\sec x$.*

We differentiate $f(x)$ as a quotient, i.e. using $d(u/v)/dx = (vu' - uv')/v^2$, and obtain

$$f(x) = \frac{1 + \sin x}{\cos x},$$

$$f'(x) = \frac{\cos x(\cos x) - (1 + \sin x)(-\sin x)}{\cos^2 x}$$

$$= \frac{1 + \sin x}{\cos^2 x}$$

$$= \frac{f(x)}{\cos x}.$$

Thus, since $\sec x = f'(x)/f(x)$, it follows that

$$\int \sec x \, dx = \ln[f(x)] + c = \ln\left(\frac{1 + \sin x}{\cos x}\right) + c = \ln(\sec x + \tan x) + c.$$

2.37 *By making the substitution $x = a\cos^2\theta + b\sin^2\theta$, evaluate the definite integrals J between limits a and b ($> a$) of the following functions:*

(a) $[(x - a)(b - x)]^{-1/2}$;
(b) $[(x - a)(b - x)]^{1/2}$;
(c) $[(x - a)/(b - x)]^{1/2}$.

Wherever the substitution $x = a\cos^2\theta + b\sin^2\theta$ is made, the terms in parentheses

31

take the following forms:

$$x - a \to a\cos^2\theta + b\sin^2\theta - a = -a\sin^2\theta + b\sin^2\theta = (b-a)\sin^2\theta,$$
$$b - x \to b - a\cos^2\theta - b\sin^2\theta = -a\cos^2\theta + b\cos^2\theta = (b-a)\cos^2\theta,$$

and dx will be given by

$$dx = [2a\cos\theta(-\sin\theta) + 2b\sin\theta(\cos\theta)]\,d\theta = 2(b-a)\cos\theta\sin\theta\,d\theta.$$

The limits a and b will be replaced by 0 and $\pi/2$, respectively. We also note that the average value of the square of a sinusoid over any whole number of quarter cycles of its argument is one-half.

(a)
$$J_a = \int_a^b \frac{dx}{[(x-a)(b-x)]^{1/2}}$$
$$= \int_0^{\pi/2} \frac{2(b-a)\cos\theta\sin\theta}{[(b-a)\sin^2\theta\,(b-a)\cos^2\theta]^{1/2}}\,d\theta$$
$$= \int_0^{\pi/2} 2\,d\theta = \pi.$$

(b)
$$J_b = \int_a^b [(x-a)(b-x)]^{1/2}\,dx$$
$$= \int_0^{\pi/2} 2(b-a)^2\cos^2\theta\sin^2\theta\,d\theta$$
$$= \frac{1}{2}(b-a)^2 \int_0^{\pi/2} \sin^2 2\theta\,d\theta$$
$$= \frac{1}{2}(b-a)^2\frac{1}{2}\frac{\pi}{2} = \frac{\pi(b-a)^2}{8}.$$

(c)
$$J_c = \int_a^b \sqrt{\frac{x-a}{b-x}}\,dx$$
$$= \int_0^{\pi/2} \sqrt{\frac{(b-a)\sin^2\theta}{(b-a)\cos^2\theta}} \times 2(b-a)\cos\theta\sin\theta\,d\theta$$
$$= \int_0^{\pi/2} 2(b-a)\sin^2\theta\,d\theta$$
$$= \frac{\pi(b-a)}{2}.$$

2.39 *Use integration by parts to evaluate the following:*

(a) $\displaystyle\int_0^y x^2 \sin x \, dx;$ (b) $\displaystyle\int_1^y x \ln x \, dx;$

(c) $\displaystyle\int_0^y \sin^{-1} x \, dx;$ (d) $\displaystyle\int_1^y \ln(a^2 + x^2)/x^2 \, dx.$

If u and v are functions of x, the general formula for integration by parts is

$$\int_a^b uv' \, dx = [\, uv \,]_a^b - \int_a^b u'v \, dx.$$

Any given integrand $w(x)$ has to be written as $w(x) = u(x)v'(x)$ with $v'(x)$ chosen so that (i) it can be integrated explicitly, and (ii) it results in a u that has u' no more complicated than u itself. There are usually several possible choices but the one that makes both u and v as simple as possible is normally the best.

(a) Here the obvious choice at the first stage is $u(x) = x^2$ and $v'(x) = \sin x$. For the second stage, $u = x$ and $v' = \cos x$ are equally clear assignments.

$$\begin{aligned}
\int_0^y x^2 \sin x dx &= \left[\, x^2(-\cos x) \,\right]_0^y - \int_0^y 2x(-\cos x) \, dx \\
&= -y^2 \cos y + [\, 2x \sin x \,]_0^y - \int_0^y 2 \sin x \, dx \\
&= -y^2 \cos y + 2y \sin y + [\, 2 \cos x \,]_0^y \\
&= (2 - y^2) \cos y + 2y \sin y - 2.
\end{aligned}$$

(b) This integration is most straightforwardly carried out by taking $v'(x) = x$ and $u(x) = \ln x$ as follows:

$$\begin{aligned}
\int_1^y x \ln x \, dx &= \left[\, \frac{x^2}{2} \ln x \,\right]_1^y - \int_1^y \frac{1}{x}\frac{x^2}{2} \, dx \\
&= \frac{y^2}{2} \ln y - \left[\, \frac{x^2}{4} \,\right]_1^y \\
&= \frac{1}{2}y^2 \ln y + \frac{1}{4}(1 - y^2).
\end{aligned}$$

However, if you know that the integral of $\ln x$ is $x \ln x - x$, then the given integral can also be found by taking $v' = \ln x$ and $u = x$:

$$\begin{aligned}
\int_1^y x \ln x \, dx &= [\, x(x \ln x - x) \,]_1^y - \int_1^y 1 \times (x \ln x - x) \, dx \\
&= y^2 \ln y - y^2 - 0 + 1 - \int_1^y x \ln x \, dx + \left[\, \frac{x^2}{2} \,\right]_1^y.
\end{aligned}$$

After the limits have been substituted, the equation can be rearranged as

$$2 \int_1^y x \ln x \, dx = y^2 \ln y - y^2 + 1 + \frac{y^2}{2} - \frac{1}{2},$$

$$\int_1^y x \ln x \, dx = \frac{1}{2} y^2 \ln y + \frac{1}{4}(1 - y^2).$$

(c) Here we do not know the integral of $\sin^{-1} x$ (that is the problem!) but we do know its derivative. Therefore consider the integrand as $1 \times \sin^{-1} x$, with $v'(x) = 1$ and $u(x) = \sin^{-1} x$.

$$\int_0^y \sin^{-1} x \, dx = \int_0^y 1 \, \sin^{-1} x \, dx$$

$$= \left[x \sin^{-1} x \right]_0^y - \int_0^y \frac{1}{\sqrt{1 - x^2}} x \, dx$$

$$= y \sin^{-1} y + \left[\sqrt{1 - x^2} \right]_0^y$$

$$= y \sin^{-1} y + \sqrt{1 - y^2} - 1.$$

(d) When the logarithm of a function of x appears as part of an integrand, it is normally helpful to remove its explicit appearance by making it the $u(x)$ part of an integration-by-parts formula. The reciprocal of the function, without any explicit logarithm, then appears in the resulting integral; this is usually easier to deal with. In this case we take $\ln(a^2 + x^2)$ as $u(x)$.

$$\int_1^y \frac{\ln(a^2 + x^2)}{x^2} \, dx = \left[-\frac{\ln(a^2 + x^2)}{x} \right]_1^y - \int_1^y \frac{2x}{a^2 + x^2} \left(-\frac{1}{x} \right) dx$$

$$= -\frac{\ln(a^2 + y^2)}{y} + \ln(a^2 + 1) + \frac{2}{a} \left[\tan^{-1} \left(\frac{x}{a} \right) \right]_1^y$$

$$= -\frac{\ln(a^2 + y^2)}{y} + \ln(a^2 + 1)$$

$$+ \frac{2}{a} \left[\tan^{-1} \left(\frac{y}{a} \right) - \tan^{-1} \left(\frac{1}{a} \right) \right].$$

2.41 *The gamma function* $\Gamma(n)$ *is defined for all* $n > -1$ *by*

$$\Gamma(n+1) = \int_0^\infty x^n e^{-x}\, dx.$$

Find a recurrence relation connecting $\Gamma(n+1)$ *and* $\Gamma(n)$.

(a) *Deduce* (i) *the value of* $\Gamma(n+1)$ *when* n *is a non-negative integer, and* (ii) *the value of* $\Gamma\left(\frac{7}{2}\right)$, *given that* $\Gamma\left(\frac{1}{2}\right) = \sqrt{\pi}$.

(b) *Now, taking factorial* m *for any* m *to be defined by* $m! = \Gamma(m+1)$, *evaluate* $\left(-\frac{3}{2}\right)!$.

Integrating the defining equation by parts,

$$\Gamma(n+1) = \int_0^\infty x^n e^{-x}\, dx = \left[-x^n e^{-x} \right]_0^\infty + \int_0^\infty n x^{n-1} e^{-x}\, dx$$

$$= 0 + n\Gamma(n), \quad \text{for } n > 0,$$

i.e. $\Gamma(n+1) = n\Gamma(n)$.

(a)(i) Clearly $\Gamma(n+1) = n(n-1)(n-2)\cdots 2\,1\,\Gamma(1)$. But

$$\Gamma(1) = \int_0^\infty e^{-x}\, dx = 1.$$

Hence $\Gamma(n+1) = n!$.

(a)(ii) Applying the recurrence relation derived above,

$$\Gamma\left(\tfrac{7}{2}\right) = \tfrac{5}{2}\tfrac{3}{2}\tfrac{1}{2}\Gamma\left(\frac{1}{2}\right) = \tfrac{15}{8}\sqrt{\pi}.$$

(b) With this general definition of a factorial, we have

$$\left(-\tfrac{3}{2}\right)! = \Gamma\left(-\tfrac{1}{2}\right) = \tfrac{1}{-\frac{1}{2}}\Gamma\left(\tfrac{1}{2}\right) = -2\sqrt{\pi}.$$

2.43 *By integrating by parts twice, prove that* I_n *as defined in the first equality below for positive integers* n *has the value given in the second equality:*

$$I_n = \int_0^{\pi/2} \sin n\theta \cos\theta\, d\theta = \frac{n - \sin(n\pi/2)}{n^2 - 1}.$$

Taking $\sin n\theta$ as u and $\cos\theta$ as v and noting that with this choice $u'' = -n^2 u$

and $v'' = -v$, we expect that after two integrations by parts we will recover (a multiple of) I_n.

$$I_n = \int_0^{\pi/2} \sin n\theta \cos \theta \, d\theta$$

$$= [\sin n\theta \sin \theta]_0^{\pi/2} - \int_0^{\pi/2} n \cos n\theta \sin \theta \, d\theta$$

$$= \sin \frac{n\pi}{2} - n \left\{ [-\cos n\theta \cos \theta]_0^{\pi/2} - \int_0^{\pi/2} (-n \sin n\theta)(-\cos \theta) \, d\theta \right\}$$

$$= \sin \frac{n\pi}{2} - n[-(-1) - nI_n].$$

Rearranging this gives

$$I_n(1 - n^2) = \sin \frac{n\pi}{2} - n,$$

and hence the stated result.

2.45 *If J_r is the integral*

$$\int_0^\infty x^r \exp(-x^2) \, dx,$$

show that

(a) $J_{2r+1} = (r!)/2$,
(b) $J_{2r} = 2^{-r}(2r - 1)(2r - 3) \cdots (5)(3)(1) J_0$.

(a) We first derive a recurrence relationship for J_{2r+1}. Since we cannot integrate $\exp(-x^2)$ explicitly but can integrate $-2x \exp(-x^2)$, we extract the factor $-2x$ from the rest of the integrand and treat what is left ($-\frac{1}{2}x^{2r}$ in this case) as $u(x)$. This is the operation that has been carried out in the second line of what follows.

$$J_{2r+1} = \int_0^\infty x^{2r+1} \exp(-x^2) \, dx$$

$$= \int_0^\infty -\frac{x^{2r}}{2}(-2x) \exp(-x^2) \, dx$$

$$= \left[-\frac{x^{2r}}{2} \exp(-x^2) \right]_0^\infty + \int_0^\infty \frac{2rx^{2r-1}}{2} \exp(-x^2) \, dx$$

$$= 0 + rJ_{2r-1}.$$

Applying the relationship r times gives

$$J_{2r+1} = r(r - 1) \cdots 1 J_1.$$

36

But

$$J_1 = \int_0^\infty x \exp(-x^2) \, dx = \left[-\frac{1}{2} \exp(-x^2) \right]_0^\infty = \frac{1}{2},$$

and so $J_{2r+1} = \frac{1}{2} r!$.

(b) Using the same method as in part (a) it can be shown that

$$J_{2r} = \frac{2r-1}{2} J_{2r-2}.$$

Hence,

$$J_{2r} = \frac{2r-1}{2} \frac{2r-3}{2} \cdots \frac{1}{2} J_0,$$

in agreement with the stated relationship.

2.47 By noting that for $0 \leq \eta \leq 1$, $\eta^{1/2} \geq \eta^{3/4} \geq \eta$, prove that

$$\frac{2}{3} \leq \frac{1}{a^{5/2}} \int_0^a (a^2 - x^2)^{3/4} \, dx \leq \frac{\pi}{4}.$$

We use the result that, if $g(x) \leq f(x) \leq h(x)$ for *all* x in the range $a \leq x \leq b$, then $\int g(x) \, dx \leq \int f(x) \, dx \leq \int h(x) \, dx$, where all integrals are between the limits a and b.

Set $\eta = 1 - (x/a)^2$ in the stated inequalities and integrate the result from 0 to a, giving

$$\int_0^a \left(1 - \frac{x^2}{a^2} \right)^{1/2} dx \geq \int_0^a \left(1 - \frac{x^2}{a^2} \right)^{3/4} dx \geq \int_0^a \left(1 - \frac{x^2}{a^2} \right) dx.$$

Substituting $x = a \sin \theta$ and $dx = a \cos \theta \, d\theta$ in the first term and carrying out the elementary integration in the third term yields

$$\int_0^{\pi/2} a \cos^2 \theta \, d\theta \geq \frac{1}{a^{3/2}} \int_0^a (a^2 - x^2)^{3/4} \, dx \geq \left[x - \frac{x^3}{3a^2} \right]_0^a,$$

$$\Rightarrow \quad a \frac{1}{2} \frac{\pi}{2} \geq \frac{1}{a^{3/2}} \int_0^a (a^2 - x^2)^{3/4} \, dx \geq \frac{2a}{3},$$

$$\Rightarrow \quad \frac{\pi}{4} \geq \frac{1}{a^{5/2}} \int_0^a (a^2 - x^2)^{3/4} \, dx \geq \frac{2}{3}.$$

2.49 *By noting that* $\sinh x < \frac{1}{2}e^x < \cosh x$, *and that* $1 + z^2 < (1 + z)^2$ *for* $z > 0$, *show that, for* $x > 0$, *the length* L *of the curve* $y = \frac{1}{2}e^x$ *measured from the origin satisfies the inequalities* $\sinh x < L < x + \sinh x$.

With $y = y' = \frac{1}{2}e^x$ and the element of curve length ds given by $ds = (1 + y'^2)^{1/2}\,dx$, the total length of the curve measured from the origin is

$$L = \int_0^x ds = \int_0^x \left(1 + \tfrac{1}{4}e^{2x}\right)^{1/2} dx.$$

But, since all quantities are positive for $x \geq 0$,

$$\sinh x < \qquad \tfrac{1}{2}e^x \qquad < \cosh x,$$
$$\Rightarrow \quad \sinh^2 x < \qquad \tfrac{1}{4}e^{2x} \qquad < \cosh^2 x,$$
$$\cosh^2 x = 1 + \sinh^2 x < \quad 1 + \tfrac{1}{4}e^{2x} \quad < 1 + \cosh^2 x < (1 + \cosh x)^2,$$
$$\Rightarrow \quad \cosh x < \left(1 + \tfrac{1}{4}e^{2x}\right)^{1/2} < 1 + \cosh x.$$

It then follows, from integrating each term in the double inequality, that

$$\int_0^x \cosh x\,dx < L < \int_0^x (1 + \cosh x)\,dx,$$
$$\Rightarrow \qquad \sinh x < L < x + \sinh x,$$

as stated in the question.

3

Complex numbers and hyperbolic functions

3.1 *Two complex numbers z and w are given by $z = 3 + 4i$ and $w = 2 - i$. On an Argand diagram, plot*

(a) $z + w$, (b) $w - z$, (c) wz, (d) z/w,
(e) $z^*w + w^*z$, (f) w^2, (g) $\ln z$, (h) $(1 + z + w)^{1/2}$.

With $z = 3 + 4i$, $w = 2 - i$ and, where needed, $i^2 = -1$:

(a) $z + w = 3 + 4i + 2 - i = 5 + 3i$;

(b) $w - z = 2 - i - 3 - 4i = -1 - 5i$;

(c) $wz = (2 - i)(3 + 4i) = 6 - 3i + 8i - 4i^2 = 10 + 5i$;

(d) $\dfrac{z}{w} = \dfrac{3 + 4i}{2 - i} = \dfrac{3 + 4i}{2 - i}\dfrac{2 + i}{2 + i} = \dfrac{6 + 8i + 3i + 4i^2}{4 - 2i + 2i - i^2} = \dfrac{2 + 11i}{5}$;

(e) $z * w + w * z = (3 - 4i)(2 - i) + (2 + i)(3 + 4i) = (2 - 11i) + (2 + 11i) = 4$;

(f) $w^2 = (2 - i)(2 - i) = 4 - 4i + i^2 = 3 - 4i$;

(g) $\ln z = \ln |z| + i \arg z$
$$= \ln(3^2 + 4^2)^{1/2} + i \tan^{-1}\left(\tfrac{4}{3}\right)$$
$$= \ln 5 + i \left[\tan^{-1}\left(\tfrac{4}{3}\right) + 2n\pi\right];$$

(h) $(1 + z + w)^{1/2} = (6 + 3i)^{1/2}$
$$= \left\{\sqrt{45}\exp\left[i\tan\left(\tfrac{3}{6}\right)\right]\right\}^{1/2}$$
$$= \pm(45)^{1/4}\exp\left[i\tfrac{1}{2}\tan^{-1}\left(\tfrac{1}{2}\right)\right]$$
$$= \pm 2.590\,(\cos 0.2318 + i \sin 0.2318)$$
$$= \pm(2.521 + 0.595i).$$

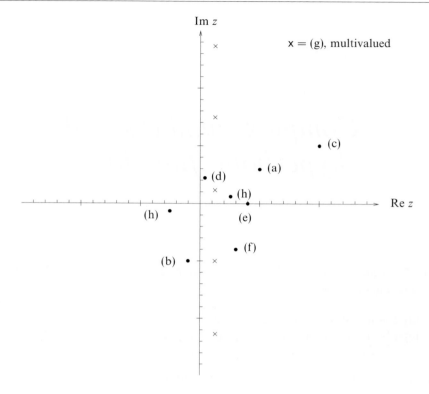

Figure 3.1   The solutions to exercise 3.1.

These results are plotted in figure 3.1. The answer to part (g) is multivalued and only five of the infinite number of possibilities are shown.

3.3 *By writing $\pi/12 = (\pi/3) - (\pi/4)$ and considering $e^{i\pi/12}$, evaluate $\cot(\pi/12)$.*

As we are expressing $\pi/12$ as the difference between two (familiar) angles, for which we know explicit formulae for their sines and cosines, namely

$$\sin \frac{\pi}{3} = \frac{\sqrt{3}}{2}, \quad \cos \frac{\pi}{3} = \frac{1}{2}, \quad \sin \frac{\pi}{4} = \cos \frac{\pi}{4} = \frac{1}{\sqrt{2}},$$

we will need the formulae for $\cos(A - B)$ and $\sin(A - B)$. They are given by

$$\cos(A - B) = \cos A \cos B + \sin A \sin B$$

$$\text{and} \quad \sin(A - B) = \sin A \cos B - \cos A \sin B.$$

Applying these with $A = \pi/3$ and $B = \pi/4$,

$$\exp\left(i\frac{\pi}{12}\right) = \exp\left[i\left(\frac{\pi}{3} - \frac{\pi}{4}\right)\right],$$

$$\cos\frac{\pi}{12} + i\sin\frac{\pi}{12} = \cos\left(\frac{\pi}{3} - \frac{\pi}{4}\right) + i\sin\left(\frac{\pi}{3} - \frac{\pi}{4}\right)$$

$$= \cos\frac{\pi}{3}\cos\frac{\pi}{4} + \sin\frac{\pi}{3}\sin\frac{\pi}{4}$$

$$+ i\left(\sin\frac{\pi}{3}\cos\frac{\pi}{4} - \cos\frac{\pi}{3}\sin\frac{\pi}{4}\right)$$

$$= \left(\frac{1}{2}\frac{1}{\sqrt{2}} + \frac{\sqrt{3}}{2}\frac{1}{\sqrt{2}}\right) + i\left(\frac{\sqrt{3}}{2}\frac{1}{\sqrt{2}} - \frac{1}{2}\frac{1}{\sqrt{2}}\right).$$

Thus

$$\cot\frac{\pi}{12} = \frac{\cos(\pi/12)}{\sin(\pi/12)} = \frac{1 + \sqrt{3}}{\sqrt{3} - 1} = 2 + \sqrt{3}.$$

3.5 *Evaluate*

(a) Re $(\exp 2iz)$, (b) Im $(\cosh^2 z)$, (c) $(-1 + \sqrt{3}i)^{1/2}$,
(d) $|\exp(i^{1/2})|$, (e) $\exp(i^3)$, (f) Im (2^{i+3}), (g) i^i, (h) $\ln[(\sqrt{3} + i)^3]$.

All of these evaluations rely directly on the definitions of the various functions involved as applied to complex numbers; these should be known to the reader. There are too many to give every one individually at each step and, if the justification for any particular step is unclear, reference should be made to a textbook.

(a) Re $(\exp 2iz) = $ Re $[\exp(2ix - 2y)] = \exp(-2y)\cos 2x$.

(b) Im $(\cosh^2 z) = $ Im $\left[\frac{1}{2}(\cosh 2z + 1)\right]$

$$= \tfrac{1}{2}\text{ Im } [\cosh(2x + 2iy)]$$

$$= \tfrac{1}{2}\text{ Im } (\cosh 2x \cosh 2iy + \sinh 2x \sinh 2iy)$$

$$= \tfrac{1}{2}\text{ Im } (\cosh 2x \cos 2y + i\sinh 2x \sin 2y)$$

$$= \tfrac{1}{2}\sinh 2x \sin 2y.$$

(c) $(-1 + \sqrt{3}i)^{1/2} = \left[(-1)^2 + (\sqrt{3})^2\right]^{1/4} \exp\left[i\tfrac{1}{2}(\tan^{-1}(\tfrac{\sqrt{3}}{-1}) + 2n\pi)\right]$

$$= \sqrt{2}\exp\left[i\tfrac{1}{2}(\tfrac{2}{3}\pi + 2n\pi)\right]$$

$$= \sqrt{2}\exp\left(\tfrac{\pi i}{3}\right) \quad\text{or}\quad \sqrt{2}\exp\left(\tfrac{4\pi i}{3}\right).$$

(d) $\left| \exp\left(i^{1/2}\right) \right| = \left| \exp\left((e^{\frac{i\pi}{2}})^{1/2}\right) \right|$

$= \left| \exp\left(e^{\frac{i\pi}{4}+in\pi}\right) \right|$

$= \left| \exp\left[\cos(n+\tfrac{1}{4})\pi + i\sin(n+\tfrac{1}{4})\pi\right] \right|$

$= \exp[\cos(n+\tfrac{1}{4})\pi]$

$= \exp\left(\tfrac{1}{\sqrt{2}}\right) \quad \text{or} \quad \exp\left(-\tfrac{1}{\sqrt{2}}\right).$

(e) $\exp\left(i^3\right) = \exp\left(e^{3(\frac{i\pi}{2})}\right) = \exp(\cos\tfrac{3\pi}{2} + i\sin\tfrac{3\pi}{2})$

$= \exp(0-i) = \cos(-1) + i\sin(-1) = 0.540 - 0.841\,i.$

(f) $\text{Im}\,(2^{i+3}) = \text{Im}\,(8\times 2^i) = 8\,\text{Im}\,(2^i) = 8\,\text{Im}\,(e^{i\ln 2}) = 8\,\sin(\ln 2) = 5.11.$

(g) $i^i = \left[\exp i(\tfrac{1}{2}\pi + 2n\pi)\right]^i = \left[\exp i^2(\tfrac{1}{2}\pi + 2n\pi)\right] = \exp[-(2n+\tfrac{1}{2})\pi].$

(h) $\ln\left[(\sqrt{3}+i)^3\right] = 3\ln(\sqrt{3}+i)$

$= 3\left(\ln 2 + i\tan^{-1}\tfrac{1}{\sqrt{3}}\right)$

$= \ln 8 + 3i(\tfrac{\pi}{6} + 2n\pi)$

$= \ln 8 + i(6n + \tfrac{1}{2})\pi.$

3.7 Show that the locus of all points $z = x+iy$ in the complex plane that satisfy

$$|z - ia| = \lambda|z + ia|, \quad \lambda > 0,$$

is a circle of radius $|2a\lambda/(1-\lambda^2)|$ centred on the point $z = ia[(1+\lambda^2)/(1-\lambda^2)]$. Sketch the circles for a few typical values of λ, including $\lambda < 1$, $\lambda > 1$ and $\lambda = 1$.

As we wish to find the locus in the x-y plane, we first express $|z \pm ia|$ explicitly in terms of x and y, remembering that a can be complex:

$$|x + iy - ia|^2 = (x+iy-ia)(x-iy+ia^*)$$
$$= x^2 + y^2 + |a|^2 - ia(x-iy) + ia^*(x+iy).$$
$$|x + iy + ia|^2 = (x+iy+ia)(x-iy-ia^*)$$
$$= x^2 + y^2 + |a|^2 + ia(x-iy) - ia^*(x+iy).$$

Substituting in

$$|x + iy - ia|^2 = \lambda^2|x + iy + ia|^2$$

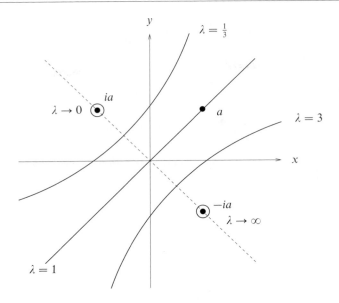

Figure 3.2   The solution to exercise 3.7.

gives, on dividing through by $1 - \lambda^2$,

$$x^2 - \frac{1 + \lambda^2}{1 - \lambda^2}(ia - ia^*)x + y^2 - \frac{1 + \lambda^2}{1 - \lambda^2}(a + a^*)y + |a|^2 = 0,$$

which can be rearranged as

$$\left(x + \frac{1 + \lambda^2}{1 - \lambda^2}\operatorname{Im} a\right)^2 + \left(y + \frac{1 + \lambda^2}{1 - \lambda^2}\operatorname{Re} a\right)^2 + |a|^2$$

$$- \left(\frac{1 + \lambda^2}{1 - \lambda^2}\right)^2 \left[(\operatorname{Im} a)^2 + (\operatorname{Re} a)^2\right] = 0.$$

This is of the form

$$(x - \alpha)^2 + (y - \beta)^2 = \left[\left(\frac{1 + \lambda^2}{1 - \lambda^2}\right)^2 - 1\right]|a|^2 = \frac{4\lambda^2}{(1 - \lambda^2)^2}|a|^2,$$

where

$$\alpha + i\beta = \frac{1 + \lambda^2}{1 - \lambda^2}(-\operatorname{Im} a + i\operatorname{Re} a) = \frac{1 + \lambda^2}{1 - \lambda^2}ia.$$

Thus it is the equation of a circle of radius $|2\lambda/(1 - \lambda^2)|a$ centred on the point $\alpha + i\beta$ as given above. See figure 3.2; note that a lies on the straight line (circle of infinite radius) corresponding to $\lambda = 1$. The circles centred on ia and $-ia$ have vanishingly small radii.

43

3.9 *For the real constant a find the loci of all points $z = x + iy$ in the complex plane that satisfy*

(a) $\text{Re} \left\{ \ln \left(\dfrac{z - ia}{z + ia} \right) \right\} = c, \qquad c > 0,$

(b) $\text{Im} \left\{ \ln \left(\dfrac{z - ia}{z + ia} \right) \right\} = k, \qquad 0 \le k \le \pi/2.$

Identify the two families of curves and verify that in case (b) all curves pass through the two points $\pm ia$.

(a) Recalling that

$$\ln z = \ln |z| + i \arg z$$

we have

$$\text{Re} \left(\ln \frac{z - ia}{z + ia} \right) = \ln \left| \frac{z - ia}{z + ia} \right| = c, \quad c > 0,$$
$$|z - ia| = e^c |z + ia|, \quad e^c > 1.$$

As in exercise 3.7, this is a circle of radius $|2ae^c/(1 - e^{2c})| = |a| \operatorname{cosech} c$ centred on the point $z = ia(1 + e^{2c})/(1 - e^{2c}) = ia \coth c$. As c varies this generates a family of circles whose centres lie on the y-axis above the point $z = ia$ (or below the point $z = ia$ if a is negative) and whose radii decrease as their centres approach that point. The curve corresponding to $c = 0$ is the x-axis.

(b) Using the principal value for the argument of a logarithm, we obtain

$$\text{Im} \left(\ln \frac{z - ia}{z + ia} \right) = \arg \frac{z - ia}{z + ia} = k, \quad 0 \le k \le \frac{\pi}{2}.$$

Now, $\dfrac{z - ia}{z + ia} = \dfrac{(z - ia)(z^* - ia)}{(z + ia)(z + ia)^*} = \dfrac{zz^* - ia(z + z^*) - a^2}{|z + ia|^2}.$

Hence, $\qquad\qquad k = \tan^{-1} \dfrac{-a(z + z^*)}{|z|^2 - a^2},$

$$a(z + z^*) = (a^2 - |z|^2) \tan k,$$
$$2ax = a^2 \tan k - (x^2 + y^2) \tan k,$$
$$(x + a \cot k)^2 + y^2 = a^2(1 + \cot^2 k).$$

This is a circle with centre $(-a \cot k, 0)$ and radius $a \operatorname{cosec} k$. As k varies the curves generate a family of circles whose centres lie on the negative x-axis (for $a > 0$) and whose radii decrease to a as their centres approach the origin. The curve corresponding to $k = 0$ is the y-axis.

The two points $z = \pm ia = (0, \pm a)$ lie on the curve if

$$(0 + a\cot k)^2 + a^2 = a^2(1 + \cot^2 k).$$

This is identically satisfied, verifying that all members of the family pass through the two points $z = \pm ia$.

3.11 *Sketch the parts of the Argand diagram in which*

   (a) *Re $z^2 < 0$, $|z^{1/2}| \leq 2$;*
   (b) *$0 \leq \arg z^* \leq \pi/2$;*
   (c) *$|\exp z^3| \to 0$ as $|z| \to \infty$.*

What is the area of the region in which all three sets of conditions are satisfied?

Since we will need to study the signs of the real parts of certain powers of z, it will be convenient to consider z as $r\,e^{i\theta}$ with $0 \leq \theta \leq 2\pi$.

Condition (a) contains two specifications. Firstly, for the real part of z^2 to be negative, its argument must be greater than $\pi/2$ but less than $3\pi/2$. The argument of z itself, which is half that of z^2 (mod 2π), must therefore lie in one of the two ranges $\pi/4 < \arg z < 3\pi/4$ and $5\pi/4 < \arg z < 7\pi/4$. Secondly, since the modulus of any complex number is real and positive, $|z^{1/2}| \leq 2$ is equivalent to $|z| \leq 4$.

Since $\arg z^* = -\arg z$, condition (b) requires $\arg z$ to lie in the range $3\pi/2 \leq \theta \leq 2\pi$, i.e z to lie in the fourth quadrant.

Condition (c) will only be satisfied if the real part of z^3 is negative. This requires

$$(4n+1)\frac{\pi}{2} < 3\theta < (4n+3)\frac{\pi}{2}, \qquad n = 0, 1, 2.$$

The allowed regions for θ are thus alternate wedges of angular size $\pi/3$ with an allowed region starting at $\theta = \pi/6$. The allowed region overlapping those specified by conditions (a) and (b) is the wedge $3\pi/2 \leq \theta \leq 11\pi/6$.

All three conditions are satisfied in the region $3\pi/2 \leq \theta \leq 7\pi/4$, $|z| \leq 4$; see figure 3.3. This wedge has an area given by

$$\frac{1}{2}r^2\theta = \frac{1}{2}16\left(\frac{7\pi}{4} - \frac{3\pi}{2}\right) = 2\pi.$$

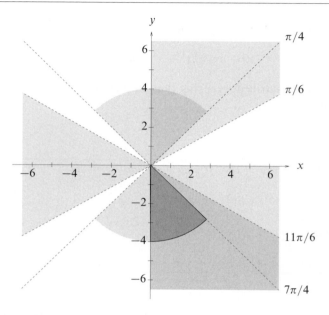

Figure 3.3 The defined region of the Argand diagram in exercise 3.11. Regions in which only one condition is satisfied are lightly shaded; those that satisfy two conditions are more heavily shaded; and the region satisfying all three conditions is most heavily shaded and outlined.

3.13 *Prove that* $x^{2m+1} - a^{2m+1}$*, where* m *is an integer* ≥ 1*, can be written as*

$$x^{2m+1} - a^{2m+1} = (x - a) \prod_{r=1}^{m} \left[x^2 - 2ax \cos\left(\frac{2\pi r}{2m+1}\right) + a^2 \right].$$

For the sake of brevity, we shall denote $x^{2m+1} - a^{2m+1}$ by $f(x)$ and the $(2m+1)$th root of unity, $\exp[2\pi i/(2m+1)]$, by Ω.

Now consider the roots of the equation $f(x) = 0$. The $2m+1$ quantities of the form $x = a\Omega^r$ with $r = 0, 1, 2, \ldots, 2m$ are all solutions of this equation and, since it is a polynomial equation of order $2m+1$, they represent all of its roots. We can therefore reconstruct the polynomial $f(x)$ (which has unity as the coefficient of its highest power) as the product of factors of the form $(x - a\Omega^r)$:

$$f(x) = (x - a)(x - a\Omega) \cdots (x - a\Omega^m)(x - a\Omega^{m+1}) \cdots (x - a\Omega^{2m}).$$

Now combine $(x - a\Omega^r)$ with $(x - a\Omega^{2m+1-r})$:

$$f(x) = (x - a) \prod_{r=1}^{m} (x - a\Omega^r)(x - a\Omega^{2m+1-r})$$

$$= (x - a) \prod_{r=1}^{m} [x^2 - ax(\Omega^r + \Omega^{2m+1-r}) + a^2\Omega^{2m+1}]$$

$$= (x - a) \prod_{r=1}^{m} [x^2 - ax(\Omega^r + \Omega^{-r}) + a^2], \quad \text{since } \Omega^{2m+1} = 1,$$

$$= (x - a) \prod_{r=1}^{m} \left[x^2 - 2ax \cos\left(\frac{2\pi r}{2m+1}\right) + a^2 \right].$$

This is the form given in the question.

3.15 *Solve the equation*

$$z^7 - 4z^6 + 6z^5 - 6z^4 + 6z^3 - 12z^2 + 8z + 4 = 0,$$

(a) *by examining the effect of setting z^3 equal to 2, and then*
(b) *by factorising and using the binomial expansion of $(z + a)^4$.*

Plot the seven roots of the equation on an Argand plot, exemplifying that complex roots of a polynomial equation always occur in conjugate pairs if the polynomial has real coefficients.

(a) Setting $z^3 = 2$ in $f(z)$ so as to leave no higher powers of z than its square, e.g. writing z^7 as $(z^3)^2 z = 4z$, gives

$$4z - 16 + 12z^2 - 12z + 12 - 12z^2 + 8z + 4 = 0,$$

which is satisfied identically. Thus $z^3 - 2$ is a factor of $f(z)$.

(b) Writing $f(z)$ as

$$f(z) = (z^3 - 2)(az^4 + bz^3 + cz^2 + dz + e) = 0$$

and equating the coefficients of the various powers of z gives $a = 1$, $b = -4$, $c = 6$, $d - 2a = -6$, $e - 2b = 6$, $-2c = -12$, $-2d = 8$ and $-2e = 4$. These imply (consistently) that $f(z)$ can be written as

$$f(z) = (z^3 - 2)(z^4 - 4z^3 + 6z^2 - 4z - 2).$$

We now note that the first four terms in the second set of parentheses are the

47

same as the corresponding terms in the expansion of $(z-1)^4$; only the constant term needs correction. Thus, we may write the original equation as

$$0 = f(z) = (z^3 - 2)[(z-1)^4 - 3],$$

with solutions

$$z = 2^{1/3} e^{2n\pi i/3} \quad n = 0, 1, 2 \quad \text{or}$$
$$z - 1 = 3^{1/4} e^{2n\pi i/4} \quad n = 0, 1, 2, 3.$$

The seven roots are therefore

$$2^{1/3}, \quad 2^{1/3} \left(\frac{-1 \pm i\sqrt{3}}{2} \right), \quad 1 \pm 3^{1/4}, \quad 1 \pm 3^{1/4} i.$$

As is to be expected, each root that has a non-zero imaginary part occurs as one of a complex conjugate pair.

3.17 *The binomial expansion of $(1+x)^n$ can be written for a positive integer n as*

$$(1+x)^n = \sum_{r=0}^{n} {}^nC_r x^r,$$

where ${}^nC_r = n!/[r!(n-r)!]$.

(a) *Use de Moivre's theorem to show that the sum*

$$S_1(n) = {}^nC_0 - {}^nC_2 + {}^nC_4 - \cdots + (-1)^m \, {}^nC_{2m}, \quad n-1 \le 2m \le n,$$

has the value $2^{n/2} \cos(n\pi/4)$.

(b) *Derive a similar result for the sum*

$$S_2(n) = {}^nC_1 - {}^nC_3 + {}^nC_5 - \cdots + (-1)^m \, {}^nC_{2m+1}, \quad n-1 \le 2m+1 \le n,$$

and verify it for the cases $n = 6, 7$ and 8.

Since we seek the sum of binomial coefficients that contain either all even or all odd indices, we need to choose a value for x such that x^r has different characteristics depending upon whether r is even or odd. The quantity i has just such a property, being purely real when r is even and purely imaginary when r is odd. We therefore take $x = i$, write $1 + i$ as $\sqrt{2} e^{i\pi/4}$ and apply de Moivre's theorem:

$$\left(\sqrt{2} e^{i\pi/4} \right)^n = (1+i)^n$$
$$= {}^nC_0 + i \, {}^nC_1 + i^2 \, {}^nC_2 + \cdots$$
$$= ({}^nC_0 - {}^nC_2 + {}^nC_4 - \cdots)$$
$$+ i ({}^nC_1 - {}^nC_3 + {}^nC_5 - \cdots).$$

Thus $S_1(n) = (\,^nC_0 - \,^nC_2 + \,^nC_4 - \cdots + (-1)^m \,^nC_{2m})$, where $n - 1 \leq 2m \leq n$, has a value equal to that of the real part of $\left(\sqrt{2}e^{i\pi/4}\right)^n$. This is the real part of $2^{n/2}e^{in\pi/4}$, which, by de Moivre's theorem, is $2^{n/2}\cos(n\pi/4)$.

(b) The corresponding result for $S_2(n)$ is that it is equal to the imaginary part of $2^{n/2}e^{in\pi/4}$, which is $2^{n/2}\sin(n\pi/4)$.

We now verify this result for $n = 6, 7$ and 8 by direct calculation:

$$S_2(6) = \,^6C_1 - \,^6C_3 + \,^6C_5 = 6 - 20 + 6 = -8 = 2^3 \sin\frac{6\pi}{4},$$

$$S_2(7) = \,^7C_1 - \,^7C_3 + \,^7C_5 - \,^7C_7$$
$$= 7 - 35 + 21 - 1 = -8 = 2^{7/2} \sin\frac{7\pi}{4},$$

$$S_2(8) = \,^8C_1 - \,^8C_3 + \,^8C_5 - \,^8C_7 = 8 - 56 + 56 - 8 = 0 = 2^4 \sin\frac{8\pi}{4}.$$

3.19 *Use de Moivre's theorem with $n = 4$ to prove that*

$$\cos 4\theta = 8\cos^4\theta - 8\cos^2\theta + 1,$$

and deduce that

$$\cos\frac{\pi}{8} = \left(\frac{2+\sqrt{2}}{4}\right)^{1/2}.$$

From de Moivre's theorem, $e^{i4\theta} = \cos 4\theta + i\sin 4\theta$. But, by the binomial theorem, we also have that

$$e^{i4\theta} = (\cos\theta + i\sin\theta)^4$$
$$= \cos^4\theta + 4i\cos^3\theta\sin\theta - 6\cos^2\theta\sin^2\theta - 4i\cos\theta\sin^3\theta + \sin^4\theta.$$

Equating the real parts of the two equal expressions and writing $\sin^2\theta$ as $1 - \cos^2\theta$,

$$\cos 4\theta = \cos^4\theta - 6\cos^2\theta(1 - \cos^2\theta) + (1 - \cos^2\theta)^2$$
$$= 8\cos^4\theta - 8\cos^2\theta + 1.$$

Now set $\theta = \pi/8$ in this result and write $\cos(\pi/8)$ as c:

$$0 = \cos\frac{4\pi}{8} = 8c^4 - 8c^2 + 1.$$

Hence, as this is a quadratic equation in c^2,

$$c^2 = \frac{4 \pm \sqrt{16 - 8}}{8} \quad \text{and} \quad c = \cos\frac{\pi}{8} = \pm\left(\frac{2 \pm \sqrt{2}}{4}\right)^{1/2}.$$

Since $0 < \pi/8 < \pi/2$, c must be positive. Further, as $\pi/8 < \pi/4$ and $\cos(\pi/4) = 1/\sqrt{2}$, c must be greater then $1/\sqrt{2}$. It is clear that the positive square roots are the appropriate ones in both cases.

3.21 *Use de Moivre's theorem to prove that*

$$\tan 5\theta = \frac{t^5 - 10t^3 + 5t}{5t^4 - 10t^2 + 1},$$

where $t = \tan\theta$. Deduce the values of $\tan(n\pi/10)$ for $n = 1, 2, 3$ and 4.

Using the binomial theorem and de Moivre's theorem to expand $(e^{i\theta})^5$ in two different ways, we have, from equating the real and imaginary parts of the two results, that

$$\cos 5\theta + i \sin 5\theta = \cos^5\theta + i5\cos^4\theta\sin\theta - 10\cos^3\theta\sin^2\theta$$
$$-i10\cos^2\theta\sin^3\theta + 5\cos\theta\sin^4\theta + i\sin^5\theta,$$
$$\cos 5\theta = \cos^5\theta - 10\cos^3\theta(1 - \cos^2\theta)$$
$$+5\cos\theta(1 - 2\cos^2\theta + \cos^4\theta)$$
$$= 16\cos^5\theta - 20\cos^3\theta + 5\cos\theta,$$
$$\sin 5\theta = 5(1 - 2\sin^2\theta + \sin^4\theta)\sin\theta$$
$$-10(1 - \sin^2\theta)\sin^3\theta + \sin^5\theta$$
$$= 16\sin^5\theta - 20\sin^3\theta + 5\sin\theta.$$

Now, writing $\cos\theta$ as c, $\sin\theta$ as s and $\tan\theta$ as t, and further recalling that $c^{-2} = 1 + t^2$, we have

$$\tan 5\theta = \frac{16s^5 - 20s^3 + 5s}{16c^5 - 20c^3 + 5c}$$

$$= \frac{16t^5 - 20t^3c^{-2} + 5tc^{-4}}{16 - 20c^{-2} + 5c^{-4}}$$

$$= \frac{16t^5 - 20t^3(1 + t^2) + 5t(1 + 2t^2 + t^4)}{16 - 20(1 + t^2) + 5(1 + 2t^2 + t^4)}$$

$$= \frac{t^5 - 10t^3 + 5t}{5t^4 - 10t^2 + 1}.$$

When θ is equal to $\dfrac{\pi}{10}$ or $\dfrac{3\pi}{10}$, $\tan 5\theta = \infty$, implying that

$$5t^4 - 10t^2 + 1 = 0 \quad\Rightarrow\quad t^2 = \frac{5 \pm \sqrt{25 - 5}}{5} \quad\Rightarrow\quad t = \pm\left(\frac{5 \pm \sqrt{20}}{5}\right)^{1/2}.$$

As both angles lie in the first quadrant the overall sign must be taken as positive in both cases, and it is clear that the positive square root in the numerator corresponds to $\theta = 3\pi/10$.

When θ is equal to $\dfrac{2\pi}{10}$ or $\dfrac{4\pi}{10}$, $\tan 5\theta = 0$, implying that

$$t^5 - 10t^3 + 5t = 0 \quad \Rightarrow \quad t^2 = 5 \pm \sqrt{25 - 5} \quad \Rightarrow \quad t = \pm \left(5 \pm \sqrt{20}\right)^{1/2}.$$

Again, as both angles lie in the first quadrant the overall sign must be taken as positive; it is also clear that the positive square root in the parentheses corresponds to $\theta = 4\pi/10$.

3.23 *Determine the conditions under which the equation*

$$a \cosh x + b \sinh x = c, \qquad c > 0,$$

has zero, one, or two real solutions for x. What is the solution if $a^2 = c^2 + b^2$?

We start by recalling that $\cosh x = \frac{1}{2}(e^x + e^{-x})$ and $\sinh x = \frac{1}{2}(e^x - e^{-x})$, and then rewrite the equation as a quadratic equation in e^x:

$$a \cosh x + b \sinh x - c = 0,$$
$$(a + b)e^x - 2c + (a - b)e^{-x} = 0,$$
$$(a + b)e^{2x} - 2ce^x + (a - b) = 0.$$

Hence,

$$e^x = \frac{c \pm \sqrt{c^2 - (a^2 - b^2)}}{a + b}.$$

For x to be real, e^x must be real and ≥ 0. Since $c > 0$, this implies that $a + b > 0$ and $c^2 + b^2 \geq a^2$. Provided these two conditions are satisfied, there are two roots if $c^2 + b^2 - a^2 < c^2$, i.e. if $b^2 < a^2$, but only one root if $c^2 + b^2 - a^2 > c^2$, i.e. if $b^2 > a^2$.

If $c^2 + b^2 = a^2$ then the double root is given by

$$e^x = \frac{c}{a + b},$$
$$e^{2x} = \frac{c^2}{(a + b)^2} = \frac{a^2 - b^2}{(a + b)^2} = \frac{a - b}{a + b},$$
$$x = \frac{1}{2} \ln \frac{a - b}{a + b}.$$

3.25 *Express* $\sinh^4 x$ *in terms of hyperbolic cosines of multiples of x, and hence find the real solutions of*

$$2\cosh 4x - 8\cosh 2x + 5 = 0.$$

In order to connect $\sinh^4 x$ to hyperbolic functions of other multiples of x, we need to express it in terms of powers of $e^{\pm x}$ and then to group the terms so as to make up those hyperbolic functions. Starting from

$$\sinh x = \tfrac{1}{2}(e^x - e^{-x}),$$

we have from the binomial theorem that

$$\sinh^4 x = \tfrac{1}{16}\left(e^{4x} - 4e^{2x} + 6 - 4e^{-2x} + e^{-4x}\right).$$

Terms containing related exponents nx and $-nx$ can now be grouped together and expressed as a linear sum of $\cosh nx$ and $\sinh nx$; here, because of the symmetry properties of the binomial coefficients, only the $\cosh nx$ combinations appear and yield

$$\sinh^4 x = \tfrac{1}{8}\cosh 4x - \tfrac{1}{2}\cosh 2x + \tfrac{3}{8}.$$

Now consider the relationship between this expression and the LHS of the given equation. They are clearly closely related; one is a multiple of the other, except in respect of the constant term. Making compensating corrections to the constant term allows us to rewrite the equation in terms of $\sinh^4 x$ as follows:

$$2\cosh 4x - 8\cosh 2x + (6-1) = 0,$$
$$16\sinh^4 x - 1 = 0,$$
$$\sinh^4 x = \tfrac{1}{16},$$
$$\sinh x = \pm\tfrac{1}{2} \quad \text{(real solutions only)}.$$

We now use the explicit expression for the inverse hyperbolic sine, namely

$$\text{If } y = \sinh^{-1} z, \text{ then } y = \ln(\sqrt{1+z^2} + z),$$

to give in this case

$$x = \ln\left(\sqrt{1+\tfrac{1}{4}} \pm \tfrac{1}{2}\right) = 0.481 \text{ or } -0.481.$$

3.27 *A closed barrel has as its curved surface the surface obtained by rotating about the x-axis the part of the curve*

$$y = a[2 - \cosh(x/a)]$$

lying in the range $-b \le x \le b$, where $b < a \cosh^{-1} 2$. Show that the total surface area, A, of the barrel is given by

$$A = \pi a[9a - 8a \exp(-b/a) + a \exp(-2b/a) - 2b].$$

If s is the length of the curve defining the surface (measured from $x = 0$) then $ds^2 = dx^2 + dy^2$ and consequently $ds/dx = (1 + y'^2)^{1/2}$.

For this particular surface,

$$y = a \left(2 - \cosh \frac{x}{a}\right)$$

$$\text{and } \frac{dy}{dx} = -\sinh \frac{x}{a}.$$

It follows that

$$\frac{ds}{dx} = \left[1 + \left(\frac{dy}{dx}\right)^2\right]^{1/2}$$

$$= \left(1 + \sinh^2 \frac{x}{a}\right)^{1/2}$$

$$= \cosh \frac{x}{a}.$$

The curved surface area, A_1, is given by

$$A_1 = 2 \int_0^b 2\pi y \, ds$$

$$= 2 \int_0^b 2\pi y \frac{ds}{dx} \, dx$$

$$= 4\pi a \int_0^b \left(2 \cosh \frac{x}{a} - \cosh^2 \frac{x}{a}\right) dx, \text{ use } \cosh^2 z = \tfrac{1}{2}(\cosh 2z + 1),$$

$$= 4\pi a \int_0^b \left(2 \cosh \frac{x}{a} - \frac{1}{2} - \frac{1}{2} \cosh \frac{2x}{a}\right) dx$$

$$= 4\pi a \left[2a \sinh \frac{x}{a} - \frac{x}{2} - \frac{a}{4} \sinh \frac{2x}{a}\right]_0^b$$

$$= \pi a \left(8a \sinh \frac{b}{a} - 2b - a \sinh \frac{2b}{a}\right).$$

The area, A_2, of the two flat ends is given by

$$A_2 = 2\pi a^2 \left(2 - \cosh \frac{b}{a} \right)^2$$

$$= 2\pi a^2 \left(4 - 4 \cosh \frac{b}{a} + \cosh^2 \frac{b}{a} \right).$$

And so the total area is

$$A = \pi a \left[4a \left(e^{b/a} - e^{-b/a} \right) - 2b - \frac{a}{2} \left(e^{2b/a} - e^{-2b/a} \right) \right.$$

$$\left. + 8a - 4a \left(e^{b/a} + e^{-b/a} \right) + \frac{2a}{4} \left(e^{2b/a} + 2 + e^{-2b/a} \right) \right]$$

$$= \pi a \left(9a - 8ae^{-b/a} + ae^{-2b/a} - 2b \right).$$

4

Series and limits

4.1 *Sum the even numbers between 1000 and 2000 inclusive.*

We must first express the given sum in terms of a summation for which we have an explicit form. The result that is needed is clearly

$$S_N = \sum_{n=1}^{N} n = \frac{1}{2}N(N+1),$$

and we must re-write the given summation in terms of sums of this form:

$$\sum_{n(\text{even})=1000}^{n=2000} n = \sum_{m=500}^{m=1000} 2m$$

$$= 2(S_{1000} - S_{499})$$

$$= 2\left(\tfrac{1}{2} \times 1000 \times 1001 - \tfrac{1}{2} \times 499 \times 500\right)$$

$$= 751\,500.$$

4.3 *How does the convergence of the series*

$$\sum_{n=r}^{\infty} \frac{(n-r)!}{n!}$$

depend on the integer r?

For $r \leq 1$, each term of the series is greater than or equal to the corresponding term of $\sum \frac{1}{n}$, which is known to be divergent (for a proof, see any standard textbook). Thus, by the comparison test, the given series is also divergent.

For $r \geq 2$, each term of the series is less than or equal to the corresponding term of $\sum\limits_{1}^{\infty} \dfrac{1}{n(n+1)}$. By writing this latter sum as

$$\sum_{n=1}^{\infty} \frac{1}{n(n+1)} = \sum_{n=1}^{\infty} \left(\frac{1}{n} - \frac{1}{n+1} \right)$$

$$= \left(1 - \tfrac{1}{2}\right) + \left(\tfrac{1}{2} - \tfrac{1}{3}\right) + \left(\tfrac{1}{3} - \tfrac{1}{4}\right) + \cdots$$

$$= 1 + \left(-\tfrac{1}{2} + \tfrac{1}{2}\right) + \left(-\tfrac{1}{3} + \tfrac{1}{3}\right) + \cdots \to 1,$$

it is shown to be convergent. Thus, by the comparison test, the given series is also convergent when $r \geq 2$.

4.5 *Find the sum, S_N, of the first N terms of the following series, and hence determine whether the series are convergent, divergent or oscillatory:*

(a) $\displaystyle\sum_{n=1}^{\infty} \ln\left(\frac{n+1}{n}\right),$ (b) $\displaystyle\sum_{n=0}^{\infty}(-2)^n,$ (c) $\displaystyle\sum_{n=1}^{\infty} \frac{(-1)^{n+1}n}{3^n}.$

(a) We express this series as the difference between two series with similar terms and find that the terms cancel in pairs, leaving an explicit expression that contains only the last term of the first series and the first term of the second:

$$\sum_{n=1}^{N} \ln \frac{n+1}{n} = \sum_{n=1}^{N} \ln(n+1) - \sum_{n=1}^{N} \ln n = \ln(N+1) - \ln 1.$$

As $\ln(N+1) \to \infty$ as $N \to \infty$, the series diverges.

(b) Applying the normal formula for a geometric sum gives

$$\sum_{n=0}^{N-1}(-2)^n = \frac{1 - (-2)^N}{3}.$$

The series therefore oscillates infinitely.

(c) Denote the partial sum by S_N. Then,

$$S_N = \sum_{n=1}^{N} \frac{(-1)^{n+1}n}{3^n},$$

$$\frac{1}{3}S_N = \sum_{n=1}^{N} \frac{(-1)^{n+1}n}{3^{n+1}} = \sum_{s=2}^{N+1} \frac{(-1)^s(s-1)}{3^s}$$

$$= \sum_{s=2}^{N+1} \frac{(-1)^s s}{3^s} - \sum_{s=2}^{N+1} \frac{(-1)^s}{3^s}.$$

Separating off the last term of the first series on the RHS and adding S_N to both sides, with the S_N added to the RHS having its $n = 1$ term written explicitly, yields

$$\frac{4}{3}S_N = \frac{(-1)^2 1}{3} + \sum_{n=2}^{N} \frac{(-1)^{n+1}n}{3^n} + \sum_{s=2}^{N} \frac{(-1)^s s}{3^s}$$
$$+ \frac{(-1)^{N+1}(N+1)}{3^{N+1}} - \sum_{s=2}^{N+1} \frac{(-1)^s}{3^s}$$
$$= \frac{1}{3} + \frac{(-1)^{N+1}(N+1)}{3^{N+1}} - \frac{1}{9} \frac{1-(-\frac{1}{3})^N}{1-(-\frac{1}{3})}.$$

To obtain the last line we note that on the RHS the second and third terms (both summations) cancel and that the final term is a geometric series (with leading term $-\frac{1}{9}$). This result can be rearranged as

$$S_N = \frac{3}{16}\left[1 - \left(-\frac{1}{3}\right)^N\right] + \frac{3N}{4}\left(-\frac{1}{3}\right)^{N+1},$$

from which it is clear that the series converges to a sum of $\frac{3}{16}$.

4.7 *Use the difference method to sum the series*

$$\sum_{n=2}^{N} \frac{2n-1}{2n^2(n-1)^2}.$$

We try to write the nth term as the difference between two consecutive values of a partial-fraction function of n. Since the second power of n appears in the denominator the function will need two terms, An^{-2} and Bn^{-1}. Hence, we must have

$$\frac{2n-1}{2n^2(n-1)^2} = \frac{A}{n^2} + \frac{B}{n} - \left[\frac{A}{(n-1)^2} + \frac{B}{n-1}\right]$$
$$= \frac{A[-2n+1] + B[n(n-1)(n-1-n)]}{n^2(n-1)^2}.$$

The powers of n in the numerators can be equated consistently if we take $A = -\frac{1}{2}$

and $B = 0$. Thus

$$\frac{2n-1}{2n^2(n-1)^2} = \frac{1}{2}\left[\frac{1}{(n-1)^2} - \frac{1}{n^2}\right].$$

We can now carry out the summation, in which the second component of each pair of terms cancels the first component of the next pair, leaving only the initial and very final components:

$$\sum_{n=2}^{N}\frac{2n-1}{2n^2(n-1)^2} = \frac{1}{2}\sum_{n=2}^{N}\left[\frac{1}{(n-1)^2} - \frac{1}{n^2}\right]$$

$$= \frac{1}{2}\left(\frac{1}{1} - \frac{1}{N^2}\right)$$

$$= \tfrac{1}{2}(1 - N^{-2}).$$

4.9 *Prove that*

$$\cos\theta + \cos(\theta + \alpha) + \cdots + \cos(\theta + n\alpha) = \frac{\sin\frac{1}{2}(n+1)\alpha}{\sin\frac{1}{2}\alpha}\cos(\theta + \tfrac{1}{2}n\alpha).$$

From de Moivre's theorem, the required sum, S, is the real part of the sum of the geometric series $\sum_{r=0}^{n} e^{i\theta}e^{ir\alpha}$. Using the formula for the partial sum of a geometric series, and multiplying by a factor that makes the denominator real, we have

$$S = \text{Re}\left(e^{i\theta}\frac{1 - e^{i(n+1)\alpha}}{1 - e^{i\alpha}}\frac{1 - e^{-i\alpha}}{1 - e^{-i\alpha}}\right)$$

$$= \frac{\cos\theta - \cos[(n+1)\alpha + \theta] - \cos(\theta - \alpha) + \cos(\theta + n\alpha)}{2 \times 2\sin^2\frac{1}{2}\alpha}$$

$$= \frac{2\sin(\theta - \tfrac{1}{2}\alpha)\sin(-\tfrac{1}{2}\alpha) + 2\sin(n\alpha + \tfrac{1}{2}\alpha + \theta)\sin\tfrac{1}{2}\alpha}{4\sin^2\frac{1}{2}\alpha}$$

$$= \frac{2\sin\tfrac{1}{2}\alpha\,2\cos(\tfrac{1}{2}n\alpha + \theta)\,\sin[\tfrac{1}{2}(n+1)\alpha]}{4\sin^2\frac{1}{2}\alpha}$$

$$= \frac{\sin\frac{1}{2}(n+1)\alpha}{\sin\frac{1}{2}\alpha}\cos(\theta + \tfrac{1}{2}n\alpha).$$

In the course of this manipulation we have used the identity $1 - \cos\theta = 2\sin^2\frac{1}{2}\theta$ and the formulae for $\cos A - \cos B$ and $\sin A - \sin B$.

4.11 *Find the real values of x for which the following series are convergent:*

$$\text{(a)} \sum_{n=1}^{\infty} \frac{x^n}{n+1}, \quad \text{(b)} \sum_{n=1}^{\infty} (\sin x)^n, \quad \text{(c)} \sum_{n=1}^{\infty} n^x,$$

$$\text{(d)} \sum_{n=1}^{\infty} e^{nx}, \quad \text{(e)} \sum_{n=2}^{\infty} (\ln n)^x.$$

(a) Using the ratio test:

$$\lim_{n \to \infty} \frac{u_{n+1}}{u_n} = \lim_{n \to \infty} \frac{x^{n+1}}{n+2} \frac{n+1}{x^n} = x.$$

Thus the series is convergent for all $|x| < 1$. At $x = 1$ the series diverges, as shown in any standard text, whilst at $x = -1$ it converges by the alternating series test. Thus we have convergence for $-1 \le x < 1$.

(b) For all x other than $x = (2m \pm \frac{1}{2})\pi$, where m is an integer, $|\sin x| < 1$ and so convergence is assured by the ratio test. At $x = (2m + \frac{1}{2})\pi$ the series diverges, whilst at $x = (2m - \frac{1}{2})\pi$ it oscillates finitely.

(c) This is the Riemann zeta series with p written as $-x$. Thus the series converges for all $x < -1$.

(d) The ratio of successive terms is e^x (independent of n) and for this to be less than unity in magnitude requires x to be negative. Thus the series is convergent when $x < 0$.

(e) The sum $S = \sum_{n=2}^{\infty} (\ln n)^x$ is clearly divergent for all $x > -1$ (by comparison with $\sum n^{-1}$). So we define a positive X by $-X = x < -1$ and consider

$$S_1 = \sum_{k=1}^{\infty} \sum_{r_k = M_{k-1}+1}^{M_k} \frac{1}{(\ln M_k)^X},$$

where M_k is the lowest integer such that $\ln M_k > k$. The notation is such that when $e^{k-1} < n < e^k$ then $n = M_{k-1} + r_k$.

For each fixed k, every term in the second (finite) summation is smaller than the corresponding term in S (because $n < M_k$). But, since all the terms in such a summation are equal, the value of the sum is simply $(M_k - M_{k-1})/(\ln M_k)^X$. Thus,

$$S_1 = \sum_{k=1}^{\infty} \frac{M_k - M_{k-1}}{(\ln M_k)^X} = \sum_{k=1}^{\infty} \frac{(1 - e^{-1})M_k}{(\ln M_k)^X}.$$

Now, the ratio of successive terms in this final summation is

$$\frac{M_{k+1}}{(\ln M_{k+1})^X} \frac{(\ln M_k)^X}{M_k} \to \frac{e}{\ln e} \quad \text{as} \quad k \to \infty.$$

This limit is > 1, and thus S_1 diverges for all X; hence, by the comparison test, so does S.

4.13 *Determine whether the following series are absolutely convergent, convergent or oscillatory:*

(a) $\displaystyle\sum_{n=1}^{\infty} \frac{(-1)^n}{n^{5/2}}$,　　(b) $\displaystyle\sum_{n=1}^{\infty} \frac{(-1)^n(2n+1)}{n}$,　　(c) $\displaystyle\sum_{n=0}^{\infty} \frac{(-1)^n|x|^n}{n!}$,

(d) $\displaystyle\sum_{n=0}^{\infty} \frac{(-1)^n}{n^2 + 3n + 2}$,　　(e) $\displaystyle\sum_{n=1}^{\infty} \frac{(-1)^n 2^n}{n^{1/2}}$.

(a) The sum $\sum n^{-5/2}$ is convergent (by comparison with $\sum n^{-2}$) and so $\sum (-1)^n n^{-5/2}$ is absolutely convergent.

(b) The magnitude of the individual terms $\to 2$ and not to zero; thus the series cannot converge. In fact it oscillates finitely about the value $-(1 + \ln 2)$.

(c) The magnitude of the successive-term ratio is

$$\left| \frac{u_{n+1}}{u_n} \right| = \frac{|x|^{n+1}}{(n+1)!} \frac{n!}{|x|^n} = \frac{|x|}{n} \to 0 \quad \text{for all } x.$$

Thus, the series is absolutely convergent for all finite x.

(d) The polynomial in the denominator has all positive signs and a non-zero constant term; it is therefore always strictly positive. Thus, to test for absolute convergence, we need to replace the numerator by its absolute value and consider $\sum_{n=0}^{N}(n^2 + 3n + 2)^{-1}$:

$$\sum_{n=0}^{N} \frac{1}{n^2 + 3n + 2} = \sum_{n=0}^{N} \left(\frac{1}{n+1} - \frac{1}{n+2} \right) = 1 - \frac{1}{N+2} \to 1 \quad \text{as} \quad N \to \infty.$$

Thus the given series is absolutely convergent.

(e) The magnitude of the individual terms does not tend to zero; in fact it grows monotonically. The effect of the alternating signs is to make the series oscillate infinitely.

4.15 *Prove that*

$$\sum_{n=2}^{\infty} \ln \left[\frac{n^r + (-1)^n}{n^r} \right]$$

is absolutely convergent for $r = 2$, but only conditionally convergent for $r = 1$.

In each case divide the sum into two sums, one for n even and one for n odd.

(i) For $r = 2$, consider first the even series:

$$\sum_{n \text{ even}} \ln \left(\frac{n^2 + 1}{n^2} \right) = \sum_{n \text{ even}} \ln \left(1 + \frac{1}{n^2} \right)$$
$$= \sum_{n \text{ even}} \left(\frac{1}{n^2} - \frac{1}{2n^4} + \cdots \right).$$

The nth logarithmic term is positive for all n but, as shown above, less than n^{-2}. It follows from the comparison test that the series is (absolutely) convergent.

For the odd series we consider

$$\ln \left[\frac{(2m+1)^2 - 1}{(2m+1)^2} \right] = \ln \frac{4m^2 + 4m}{4m^2 + 4m + 1}$$
$$= -\ln \left(1 + \frac{1}{4m(m+1)} \right).$$

By a similar argument to that above, each term is negative but greater than $-[4m(m+1)]^{-1}$. Again, the comparison test shows that the series is (absolutely) convergent.

Thus the original series, being the sum of two absolutely convergent series, is also absolutely convergent.

(ii) For $r = 1$ we have to consider $\ln[(n \pm 1)/n]$, whose expansion contains a term $\pm n^{-1}$ and other inverse powers of n. The summations over the other powers converge and cannot cancel the divergence arising from $\sum \pm n^{-1}$. Thus both the even and odd series diverge; consequently the original series cannot be absolutely convergent.

However, if we group together consecutive pairs of terms, $n = 2m$ and $n = 2m+1$, then we see that

$$\sum_{n=2}^{\infty} \ln \left[\frac{n + (-1)^n}{n} \right] = \sum_{m=1}^{\infty} \left[\ln \frac{2m+1}{2m} + \ln \frac{2m+1-1}{2m+1} \right]$$
$$= \sum_{m=1}^{\infty} \ln 1 = \sum_{m=1}^{\infty} 0 = 0,$$

i.e. the terms cancel in pairs and the series is conditionally convergent to zero.

4.17 *Demonstrate that rearranging the order of its terms can make a conditionally convergent series converge to a different limit by considering the series* $\sum(-1)^{n+1}n^{-1} = \ln 2 = 0.693$. *Rearrange the series as*

$$S = \tfrac{1}{1} + \tfrac{1}{3} - \tfrac{1}{2} + \tfrac{1}{5} + \tfrac{1}{7} - \tfrac{1}{4} + \tfrac{1}{9} + \tfrac{1}{11} - \tfrac{1}{6} + \tfrac{1}{13} + \cdots$$

and group each set of three successive terms. Show that the series can then be written

$$\sum_{m=1}^{\infty} \frac{8m-3}{2m(4m-3)(4m-1)},$$

which is convergent (by comparison with $\sum n^{-2}$*) and contains only positive terms. Evaluate the first of these and hence deduce that* S *is not equal to* $\ln 2$.

Proceeding as indicated, we have

$$\begin{aligned} S &= \left(\frac{1}{1} + \frac{1}{3} - \frac{1}{2}\right) + \left(\frac{1}{5} + \frac{1}{7} - \frac{1}{4}\right) + \left(\frac{1}{9} + \frac{1}{11} - \frac{1}{6}\right) + \cdots \\ &= \sum_{m=1}^{\infty} \left(\frac{1}{4m-3} + \frac{1}{4m-1} - \frac{1}{2m}\right) \\ &= \sum_{m=1}^{\infty} \frac{(8m^2 - 2m) + (8m^2 - 6m) - (16m^2 - 16m + 3)}{2m(4m-3)(4m-1)} \\ &= \sum_{m=1}^{\infty} \frac{8m-3}{2m(4m-3)(4m-1)}. \end{aligned}$$

As noted, this series is convergent and contains only positive terms. The first of these terms ($m = 1$) is $5/6 = 0.833$. This, by itself, is greater than the known sum (0.693) of the original series. Thus S cannot be equal to $\ln 2$.

4.19 *A Fabry–Pérot interferometer consists of two parallel heavily silvered glass plates; light enters normally to the plates, and undergoes repeated reflections between them, with a small transmitted fraction emerging at each reflection. Find the intensity* $|B|^2$ *of the emerging wave, where*

$$B = A(1-r)\sum_{n=0}^{\infty} r^n e^{in\phi},$$

with r *and* ϕ *real.*

This is a simple geometric series but with a complex common ratio $re^{i\phi}$. Thus we have

$$B = A(1-r)\sum_{n=0}^{\infty} r^n e^{in\phi}$$

$$= A\frac{1-r}{1-re^{i\phi}}.$$

To obtain the intensity $|B|^2$ we multiply this result by its complex conjugate, recalling that r and ϕ are real, but A may not be:

$$|B|^2 = \frac{|A|^2(1-r)^2}{(1-re^{i\phi})(1-re^{-i\phi})}$$

$$= \frac{|A|^2(1-r)^2}{1-2r\cos\phi+r^2}.$$

4.21 *Starting from the Maclaurin series for* $\cos x$, *show that*

$$(\cos x)^{-2} = 1 + x^2 + \frac{2x^4}{3} + \cdots .$$

Deduce the first three terms in the Maclaurin series for $\tan x$.

From the Maclaurin series for (or definition of) $\cos x$,

$$\cos x = 1 - \frac{x^2}{2!} + \frac{x^4}{4!} + \cdots .$$

Using the binomial expansion of $(1+z)^{-2}$, we have

$$(\cos x)^{-2} = \left(1 - \frac{x^2}{2!} + \frac{x^4}{4!} + \cdots\right)^{-2}$$

$$= 1 - 2\left(-\frac{x^2}{2!} + \frac{x^4}{4!} + \cdots\right) + \frac{2\,3}{2!}\left(-\frac{x^2}{2!} + \frac{x^4}{4!} + \cdots\right)^2 + \cdots$$

$$= 1 + x^2 + x^4\left(-\frac{2}{4!} + \frac{2\,3}{2!\,2!\,2!}\right) + O(x^6)$$

$$= 1 + x^2 + \tfrac{2}{3}x^4 + \cdots .$$

We now integrate both sides of the expansion from 0 to x, noting that $(\cos x)^{-2} \equiv \sec^2 x$ and that this integrates to $\tan x$. Thus

$$\tan x = \int_0^x \sec^2 u\, du = x + \frac{x^3}{3} + \frac{2x^5}{15} + \cdots .$$

4.23 *If $f(x) = \sinh^{-1} x$, and its nth derivative $f^{(n)}(x)$ is written as*

$$f^{(n)} = \frac{P_n}{(1 + x^2)^{n-1/2}},$$

where $P_n(x)$ is a polynomial (of order $n - 1$), show that the $P_n(x)$ satisfy the recurrence relation

$$P_{n+1}(x) = (1 + x^2)P_n'(x) - (2n - 1)xP_n(x).$$

Hence generate the coefficients necessary to express $\sinh^{-1} x$ as a Maclaurin series up to terms in x^5.

With $f(x) = \sinh^{-1} x$,

$$x = \sinh f \quad \Rightarrow \quad \frac{dx}{df} = \cosh f \quad \Rightarrow \quad \frac{df}{dx} = \frac{1}{\cosh f} = \frac{1}{(1 + x^2)^{1/2}}.$$

Thus $P_1(x) = 1$; we will need this as a starting value for the recurrence relation.

With the definition of $P_n(x)$ given,

$$f^{(n)} = \frac{P_n}{(1 + x^2)^{n-1/2}},$$

$$f^{(n+1)} = \frac{P_n'}{(1 + x^2)^{n-1/2}} - \frac{(n - \frac{1}{2})\, 2x\, P_n}{(1 + x^2)^{n+1/2}}$$

$$= \frac{(1 + x^2)P_n' - (2n - 1)xP_n}{(1 + x^2)^{n+1-1/2}}.$$

It then follows that

$$P_{n+1}(x) = (1 + x^2)P_n'(x) - (2n - 1)xP_n(x).$$

With $P_1 = 1$, as shown,

$$P_2 = (1 + x^2)0 - (2 - 1)x\,1 = -x,$$
$$P_3 = (1 + x^2)(-1) - (4 - 1)x(-x) = 2x^2 - 1,$$
$$P_4 = (1 + x^2)(4x) - (6 - 1)x(2x^2 - 1) = 9x - 6x^3,$$
$$P_5 = (1 + x^2)(9 - 18x^2) - (8 - 1)x(9x - 6x^3) = 24x^4 - 72x^2 + 9.$$

The corresponding values of $f^{(n)}(0) = P_n(0)/(1 + 0^2)^{n-1/2}$ can then be used to express the Maclaurin series for $\sinh^{-1} x$ as

$$\sinh^{-1} x = f(0) + \sum_{n=1}^{\infty} \frac{f^n(0)x^n}{n!} = x - \frac{x^3}{3!} + \frac{9x^5}{5!} - \cdots .$$

4.25 By using the logarithmic series, prove that if a and b are positive and nearly equal then

$$\ln\frac{a}{b} \simeq \frac{2(a-b)}{a+b}.$$

Show that the error in this approximation is about $2(a-b)^3/[3(a+b)^3]$.

Write $a+b = 2c$ and $a - b = 2\delta$. Then

$$\ln\frac{a}{b} = \ln a - \ln b$$

$$= \ln(c+\delta) - \ln(c-\delta)$$

$$= \ln c + \ln\left(1+\frac{\delta}{c}\right) - \ln c - \ln\left(1-\frac{\delta}{c}\right)$$

$$= \left(\frac{\delta}{c} - \frac{\delta^2}{2c^2} + \frac{\delta^3}{3c^3} - \cdots\right) - \left(-\frac{\delta}{c} - \frac{\delta^2}{2c^2} - \frac{\delta^3}{3c^3} - \cdots\right)$$

$$= \frac{2\delta}{c} + \frac{2}{3}\left(\frac{\delta}{c}\right)^3 + \cdots$$

$$= \frac{2(a-b)}{a+b} + \frac{2}{3}\left(\frac{a-b}{a+b}\right)^3 + \cdots,$$

i.e. as stated in the question.

We note that other approximations are possible, and equally valid, e.g. setting $b = a+\epsilon$ leading to $-(\epsilon/a)[1-\epsilon/2a+\epsilon^2/3a^2 - \cdots]$, but the given one, expanding symmetrically about $c = (a+b)/2$, contains no quadratic terms in $(a-b)$, only cubic and higher terms.

4.27 Find the limit as $x \to 0$ of $[\sqrt{1+x^m} - \sqrt{1-x^m}]/x^n$, in which m and n are positive integers.

Using the binomial expansions of the terms in the numerator,

$$\frac{\sqrt{1+x^m} - \sqrt{1-x^m}}{x^n} = \frac{1 + \frac{1}{2}x^m + \cdots - (1 - \frac{1}{2}x^m + \cdots)}{x^n}$$

$$= \frac{x^m + \cdots}{x^n}$$

$$= x^{m-n} + \cdots.$$

Thus the limit of the function as $x \to 0$ is 0 for $m > n$, 1 for $m = n$ and ∞ for $m < n$.

4.29 *Find the limits of the following functions:*

(a) $\dfrac{x^3 + x^2 - 5x - 2}{2x^3 - 7x^2 + 4x + 4}$, *as* $x \to 0$, $x \to \infty$ *and* $x \to 2$;

(b) $\dfrac{\sin x - x \cosh x}{\sinh x - x}$, *as* $x \to 0$;

(c) $\displaystyle\int_x^{\pi/2} \left(\dfrac{y \cos y - \sin y}{y^2} \right) dy$, *as* $x \to 0$.

(a) Denote the ratio of polynomials by $f(x)$. Then

$$\lim_{x \to 0} f(x) = \lim_{x \to 0} \frac{x^3 + x^2 - 5x - 2}{2x^3 - 7x^2 + 4x + 4} = \frac{-2}{4} = -\frac{1}{2};$$

$$\lim_{x \to \infty} f(x) = \lim_{x \to \infty} \frac{1 + x^{-1} - 5x^{-2} - 2x^{-3}}{2 - 7x^{-1} + 4x^{-2} + 4x^{-3}} = \frac{1}{2};$$

$$\lim_{x \to 2} f(x) = \lim_{x \to 2} \frac{x^3 + x^2 - 5x - 2}{2x^3 - 7x^2 + 4x + 4} = \frac{0}{0}.$$

This final value is indeterminate and so, using l'Hôpital's rule, consider instead

$$\lim_{x \to 2} f(x) = \lim_{x \to 2} \frac{3x^2 + 2x - 5}{6x^2 - 14x + 4} = \frac{11}{0} = \infty.$$

(b) Using l'Hôpital's rule repeatedly,

$$\lim_{x \to 0} \frac{\sin x - x \cosh x}{\sinh x - x} = \lim_{x \to 0} \frac{\cos x - \cosh x - x \sinh x}{\cosh x - 1}$$

$$= \lim_{x \to 0} \frac{- \sin x - \sinh x - \sinh x - x \cosh x}{\sinh x}$$

$$= \lim_{x \to 0} \frac{- \cos x - 2 \cosh x - \cosh x - x \sinh x}{\cosh x} = -4.$$

(c) Before taking the limit we need to find a closed form for the integral. So,

$$\lim_{x \to 0} \int_x^{\pi/2} \left(\frac{y \cos y - \sin y}{y^2} \right) dy = \lim_{x \to 0} \int_x^{\pi/2} \frac{d}{dy} \left(\frac{\sin y}{y} \right) dy$$

$$= \lim_{x \to 0} \left[\frac{\sin y}{y} \right]_x^{\pi/2}$$

$$= \lim_{x \to 0} \left(\frac{2}{\pi} - \frac{\sin x}{x} \right)$$

$$= \lim_{x \to 0} \left[\frac{2}{\pi} - \frac{1}{x} \left(x - \frac{x^3}{3!} + \cdots \right) \right]$$

$$= \frac{2}{\pi} - 1.$$

4.31 *Using a first-order Taylor expansion about* $x = x_0$, *show that a better approximation than* x_0 *to the solution of the equation*

$$f(x) = \sin x + \tan x = 2$$

is given by $x = x_0 + \delta$, *where*

$$\delta = \frac{2 - f(x_0)}{\cos x_0 + \sec^2 x_0}.$$

(a) *Use this procedure twice to find the solution of* $f(x) = 2$ *to six significant figures, given that it is close to* $x = 0.9$.

(b) *Use the result in (a) to deduce, to the same degree of accuracy, one solution of the quartic equation*

$$y^4 - 4y^3 + 4y^2 + 4y - 4 = 0.$$

(a) We write the solution to $f(x) = \sin x + \tan x = 2$ as $x = x_0 + \delta$. Substituting this form and retaining the first-order terms in δ in the Taylor expansions of $\sin x$ and $\tan x$ we obtain

$$\sin x_0 + \delta \cos x_0 + \cdots + \tan x_0 + \delta \sec^2 x_0 + \cdots = 2$$

$$\delta = \frac{2 - \sin x_0 - \tan x_0}{\cos x_0 + \sec^2 x_0}.$$

With $x_0 = 0.9$,

$$\delta_1 = \frac{2 - 0.783327 - 1.260158}{0.621610 + 2.587999} = \frac{-0.043485}{3.209609} = -0.013548,$$

making the first improved approximation $x_1 = x_0 + \delta_1 = 0.886452$.

Now, using x_1 instead of x_0 and repeating the process gives

$$\delta_2 = \frac{2 - 0.774833 - 1.225682}{0.632165 + 2.502295} = \frac{-5.15007 \times 10^{-4}}{3.13446} = -1.6430 \times 10^{-4},$$

making the second improved approximation $x_2 = x_1 + \delta_2 = 0.886287$. The method used up to here does not *prove* that this latest answer is accurate to six significant figures, but a further application of the procedure shows that $\delta_3 \approx 3 \times 10^{-7}$.

(b) In order to make use of the result in part (a) we need to make a change of variable that converts the geometric equation into an algebraic one. Since $\tan x$ can be expressed in terms of $\sin x$, if we set $y = \sin x$ in the equation

$\sin x + \tan x = 2$, it will become an algebraic equation:

$$\sin x + \tan x = \sin x + \frac{\sin x}{\cos x} = 2,$$

$$\Rightarrow \quad y + \frac{y}{\sqrt{1 - y^2}} = 2,$$

$$\frac{y^2}{1 - y^2} = (2 - y)^2,$$

$$y^2 = (1 - y^2)(4 - 4y + y^2)$$

$$= -y^4 + 4y^3 - 3y^2 - 4y + 4,$$

$$0 = y^4 - 4y^3 + 4y^2 + 4y - 4.$$

This is the equation that is to be solved. Thus, since $x = 0.886287$ is an approximation to the solution of $\sin x + \tan x = 2$, $y = \sin x = 0.774730$ is an approximation to one of the solutions of $y^4 - 4y^3 + 4y^2 + 4y - 4 = 0$ to the same degree of accuracy.

We note that an equally plausible change of variable is to set $y = \tan x$, with $\sin x$ expressed as $\tan x / \sec x$, i.e. as $y / \sqrt{1 + y^2}$. With this substitution the resulting algebraic equation is the quartic $y^4 - 4y^3 + 4y^2 - 4y + 4 = 0$ (very similar to, but not exactly the same as, the given quartic equation). The reader may wish to verify this. By a parallel argument to that above, $y = \tan 0.886287 = 1.225270$ is an approximate solution of this second quartic equation.

4.33 *In quantum theory, a system of oscillators, each of fundamental frequency v and interacting at temperature T, has an average energy $\bar{E}$ given by*

$$\bar{E} = \frac{\sum_{n=0}^{\infty} nhv e^{-nx}}{\sum_{n=0}^{\infty} e^{-nx}},$$

where $x = hv/kT$, h and k being the Planck and Boltzmann constants, respectively. Prove that both series converge, evaluate their sums, and show that at high temperatures $\bar{E} \approx kT$, whilst at low temperatures $\bar{E} \approx hv \exp(-hv/kT)$.

In the expression

$$\bar{E} = \frac{\sum_{n=0}^{\infty} nhv e^{-nx}}{\sum_{n=0}^{\infty} e^{-nx}},$$

the ratio of successive terms in the series in the numerator is given by

$$\left| \frac{a_{n+1}}{a_n} \right| = \left| \frac{(n+1)hv e^{-(n+1)x}}{nhv e^{-nx}} \right| = \left| \frac{n+1}{n} e^{-x} \right| \to e^{-x} \quad \text{as} \quad n \to \infty,$$

where $x = hv/kT$. Since $x > 0$, $e^{-x} < 1$, and the series is convergent by the ratio test.

The series in the denominator is a geometric series with common ratio $r = e^{-x}$. This is < 1 and so the series converges with sum

$$S(x) = 1 + e^{-x} + e^{-2x} + \cdots + e^{-nx} + \cdots = \frac{1}{1 - e^{-x}}.$$

Now consider

$$-\frac{dS(x)}{dx} = e^{-x} + 2e^{-2x} + \cdots + ne^{-nx} + \cdots.$$

The series on the RHS, when multiplied by hv, gives the numerator in the expression for $\bar{E}$; the numerator therefore has the value

$$-\frac{dS(x)}{dx} = -\frac{d}{dx}\left(\frac{1}{1 - e^{-x}}\right) = \frac{e^{-x}}{(1 - e^{-x})^2}.$$

Hence,

$$\bar{E} = \frac{hv\, e^{-x}}{(1 - e^{-x})^2} \frac{1 - e^{-x}}{1} = \frac{hv}{e^x - 1}.$$

At high temperatures, $x \ll 1$ and

$$\bar{E} = \frac{hv}{\left(1 + \frac{hv}{kT} + \cdots\right) - 1} \approx kT.$$

At low temperatures, $x \gg 1$ and $e^x \gg 1$. Thus the -1 in the denominator can be neglected and $\bar{E} \approx hv \exp(-hv/kT)$.

4.35 *One of the factors contributing to the high relative permittivity of water to static electric fields is the permanent electric dipole moment, p, of the water molecule. In an external field E the dipoles tend to line up with the field, but they do not do so completely because of thermal agitation corresponding to the temperature, T, of the water. A classical (non-quantum) calculation using the Boltzmann distribution shows that the average polarisability per molecule, α, is given by*

$$\alpha = \frac{p}{E}(\coth x - x^{-1}),$$

where $x = pE/(kT)$ and k is the Boltzmann constant.

At ordinary temperatures, even with high field strengths ($10^4\,\mathrm{V\,m^{-1}}$ or more), $x \ll 1$. By making suitable series expansions of the hyperbolic functions involved, show that $\alpha = p^2/(3kT)$ to an accuracy of about one part in $15x^{-2}$.

As $x \ll 1$, we have to deal with a function that is the difference between two terms that individually tend to infinity as $x \to 0$. We will need to expand each in a series and consider the leading non-cancelling terms. The coth function will have to be expressed in terms of the series for the sinh and cosh functions, as follows:

$$\alpha = \frac{p}{E}\left(\coth x - \frac{1}{x}\right), \quad \text{with} \quad x = \frac{pE}{kT},$$

$$= \frac{p}{E}\left(\frac{\cosh x}{\sinh x} - \frac{1}{x}\right)$$

$$= \frac{p}{E}\left[\frac{1 + \frac{x^2}{2!} + \frac{x^4}{4!} + \cdots}{x\left(1 + \frac{x^2}{3!} + \frac{x^4}{5!} + \cdots\right)} - \frac{1}{x}\right]$$

$$= \frac{p}{Ex}\left\{\left(1 + \frac{x^2}{2!} + \frac{x^4}{4!} + \cdots\right)\left[1 - \left(\frac{x^2}{3!} + \frac{x^4}{5!} + \cdots\right)\right.\right.$$

$$\left.\left. + \left(\frac{x^2}{3!} + \frac{x^4}{5!} + \cdots\right)^2 + \cdots\right] - 1\right\}$$

$$= \frac{p}{Ex}\left[0 + x^2\left(\frac{1}{2!} - \frac{1}{3!}\right) + x^4\left(-\frac{1}{5!} + \frac{1}{(3!)^2} - \frac{1}{2!\,3!} + \frac{1}{4!}\right) + \cdots\right]$$

$$= \frac{px}{E}\left(\frac{1}{3} - \frac{x^2}{45} + \cdots\right).$$

Thus the polarisability $\approx px/3E = p^2/3kT$, with the correction term being a factor of about $x^2/15$ smaller.

Partial differentiation

5.1 *Using the appropriate properties of ordinary derivatives, perform the following.*

(a) *Find all the first partial derivatives of the following functions $f(x, y)$:*
   (i) $x^2 y$, (ii) $x^2 + y^2 + 4$, (iii) $\sin(x/y)$, (iv) $\tan^{-1}(y/x)$,
   (v) $r(x, y, z) = (x^2 + y^2 + z^2)^{1/2}$.
(b) *For (i), (ii) and (v), find $\partial^2 f/\partial x^2$, $\partial^2 f/\partial y^2$ and $\partial^2 f/\partial x \partial y$.*
(c) *For (iv) verify that $\partial^2 f/\partial x \partial y = \partial^2 f/\partial y \partial x$.*

These are all straightforward applications of the definitions of partial derivatives.

(a) (i) $\dfrac{\partial f}{\partial x} = \dfrac{\partial(x^2 y)}{\partial x} = 2xy$; $\quad \dfrac{\partial f}{\partial y} = \dfrac{\partial(x^2 y)}{\partial y} = x^2$.

(ii) $\dfrac{\partial f}{\partial x} = \dfrac{\partial(x^2 + y^2 + 4)}{\partial x} = 2x$; $\quad \dfrac{\partial f}{\partial y} = \dfrac{\partial(x^2 + y^2 + 4)}{\partial y} = 2y$.

(iii) $\dfrac{\partial f}{\partial x} = \dfrac{\partial}{\partial x} \sin\left(\dfrac{x}{y}\right) = \cos\left(\dfrac{x}{y}\right) \dfrac{1}{y}$;

$\dfrac{\partial f}{\partial y} = \dfrac{\partial}{\partial y} \sin\left(\dfrac{x}{y}\right) = \cos\left(\dfrac{x}{y}\right) \dfrac{-x}{y^2}$.

(iv) $\dfrac{\partial f}{\partial x} = \dfrac{\partial}{\partial x} \left[\tan^{-1}\left(\dfrac{y}{x}\right)\right] = \dfrac{1}{1 + \frac{y^2}{x^2}} \dfrac{-y}{x^2} = -\dfrac{y}{x^2 + y^2}$;

$\dfrac{\partial f}{\partial y} = \dfrac{\partial}{\partial y} \left[\tan^{-1}\left(\dfrac{y}{x}\right)\right] = \dfrac{1}{1 + \frac{y^2}{x^2}} \dfrac{1}{x} = \dfrac{x}{x^2 + y^2}$.

(v) $\dfrac{\partial r}{\partial x} = \dfrac{\partial(x^2 + y^2 + z^2)^{1/2}}{\partial x} = \dfrac{\frac{1}{2} \times 2x}{(x^2 + y^2 + z^2)^{1/2}} = \dfrac{x}{r}$;

similarly for $\dfrac{\partial r}{\partial y}$ and $\dfrac{\partial r}{\partial z}$.

(b) (i) $\dfrac{\partial^2(x^2y)}{\partial x^2} = \dfrac{\partial(2xy)}{\partial x} = 2y; \quad \dfrac{\partial^2(x^2y)}{\partial y^2} = \dfrac{\partial(x^2)}{\partial y} = 0;$

$\dfrac{\partial^2(x^2y)}{\partial x \partial y} = \dfrac{\partial(x^2)}{\partial x} = 2x.$

(ii) $\dfrac{\partial^2(x^2 + y^2 + 4)}{\partial x^2} = \dfrac{\partial(2x)}{\partial x} = 2; \quad \dfrac{\partial^2(x^2 + y^2 + 4)}{\partial y^2} = \dfrac{\partial(2y)}{\partial y} = 2;$

$\dfrac{\partial^2(x^2 + y^2 + 4)}{\partial x \partial y} = \dfrac{\partial(2y)}{\partial x} = 0.$

(v) $\dfrac{\partial^2(x^2 + y^2 + z^2)^{1/2}}{\partial x^2} = \dfrac{\partial}{\partial x}\left(\dfrac{x}{r}\right) = \dfrac{1}{r} - \dfrac{x}{r^2}\dfrac{\partial r}{\partial x}$

$= \dfrac{1}{r} - \dfrac{x}{r^2}\dfrac{x}{r} = \dfrac{y^2 + z^2}{r^3};$

similarly for $\dfrac{\partial^2 r}{\partial y^2};$

$\dfrac{\partial^2(x^2 + y^2 + z^2)^{1/2}}{\partial x \partial y} = \dfrac{\partial}{\partial x}\left(\dfrac{y}{r}\right) = -\dfrac{y}{r^2}\dfrac{x}{r} = -\dfrac{xy}{r^3}.$

(c) $\dfrac{\partial^2 f}{\partial y \partial x} = \dfrac{\partial}{\partial y}\left(\dfrac{-y}{x^2 + y^2}\right) = -\dfrac{(x^2 + y^2) - y\,2y}{(x^2 + y^2)^2} = \dfrac{y^2 - x^2}{(x^2 + y^2)^2}$

and

$\dfrac{\partial^2 f}{\partial x \partial y} = \dfrac{\partial}{\partial x}\left(\dfrac{x}{x^2 + y^2}\right) = \dfrac{(x^2 + y^2) - x\,2x}{(x^2 + y^2)^2} = \dfrac{y^2 - x^2}{(x^2 + y^2)^2},$

thus verifying the general result for this particular case.

5.3 *Show that the differential*

$$df = x^2\,dy - (y^2 + xy)\,dx$$

is not exact, but that $dg = (xy^2)^{-1}df$ *is exact.*

If $df = A\,dx + B\,dy$ then a necessary and sufficient condition for df to be exact is

$$\dfrac{\partial A(x, y)}{\partial y} = \dfrac{\partial B(x, y)}{\partial x}.$$

Here $A = -(y^2 + xy)$ and $B = x^2$, and so we calculate

$$\dfrac{\partial(x^2)}{\partial x} = 2x \quad \text{and} \quad \dfrac{\partial(-y^2 - xy)}{\partial y} = -2y - x.$$

These are not equal and so df is *not* an exact differential.

However, for dg, $A = -(y^2 + xy)/(xy^2)$ and $B = x^2/(xy^2)$. Taking the appropriate partial derivatives gives

$$\frac{\partial}{\partial x}\left(\frac{x^2}{xy^2}\right) = \frac{1}{y^2} \quad \text{and} \quad \frac{\partial}{\partial y}\left(\frac{-y^2 - xy}{xy^2}\right) = 0 + \frac{1}{y^2}.$$

These are equal, implying that dg *is* an exact differential and that the original inexact differential has $1/xy^2$ as its integrating factor.

5.5 *The equation* $3y = z^3 + 3xz$ *defines z implicitly as a function of x and y. Evaluate all three second partial derivatives of z with respect to x and/or y. Verify that z is a solution of*

$$x\frac{\partial^2 z}{\partial y^2} + \frac{\partial^2 z}{\partial x^2} = 0.$$

By successive partial differentiations of

$$3y = z^3 + 3xz \qquad (*)$$

and its derivatives with respect to (wrt) x and y, we obtain the following.

Of (*) wrt x
$$0 = 3z^2\frac{\partial z}{\partial x} + 3z + 3x\frac{\partial z}{\partial x},$$

(i)
$$\Rightarrow \quad \frac{\partial z}{\partial x} = -\frac{z}{x + z^2}.$$

Of (*) wrt y
$$3 = 3z^2\frac{\partial z}{\partial y} + 3x\frac{\partial z}{\partial y},$$

(ii)
$$\Rightarrow \quad \frac{\partial z}{\partial y} = \frac{1}{x + z^2}.$$

For the second derivatives:

differentiating (i) wrt x
$$\frac{\partial^2 z}{\partial x^2} = -\frac{(x + z^2)\frac{\partial z}{\partial x} - z\left(1 + 2z\frac{\partial z}{\partial x}\right)}{(x + z^2)^2}$$

$$= \frac{(z^2 - x)\frac{\partial z}{\partial x} + z}{(x + z^2)^2}$$

$$= \frac{(z^2 - x)(-z) + z(x + z^2)}{(x + z^2)^3}, \quad \text{using (i)},$$

$$= \frac{2xz}{(x + z^2)^3};$$

differentiating (i) wrt y

$$\frac{\partial^2 z}{\partial y \partial x} = -\frac{(x+z^2)\frac{\partial z}{\partial y} - z \, 2z \frac{\partial z}{\partial y}}{(x+z^2)^2}$$

$$= \frac{(z^2 - x)\frac{\partial z}{\partial y}}{(x+z^2)^2}$$

$$= \frac{z^2 - x}{(x+z^2)^3}, \qquad \text{using (ii);}$$

differentiating (ii) wrt y

$$\frac{\partial^2 z}{\partial y^2} = \frac{-1}{(x+z^2)^2} 2z \frac{\partial z}{\partial y}$$

$$= \frac{-2z}{(x+z^2)^3}, \qquad \text{using (ii).}$$

We now have that

$$x\frac{\partial^2 z}{\partial y^2} + \frac{\partial^2 z}{\partial x^2} = \frac{-2zx}{(x+z^2)^3} + \frac{2zx}{(x+z^2)^3} = 0,$$

i.e. z is a solution of the given partial differential equation.

5.7 *The function $G(t)$ is defined by*

$$G(t) = F(x, y) = x^2 + y^2 + 3xy,$$

where $x(t) = at^2$ and $y(t) = 2at$. Use the chain rule to find the values of (x, y) at which $G(t)$ has stationary values as a function of t. Do any of them correspond to the stationary points of $F(x, y)$ as a function of x and y?

Using the chain rule,

$$\frac{dG}{dt} = \frac{\partial F}{\partial x}\frac{dx}{dt} + \frac{\partial F}{\partial y}\frac{dy}{dt}$$

$$= (2x + 3y)2at + (2y + 3x)2a$$

$$= 2at(2at^2 + 6at) + 2a(4at + 3at^2)$$

$$= 2a^2 t(2t^2 + 9t + 4)$$

$$= 2a^2 t(2t + 1)(t + 4).$$

Thus dG/dt has zeroes at $t = 0$, $t = -\frac{1}{2}$ and $t = -4$; the corresponding values of (x, y) are $(0, 0)$, $(\frac{1}{4}a, -a)$ and $(16a, -8a)$.

Considered as a function of x and y, $F(x, y)$ has stationary points when

$$\frac{\partial F}{\partial x} = 2x + 3y = 0,$$

$$\frac{\partial F}{\partial y} = 3x + 2y = 0.$$

The *only* solution to this pair of equations is $(x, y) = (0, 0)$, which corresponds to (only) one of the points found previously. This stationary point is a saddle point at the origin and is the only stationary point of $F(x, y)$.

The stationary points of $G(t)$ as a function of t are a maximum of $5a^2/16$ at $(\frac{1}{4}a, -a)$, a minimum of $-64a^2$ at $(16a, -8a)$, and a point of inflection at the origin. The first two are not stationary points of $F(x, y)$ for general values of x and y. They only appear to be so because the parameterisation, which restricts the search to the (one-dimensional) line defined by the parabola $y^2 = 4ax$, does not take into account the values of $F(x, y)$ at points close to, but not on, the line.

5.9 *The function $f(x, y)$ satisfies the differential equation*

$$y \frac{\partial f}{\partial x} + x \frac{\partial f}{\partial y} = 0.$$

By changing to new variables $u = x^2 - y^2$ and $v = 2xy$, show that f is, in fact, a function of $x^2 - y^2$ only.

In order to use the equations

$$\frac{\partial f}{\partial x_j} = \sum_{i=1}^{n} \frac{\partial f}{\partial u_i} \frac{\partial u_i}{\partial x_j}$$

that govern a change of variables, we need the partial derivatives

$$\frac{\partial u}{\partial x} = 2x, \quad \frac{\partial u}{\partial y} = -2y, \quad \frac{\partial v}{\partial x} = 2y, \quad \frac{\partial v}{\partial y} = 2x.$$

Then, with $f(x, y)$ written as $g(u, v)$,

$$\frac{\partial f}{\partial x} = 2x \frac{\partial g}{\partial u} + 2y \frac{\partial g}{\partial v},$$
$$\frac{\partial f}{\partial y} = -2y \frac{\partial g}{\partial u} + 2x \frac{\partial g}{\partial v}.$$

Thus,

$$y \frac{\partial f}{\partial x} + x \frac{\partial f}{\partial y} = (2xy - 2xy) \frac{\partial g}{\partial u} + 2(y^2 + x^2) \frac{\partial g}{\partial v}$$

and the equation reduces to

$$\frac{\partial g}{\partial v} = 0 \quad \Rightarrow \quad g = g(u), \text{ i.e. } f(x, y) = g(x^2 - y^2) \text{ only.}$$

> **5.11** *Find and evaluate the maxima, minima and saddle points of the function*
> $$f(x, y) = xy(x^2 + y^2 - 1).$$

The required derivatives are given by

$$\frac{\partial f}{\partial x} = 3x^2 y + y^3 - y, \qquad \frac{\partial f}{\partial y} = x^3 + 3y^2 x - x,$$

$$\frac{\partial^2 f}{\partial x^2} = 6xy, \qquad \frac{\partial^2 f}{\partial x \partial y} = 3x^2 + 3y^2 - 1, \qquad \frac{\partial^2 f}{\partial y^2} = 6xy.$$

Any stationary points must satisfy both of the equations

$$\frac{\partial f}{\partial x} = y(3x^2 + y^2 - 1) = 0,$$
$$\frac{\partial f}{\partial y} = x(x^2 + 3y^2 - 1) = 0.$$

If $x = 0$ then $y = 0$ or ± 1. If $y = 0$ then $x = 0$ or ± 1.

Otherwise, adding and subtracting the factors in parentheses gives

$$4(x^2 + y^2) = 2,$$
$$2(x^2 - y^2) = 0.$$

These have the solutions $x = \pm\frac{1}{2}$, $y = \pm\frac{1}{2}$.

Thus the nine stationary points are $(0,0)$, $(0,\pm 1)$, $(\pm 1, 0)$, $\pm(\frac{1}{2}, \frac{1}{2})$ and $\pm(\frac{1}{2}, -\frac{1}{2})$. The corresponding values for $f(x, y)$ are 0 for the first five, $-\frac{1}{8}$ for the next two and $\frac{1}{8}$ for the final two.

For the first five cases, $\partial^2 f / \partial^2 x = \partial^2 f / \partial^2 y = 0$, whilst $\partial^2 f / \partial x \partial y = -1$ or 2. Since $(-1)^2 > 0 \times 0$ and $2^2 > 0 \times 0$, these points are all saddle points.

At $\pm(\frac{1}{2}, \frac{1}{2})$, $\partial^2 f / \partial^2 x = \partial^2 f / \partial^2 y = \frac{3}{2}$, whilst $\partial^2 f / \partial x \partial y = \frac{1}{2}$. Since $(\frac{1}{2})^2 < \frac{3}{2} \times \frac{3}{2}$, these two points are either maxima or minima (i.e. not saddle points) and the positive signs for $\partial^2 f / \partial^2 x$ and $\partial^2 f / \partial^2 y$ indicate that they are, in fact, minima.

At $\pm(\frac{1}{2}, -\frac{1}{2})$, $\partial^2 f / \partial^2 x = \partial^2 f / \partial^2 y = -\frac{3}{2}$, whilst $\partial^2 f / \partial x \partial y = \frac{1}{2}$. Since $(\frac{1}{2})^2 < -\frac{3}{2} \times -\frac{3}{2}$, these two points are also either maxima or minima; the common negative sign for $\partial^2 f / \partial^2 x$ and $\partial^2 f / \partial^2 y$ indicates that they are maxima.

5.13 *Locate the stationary points of the function*

$$f(x,y) = (x^2 - 2y^2)\exp[-(x^2 + y^2)/a^2],$$

where a is a non-zero constant.

Sketch the function along the x- and y-axes and hence identify the nature and values of the stationary points.

To find the stationary points, we set each of the two first partial derivatives,

$$\frac{\partial f}{\partial x} = \left[2x - \frac{2x}{a^2}(x^2 - 2y^2)\right]\exp\left(-\frac{x^2 + y^2}{a^2}\right),$$

$$\frac{\partial f}{\partial y} = \left[-4y - \frac{2y}{a^2}(x^2 - 2y^2)\right]\exp\left(-\frac{x^2 + y^2}{a^2}\right),$$

equal to zero:

$$\frac{\partial f}{\partial x} = 0 \quad \Rightarrow \quad x = 0 \text{ or } x^2 - 2y^2 = a^2;$$

$$\frac{\partial f}{\partial y} = 0 \quad \Rightarrow \quad y = 0 \text{ or } x^2 - 2y^2 = -2a^2.$$

Since $a \neq 0$, possible solutions for (x,y) are $(0,0)$, $(0,\pm a)$ and $(\pm a, 0)$. The corresponding values are $f(0,0) = 0$, $f(0,\pm a) = -2a^2 e^{-1}$ and $f(\pm a, 0) = a^2 e^{-1}$. These results, taken together with the observation that $|f(x,y)| \to 0$ as either or both of $|x|$ and $|y| \to \infty$, show that $f(x,y)$ has maxima at $(\pm a, 0)$, minima at $(0, \pm a)$ and a saddle point at the origin.

Sketches of $f(x,0)$ and $f(0,y)$, whilst hardly necessary, illustrate rather than confirm these conclusions.

5.15 *Find the stationary values of*

$$f(x,y) = 4x^2 + 4y^2 + x^4 - 6x^2y^2 + y^4$$

and classify them as maxima, minima or saddle points. Make a rough sketch of the contours of f in the quarter plane x, y ≥ 0.

The required derivatives are as follows:

$$\frac{\partial f}{\partial x} = 8x + 4x^3 - 12xy^2, \qquad \frac{\partial f}{\partial y} = 8y - 12x^2y + 4y^3,$$

$$\frac{\partial^2 f}{\partial x^2} = 8 + 12x^2 - 12y^2, \qquad \frac{\partial^2 f}{\partial x \partial y} = -24xy, \qquad \frac{\partial^2 f}{\partial y^2} = 8 - 12x^2 + 12y^2.$$

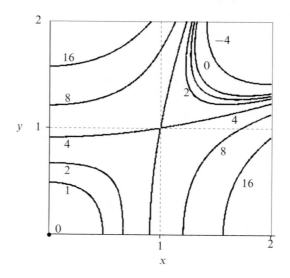

Figure 5.1   The contours found in exercise 5.15.

Any stationary points must satisfy both of the equations

$$\frac{\partial f}{\partial x} = 4x(2 + x^2 - 3y^2) = 0,$$

$$\frac{\partial f}{\partial y} = 4y(2 - 3x^2 + y^2) = 0.$$

If $x = 0$ then $4y(2 + y^2) = 0$, implying that $y = 0$ also, since $2 + y^2 = 0$ has no real solutions. Conversely, $y = 0$ implies $x = 0$. Further solutions exist if both expressions in parentheses equal zero; this requires $x^2 = y^2 = 1$.

Thus the stationary points are $(0,0)$, $(1,1)$, $(-1,1)$, $(1,-1)$ and $(-1,-1)$, with corresponding values 0, 4, 4, 4 and 4.

At $(0,0)$, $\partial^2 f/\partial^2 x = \partial^2 f/\partial^2 y = 8$, whilst $\partial^2 f/\partial x \partial y = 0$. Since $0^2 < 8 \times 8$, this point is a minimum.

In the other four cases, $\partial^2 f/\partial^2 x = \partial^2 f/\partial^2 y = 8$, whilst $\partial^2 f/\partial x \partial y = \pm 24$. Since $(24)^2 > 8 \times 8$, these four points are all saddle points.

It will probably be helpful when sketching the contours (figure 5.1) to determine the behaviour of $f(x, y)$ along the line $x = y$ and to note the symmetry about it. In particular, note that $f(x, y) = 0$ at both the origin and the point $(\sqrt{2}, \sqrt{2})$.

78

> **5.17** *A rectangular parallelepiped has all eight vertices on the ellipsoid*
>
> $$x^2 + 3y^2 + 3z^2 = 1.$$
>
> *Using the symmetry of the parallelepiped about each of the planes $x = 0$, $y = 0$, $z = 0$, write down the surface area of the parallelepiped in terms of the coordinates of the vertex that lies in the octant $x, y, z \geq 0$. Hence find the maximum value of the surface area of such a parallelepiped.*

Let S be the surface area and (x, y, z) the coordinates of one of the corners of the parallelepiped with x, y and z all positive. Then we need to maximise $S = 8(xy + yz + zx)$ subject to x, y and z satisfying $x^2 + 3y^2 + 3z^2 = 1$.

Consider

$$f(x, y, z) = 8(xy + yz + zx) + \lambda(x^2 + 3y^2 + 3z^2),$$

where λ is a Lagrange undetermined multipier. Then, setting each of the first partial derivatives separately to zero, we have the simultaneous equations

$$0 = \frac{\partial f}{\partial x} = 8y + 8z + 2\lambda x,$$

$$0 = \frac{\partial f}{\partial y} = 8x + 8z + 6\lambda y,$$

$$0 = \frac{\partial f}{\partial z} = 8x + 8y + 6\lambda z.$$

From symmetry, $y = z$, leading to

$$0 = 16y + 2\lambda x,$$

$$0 = 8x + 8y + 6\lambda y.$$

Thus, rejecting the trivial solution $x = 0$, $y = 0$, we conclude that $\lambda = -8y/x$, leading to $x^2 + xy - 6y^2 = (x - 2y)(x + 3y) = 0$. The only solution to this quadratic equation with x, y and z all positive is $x = 2y = 2z$. Substituting this into the equation of the ellipse gives

$$(2y)^2 + 3y^2 + 3y^2 = 1 \qquad \Rightarrow \qquad y = \frac{1}{\sqrt{10}}.$$

The value of S is then given by

$$S = 8\left(\frac{2}{10} + \frac{1}{10} + \frac{2}{10}\right) = 4.$$

> **5.19** *A barn is to be constructed with a uniform cross-sectional area A throughout its length. The cross-section is to be a rectangle of wall height h (fixed) and width w, surmounted by an isosceles triangular roof that makes an angle θ with the horizontal. The cost of construction is α per unit height of wall and β per unit (slope) length of roof. Show that, irrespective of the values of α and β, to minimise costs w should be chosen to satisfy the equation*
>
> $$w^4 = 16A(A - wh),$$
>
> *and θ made such that* $2 \tan 2\theta = w/h$.

The cost *always* includes $2\alpha h$ for the vertical walls, which can therefore be ignored in the minimisation procedure. The rest of the calculation will be solely concerned with minimising the roof area, and the optimum choices for w and θ will be independent of β, the actual cost per unit length of the roof.

The cost of the roof is $2\beta \times \frac{1}{2}w \sec \theta$, but w and θ are constrained by the requirement that

$$A = wh + \frac{1}{2}w\frac{w}{2}\tan \theta.$$

So we consider $G(w, \theta)$, where

$$G(w, \theta) = \beta w \sec \theta - \lambda(wh + \tfrac{1}{4}w^2 \tan \theta),$$

and the implications of equating its partial derivatives to zero. The first derivative to be set to zero is

$$\frac{\partial G}{\partial \theta} = \beta w \sec \theta \tan \theta - \frac{\lambda}{4}w^2 \sec^2 \theta,$$
$$\Rightarrow \quad 0 = \beta \sin \theta - \tfrac{1}{4}\lambda w,$$
$$\Rightarrow \quad \lambda = \frac{4\beta \sin \theta}{w}.$$

A second equation is provided by differentiation with respect to w and yields

$$\frac{\partial G}{\partial w} = \beta \sec \theta - \lambda h - \tfrac{1}{2}\lambda w \tan \theta.$$

Setting $\partial G/\partial w = 0$, multiplying through by $\cos \theta$ and substituting for λ, we obtain

$$\beta - 2\beta \sin^2 \theta = \frac{4\beta \sin \theta\, h \cos \theta}{w},$$
$$w \cos 2\theta = 2h \sin 2\theta,$$
$$\tan 2\theta = \frac{w}{2h}.$$

This is the second result quoted.

The overall area constraint can be written

$$\tan\theta = \frac{4(A - wh)}{w^2}.$$

From these two results and the double angle formula $\tan 2\phi = 2\tan\phi/(1-\tan^2\phi)$, it follows that

$$\frac{w}{2h} = \tan 2\theta$$

$$= \frac{\dfrac{8(A - wh)}{w^2}}{1 - \dfrac{16(A - wh)^2}{w^4}},$$

$$16wh(A - wh) = w^4 - 16(A - wh)^2,$$

$$w^4 = 16A(A - wh).$$

This is the first quoted result, and we note that, as expected, both optimum values are independent of β.

5.21 *Find the area of the region covered by points on the lines*

$$\frac{x}{a} + \frac{y}{b} = 1,$$

where the sum of any line's intercepts on the coordinate axes is fixed and equal to c.

The equation of a typical line with intercept a on the x-axis is

$$f(x, y, a) = \frac{x}{a} + \frac{y}{c - a} - 1 = 0.$$

To find the envelope of the lines we set $\partial f/\partial a = 0$. This gives

$$\frac{\partial f}{\partial a} = -\frac{x}{a^2} + \frac{y}{(c - a)^2} = 0.$$

Hence,

$$(c - a)\sqrt{x} = a\sqrt{y},$$

$$a = \frac{c\sqrt{x}}{\sqrt{x} + \sqrt{y}}.$$

Substituting this value into $f(x, y, a) = 0$ gives the equation of the envelope as

$$\frac{x(\sqrt{x} + \sqrt{y})}{c\sqrt{x}} + \frac{y}{c - \dfrac{c\sqrt{x}}{\sqrt{x} + \sqrt{y}}} = 1,$$

$$\sqrt{x}(\sqrt{x} + \sqrt{y}) + \sqrt{y}(\sqrt{x} + \sqrt{y}) = c,$$

$$\sqrt{x} + \sqrt{y} = \sqrt{c}.$$

This is a curve (not a straight line) whose end-points are $(c, 0)$ on the x-axis and $(0, c)$ on the y-axis. All points on lines with the given property lie below this envelope curve (except for one point on each line, which lies on the curve). Consequently, the area covered by the points is that bounded by the envelope and the two axes. It has the value

$$\int_0^c y \, dx = \int_0^c (\sqrt{c} - \sqrt{x})^2 \, dx$$

$$= \int_0^c (c - 2\sqrt{c}\sqrt{x} + x) \, dx$$

$$= c^2 - \tfrac{4}{3}\sqrt{c}\, c^{3/2} + \tfrac{1}{2}c^2 = \tfrac{1}{6}c^2.$$

5.23 *A water feature contains a spray head at water level at the centre of a round basin. The head is in the form of a small hemisphere perforated by many evenly distributed small holes, through which water spurts out at the same speed, v_0, in all directions.*

    (a) *What is the shape of the 'water bell' so formed?*
    (b) *What must be the minimum diameter of the bowl if no water is to be lost?*

The system has cylindrical symmetry and so we work with cylindrical polar coordinates ρ and z.

For a jet of water emerging from the spray head at an angle θ to the vertical, the equations of motion are

$$z = v_0 \cos\theta\, t - \tfrac{1}{2}gt^2,$$
$$\rho = v_0 \sin\theta\, t.$$

Eliminating the time, t, and writing $\cot\theta = \alpha$, we have

$$z = \frac{\rho\, v_0 \cos\theta}{v_0 \sin\theta} - \frac{1}{2}g\frac{\rho^2}{v_0^2 \sin^2\theta},$$

$$\Rightarrow \quad 0 = z - \rho\cot\theta + \frac{g\rho^2}{2v_0^2}\operatorname{cosec}^2\theta,$$

i.e. the trajectory of this jet is given by

$$f(\rho, z, \alpha) = z - \rho\alpha + \frac{g\rho^2}{2v_0^2}(1 + \alpha^2) = 0.$$

To find the envelope of all these trajectories as θ (and hence α) is varied, we set

$\partial f / \partial \alpha$ equal to zero:

$$0 = \frac{\partial f}{\partial \alpha} = 0 - \rho + \frac{2\alpha g \rho^2}{2v_0^2},$$

$$\Rightarrow \quad \alpha = \frac{v_0^2}{g\rho}.$$

Hence, the equation of the envelope, and thus of the water bell, is

$$g(\rho, z) = z - \frac{v_0^2}{g} + \frac{g\rho^2}{2v_0^2}\left(1 + \frac{v_0^4}{g^2\rho^2}\right) = 0,$$

$$\Rightarrow \quad z = \frac{v_0^2}{2g} - \frac{g\rho^2}{2v_0^2}.$$

(a) This is the equation of a parabola whose apex is at $z = v_0^2/2g$, $\rho = 0$. It follows that the water bell has the shape of an inverted paraboloid of revolution.

(b) When $z = 0$, ρ has the value v_0^2/g, and hence the minimum value needed for the diameter of the bowl is given by $2\rho = 2v_0^2/g$.

5.25 *By considering the differential*

$$dG = d(U + PV - ST),$$

where G is the Gibbs free energy, P the pressure, V the volume, S the entropy and T the temperature of a system, and given further that U, the internal energy, satisfies

$$dU = T dS - P dV,$$

derive a Maxwell relation connecting $(\partial V / \partial T)_P$ and $(\partial S / \partial P)_T$.

Given that $dU = T dS - P dV$, we have that

$$dG = d(U + PV - ST)$$
$$= dU + P dV + V dP - S dT - T dS$$
$$= V dP - S dT.$$

Hence,

$$\left(\frac{\partial G}{\partial P}\right)_T = V \quad \text{and} \quad \left(\frac{\partial G}{\partial T}\right)_P = -S.$$

It follows that

$$\left(\frac{\partial V}{\partial T}\right)_P = \frac{\partial^2 G}{\partial T \partial P} = \frac{\partial^2 G}{\partial P \partial T} = -\left(\frac{\partial S}{\partial P}\right)_T.$$

This is the required Maxwell thermodynamic relation.

> **5.27** *As implied in exercise 5.25 on the thermodynamics of a simple gas, the quantity $dS = T^{-1}(dU + P\,dV)$ is an exact differential. Use this to prove that*
>
> $$\left(\frac{\partial U}{\partial V}\right)_T = T\left(\frac{\partial P}{\partial T}\right)_V - P.$$
>
> *In the van der Waals model of a gas, P obeys the equation*
>
> $$P = \frac{RT}{V-b} - \frac{a}{V^2},$$
>
> *where R, a and b are constants. Further, in the limit $V \to \infty$, the form of U becomes $U = cT$, where c is another constant. Find the complete expression for $U(V,T)$.*

Writing the total differentials in $dS = T^{-1}(dU + P\,dV)$ in terms of partial derivatives with respect to V and T gives

$$T\left(\frac{\partial S}{\partial V}\right)_T dV + T\left(\frac{\partial S}{\partial T}\right)_V dT = \left(\frac{\partial U}{\partial V}\right)_T dV + \left(\frac{\partial U}{\partial T}\right)_V dT + P\,dV,$$

from which it follows that

$$T\left(\frac{\partial S}{\partial V}\right)_T = \left(\frac{\partial U}{\partial V}\right)_T + P \quad (*) \quad \text{and} \quad T\left(\frac{\partial S}{\partial T}\right)_V = \left(\frac{\partial U}{\partial T}\right)_V.$$

Differentiating the first of these with respect to T and the second with respect to V, and then combining the two equations so obtained, gives

$$\left(\frac{\partial S}{\partial V}\right)_T + T\frac{\partial^2 S}{\partial T \partial V} = \frac{\partial^2 U}{\partial T \partial V} + \left(\frac{\partial P}{\partial T}\right)_V,$$

$$T\frac{\partial^2 S}{\partial V \partial T} = \frac{\partial^2 U}{\partial V \partial T},$$

$$\Rightarrow \quad \left(\frac{\partial S}{\partial V}\right)_T = \left(\frac{\partial P}{\partial T}\right)_V.$$

The equation (*) can now be written in the required form:

$$\left(\frac{\partial U}{\partial V}\right)_T = T\left(\frac{\partial P}{\partial T}\right)_V - P.$$

For the van der Waals model gas,

$$P = \frac{RT}{V-b} - \frac{a}{V^2},$$

and we can substitute for P in the previous result to give

$$\left(\frac{\partial U}{\partial V}\right)_T = T\left(\frac{R}{V-b}\right) - \left(\frac{RT}{V-b} - \frac{a}{V^2}\right) = \frac{a}{V^2},$$

which integrates to

$$U(V,T) = -\frac{a}{V} + f(T).$$

Since $U \to cT$ as $V \to \infty$ for all T, the unknown function, $f(T)$, must be simply $f(T) = cT$. Thus, the full expression for $U(V,T)$ is

$$U(V,T) = cT - \frac{a}{V}.$$

We note that, in the limit $V \to \infty$, van der Waals' equation becomes $PV = RT$ and thus recognise c as the specific heat at constant volume of a perfect gas.

5.29 *By finding dI/dy, evaluate the integral*

$$I(y) = \int_0^\infty \frac{e^{-xy}\sin x}{x}\, dx.$$

Hence show that

$$J = \int_0^\infty \frac{\sin x}{x}\, dx = \frac{\pi}{2}.$$

Since the integral is over positive values of x, its convergence requires that $y \geq 0$. We first express the $\sin x$ factor as a complex exponential:

$$I(y) = \int_0^\infty \frac{e^{-xy}\sin x}{x}\, dx$$

$$= \mathrm{Im}\, \int_0^\infty \frac{e^{-xy+ix}}{x}\, dx.$$

And now differentiate under the integral sign:

$$\frac{dI}{dy} = \mathrm{Im}\, \int_0^\infty \frac{(-x)e^{-xy+ix}}{x}\, dx$$

$$= \mathrm{Im}\, \left[\frac{-e^{-xy+ix}}{-y+i}\right]_0^\infty$$

$$= \mathrm{Im}\, \left(\frac{1}{-y+i}\right)$$

$$= -\frac{1}{1+y^2}.$$

This differential equation expresses how the integral varies as a function of y.

But, as we can see immediately that for $y = \infty$ the integral must be zero, we can find its value for non-infinite y by integrating the differential equation:

$$I(y) - I(\infty) = \int_\infty^y \frac{-1}{1+y^2}\,dy = -\tan^{-1} y + \tan^{-1}\infty = \frac{\pi}{2} - \tan^{-1} y.$$

In the limit $y \to 0$ this becomes

$$J = \int_0^\infty \frac{\sin x}{x}\,dx = I(0) = \frac{\pi}{2} - 0 = \frac{\pi}{2}.$$

5.31 *The function $f(x)$ is differentiable and $f(0) = 0$. A second function $g(y)$ is defined by*

$$g(y) = \int_0^y \frac{f(x)\,dx}{\sqrt{y-x}}.$$

Prove that

$$\frac{dg}{dy} = \int_0^y \frac{df}{dx}\frac{dx}{\sqrt{y-x}}.$$

For the case $f(x) = x^n$, prove that

$$\frac{d^n g}{dy^n} = 2(n!)\sqrt{y}.$$

Integrating the definition of $g(y)$ by parts:

$$g(y) = \int_0^y \frac{f(x)\,dx}{\sqrt{y-x}}$$

$$= \left[-2f(x)\sqrt{y-x}\right]_0^y + \int_0^y 2\frac{df}{dx}\sqrt{y-x}\,dx$$

$$= 2\int_0^y \frac{df}{dx}\sqrt{y-x},$$

where we have used $f(0) = 0$ in setting the definite integral to zero.

Now, differentiating $g(y)$ with respect to both its upper limit and its integrand, we obtain

$$\frac{dg}{dy} = 2\frac{df}{dx}\sqrt{y-y} + 2\int_0^y \frac{1}{2}\frac{df}{dx}\frac{1}{\sqrt{y-x}} = \int_0^y \frac{df}{dx}\frac{1}{\sqrt{y-x}}.$$

This result, showing that the construction of the derivative of g from the derivative of f is the same as that of g from f, applies to any function that satisfies $f(0) = 0$

and so applies to x^n and all of its derivatives. It follows that

$$\frac{d^n g}{dy^n} = \int_0^y \frac{d^n f}{dx^n} \frac{1}{\sqrt{y-x}}\, dx$$

$$= \int_0^y \frac{n!}{\sqrt{y-x}}\, dx$$

$$= \left[\frac{n!(-1)\sqrt{y-x}}{\frac{1}{2}} \right]_0^y$$

$$= 2(n!)\sqrt{y}.$$

5.33 *If*

$$I(\alpha) = \int_0^1 \frac{x^\alpha - 1}{\ln x}\, dx, \qquad \alpha > -1,$$

what is the value of $I(0)$? *Show that*

$$\frac{d}{d\alpha} x^\alpha = x^\alpha \ln x,$$

and deduce that

$$\frac{d}{d\alpha} I(\alpha) = \frac{1}{\alpha + 1}.$$

Hence prove that $I(\alpha) = \ln(1 + \alpha)$.

Since the integrand is singular at $x = 1$, we need to define $I(0)$ as a limit:

$$I(0) = \lim_{y\to 1} \int_0^y \frac{x^0 - 1}{\ln x}\, dx = \lim_{y\to 1} \int_0^y 0\, dx = \lim_{y\to 1} 0 = 0,$$

i.e. $I(0) = 0$.

With $z = x^\alpha$, we have

$$\ln z = \alpha \ln x \quad \Rightarrow \quad \frac{1}{z}\frac{dz}{d\alpha} = \ln x$$

$$\Rightarrow \quad \frac{dz}{d\alpha} = z \ln x \quad \Rightarrow \quad \frac{d}{d\alpha} x^\alpha = x^\alpha \ln x.$$

The derivative of $I(\alpha)$ is then

$$\frac{dI}{d\alpha} = \int_0^1 \frac{1}{\ln x} x^\alpha \ln x\, dx$$

$$= \left[\frac{x^{\alpha+1}}{\alpha + 1} \right]_0^1$$

$$= \frac{1}{\alpha + 1}.$$

87

Finally, intergation gives

$$I(\alpha) - I(0) = \int_0^\alpha \frac{d\beta}{\beta + 1},$$
$$I(\alpha) - 0 = \ln(1 + \alpha).$$

To obtain this final line we have used our first result that $I(0) = 0$.

5.35   *The function $G(t, \xi)$ is defined for $0 \leq t \leq \pi$ by*

$$G(t, \xi) = -\cos t \sin \xi \quad \text{for } \xi \leq t,$$
$$= -\sin t \cos \xi \quad \text{for } \xi > t.$$

Show that the function $x(t)$ defined by

$$x(t) = \int_0^\pi G(t, \xi) f(\xi) \, d\xi$$

satisfies the equation

$$\frac{d^2 x}{dt^2} + x = f(t),$$

where $f(t)$ can be any arbitrary (continuous) function. Show further that $x(0) = [dx/dt]_{t=\pi} = 0$, again for any $f(t)$, but that the value of $x(\pi)$ does depend upon the form of $f(t)$.

[The function $G(t, \xi)$ is an example of a Green's function, an important concept in the solution of differential equations.]

The explicit integral expression for $x(t)$ is

$$x(t) = \int_0^\pi G(t, \xi) f(\xi) \, d\xi$$
$$= -\int_0^t \cos t \sin \xi \, f(\xi) \, d\xi - \int_t^\pi \sin t \cos \xi \, f(\xi) \, d\xi.$$

We now form its first two derivatives using Leibnitz' rule:

$$\frac{dx}{dt} = -\cos t [\sin t \, f(t)] + \sin t \int_0^t \sin \xi \, f(\xi) \, d\xi$$
$$+ \sin t [\cos t \, f(t)] - \cos t \int_t^\pi \cos \xi \, f(\xi) \, d\xi$$
$$= \sin t \int_0^t \sin \xi \, f(\xi) \, d\xi - \cos t \int_t^\pi \cos \xi \, f(\xi) \, d\xi.$$

$$\frac{d^2x}{dt^2} = \cos t \int_0^t \sin \xi \, f(\xi) \, d\xi + \sin t [\sin t \, f(t)]$$

$$+ \sin t \int_t^\pi \cos \xi \, f(\xi) \, d\xi + \cos t [\cos t \, f(t)]$$

$$= -x(t) + f(t)(\sin^2 t + \cos^2 t).$$

This shows that

$$\frac{d^2x}{dt^2} + x = f(t)$$

for *any* continuous function $f(x)$.

When $t = 0$ the first integral in the expression for $x(t)$ has zero range and the second is multiplied by $\sin 0$; consequently $x(0) = 0$.

When $t = \pi$ the second integral in the expression for dx/dt has zero range and the first is multiplied by $\sin \pi$; consequently $[dx/dt]_{t=\pi} = 0$.

However, when $t = \pi$, although the second integral in the expression for $x(t)$ is multiplied by $\sin \pi$ and contributes nothing, the first integral is not zero in general and its value *will* depend upon the form of $f(t)$.

6

Multiple integrals

6.1 *Identify the curved wedge bounded by the surfaces $y^2 = 4ax$, $x + z = a$ and $z = 0$, and hence calculate its volume V.*

As will readily be seen from a rough sketch, the wedge consists of that part of a parabolic cylinder, parallel to the z-axis, that is cut off by two planes, one parallel to the y-axis and the other the coordinate plane $z = 0$.

For the first stage of the multiple integration, the volume can be divided equally easily into 'vertical columns' or into horizontal strips parallel to the y-axis. Thus there are two equivalent and equally obvious ways of proceeding.

Either

$$V = \int_0^a dx \int_{-\sqrt{4ax}}^{\sqrt{4ax}} dy \int_0^{a-x} dz$$

$$= \int_0^a 2\sqrt{4ax}(a - x)\, dx$$

$$= 4\sqrt{a}\left[\tfrac{2}{3}ax^{3/2} - \tfrac{2}{5}x^{5/2}\right]_0^a = \tfrac{16}{15}a^3;$$

or

$$V = \int_0^a dz \int_0^{a-z} dx \int_{-\sqrt{4ax}}^{\sqrt{4ax}} dy$$

$$= \int_0^a dz \int_0^{a-z} 2\sqrt{4ax}\, dx$$

$$= 4\sqrt{a}\int \tfrac{2}{3}(a - z)^{3/2}\, dz$$

$$= \frac{8\sqrt{a}}{3}\left[-\tfrac{2}{5}(a - z)^{5/2}\right]_0^a = \tfrac{16}{15}a^3.$$

90

> **6.3** *Find the volume integral of x^2y over the tetrahedral volume bounded by the planes $x = 0$, $y = 0$, $z = 0$ and $x + y + z = 1$.*

The bounding surfaces of the integration volume are symmetric in x, y and z and, on these grounds, there is nothing to choose between the various possible orders of integration.

However, the integrand does not contain z and so there is some advantage in carrying out the z-integration first. Its value can simply be set equal to the length of the z-interval and the dimension of the integral will have been reduced by one 'at a stroke'.

$$
\begin{aligned}
I &= \int_0^1 dx \int_0^{1-x} dy \int_0^{1-x-y} x^2 y \, dz \\
&= \int_0^1 dx \int_0^{1-x} x^2 y (1 - x - y) \, dy \\
&= \int_0^1 \left[x^2(1-x)\frac{(1-x)^2}{2} - x^2\frac{(1-x)^3}{3} \right] dx \\
&= \frac{1}{6} \int_0^1 x^2(1 - 3x + 3x^2 - x^3) \, dx \\
&= \frac{1}{6} \left(\frac{1}{3} - \frac{3}{4} + \frac{3}{5} - \frac{1}{6} \right) \\
&= \frac{1}{6} \frac{20 - 45 + 36 - 10}{60} = \frac{1}{360}.
\end{aligned}
$$

> **6.5** *Calculate the volume of an ellipsoid as follows:*
>
> (a) *Prove that the area of the ellipse*
>
> $$\frac{x^2}{a^2} + \frac{y^2}{b^2} = 1$$
>
> *is πab.*
>
> (b) *Use this result to obtain an expression for the volume of a slice of thickness dz of the ellipsoid*
>
> $$\frac{x^2}{a^2} + \frac{y^2}{b^2} + \frac{z^2}{c^2} = 1.$$
>
> *Hence show that the volume of the ellipsoid is $4\pi abc/3$.*

(a) Dividing the ellipse into thin strips parallel to the y-axis, we may write its area as

$$\text{area} = 2 \int_{-a}^{a} y \, dx = 2 \int_{-a}^{a} b \sqrt{1 - \left(\frac{x}{a}\right)^2} \, dx.$$

Set $x = a \cos \phi$ with $dx = -a \sin \phi \, d\phi$. Then

$$\text{area} = 2b \int_{\pi}^{0} \sin \phi (-a \sin \phi) \, d\phi = 2ab \int_{0}^{\pi} \sin^2 \phi \, d\phi = 2ab \frac{\pi}{2} = \pi ab.$$

(b) Consider slices of the ellipsoid, of thickness dz, taken perpendicular to the z-axis. Each is an ellipse whose bounding curve is given by the equation

$$\frac{x^2}{a^2} + \frac{y^2}{b^2} = 1 - \frac{z^2}{c^2}$$

and is thus a scaled-down version of the ellipse considered in part (a) with semi-axes $a(1-(z/c)^2)^{1/2}$ and $b(1-(z/c)^2)^{1/2}$. Its area is therefore $\pi a(1-(z/c)^2)^{1/2}b(1-(z/c)^2)^{1/2}$ and its volume dV is this multiplied by dz. Thus, the total volume V of the ellipsoid is given by

$$\int_{-c}^{c} \pi ab \left(1 - \frac{z^2}{c^2}\right) dz = \pi ab \left[z - \frac{1}{3} \frac{z^3}{c^2}\right]_{-c}^{c} = \frac{4\pi abc}{3}.$$

6.7 *In quantum mechanics the electron in a hydrogen atom in some particular state is described by a wavefunction Ψ, which is such that $|\Psi|^2 \, dV$ is the probability of finding the electron in the infinitesimal volume dV. In spherical polar coordinates $\Psi = \Psi(r, \theta, \phi)$ and $dV = r^2 \sin \theta \, dr \, d\theta \, d\phi$. Two such states are described by*

$$\Psi_1 = \left(\frac{1}{4\pi}\right)^{1/2} \left(\frac{1}{a_0}\right)^{3/2} 2e^{-r/a_0},$$

$$\Psi_2 = -\left(\frac{3}{8\pi}\right)^{1/2} \sin \theta \, e^{i\phi} \left(\frac{1}{2a_0}\right)^{3/2} \frac{re^{-r/2a_0}}{a_0\sqrt{3}}.$$

(a) *Show that each Ψ_i is normalised, i.e. the integral over all space $\int |\Psi|^2 \, dV$ is equal to unity – physically, this means that the electron must be somewhere.*

(b) *The (so-called) dipole matrix element between the states 1 and 2 is given by the integral*

$$p_x = \int \Psi_1^* qr \sin \theta \cos \phi \, \Psi_2 \, dV,$$

where q is the charge on the electron. Prove that p_x has the value $-2^7 q a_0 / 3^5$.

We need to show that the volume integral of $|\Psi_i|^2$ is equal to unity, and begin by noting that, since ϕ is not explicitly mentioned, or appears only in the form $e^{i\phi}$, the ϕ integration of $|\Psi|^2$ yields a factor of 2π in each case. For Ψ_1 we have

$$
\begin{aligned}
\int |\Psi_1|^2 \, dV &= \int |\Psi_1|^2 r^2 \sin\theta \, d\theta \, d\phi \, dr \\
&= \frac{1}{4\pi} \frac{4}{a_0^3} 2\pi \int_0^\infty r^2 e^{-2r/a_0} \, dr \int_0^\pi \sin\theta \, d\theta \\
&= \frac{2}{a_0^3} \int_0^\infty 2r^2 e^{-2r/a_0} \, dr \\
&= \frac{4}{a_0^3} \frac{a_0}{2} 2 \frac{a_0}{2} 1 \frac{a_0}{2} = 1.
\end{aligned}
$$

The last line has been obtained using repeated integration by parts.

For Ψ_2, the corresponding calculation is

$$
\begin{aligned}
\int |\Psi_2|^2 \, dV &= \int |\Psi_2|^2 r^2 \sin\theta \, d\theta \, d\phi \, dr \\
&= \frac{2\pi}{64\pi \, a_0^5} \int_0^\infty r^4 e^{-r/a_0} \, dr \int_0^\pi \sin^3\theta \, d\theta \\
&= \frac{1}{32 \, a_0^5} \int_0^\infty r^4 e^{-r/a_0} \, dr \int_0^\pi (1 - \cos^2\theta) \sin\theta \, d\theta \\
&= \frac{1}{32 \, a_0^5} 4! \, a_0^5 \left(2 - \frac{2}{3} \right) = 1.
\end{aligned}
$$

Again, the r-integral was calculated using integration by parts. In summary, both functions are correctly normalised.

(b) The dipole matrix element has important physical properties, but for the purposes of this exercise it is simply an integral to be evaluated according to a formula, as follows:

$$
\begin{aligned}
p_x &= \int \Psi_1^* q r \sin\theta \cos\phi \, \Psi_2 \, r^2 \sin\theta \, d\theta \, d\phi \, dr \\
&= \frac{-q}{8\pi a_0^4} \int_0^\pi \sin^3\theta \, d\theta \int_0^{2\pi} \cos\phi(\cos\phi + i\sin\phi) \, d\phi \int_0^\infty r^4 e^{-3r/2a_0} \, dr \\
&= -\frac{q}{8\pi a_0^4} \left(2 - \frac{2}{3} \right) (\pi + i0) 4! \left(\frac{2a_0}{3} \right)^5 \\
&= -\frac{2^7}{3^5} q a_0.
\end{aligned}
$$

6.9 *A certain torus has a circular vertical cross-section of radius a centred on a horizontal circle of radius c ($> a$).*

    (a) *Find the volume V and surface area A of the torus, and show that they can be written as*

$$V = \frac{\pi^2}{4}(r_o^2 - r_i^2)(r_o - r_i), \qquad A = \pi^2(r_o^2 - r_i^2),$$

    where r_o and r_i are, respectively, the outer and inner radii of the torus.

    (b) *Show that a vertical circular cylinder of radius c, coaxial with the torus, divides A in the ratio*

$$\pi c + 2a \; : \; \pi c - 2a.$$

(a) The inner and outer radii of the torus are $r_i = c - a$ and $r_o = c + a$, from which it follows that $r_o^2 - r_i^2 = 4ac$ and that $r_o - r_i = 2a$.

The torus is generated by sweeping the centre of a circle of radius a, area πa^2 and circumference $2\pi a$ around a circle of radius c. Therefore, by Pappus' first theorem, the volume of the torus is given by

$$V = \pi a^2 \times 2\pi c = 2\pi^2 a^2 c = \frac{\pi^2}{4}(r_o^2 - r_i^2)(r_o - r_i),$$

whilst, by his second theorem, its surface area is

$$A = 2\pi a \times 2\pi c = 4\pi^2 ac = \pi^2(r_o^2 - r_i^2).$$

(b) The vertical cylinder divides the perimeter of a cross-section of the torus into two equal parts. The distance from the cylinder of the centroid of either half is given by

$$\bar{x} = \frac{\int x\, ds}{\int ds} = \frac{\int_{-\pi/2}^{\pi/2} a\cos\phi\, a\, d\phi}{\int_{-\pi/2}^{\pi/2} a\, d\phi} = \frac{2a}{\pi}.$$

It therefore follows from Pappus' second theorem that

$$A_o = \pi a \times 2\pi \left(c + \frac{2a}{\pi}\right) \quad \text{and} \quad A_i = \pi a \times 2\pi \left(c - \frac{2a}{\pi}\right),$$

leading to the stated result.

6.11 *In some applications in mechanics the moment of inertia of a body about a single point (as opposed to about an axis) is needed. The moment of inertia, I, about the origin of a uniform solid body of density ρ is given by the volume integral*

$$I = \int_V (x^2 + y^2 + z^2)\rho \, dV.$$

Show that the moment of inertia of a right circular cylinder of radius a, length 2b and mass M about its centre is given by

$$M \left(\frac{a^2}{2} + \frac{b^2}{3} \right).$$

Since the cylinder is easily described in cylindrical polar coordinates (ρ, ϕ, z), we convert the calculation to one using those coordinates and denote the density by ρ_0 to avoid confusion:

$$I = \int_V (x^2 + y^2 + z^2)\rho_0 \, dV$$

$$= \rho_0 \int_V (\rho^2 + z^2)\rho \, d\phi \, d\phi \, dz$$

$$= \rho_0 \int_0^{2\pi} d\phi \int_0^a \rho \, d\rho \int_{-b}^b (\rho^2 + z^2) \, dz$$

$$= 2\pi\rho_0 \int_0^a \rho \left(2b\rho^2 + \frac{2b^3}{3} \right) d\rho$$

$$= 2\pi\rho_0 \left(2b\frac{a^4}{4} + \frac{2b^3}{3}\frac{a^2}{2} \right).$$

Now $M = \pi a^2 \times 2b \times \rho_0$, and so the moment of inertia about the origin can be expressed as

$$I = M \left(\frac{a^2}{2} + \frac{b^2}{3} \right).$$

6.13 *In spherical polar coordinates r, θ, φ the element of volume for a body that is symmetrical about the polar axis is $dV = 2\pi r^2 \sin\theta \, dr \, d\theta$, whilst its element of surface area is $2\pi r \sin\theta[(dr)^2 + r^2(d\theta)^2]^{1/2}$. A particular surface is defined by $r = 2a\cos\theta$, where a is a constant and $0 \le \theta \le \pi/2$. Find its total surface area and the volume it encloses, and hence identify the surface.*

With the surface of the body defined by $r = 2a \cos \theta$, for calculating its total volume the radial integration variable r' lies in the range $0 \le r' \le 2a \cos \theta$. Hence

$$V = \int_0^{\pi/2} 2\pi \sin \theta \, d\theta \int_0^{2a \cos \theta} r'^2 \, dr'$$

$$= 2\pi \int_0^{\pi/2} \sin \theta \frac{(2a \cos \theta)^3}{3} \, d\theta$$

$$= \frac{16\pi a^3}{3} \int_0^{\pi/2} \cos^3 \theta \sin \theta \, d\theta$$

$$= \frac{16\pi a^3}{3} \left[-\frac{\cos^4 \theta}{4} \right]_0^{\pi/2}$$

$$= \tfrac{4}{3} \pi a^3.$$

The additional strip of surface area resulting from a change from θ to $\theta + d\theta$ is $2\pi r \sin \theta \, d\ell$, where $d\ell$ is the length of the generating curve that lies in this infinitesimal range of θ. This is given by

$$(d\ell)^2 = (dr)^2 + (r \, d\theta)^2$$

$$= (-2a \sin \theta \, d\theta)^2 + (2a \cos \theta \, d\theta)^2$$

$$= 4a^2 \, (d\theta)^2$$

The integral becomes one-dimensional with

$$S = 2\pi \int_0^{\pi/2} 2a \cos \theta \sin \theta \, 2a \, d\theta$$

$$= 8\pi a^2 \left[\frac{\sin^2 \theta}{2} \right]_0^{\pi/2}$$

$$= 4\pi a^2.$$

With a volume of $\tfrac{4}{3} \pi a^3$ and a surface area of $4\pi a^2$, the surface is probably that of a sphere of radius a, with the origin at the 'lowest' point of the sphere. This conclusion is confirmed by the fact that the triangle formed by the two ends of the vertical diameter of the sphere and any point on its surface is a right-angled triangle in which $r/2a = \cos \theta$.

6.15 *By transforming to cylindrical polar coordinates, evaluate the integral*

$$I = \int \int \int \ln(x^2 + y^2) \, dx \, dy \, dz$$

over the interior of the conical region $x^2 + y^2 \le z^2$, $0 \le z \le 1$.

The volume element $dx \, dy \, dz$ becomes $\rho \, d\rho \, d\phi \, dz$ in cylindrical polar coordinates and the integrand contains a factor $\rho \ln \rho^2 = 2\rho \ln \rho$. This is dealt with using integration by parts and the integral becomes

$$I = \int \int \int 2\rho \, \ln \rho \, d\rho \, d\phi \, dz \quad \text{over} \quad \rho \le z, \; 0 \le z \le 1,$$

$$= 2 \int_0^{2\pi} d\phi \int_0^1 dz \int_0^z \rho \ln \rho \, d\rho$$

$$= 2 \cdot 2\pi \int_0^1 \left(\left[\frac{\rho^2 \ln \rho}{2} \right]_0^z - \int_0^z \frac{1}{\rho} \frac{\rho^2}{2} \, d\rho \right) dz$$

$$= 4\pi \int_0^1 \left(\frac{1}{2} z^2 \ln z - \frac{1}{4} z^2 \right) dz$$

$$= 2\pi \left(\left[\frac{z^3 \ln z}{3} \right]_0^1 - \int_0^1 \frac{1}{z} \frac{z^3}{3} \, dz \right) - \pi \left[\frac{z^3}{3} \right]_0^1$$

$$= 2\pi \left(0 - \left[\frac{z^3}{9} \right]_0^1 \right) - \frac{\pi}{3}$$

$$= -\frac{2\pi}{9} - \frac{\pi}{3} = -\frac{5\pi}{9}.$$

Although the integrand contains no explicit minus signs, a negative value for the integral is to be expected, since $1 \ge z^2 \ge x^2 + y^2$ and $\ln(x^2 + y^2)$ is therefore negative.

6.17 *By making two successive simple changes of variables, evaluate*

$$I = \int \int \int x^2 \, dx \, dy \, dz$$

over the ellipsoidal region

$$\frac{x^2}{a^2} + \frac{y^2}{b^2} + \frac{z^2}{c^2} \le 1.$$

We start by making a scaling change aimed at producing an integration volume that has more amenable properties than an ellipsoid, namely a sphere. To do this, set $\xi = x/a$, $\eta = y/b$ and $\zeta = z/c$; the integral then becomes

$$I = \int \int \int a^2 \xi^2 \, a \, d\xi \, b \, d\eta \, c \, d\zeta \quad \text{over} \quad \xi^2 + \eta^2 + \zeta^2 \le 1$$

$$= a^3 bc \int \int \int \xi^2 \, d\xi \, d\eta \, d\zeta.$$

With the integration volume now a sphere it is sensible to change to spherical polar variables: $\xi = r \cos \theta$, $\eta = r \sin \theta \cos \phi$ and $\zeta = r \sin \theta \sin \phi$, with volume

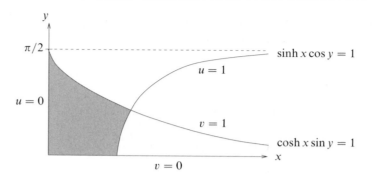

Figure 6.1   The integration area for exercise 6.19.

element $d\xi\, d\eta\, d\zeta = r^2 \sin\theta\, dr\, d\theta\, d\phi$. Note that we have chosen to orientate the polar axis along the old x-axis, rather than along the more conventional z-axis.

$$I = a^3bc \int_0^{2\pi} d\phi \int_0^{\pi} \cos^2\theta \sin\theta\, d\theta \int_0^1 r^4\, dr$$

$$= a^3bc\, 2\pi\, \frac{2}{3}\frac{1}{5}$$

$$= \tfrac{4}{15}\pi a^3bc.$$

6.19 *Sketch that part of the region $0 \le x$, $0 \le y \le \pi/2$ which is bounded by the curves $x = 0$, $y = 0$, $\sinh x \cos y = 1$ and $\cosh x \sin y = 1$. By making a suitable change of variables, evaluate the integral*

$$I = \int\int (\sinh^2 x + \cos^2 y)\sinh 2x \sin 2y\, dx\, dy$$

over the bounded subregion.

The integration area is shaded in figure 6.1. We are guided in making a choice of new variables by the equations defining the 'awkward' parts of the subregion's boundary curve. Ideally, the new variables should each be constant along one or more of the curves making up the boundary. This consideration leads us to make a change to new variables, $u = \sinh x \cos y$ and $v = \cosh x \sin y$. We then find the following.

(i) The boundary $y = 0$ becomes $v = 0$.
(ii) The boundary $x = 0$ becomes $u = 0$.
(iii) The boundary $\sinh x \cos y = 1$ becomes $u = 1$.
(iv) The boundary $\cosh x \sin y = 1$ becomes $v = 1$.

With this choice for the change, all four parts of the boundary can be characterised as being lines along which one of the coordinates is constant.

The Jacobian relating $dx\,dy$ to $du\,dv$, i.e. $du\,dv = \dfrac{\partial(u,v)}{\partial(x,y)}\,dx\,dy$, is

$$\frac{\partial(u,v)}{\partial(x,y)} = \frac{\partial u}{\partial x}\frac{\partial v}{\partial y} - \frac{\partial u}{\partial y}\frac{\partial v}{\partial x}$$
$$= (\cosh x \cos y)(\cosh x \cos y) - (-\sinh x \sin y)(\sinh x \sin y)$$
$$= (\sinh^2 x + 1)\cos^2 y + \sinh^2 x \sin^2 y$$
$$= \sinh^2 x + \cos^2 y.$$

The Jacobian required for the change of variables in the current case is the inverse of this.

Making the change of variables, and recalling that $\sin 2z = 2\sin z \cos z$, and similarly for $\sinh 2z$, gives

$$I = \int\int (\sinh^2 x + \cos^2 y)\sinh 2x \sin 2y\,dx\,dy$$
$$= \int_0^1\int_0^1 (\sinh^2 x + \cos^2 y)(4uv)\,\frac{du\,dv}{\sinh^2 x + \cos^2 y}$$
$$= 4\int_0^1 u\,du\int_0^1 v\,dv$$
$$= 4\left[\frac{u^2}{2}\right]_0^1\left[\frac{v^2}{2}\right]_0^1 = 1.$$

This is the simple answer to a superficially difficult integral!

6.21 *As stated in some of the exercises in chapter 5, the first law of thermodynamics can be expressed as*

$$dU = T\,dS - P\,dV.$$

By calculating and equating $\partial^2 U/\partial Y\,\partial X$ and $\partial^2 U/\partial X\partial Y$, where X and Y are an unspecified pair of variables (drawn from P, V, T and S), prove that

$$\frac{\partial(S,T)}{\partial(X,Y)} = \frac{\partial(V,P)}{\partial(X,Y)}.$$

Using the properties of Jacobians, deduce that

$$\frac{\partial(S,T)}{\partial(V,P)} = 1.$$

Starting from

$$dU = T\,dS - P\,dV,$$

99

the partial derivatives of U with respect to X and Y are

$$\frac{\partial U}{\partial X} = T\frac{\partial S}{\partial X} - P\frac{\partial V}{\partial X} \quad \text{and} \quad \frac{\partial U}{\partial Y} = T\frac{\partial S}{\partial Y} - P\frac{\partial V}{\partial Y}.$$

We next differentiate these two expressions to obtain two (equal) second derivatives. Note that, since X and Y can be any pair drawn from P, V, T and S, we must differentiate all four terms on the RHS as products, giving rise to two terms each. The resulting equations are

$$\frac{\partial^2 U}{\partial Y \partial X} = T\frac{\partial^2 S}{\partial Y \partial X} + \frac{\partial T}{\partial Y}\frac{\partial S}{\partial X} - P\frac{\partial^2 V}{\partial Y \partial X} - \frac{\partial P}{\partial Y}\frac{\partial V}{\partial X},$$

$$\frac{\partial^2 U}{\partial X \partial Y} = T\frac{\partial^2 S}{\partial X \partial Y} + \frac{\partial T}{\partial X}\frac{\partial S}{\partial Y} - P\frac{\partial^2 V}{\partial X \partial Y} - \frac{\partial P}{\partial X}\frac{\partial V}{\partial Y}.$$

Equating the two expressions, and then cancelling the terms that appear on both side of the equality, yields

$$\frac{\partial T}{\partial Y}\frac{\partial S}{\partial X} - \frac{\partial P}{\partial Y}\frac{\partial V}{\partial X} = \frac{\partial T}{\partial X}\frac{\partial S}{\partial Y} - \frac{\partial P}{\partial X}\frac{\partial V}{\partial Y},$$

$$\Rightarrow \quad \frac{\partial T}{\partial Y}\frac{\partial S}{\partial X} - \frac{\partial T}{\partial X}\frac{\partial S}{\partial Y} = \frac{\partial P}{\partial Y}\frac{\partial V}{\partial X} - \frac{\partial P}{\partial X}\frac{\partial V}{\partial Y},$$

$$\Rightarrow \quad \frac{\partial(S, T)}{\partial(X, Y)} = \frac{\partial(V, P)}{\partial(X, Y)}.$$

Now, using this result and the properties of Jacobians ($J_{pr} = J_{pq}J_{qr}$ and $J_{pq} = [J_{qp}]^{-1}$), we can write

$$\frac{\partial(S, T)}{\partial(V, P)} = \frac{\partial(S, T)}{\partial(X, Y)}\frac{\partial(X, Y)}{\partial(V, P)}$$

$$= \frac{\partial(S, T)}{\partial(X, Y)}\left[\frac{\partial(V, P)}{\partial(X, Y)}\right]^{-1}$$

$$= \frac{\partial(S, T)}{\partial(X, Y)}\left[\frac{\partial(S, T)}{\partial(X, Y)}\right]^{-1}$$

$$= 1.$$

6.23 *This is a more difficult question about 'volumes' in an increasing number of dimensions.*

(a) *Let R be a real positive number and define K_m by*

$$K_m = \int_{-R}^{R} \left(R^2 - x^2\right)^m dx.$$

Show, using integration by parts, that K_m satisfies the recurrence relation

$$(2m + 1)K_m = 2mR^2 K_{m-1}.$$

(b) *For integer n, define $I_n = K_n$ and $J_n = K_{n+1/2}$. Evaluate I_0 and J_0 directly and hence prove that*

$$I_n = \frac{2^{2n+1}(n!)^2 R^{2n+1}}{(2n + 1)!} \qquad and \qquad J_n = \frac{\pi(2n + 1)! R^{2n+2}}{2^{2n+1} n!(n + 1)!}.$$

(c) *A sequence of functions $V_n(R)$ is defined by*

$$V_0(R) = 1,$$
$$V_n(R) = \int_{-R}^{R} V_{n-1}\left(\sqrt{R^2 - x^2}\right) dx, \qquad n \geq 1.$$

Prove by induction that

$$V_{2n}(R) = \frac{\pi^n R^{2n}}{n!}, \qquad V_{2n+1}(R) = \frac{\pi^n 2^{2n+1} n! R^{2n+1}}{(2n + 1)!}.$$

(d) *For interest,*

   (i) *show that $V_{2n+2}(1) < V_{2n}(1)$ and $V_{2n+1}(1) < V_{2n-1}(1)$ for all $n \geq 3$;*
   (ii) *hence, by explicitly writing out $V_k(R)$ for $1 \leq k \leq 8$ (say), show that the 'volume' of the totally symmetric solid of unit radius is a maximum in five dimensions.*

(a) Taking the second factor in the integrand to be unity and integrating by parts, we have

$$K_m = \int_{-R}^{R} \left(R^2 - x^2\right)^m dx$$

$$= \left[x(R^2 - x^2)^m\right]_{-R}^{R} - \int_{-R}^{R} mx(R^2 - x^2)^{m-1}(-2x)\, dx$$

$$= 0 + 2m \int_{-R}^{R} (R^2 - x^2)^{m-1}(x^2 - R^2 + R^2)\, dx$$

$$= -2mK_m + 2mR^2 K_{m-1},$$

i.e. $\qquad (2m+1)K_m = 2mR^2 K_{m-1}. \qquad (*)$

(b) With $I_n = K_n$ and $J_N = K_{n+1/2}$,

$$I_0 = \int_{-R}^{R} 1 \, dx = 2R \quad \text{and}$$

$$J_0 = \int_{-R}^{R} (R^2 - x^2)^{1/2} \, dx, \qquad \text{(now set } x = R\cos\theta)$$

$$= \int_0^\pi R^2 \sin\theta \sin\theta \, d\theta$$

$$= \tfrac{1}{2}\pi R^2.$$

Using the recurrence relation ($*$) then gives

$$I_n = \frac{2n}{2n+1} \frac{2n-2}{2n-1} \cdots \frac{2}{3} R^{2n} I_0$$

$$= \frac{2^{n+1} \, n! \, (2^n \, n!)}{(2n+1)!} R^{2n+1}$$

$$= \frac{2^{2n+1} (n!)^2 R^{2n+1}}{(2n+1)!}.$$

Here, and below, we have written $(2n+1)(2n-1)\cdots 3$ in the form $(2n+1)!/(2^n \, n!)$. For J_n the corresponding calculation is

$$J_n = \frac{2n+1}{2n+2} \frac{2n-1}{2n} \cdots \frac{3}{4} R^{2n} J_0$$

$$= \frac{(2n+1)!}{(2^{n+1}/2)(n+1)! \, (2^n \, n!)} \frac{R^{2n}}{2} \frac{\pi R^2}{2}$$

$$= \frac{\pi (2n+1)! \, R^{2n+2}}{2^{2n+1} \, n! \, (n+1)!}.$$

(c) This is the most difficult part of the question as, although we proceed by induction on n, the general form of the expression for $n = N + 1$ is not the same as that for $n = N$. In fact it is the same as that for $n = N - 1$. Thus we will find two interleaving series of forms and have to prove the induction procedure for even and odd values of N separately. We start by assuming that

$$V_{2n}(R) = \frac{\pi^n R^{2n}}{n!}, \qquad V_{2n+1}(R) = \frac{\pi^n 2^{2n+1} n! R^{2n+1}}{(2n+1)!}.$$

For $n = 0$, the second expression gives $V_1(R) = (\pi^0 \, 2 \, 0! \, R)/1! = 2R$, whilst, for $n = 1$, the first gives $V_2(R) = \pi^1 R^2/1! = \pi R^2$; both of these are clearly valid.

Now, taking $n = 2N$, we compute $V_{2N+1}(R)$ from $V_{2N}(R)$ as

$$
\begin{aligned}
V_{2N+1}(R) &= \int_{-R}^{R} V_{2N}(\sqrt{R^2 - x^2})\, dx \\
&= \int_{-R}^{R} \frac{\pi^N}{N!}(R^2 - x^2)^{2N/2}\, dx \\
&= \frac{\pi^N}{N!} I_N \\
&= \frac{\pi^N\, 2^{2N+1}(N!)^2 R^{2N+1}}{N!\,(2N+1)!},
\end{aligned}
$$

i.e. in agreement with the assumption about $V_{2n+1}(R)$.

Next, taking $n = 2N + 1$ we compute $V_{2N+2}(R)$ from $V_{2N+1}(R)$ as

$$
\begin{aligned}
V_{2N+2}(R) &= \int_{-R}^{R} V_{2N+1}(\sqrt{R^2 - x^2})\, dx \\
&= \frac{\pi^N 2^{2N+1} N!}{(2N+1)!} \int_{-R}^{R}(\sqrt{R^2 - x^2})^{2N+1}\, dx \\
&= \frac{\pi^N 2^{2N+1} N!}{(2N+1)!} J_N \\
&= \frac{\pi^N 2^{2N+1} N!}{(2N+1)!}\, \frac{\pi\,(2N+1)!\, R^{2N+2}}{2^{2N+1}\, N!\,(N+1)!} \\
&= \frac{\pi^{N+1} R^{2N+2}}{(N+1)!},
\end{aligned}
$$

i.e. in agreement with the assumption about $V_{2n}(R)$.

Thus the two definitions generate each other consistently and, as has been shown, are directly verifiable for $N = 1$ and $N = 2$. This completes the proof.

(d)(i) Using the formulae just proved

$$
\frac{V_{2n+2}(1)}{V_{2n}(1)} = \frac{\pi^{n+1} n!}{(n+1)!\, \pi^n} = \frac{\pi}{n+1} < 1 \quad \text{for} \quad n \geq 3,
$$

$$
\begin{aligned}
\frac{V_{2n+1}(1)}{V_{2n-1}(1)} &= \frac{\pi^n 2^{2n+1}\, n!}{(2n+1)!}\, \frac{(2n-1)!}{\pi^{n-1} 2^{2n-1}(n-1)!} \\
&= \frac{2\pi}{2n+1} < 1 \quad \text{for} \quad n \geq 3.
\end{aligned}
$$

(ii) These two results show that the 'volumes' of all totally symmetric solids of unit radius in n dimensions are smaller than those in five or six dimensions if $n > 6$. Explicit calculations give the following for the first eight:

$$2, \quad \pi, \quad 4\pi/3, \quad \pi^2/2, \quad 8\pi^2/15, \quad \pi^3/6, \quad 16\pi^3/105, \quad \pi^4/24.$$

The largest of these is $V_5(1) = 8\pi^2/15 = 5.26$.

7

Vector algebra

7.1 *Which of the following statements about general vectors* **a**, **b** *and* **c** *are true?*

(a) $\mathbf{c} \cdot (\mathbf{a} \times \mathbf{b}) = (\mathbf{b} \times \mathbf{a}) \cdot \mathbf{c}$;

(b) $\mathbf{a} \times (\mathbf{b} \times \mathbf{c}) = (\mathbf{a} \times \mathbf{b}) \times \mathbf{c}$;

(c) $\mathbf{a} \times (\mathbf{b} \times \mathbf{c}) = (\mathbf{a} \cdot \mathbf{c})\mathbf{b} - (\mathbf{a} \cdot \mathbf{b})\mathbf{c}$;

(d) $\mathbf{d} = \lambda\mathbf{a} + \mu\mathbf{b}$ *implies* $(\mathbf{a} \times \mathbf{b}) \cdot \mathbf{d} = 0$;

(e) $\mathbf{a} \times \mathbf{c} = \mathbf{b} \times \mathbf{c}$ *implies* $\mathbf{c} \cdot \mathbf{a} - \mathbf{c} \cdot \mathbf{b} = c\,|\mathbf{a} - \mathbf{b}|$;

(f) $(\mathbf{a} \times \mathbf{b}) \times (\mathbf{c} \times \mathbf{b}) = \mathbf{b}[\mathbf{b} \cdot (\mathbf{c} \times \mathbf{a})]$.

All of the tests below are made using combinations of the common properties of the various types of vector products and justifications for individual steps are therefore not given. If the properties used are not recognised, they can be found in and learned from almost any standard textbook.

(a) $\mathbf{c} \cdot (\mathbf{a} \times \mathbf{b}) = -\mathbf{c} \cdot (\mathbf{b} \times \mathbf{a}) = -(\mathbf{b} \times \mathbf{a}) \cdot \mathbf{c} \neq (\mathbf{b} \times \mathbf{a}) \cdot \mathbf{c}$.

(b) $\mathbf{a} \times (\mathbf{b} \times \mathbf{c}) = \mathbf{b}(\mathbf{a} \cdot \mathbf{c}) - \mathbf{c}(\mathbf{a} \cdot \mathbf{b}) \neq \mathbf{b}(\mathbf{a} \cdot \mathbf{c}) - \mathbf{a}(\mathbf{b} \cdot \mathbf{c}) = (\mathbf{a} \times \mathbf{b}) \times \mathbf{c}$.

(c) $\mathbf{a} \times (\mathbf{b} \times \mathbf{c}) = (\mathbf{a} \cdot \mathbf{c}\mathbf{b}) - (\mathbf{a} \cdot \mathbf{b})\mathbf{c}$, a standard result.

(d) $(\mathbf{a} \times \mathbf{b}) \cdot \mathbf{d} = (\mathbf{a} \times \mathbf{b}) \cdot (\lambda\mathbf{a} + \mu\mathbf{b}) = \lambda(\mathbf{a} \times \mathbf{b}) \cdot \mathbf{a} + \mu(\mathbf{a} \times \mathbf{b}) \cdot \mathbf{b}$
$= \lambda 0 + \mu 0 = 0$.

(e) $\mathbf{a} \times \mathbf{c} = \mathbf{b} \times \mathbf{c} \Rightarrow (\mathbf{a} - \mathbf{b}) \times \mathbf{c} = 0 \Rightarrow \mathbf{a} - \mathbf{b} \parallel \mathbf{c}$
$\Rightarrow (\mathbf{a} - \mathbf{b}) \cdot \mathbf{c} = c\,|\mathbf{a} - \mathbf{b}| \Rightarrow \mathbf{c} \cdot \mathbf{a} - \mathbf{c} \cdot \mathbf{b} = c\,|\mathbf{a} - \mathbf{b}|$.

(f) $(\mathbf{a} \times \mathbf{b}) \times (\mathbf{c} \times \mathbf{b}) = \mathbf{b}\,[\mathbf{a} \cdot (\mathbf{c} \times \mathbf{b})] - \mathbf{a}\,[\mathbf{b} \cdot (\mathbf{c} \times \mathbf{b})]$
$= \mathbf{b}\,[\mathbf{a} \cdot (\mathbf{c} \times \mathbf{b})] - 0 = \mathbf{b}\,[\mathbf{b} \cdot (\mathbf{a} \times \mathbf{c})] = -\mathbf{b}\,[\mathbf{b} \cdot (\mathbf{c} \times \mathbf{a})]$
$\neq \mathbf{b}\,[\mathbf{b} \cdot (\mathbf{c} \times \mathbf{a})]$.

Thus only (c), (d) and (e) are true.

7.3 *Identify the following surfaces:*

  (a) $|\mathbf{r}| = k$; (b) $\mathbf{r} \cdot \mathbf{u} = l$; (c) $\mathbf{r} \cdot \mathbf{u} = m|\mathbf{r}|$ *for* $-1 \le m \le +1$;
  (d) $|\mathbf{r} - (\mathbf{r} \cdot \mathbf{u})\mathbf{u}| = n$.

Here k, l, m *and* n *are fixed scalars and* $\mathbf{u}$ *is a fixed unit vector.*

(a) All points on the surface are a distance k from the origin. The surface is therefore a sphere of radius k centred on the origin.

(b) This is the standard vector equation of a plane whose normal is in the direction $\mathbf{u}$ and whose distance from the origin is l.

(c) This is the surface generated by all vectors that make an angle $\cos^{-1} m$ with the fixed unit vector $\mathbf{u}$. The surface is therefore the cone of semi-angle $\cos^{-1} m$ that has the direction of $\mathbf{u}$ as its axis and the origin as its vertex.

(d) Since $(\mathbf{r} \cdot \mathbf{u})\mathbf{u}$ is the component of $\mathbf{r}$ that is parallel to $\mathbf{u}$, $\mathbf{r} - (\mathbf{r} \cdot \mathbf{u})\mathbf{u}$ is the component perpendicular to $\mathbf{u}$. As this latter component is constant for all points on the surface, the surface must be a circular cylinder of radius n that has its axis parallel to $\mathbf{u}$.

7.5 *A, B, C and D are the four corners, in order, of one face of a cube of side 2 units. The opposite face has corners E, F, G and H, with AE, BF, CG and DH as parallel edges of the cube. The centre O of the cube is taken as the origin and the x-, y- and z- axes are parallel to AD, AE and AB, respectively. Find the following:*

  (a) *the angle between the face diagonal AF and the body diagonal AG;*
  (b) *the equation of the plane through B that is parallel to the plane CGE;*
  (c) *the perpendicular distance from the centre J of the face $BCGF$ to the plane OCG;*
  (d) *the volume of the tetrahedron $JOCG$.*

(a) Unit vectors in the directions of the two diagonals have components

$$\mathbf{f} - \mathbf{a} = \frac{(0, 2, 2)}{\sqrt{8}} \quad \text{and} \quad \mathbf{g} - \mathbf{a} = \frac{(2, 2, 2)}{\sqrt{12}}.$$

Taking the scalar product of these two unit vectors gives the angle between them as

$$\theta = \cos^{-1} \frac{0 + 4 + 4}{\sqrt{96}} = \cos^{-1} \sqrt{\frac{2}{3}}.$$

105

(b) The direction of a normal **n** to the plane CGE is in the direction of the cross product of any two non-parallel vectors that lie in the plane. These can be taken as those from C to G and from C to E:

$$(\mathbf{g} - \mathbf{c}) \times (\mathbf{e} - \mathbf{c}) = (0, 2, 0) \times (-2, 2, -2) = (-4, 0, 4).$$

The equation of the plane is therefore of the form

$$c = \mathbf{n} \cdot \mathbf{r} = -4x + 0y + 4z = -4x + 4z.$$

Since it passes through $\mathbf{b} = (-1, -1, 1)$, the value of c must be 8 and the equation of the plane is $z - x = 2$.

(c) The direction of a normal **n** to the plane OCG is given by

$$\mathbf{c} \times \mathbf{g} = (1, -1, 1) \times (1, 1, 1) = (-2, 0, 2).$$

The equation of the plane is therefore of the form

$$c = \mathbf{n} \cdot \mathbf{r} = -2x + 0y + 2z = -2x + 2z.$$

Since it passes through the origin, the value of c must be 0 and the equation of the plane written in the form $\hat{\mathbf{n}} \cdot \mathbf{r} = p$ is

$$-\frac{x}{\sqrt{2}} + \frac{z}{\sqrt{2}} = 0.$$

The distance of J from this plane is $\hat{\mathbf{n}} \cdot \mathbf{j}$, where $\mathbf{j} = (0, 0, 1)$. The distance is thus $-0 + (1/\sqrt{2}) = 1/\sqrt{2}$.

(d) The volume of the tetrahedron $= \frac{1}{3}$(base area $\times$ height perpendicular to the base). The area of triangle OCG is $\frac{1}{2}|\mathbf{c} \times \mathbf{g}|$ and the perpendicular height of the tetrahedron is the component of $\mathbf{j}$ in the direction of $\mathbf{c} \times \mathbf{g}$. Thus the volume is

$$V = \frac{1}{3}\left|\frac{1}{2}(\mathbf{c} \times \mathbf{g}) \cdot \mathbf{j}\right| = \frac{1}{6}|(-2, 0, 2) \cdot (0, 0, 1)| = \frac{1}{3}.$$

7.7 *The edges OP, OQ and OR of a tetrahedron $OPQR$ are vectors* **p**, **q** *and* **r**, *respectively, where* $\mathbf{p} = 2\mathbf{i} + 4\mathbf{j}$, $\mathbf{q} = 2\mathbf{i} - \mathbf{j} + 3\mathbf{k}$ *and* $\mathbf{r} = 4\mathbf{i} - 2\mathbf{j} + 5\mathbf{k}$. *Show that OP is perpendicular to the plane containing OQR. Express the volume of the tetrahedron in terms of* **p**, **q** *and* **r** *and hence calculate the volume.*

The plane containing OQR has a normal in the direction $\mathbf{q} \times \mathbf{r} = (2, -1, 3) \times (4, -2, 5) = (1, 2, 0)$. This is parallel to **p** since $\mathbf{q} \times \mathbf{r} = \frac{1}{2}\mathbf{p}$. The volume of the tetrahedron is therefore one-third times $\frac{1}{2}|\mathbf{q} \times \mathbf{r}|$ times $|\mathbf{p}|$, i.e. $\frac{1}{6}|(1, 2, 0)|\sqrt{20} = \frac{5}{3}$.

7.9 *Prove Lagrange's identity, i.e.*

$$(\mathbf{a} \times \mathbf{b}) \cdot (\mathbf{c} \times \mathbf{d}) = (\mathbf{a} \cdot \mathbf{c})(\mathbf{b} \cdot \mathbf{d}) - (\mathbf{a} \cdot \mathbf{d})(\mathbf{b} \cdot \mathbf{c}).$$

We treat the expression on the LHS as the triple scalar product of the three vectors $\mathbf{a} \times \mathbf{b}$, $\mathbf{c}$ and $\mathbf{d}$ and use the cyclic properties of triple scalar products:

$$(\mathbf{a} \times \mathbf{b}) \cdot (\mathbf{c} \times \mathbf{d}) = \mathbf{d} \cdot [(\mathbf{a} \times \mathbf{b}) \times \mathbf{c}]$$
$$= \mathbf{d} \cdot [(\mathbf{a} \cdot \mathbf{c})\mathbf{b} - (\mathbf{b} \cdot \mathbf{c})\mathbf{a}]$$
$$= (\mathbf{a} \cdot \mathbf{c})(\mathbf{d} \cdot \mathbf{b}) - (\mathbf{b} \cdot \mathbf{c})(\mathbf{d} \cdot \mathbf{a}).$$

In going from the first to the second line we used the standard result

$$(\mathbf{a} \times \mathbf{b}) \times \mathbf{c} = (\mathbf{a} \cdot \mathbf{c})\mathbf{b} - (\mathbf{b} \cdot \mathbf{c})\mathbf{a}$$

to replace $(\mathbf{a} \times \mathbf{b}) \times \mathbf{c}$. This result, if not known, can be proved by writing it out in component form as follows.

Consider only the x-component of each side of the equation. The corresponding results for other components can be obtained by cyclic permutation of x, y and z.

$$\mathbf{a} \times \mathbf{b} = (a_y b_z - a_z b_y, \; a_z b_x - a_x b_z, \; a_x b_y - a_y b_x)$$
$$[(\mathbf{a} \times \mathbf{b}) \times \mathbf{c}]_x = (a_z b_x - a_x b_z)c_z - (a_x b_y - a_y b_x)c_y$$
$$= b_x(a_z c_z + a_y c_y) - a_x(b_z c_z + b_y c_y)$$
$$= b_x(a_z c_z + a_y c_y + a_x c_x) - a_x(b_x c_x + b_z c_z + b_y c_y)$$
$$= [(\mathbf{a} \cdot \mathbf{c})\mathbf{b} - (\mathbf{b} \cdot \mathbf{c})\mathbf{a}]_x.$$

To obtain the penultimate line we both added and subtracted $a_x b_x c_x$ on the RHS. This establishes the result for the x-component and hence for all three components.

7.11 *Show that the points* $(1, 0, 1)$, $(1, 1, 0)$ *and* $(1, -3, 4)$ *lie on a straight line. Give the equation of the line in the form*

$$\mathbf{r} = \mathbf{a} + \lambda \mathbf{b}.$$

To show that the points lie on a line, we need to show that their position vectors are linearly dependent. That this is so follows from noting that

$$(1, -3, 4) = 4(1, 0, 1) - 3(1, 1, 0).$$

This can also be written

$$(1, -3, 4) = (1, 0, 1) + 3[(1, 0, 1) - (1, 1, 0)] = (1, 0, 1) + 3(0, -1, 1).$$

The equation of the line is therefore

$$\mathbf{r} = \mathbf{a} + \lambda(-\mathbf{j} + \mathbf{k}),$$

where $\mathbf{a}$ is the vector position of *any* point on the line, e.g. $\mathbf{i} + \mathbf{k}$ or $\mathbf{i} + \mathbf{j}$ or $\mathbf{i} - 3\mathbf{j} + 4\mathbf{k}$ or many others. Of course, choosing different points for $\mathbf{a}$ will entail using different values of λ to describe the same point $\mathbf{r}$ on the line. For example,

$$
\begin{aligned}
(1, -5, 6) &= (1, 0, 1) + 5(0, -1, 1) \\
\text{or} \ &= (1, 1, 0) + 6(0, -1, 1) \\
\text{or} \ &= (1, -3, 4) + 2(0, -1, 1).
\end{aligned}
$$

7.13 *Two planes have non-parallel unit normals $\hat{\mathbf{n}}$ and $\hat{\mathbf{m}}$ and their closest distances from the origin are λ and μ, respectively. Find the vector equation of their line of intersection in the form $\mathbf{r} = v\mathbf{p} + \mathbf{a}$.*

The equations of the two planes are

$$\hat{\mathbf{n}} \cdot \mathbf{r} = \lambda \qquad \text{and} \qquad \hat{\mathbf{m}} \cdot \mathbf{r} = \mu.$$

The line of intersection lies in both planes and is thus perpendicular to both normals; it therefore has direction $\mathbf{p} = \hat{\mathbf{n}} \times \hat{\mathbf{m}}$. Consequently the equation of the line takes the form $\mathbf{r} = v\mathbf{p} + \mathbf{a}$, where $\mathbf{a}$ is any one point lying on it. One such point is the one in which the line meets the plane containing $\hat{\mathbf{n}}$ and $\hat{\mathbf{m}}$; we take this point as $\mathbf{a}$. Since $\mathbf{a}$ also lies in both of the original planes, we must have

$$\hat{\mathbf{n}} \cdot \mathbf{a} = \lambda \quad \text{and} \quad \hat{\mathbf{m}} \cdot \mathbf{a} = \mu.$$

If we now write $\mathbf{a} = x\hat{\mathbf{n}} + y\hat{\mathbf{m}}$, these two conditions become

$$\lambda = \hat{\mathbf{n}} \cdot \mathbf{a} = x + y(\hat{\mathbf{n}} \cdot \hat{\mathbf{m}}),$$
$$\mu = \hat{\mathbf{m}} \cdot \mathbf{a} = x(\hat{\mathbf{n}} \cdot \hat{\mathbf{m}}) + y.$$

It then follows that

$$x = \frac{\lambda - \mu(\hat{\mathbf{n}} \cdot \hat{\mathbf{m}})}{1 - (\hat{\mathbf{n}} \cdot \hat{\mathbf{m}})^2} \quad \text{and} \quad y = \frac{\mu - \lambda(\hat{\mathbf{n}} \cdot \hat{\mathbf{m}})}{1 - (\hat{\mathbf{n}} \cdot \hat{\mathbf{m}})^2},$$

thus determining $\mathbf{a}$. Both $\mathbf{p}$ and $\mathbf{a}$ are therefore determined in terms of λ, μ, $\hat{\mathbf{n}}$ and $\hat{\mathbf{m}}$, and so consequently is the line of intersection of the planes.

> **7.15** *Let O, A, B and C be four points with position vectors **0**, **a**, **b** and **c**, and denote by **g** = λ**a** + μ**b** + ν**c** the position of the centre of the sphere on which they all lie.*
>
> (a) *Prove that λ, μ and ν simultaneously satisfy*
>
> $$(\mathbf{a} \cdot \mathbf{a})\lambda + (\mathbf{a} \cdot \mathbf{b})\mu + (\mathbf{a} \cdot \mathbf{c})\nu = \tfrac{1}{2}a^2$$
>
> *and two other similar equations.*
>
> (b) *By making a change of origin, find the centre and radius of the sphere on which the points **p** = 3**i**+**j**−2**k**, **q** = 4**i**+3**j**−3**k**, **r** = 7**i**−3**k** and **s** = 6**i**+**j**−**k** all lie.*

(a) Each of the points O, A, B and C is the same distance from the centre G of the sphere. In particular, $OG = OA$, i.e.

$$|\mathbf{g} - \mathbf{0}|^2 = |\mathbf{a} - \mathbf{g}|^2,$$
$$g^2 = a^2 - 2\mathbf{a} \cdot \mathbf{g} + g^2,$$
$$\mathbf{a} \cdot \mathbf{g} = \tfrac{1}{2}a^2,$$
$$\mathbf{a} \cdot (\lambda\mathbf{a} + \mu\mathbf{b} + \nu\mathbf{c}) = \tfrac{1}{2}a^2,$$
$$(\mathbf{a} \cdot \mathbf{a})\lambda + (\mathbf{a} \cdot \mathbf{b})\mu + (\mathbf{a} \cdot \mathbf{c})\nu = \tfrac{1}{2}a^2.$$

Two similar equations can be obtained from $OG = OB$ and $OG = OC$.

(b) To use the previous result we make P, say, the origin of a new coordinate system in which

$$\mathbf{p}' = \mathbf{p} - \mathbf{p} = (0,0,0),$$
$$\mathbf{q}' = \mathbf{q} - \mathbf{p} = (1,2,-1),$$
$$\mathbf{r}' = \mathbf{r} - \mathbf{p} = (4,-1,-1),$$
$$\mathbf{s}' = \mathbf{s} - \mathbf{p} = (3,0,1).$$

The centre, G, of the sphere on which P, Q, R and S lie is then given by

$$\mathbf{g}' = \lambda\mathbf{q}' + \mu\mathbf{r}' + \nu\mathbf{s}',$$

where
$$(\mathbf{q}' \cdot \mathbf{q}')\lambda + (\mathbf{q}' \cdot \mathbf{r}')\mu + (\mathbf{q}' \cdot \mathbf{s}')\nu = \tfrac{1}{2}\mathbf{q}' \cdot \mathbf{q}',$$
$$(\mathbf{r}' \cdot \mathbf{q}')\lambda + (\mathbf{r}' \cdot \mathbf{r}')\mu + (\mathbf{r}' \cdot \mathbf{s}')\nu = \tfrac{1}{2}\mathbf{r}' \cdot \mathbf{r}',$$
$$(\mathbf{s}' \cdot \mathbf{q}')\lambda + (\mathbf{s}' \cdot \mathbf{r}')\mu + (\mathbf{s}' \cdot \mathbf{s}')\nu = \tfrac{1}{2}\mathbf{s}' \cdot \mathbf{s}',$$

i.e.
$$6\lambda + 3\mu + 2\nu = 3,$$
$$3\lambda + 18\mu + 11\nu = 9,$$
$$2\lambda + 11\mu + 10\nu = 5.$$

These equations have the solution

$$\lambda = \frac{5}{18}, \qquad \mu = \frac{5}{9}, \qquad \nu = -\frac{1}{6}.$$

Thus, the centre of the sphere can be calculated as

$$\mathbf{g}' = \frac{5}{18}(1, 2, -1) + \frac{5}{9}(4, -1, -1) - \frac{1}{6}(3, 0, 1) = (2, 0, -1).$$

Its radius is therefore $|G'O'| = |\mathbf{g}'| = \sqrt{5}$ and its centre in the original coordinate system is at $\mathbf{g}' + \mathbf{p} = (5, 1, -3)$.

7.17 *Using vector methods:*

    (a) *Show that the line of intersection of the planes $x + 2y + 3z = 0$ and $3x + 2y + z = 0$ is equally inclined to the x- and z-axes and makes an angle $\cos^{-1}(-2/\sqrt{6})$ with the y-axis.*

    (b) *Find the perpendicular distance between one corner of a unit cube and the major diagonal not passing through it.*

(a) The origin O is clearly in both planes. A second such point can be found by setting $z = 1$, say, and solving the pair of simultaneous equations to give $x = 1$ and $y = -2$, i.e. $(1, -2, 1)$ is in both planes. The direction cosines of the line of intersection, OP, are therefore

$$\left(\frac{1}{\sqrt{6}}, -\frac{2}{\sqrt{6}}, \frac{1}{\sqrt{6}} \right),$$

i.e. the line is equally inclined to the x- and z-axes and makes an angle $\cos^{-1}(-2/\sqrt{6})$ with the y-axis.

The same conclusion can be reached by reasoning as follows. The line of intersection of the two planes must be orthogonal to the normal of either plane. Therefore it is in the direction of the cross product of the two normals and is given by

$$(1, 2, 3) \times (3, 2, 1) = (-4, 8, -4) = -4\sqrt{6} \left(\frac{1}{\sqrt{6}}, -\frac{2}{\sqrt{6}}, \frac{1}{\sqrt{6}} \right).$$

(b) We first note that all three major diagonals not passing through a corner come equally close to it. Taking the corner to be at the origin and the diagonal to be the one that passes through $(0, 1, 1)$ [and $(1, 0, 0)$], the equation of the diagonal is

$$(x, y, z) = (0, 1, 1) + \frac{\lambda}{\sqrt{3}}(1, -1, -1).$$

110

Using the result that the distance d of the point $\mathbf{p}$ from the line $\mathbf{r} = \mathbf{a} + \lambda\hat{\mathbf{b}}$ is given by

$$d = |(\mathbf{p} - \mathbf{a}) \times \hat{\mathbf{b}}|,$$

the distance of $(0, 0, 0)$ from the line of the diagonal is

$$\left| [(0, 0, 0) - (0, 1, 1)] \times \frac{1}{\sqrt{3}}(1, -1, -1) \right| = \frac{1}{\sqrt{3}}|(0, -1, 1)| = \sqrt{\frac{2}{3}}.$$

7.19 *The vectors* $\mathbf{a}$, $\mathbf{b}$ *and* $\mathbf{c}$ *are not coplanar. Verify that the expressions*

$$\mathbf{a}' = \frac{\mathbf{b} \times \mathbf{c}}{[\mathbf{a}, \mathbf{b}, \mathbf{c}]}, \quad \mathbf{b}' = \frac{\mathbf{c} \times \mathbf{a}}{[\mathbf{a}, \mathbf{b}, \mathbf{c}]}, \quad \mathbf{c}' = \frac{\mathbf{a} \times \mathbf{b}}{[\mathbf{a}, \mathbf{b}, \mathbf{c}]}$$

define a set of reciprocal vectors $\mathbf{a}'$, $\mathbf{b}'$ *and* $\mathbf{c}'$ *with the following properties:*

(a) $\mathbf{a}' \cdot \mathbf{a} = \mathbf{b}' \cdot \mathbf{b} = \mathbf{c}' \cdot \mathbf{c} = 1$;
(b) $\mathbf{a}' \cdot \mathbf{b} = \mathbf{a}' \cdot \mathbf{c} = \mathbf{b}' \cdot \mathbf{a}$ *etc* $= 0$;
(c) $[\mathbf{a}', \mathbf{b}', \mathbf{c}'] = 1/[\mathbf{a}, \mathbf{b}, \mathbf{c}]$;
(d) $\mathbf{a} = (\mathbf{b}' \times \mathbf{c}')/[\mathbf{a}', \mathbf{b}', \mathbf{c}']$.

Direct substitutions and the expansion formula for a triple vector product (proved in 7.9) enable the verifications to be made as follows. We make repeated use of the general result $(\mathbf{p} \times \mathbf{q}) \cdot \mathbf{p} = 0 = (\mathbf{p} \times \mathbf{q}) \cdot \mathbf{q}$.

(a) $\quad \mathbf{a}' \cdot \mathbf{a} = \dfrac{(\mathbf{b} \times \mathbf{c}) \cdot \mathbf{a}}{[\mathbf{a}, \mathbf{b}, \mathbf{c}]} = 1$. Similarly for $\mathbf{b}' \cdot \mathbf{b}$ and $\mathbf{c}' \cdot \mathbf{c}$.

(b) $\quad \mathbf{a}' \cdot \mathbf{b} = \dfrac{(\mathbf{b} \times \mathbf{c}) \cdot \mathbf{b}}{[\mathbf{a}, \mathbf{b}, \mathbf{c}]} = 0$. Similarly for $\mathbf{a}' \cdot \mathbf{c}$, $\mathbf{b}' \cdot \mathbf{a}$ etc.

(c) $\quad [\mathbf{a}', \mathbf{b}', \mathbf{c}'] = \dfrac{\mathbf{a}' \cdot \{(\mathbf{c} \times \mathbf{a}) \times (\mathbf{a} \times \mathbf{b})\}}{[\mathbf{a}, \mathbf{b}, \mathbf{c}]^2}$

$$= \frac{\mathbf{a}' \cdot \{[\mathbf{b} \cdot (\mathbf{c} \times \mathbf{a})]\,\mathbf{a} - [\mathbf{a} \cdot (\mathbf{c} \times \mathbf{a})]\,\mathbf{b}\}}{[\mathbf{a}, \mathbf{b}, \mathbf{c}]^2}$$

$$= \frac{1\,[\mathbf{b}, \mathbf{c}, \mathbf{a}] - 0\,(\mathbf{a}' \cdot \mathbf{b})}{[\mathbf{a}, \mathbf{b}, \mathbf{c}]^2}, \text{ using results (a) and (b)},$$

$$= \frac{1}{[\mathbf{a}, \mathbf{b}, \mathbf{c}]}.$$

(d) $\quad \dfrac{\mathbf{b}' \times \mathbf{c}'}{[\mathbf{a}', \mathbf{b}', \mathbf{c}']} = \dfrac{[\mathbf{b}, \mathbf{c}, \mathbf{a}]\,\mathbf{a} - 0\,\mathbf{b}}{[\mathbf{a}, \mathbf{b}, \mathbf{c}]^2\,[\mathbf{a}', \mathbf{b}', \mathbf{c}']}, \qquad$ as in part (c),

$$= \mathbf{a}, \qquad\qquad\qquad \text{from result (c)}.$$

7.21 *In a crystal with a face-centred cubic structure, the basic cell can be taken as a cube of edge a with its centre at the origin of coordinates and its edges parallel to the Cartesian coordinate axes; atoms are sited at the eight corners and at the centre of each face. However, other basic cells are possible. One is the rhomboid shown in figure 7.1, which has the three vectors **b**, **c** and **d** as edges.*

(a) *Show that the volume of the rhomboid is one-quarter that of the cube.*
(b) *Show that the angles between pairs of edges of the rhomboid are 60° and that the corresponding angles between pairs of edges of the rhomboid defined by the reciprocal vectors to **b**, **c**, **d** are each 109.5°. (This rhomboid can be used as the basic cell of a body-centred cubic structure, more easily visualised as a cube with an atom at each corner and one at its centre.)*
(c) *In order to use the Bragg formula, $2d\sin\theta = n\lambda$, for the scattering of X-rays by a crystal, it is necessary to know the perpendicular distance d between successive planes of atoms; for a given crystal structure, d has a particular value for each set of planes considered. For the face-centred cubic structure find the distance between successive planes with normals in the **k**, **i** + **j** and **i** + **j** + **k** directions.*

(a) From the figure it is easy to see that the edges of the rhomboid are the vectors $\mathbf{b} = \frac{1}{2}a(0,1,1)$, $\mathbf{c} = \frac{1}{2}a(1,0,1)$, and $\mathbf{d} = \frac{1}{2}a(1,1,0)$. The volume V of the rhomboid is therefore given by

$$V = |\,[\mathbf{b},\mathbf{c},\mathbf{d}]\,|$$
$$= |\mathbf{b}\cdot(\mathbf{c}\times\mathbf{d})|$$
$$= \tfrac{1}{8}a^3\,|(0,1,1)\cdot(-1,1,1)|$$
$$= \tfrac{1}{4}a^3,$$

i.e. one-quarter that of the cube.

(b) To find the angle between two edges of the rhomboid we calculate the scalar product of two unit vectors, one along each edge; its value is $1\times1\times\cos\phi$, where ϕ is the angle between the edges. Unit vectors along the edges of the rhomboid are

$$\hat{\mathbf{b}} = \frac{1}{\sqrt{2}}(0,1,1), \quad \hat{\mathbf{c}} = \frac{1}{\sqrt{2}}(1,0,1), \quad \hat{\mathbf{d}} = \frac{1}{\sqrt{2}}(1,1,0).$$

The scalar product of *any* pair of these particular vectors has the value $\frac{1}{2}$, e.g.

$$\hat{\mathbf{b}}\cdot\hat{\mathbf{c}} = \tfrac{1}{2}(0+0+1) = \tfrac{1}{2}.$$

Thus the angle between any pair of edges is $\cos^{-1}(\tfrac{1}{2}) = 60°$.

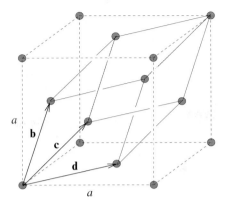

Figure 7.1   A face-centred cubic crystal.

The reciprocal vectors are, for example,

$$\mathbf{b}' = \frac{\mathbf{c} \times \mathbf{d}}{[\mathbf{b}, \mathbf{c}, \mathbf{d}]} = \frac{a^2}{4} \frac{(-1, 1, 1)}{(a^3/4)} = \frac{1}{a}(-1, 1, 1) = \frac{1}{a}(-\mathbf{i} + \mathbf{j} + \mathbf{k}),$$

where in the second equality we have used the result of part (a). Similarly, or by cyclic permutation, $\mathbf{c}' = a^{-1}(\mathbf{i} - \mathbf{j} + \mathbf{k})$ and $\mathbf{d}' = a^{-1}(\mathbf{i} + \mathbf{j} - \mathbf{k})$.

The angle between any pair of reciprocal vectors has the value 109.5°, e.g.

$$\theta = \cos^{-1}\left(\frac{\mathbf{b}' \cdot \mathbf{c}'}{|\mathbf{b}'||\mathbf{c}'|}\right) = \cos^{-1}\left(\frac{a^{-2}(-1 - 1 + 1)}{(\sqrt{3}\, a^{-1})^2}\right) = \cos^{-1}(-\tfrac{1}{3}) = 109.5°.$$

Other pairs yield the same value.

(c) Planes with normals in the $\mathbf{k}$ direction are clearly separated by $\frac{1}{2}a$.

A plane with its normal in the direction $\mathbf{i} + \mathbf{j}$ has an equation of the form

$$\frac{1}{\sqrt{2}}(1, 1, 0) \cdot (x, y, z) = p,$$

where p is the perpendicular distance of the origin from the plane. Since the plane with the smallest positive value of p passes through $(\frac{1}{2}a, 0, \frac{1}{2}a)$, p has the value $a/\sqrt{8}$, which is therefore the distance between successive planes with normals in the direction $\mathbf{i} + \mathbf{j}$.

Planes with their normals in the direction $\mathbf{i} + \mathbf{j} + \mathbf{k}$ have equations of the form

$$\frac{1}{\sqrt{3}}(1, 1, 1) \cdot (x, y, z) = p.$$

For the plane P_1 containing $\mathbf{b}$, $\mathbf{c}$ and $\mathbf{d}$ we have (for $\mathbf{b}$, say)

$$\frac{1}{\sqrt{3}}(1, 1, 1) \cdot (0, \tfrac{1}{2}a, \tfrac{1}{2}a) = p_1,$$

giving $p_1 = a/\sqrt{3}$. Similarly for the plane P_2 containing $\mathbf{c}+\mathbf{d}$, $\mathbf{b}+\mathbf{d}$ and $\mathbf{b}+\mathbf{c}$ we have (for $\mathbf{c}+\mathbf{d}$, say)

$$\frac{1}{\sqrt{3}}(1,1,1)\cdot(a,\tfrac{1}{2}a,\tfrac{1}{2}a) = p_2,$$

giving $p_2 = 2a/\sqrt{3}$. Thus the distance, d, between successive planes with normals in the direction $\mathbf{i}+\mathbf{j}+\mathbf{k}$ is the difference between these two values, i.e. $d = p_2 - p_1 = a/\sqrt{3}$.

7.23 *By proceeding as indicated below, prove the* parallel axis theorem, *which states that, for a body of mass M, the moment of inertia I about any axis is related to the corresponding moment of inertia I_0 about a parallel axis that passes through the centre of mass of the body by*

$$I = I_0 + Ma_\perp^2,$$

where $a_\perp$ is the perpendicular distance between the two axes. Note that I_0 can be written as

$$\int (\hat{\mathbf{n}} \times \mathbf{r})\cdot(\hat{\mathbf{n}} \times \mathbf{r})\,dm,$$

where $\mathbf{r}$ is the vector position, relative to the centre of mass, of the infinitesimal mass dm and $\hat{\mathbf{n}}$ is a unit vector in the direction of the axis of rotation. Write a similar expression for I in which $\mathbf{r}$ is replaced by $\mathbf{r}' = \mathbf{r}-\mathbf{a}$, where $\mathbf{a}$ is the vector position of any point on the axis to which I refers. Use Lagrange's identity and the fact that $\int \mathbf{r}\,dm = \mathbf{0}$ (by the definition of the centre of mass) to establish the result.

Figure 7.2 shows the vectors involved in describing the physical arrangement. With

$$I_0 = \int (\hat{\mathbf{n}} \times \mathbf{r})\cdot(\hat{\mathbf{n}} \times \mathbf{r})\,dm$$
$$= \int \left[(\hat{\mathbf{n}}\cdot\hat{\mathbf{n}})(\mathbf{r}\cdot\mathbf{r}) - (\hat{\mathbf{n}}\cdot\mathbf{r})^2\right]\,dm,$$

the moment of inertia of the same mass distribution about a parallel axis passing

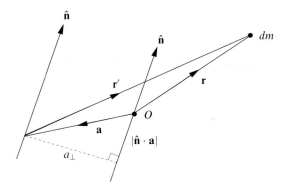

Figure 7.2 The vectors used in the proof of the parallel axis theorem in exercise 7.23.

through **a** is given by

$$I = \int (\hat{\mathbf{n}} \times \mathbf{r}') \cdot (\hat{\mathbf{n}} \times \mathbf{r}') \, dm$$

$$= \int [\hat{\mathbf{n}} \times (\mathbf{r} - \mathbf{a})] \cdot [\hat{\mathbf{n}} \times (\mathbf{r} - \mathbf{a})] \, dm$$

$$= \int \left\{ (\hat{\mathbf{n}} \cdot \hat{\mathbf{n}})[(\mathbf{r} - \mathbf{a}) \cdot (\mathbf{r} - \mathbf{a})] - [\hat{\mathbf{n}} \cdot (\mathbf{r} - \mathbf{a})]^2 \right\} \, dm,$$

$$= \int \left[r^2 - 2\mathbf{a} \cdot \mathbf{r} + a^2 - (\hat{\mathbf{n}} \cdot \mathbf{r})^2 + 2(\hat{\mathbf{n}} \cdot \mathbf{r})(\hat{\mathbf{n}} \cdot \mathbf{a}) - (\hat{\mathbf{n}} \cdot \mathbf{a})^2 \right] \, dm$$

$$= I_0 - 2\mathbf{a} \cdot \mathbf{0} + 2(\hat{\mathbf{n}} \cdot \mathbf{a})(\hat{\mathbf{n}} \cdot \mathbf{0}) + \int \left[a^2 - (\hat{\mathbf{n}} \cdot \mathbf{a})^2 \right] \, dm$$

$$= I_0 + a_\perp^2 M.$$

When obtaining the penultimate line we (twice) used the fact that O is the centre of mass of the body and so, by definition, $\int \mathbf{r} \, dm = \mathbf{0}$. To obtain the final line we noted that $\hat{\mathbf{n}} \cdot \mathbf{a}$ is the component of **a** parallel to $\hat{\mathbf{n}}$ and so $a^2 - (\hat{\mathbf{n}} \cdot \mathbf{a})^2$ is the square of the component of **a** perpendicular to $\hat{\mathbf{n}}$.

7.25 *Define a set of (non-orthogonal) base vectors* $\mathbf{a} = \mathbf{j} + \mathbf{k}$, $\mathbf{b} = \mathbf{i} + \mathbf{k}$ *and* $\mathbf{c} = \mathbf{i} + \mathbf{j}$.

(a) *Establish their reciprocal vectors and hence express the vectors* $\mathbf{p} = 3\mathbf{i} - 2\mathbf{j} + \mathbf{k}$, $\mathbf{q} = \mathbf{i} + 4\mathbf{j}$ *and* $\mathbf{r} = -2\mathbf{i} + \mathbf{j} + \mathbf{k}$ *in terms of the base vectors* **a**, **b** *and* **c**.

(b) *Verify that the scalar product* $\mathbf{p} \cdot \mathbf{q}$ *has the same value, -5, when evaluated using either set of components.*

The new base vectors are $\mathbf{a} = (0, 1, 1)$, $\mathbf{b} = (1, 0, 1)$ and $\mathbf{c} = (1, 1, 0)$.

(a) The corresponding reciprocal vectors are thus

$$\mathbf{a}' = \frac{\mathbf{b} \times \mathbf{c}}{[\mathbf{a}, \mathbf{b}, \mathbf{c}]} = \frac{(-1, 1, 1)}{2} = \tfrac{1}{2}(-1, 1, 1),$$

and similarly for $\mathbf{b}' = \tfrac{1}{2}(1, -1, 1)$ and $\mathbf{c}' = \tfrac{1}{2}(1, 1, -1)$.

The coefficient of (say) $\mathbf{a}$ in the expression for (say) $\mathbf{p}$ is $\mathbf{a}' \cdot \mathbf{p} = -2$. The coefficient of $\mathbf{b}$ is $\mathbf{b}' \cdot \mathbf{p} = 3$, etc. Building up each of $\mathbf{p}$, $\mathbf{q}$ and $\mathbf{r}$ in this way, we find that their coordinates in terms of the new basis $\{\mathbf{a}, \mathbf{b}, \mathbf{c}\}$ are $\mathbf{p} = (-2, 3, 0)$, $\mathbf{q} = (\tfrac{3}{2}, -\tfrac{3}{2}, \tfrac{5}{2})$ and $\mathbf{r} = (2, -1, -1)$.

(b) The new basis vectors, which are neither orthogonal nor normalised, have the properties $\mathbf{a} \cdot \mathbf{a} = \mathbf{b} \cdot \mathbf{b} = \mathbf{c} \cdot \mathbf{c} = 2$ and $\mathbf{b} \cdot \mathbf{c} = \mathbf{c} \cdot \mathbf{a} = \mathbf{a} \cdot \mathbf{b} = 1$. Thus the scalar product $\mathbf{p} \cdot \mathbf{q}$, calculated in the new basis, has the value

$$2\left(-3 - \tfrac{9}{2} + 0\right) + 1\left(3 - 5 + \tfrac{9}{2} + \tfrac{15}{2} + 0 + 0\right) = -15 + 10 = -5.$$

Using the original basis, $\mathbf{p} \cdot \mathbf{q} = 3 - 8 + 0 = -5$, verifying that the scalar product has the same value in both sets of coordinates.

7.27 *According to alternating current theory, the currents and potential differences in the components of the circuit shown in figure 7.3 are determined by Kirchhoff's laws and the relationships*

$$I_1 = \frac{V_1}{R_1}, \qquad I_2 = \frac{V_2}{R_2}, \qquad I_3 = i\omega C V_3, \qquad V_4 = i\omega L I_2.$$

The factor $i = \sqrt{-1}$ in the expression for I_3 indicates that the phase of I_3 is $90°$ ahead of V_3. Similarly the phase of V_4 is $90°$ ahead of I_2.

Measurement shows that V_3 has an amplitude of $0.661 V_0$ and a phase of $+13.4°$ relative to that of the power supply. Taking $V_0 = 1\,\mathrm{V}$ and using a series of vector plots for potential differences and currents (they could all be on the same plot if suitable scales were chosen), determine all unknown currents and potential differences and find values for the inductance of L and the resistance of R_2.

[Scales of $1\,\mathrm{cm} = 0.1\,\mathrm{V}$ for potential differences and $1\,\mathrm{cm} = 1\,\mathrm{mA}$ for currents are convenient.]

Using the suggested scales, we construct the vectors shown in figure 7.4 in the following order:

(1) V_0 joining $(0, 0)$ to $(10, 0)$;
(2) V_3 of length 6.61 and phase $+13.4°$;
(3) $V_1 = V_0 - V_3$;

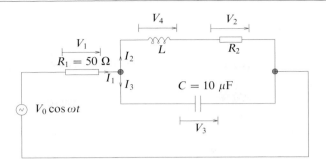

Figure 7.3   The oscillatory electric circuit in exercise 7.27. The power supply has angular frequency $\omega = 2\pi f = 400\pi\ \text{s}^{-1}$.

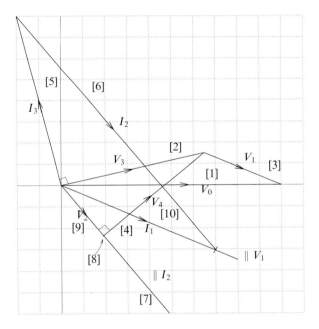

Figure 7.4   The vector solution to exercise 7.27.

(4) I_1 parallel to V_1 and $(0.1 \times 1000)/50 = 2$ times as long;
(5) I_3, 90° ahead of V_3 in phase and $(0.1 \times 1000) \times 400\pi \times 10^{-5} = 1.26$ times as long;
(6) $I_2 = I_1 - I_3$;
(7) draw a parallel to I_2 through the origin;
(8) drop a perpendicular from V_3 onto this parallel to I_2;
(9) since $V_3 = V_2 + V_4$ and $V_2 \parallel I_2$, whilst $V_4 \perp I_2$, the foot of the perpendicular

117

gives V_2;

(10) $V_4 = V_3 - V_2$.

The corresponding steps are labelled in the figure, which is somewhat reduced from its actual size.

Finally, $R_2 = V_2/I_2$ and $L = (V_4 \times 0.1 \times 1000)/(400\pi \times I_2)$.

The accurate solutions (obtained by calculation rather than drawing) are:

$I_1 = (7.76, -23.2°)$, $I_2 = (14.36, -50.8°)$, $I_3 = (8.30, 103.4°)$;

$V_1 = (0.388, -23.2°)$, $V_2 = (0.287, -50.8°)$, $V_4 = (0.596, 39.2°)$;

$L = 33$ mH, $R_2 = 20 \ \Omega$.

8

Matrices and vector spaces

8.1 *Which of the following statements about linear vector spaces are true? Where a statement is false, give a counter-example to demonstrate this.*

(a) *Non-singular $N \times N$ matrices form a vector space of dimension N^2.*

(b) *Singular $N \times N$ matrices form a vector space of dimension N^2.*

(c) *Complex numbers form a vector space of dimension 2.*

(d) *Polynomial functions of x form an infinite-dimensional vector space.*

(e) *Series $\{a_0, a_1, a_2, \ldots, a_N\}$ for which $\sum_{n=0}^{N} |a_n|^2 = 1$ form an N-dimensional vector space.*

(f) *Absolutely convergent series form an infinite-dimensional vector space.*

(g) *Convergent series with terms of alternating sign form an infinite-dimensional vector space.*

We first remind ourselves that for a set of entities to form a vector space, they must pass five tests: (i) closure under commutative and associative addition; (ii) closure under multiplication by a scalar; (iii) the existence of a null vector in the set; (iv) multiplication by unity leaves any vector unchanged; (v) each vector has a corresponding negative vector.

(a) False. The matrix O_N, the $N \times N$ null matrix, required by (iii) is *not* non-singular and is therefore not in the set.

(b) Consider the sum of $\begin{pmatrix} 1 & 0 \\ 0 & 0 \end{pmatrix}$ and $\begin{pmatrix} 0 & 0 \\ 0 & 1 \end{pmatrix}$. The sum is the unit matrix which is not singular and so the set is not closed; this violates requirement (i). The statement is false.

(c) The space is closed under addition and multiplication by a scalar; multiplication by unity leaves a complex number unchanged; there is a null vector ($= 0 + i0$)

and a negative complex number for each vector. All the necessary conditions are satisfied and the statement is true.

(d) As in the previous case, all the conditions are satisfied and the statement is true.

(e) This statement is false. To see why, consider $b_n = a_n + a_n$ for which $\sum_{n=0}^{N} |b_n|^2 = 4 \neq 1$, i.e. the set is not closed (violating (i)), or note that there is no zero vector with unit norm (violating (iii)).

(f) True. Note that an absolutely convergent series remains absolutely convergent when the signs of all of its terms are reversed.

(g) False. Consider the two series defined by

$$a_0 = \tfrac{1}{2}, \quad a_n = 2(-\tfrac{1}{2})^n \text{ for } n \geq 1; \quad b_n = -(-\tfrac{1}{2})^n \text{ for } n \geq 0.$$

The series that is the sum of $\{a_n\}$ and $\{b_n\}$ does not have alternating signs and so closure (required by (i)) does not hold.

8.3 *Using the properties of determinants, solve with a minimum of calculation the following equations for x:*

(a) $\begin{vmatrix} x & a & a & 1 \\ a & x & b & 1 \\ a & b & x & 1 \\ a & b & c & 1 \end{vmatrix} = 0,$ (b) $\begin{vmatrix} x+2 & x+4 & x-3 \\ x+3 & x & x+5 \\ x-2 & x-1 & x+1 \end{vmatrix} = 0.$

(a) In view of the similarities between some rows and some columns, the property most likely to be useful here is that if a determinant has two rows/columns equal (or multiples of each other) then its value is zero.

(i) We note that setting $x = a$ makes the first and fourth columns multiples of each other and hence makes the value of the determinant 0; thus $x = a$ is one solution to the equation.

(ii) Setting $x = b$ makes the second and third rows equal, and again the determinant vanishes; thus b is another root of the equation.

(iii) Setting $x = c$ makes the third and fourth rows equal, and yet again the determinant vanishes; thus c is also a root of the equation.

Since the determinant contains no x in its final column, it is a cubic polynomial in x and there will be exactly three roots to the equation. We have already found all three!

(b) Here, the presence of x multiplied by unity in every entry means that subtracting rows/columns will lead to a simplification. After (i) subtracting the first column from each of the others, and then (ii) subtracting the first row from each of the others, the determinant becomes

$$
\begin{vmatrix} x+2 & 2 & -5 \\ x+3 & -3 & 2 \\ x-2 & 1 & 3 \end{vmatrix} = \begin{vmatrix} x+2 & 2 & -5 \\ 1 & -5 & 7 \\ -4 & -1 & 8 \end{vmatrix}
$$

$$
= (x+2)(-40+7) + 2(-28-8) - 5(-1-20)
$$

$$
= -33(x+2) - 72 + 105
$$

$$
= -33x - 33.
$$

Thus $x = -1$ is the only solution to the original (linear!) equation.

8.5 *By considering the matrices*

$$
A = \begin{pmatrix} 1 & 0 \\ 0 & 0 \end{pmatrix}, \qquad B = \begin{pmatrix} 0 & 0 \\ 3 & 4 \end{pmatrix},
$$

show that $AB = 0$ *does* not *imply that either* A *or* B *is the zero matrix but that it does imply that at least one of them is singular.*

We have

$$
AB = \begin{pmatrix} 1 & 0 \\ 0 & 0 \end{pmatrix}\begin{pmatrix} 0 & 0 \\ 3 & 4 \end{pmatrix} = \begin{pmatrix} 0 & 0 \\ 0 & 0 \end{pmatrix}.
$$

Thus AB is the zero matrix O without either A = O or B = O.

However, $AB = O \Rightarrow |A||B| = |O| = 0$ and therefore either $|A| = 0$ or $|B| = 0$ (or both).

8.7 *Prove the following results involving Hermitian matrices:*

   (a) *If* A *is Hermitian and* U *is unitary then* $U^{-1}AU$ *is Hermitian.*
   (b) *If* A *is anti-Hermitian then* iA *is Hermitian.*
   (c) *The product of two Hermitian matrices* A *and* B *is Hermitian if and only if* A *and* B *commute.*
   (d) *If* S *is a real antisymmetric matrix then* $A = (I - S)(I + S)^{-1}$ *is orthogonal. If* A *is given by*

$$
A = \begin{pmatrix} \cos\theta & \sin\theta \\ -\sin\theta & \cos\theta \end{pmatrix}
$$

   then find the matrix S *that is needed to express* A *in the above form.*
   (e) *If* K *is skew-hermitian, i.e.* $K^\dagger = -K$, *then* $V = (I + K)(I - K)^{-1}$ *is unitary.*

121

The general properties of matrices that we will need are $(A^\dagger)^{-1} = (A^{-1})^\dagger$ and

$$(AB \cdots C)^T = C^T \cdots B^T A^T, \qquad (AB \cdots C)^\dagger = C^\dagger \cdots B^\dagger A^\dagger.$$

(a) Given that $A = A^\dagger$ and $U^\dagger U = I$, consider

$$(U^{-1}AU)^\dagger = U^\dagger A^\dagger (U^{-1})^\dagger = U^{-1}A(U^\dagger)^{-1} = U^{-1}A(U^{-1})^{-1} = U^{-1}AU,$$

i.e. $U^{-1}AU$ is Hermitian.

(b) Given $A^\dagger = -A$, consider

$$(iA)^\dagger = -iA^\dagger = -i(-A) = iA,$$

i.e. iA is Hermitian.

(c) Given $A = A^\dagger$ and $B = B^\dagger$.

(i) Suppose $AB = BA$, then

$$(AB)^\dagger = B^\dagger A^\dagger = BA = AB,$$

i.e. AB is Hermitian.

(ii) Now suppose that $(AB)^\dagger = AB$. Then

$$BA = B^\dagger A^\dagger = (AB)^\dagger = AB,$$

i.e. A and B commute.

Thus, AB is Hermitian $\iff$ A and B commute.

(d) Given that S is real and $S^T = -S$ with $A = (I - S)(I + S)^{-1}$, consider

$$
\begin{aligned}
A^T A &= [(I - S)(I + S)^{-1}]^T [(I - S)(I + S)^{-1}] \\
&= [(I + S)^{-1}]^T (I + S)(I - S)(I + S)^{-1} \\
&= (I - S)^{-1}(I + S - S - S^2)(I + S)^{-1} \\
&= (I - S)^{-1}(I - S)(I + S)(I + S)^{-1} \\
&= I\,I = I,
\end{aligned}
$$

i.e. A is orthogonal.

If $A = (I - S)(I + S)^{-1}$, then $A + AS = I - S$ and $(A + I)S = I - A$, giving

$$
\begin{aligned}
S &= (A + I)^{-1}(I - A) \\
&= \begin{pmatrix} 1 + \cos\theta & \sin\theta \\ -\sin\theta & 1 + \cos\theta \end{pmatrix}^{-1} \begin{pmatrix} 1 - \cos\theta & -\sin\theta \\ \sin\theta & 1 - \cos\theta \end{pmatrix} \\
&= \frac{1}{2 + 2\cos\theta} \begin{pmatrix} 1 + \cos\theta & -\sin\theta \\ \sin\theta & 1 + \cos\theta \end{pmatrix} \begin{pmatrix} 1 - \cos\theta & -\sin\theta \\ \sin\theta & 1 - \cos\theta \end{pmatrix} \\
&= \frac{1}{4\cos^2(\theta/2)} \begin{pmatrix} 0 & -2\sin\theta \\ 2\sin\theta & 0 \end{pmatrix} \\
&= \begin{pmatrix} 0 & -\tan(\theta/2) \\ \tan(\theta/2) & 0 \end{pmatrix}.
\end{aligned}
$$

(e) This proof is almost identical to the first section of part (d) but with S replaced by −K and transposed matrices replaced by hermitian conjugate matrices.

8.9 *The* commutator $[X, Y]$ *of two matrices is defined by the equation*

$$[X, Y] = XY − YX.$$

Two anticommuting matrices A *and* B *satisfy*

$$A^2 = I, \qquad B^2 = I, \qquad [A, B] = 2iC.$$

(a) *Prove that* $C^2 = I$ *and that* $[B, C] = 2iA$.
(b) *Evaluate* $[[[A, B], [B, C]], [A, B]]$.

(a) From $AB − BA = 2iC$ and $AB = −BA$ it follows that $AB = iC$. Thus,

$$−C^2 = iCiC = ABAB = A(−AB)B = −(AA)(BB) = −II = −I,$$

i.e. $C^2 = I$. In deriving the above result we have used the associativity of matrix multiplication.

For the commutator of B and C,

$$
\begin{aligned}
[B, C] &= BC − CB \\
&= B(−iAB) − (−i)ABB \\
&= −i(BA)B + iAI \\
&= −i(−AB)B + iA \\
&= iA + iA = 2iA.
\end{aligned}
$$

(b) To evaluate this multiple-commutator expression we must work outwards from the innermost 'explicit' commutators. There are three such commutators at the first stage. We also need the result that $[C, A] = 2iB$; this can be proved in the same way as that for $[B, C]$ in part (a), or by making the cyclic replacements $A \to B \to C \to A$ in the assumptions and their consequences, as proved in part (a). Then we have

$$
\begin{aligned}
[[[A, B], [B, C]], [A, B]] &= [[2iC, 2iA], 2iC] \\
&= −4[[C, A], 2iC] \\
&= −4[2iB, 2iC] \\
&= (−4)(−4)[B, C] = 32iA.
\end{aligned}
$$

> **8.11** *A general triangle has angles* α, β *and* γ *and corresponding opposite sides* a, b *and* c. *Express the length of each side in terms of the lengths of the other two sides and the relevant cosines, writing the relationships in matrix and vector form, using the vectors having components* a, b, c *and* $\cos \alpha, \cos \beta, \cos \gamma$. *Invert the matrix and hence deduce the cosine-law expressions involving* α, β *and* γ.

By considering each side of the triangle as the sum of the projections onto it of the other two sides, we have the three simultaneous equations:

$$a = b \cos \gamma + c \cos \beta,$$
$$b = c \cos \alpha + a \cos \gamma,$$
$$c = b \cos \alpha + a \cos \beta.$$

Written in matrix and vector form, $\mathbf{Ax} = \mathbf{y}$, they become

$$\begin{pmatrix} 0 & c & b \\ c & 0 & a \\ b & a & 0 \end{pmatrix} \begin{pmatrix} \cos \alpha \\ \cos \beta \\ \cos \gamma \end{pmatrix} = \begin{pmatrix} a \\ b \\ c \end{pmatrix}.$$

The matrix $\mathbf{A}$ is non-singular, since $|\mathbf{A}| = 2abc \neq 0$, and therefore has an inverse given by

$$\mathbf{A}^{-1} = \frac{1}{2abc} \begin{pmatrix} -a^2 & ab & ac \\ ab & -b^2 & bc \\ ac & bc & -c^2 \end{pmatrix}.$$

And so, writing $\mathbf{x} = \mathbf{A}^{-1}\mathbf{y}$, we have

$$\begin{pmatrix} \cos \alpha \\ \cos \beta \\ \cos \gamma \end{pmatrix} = \frac{1}{2abc} \begin{pmatrix} -a^2 & ab & ac \\ ab & -b^2 & bc \\ ac & bc & -c^2 \end{pmatrix} \begin{pmatrix} a \\ b \\ c \end{pmatrix}.$$

From this we can read off the cosine-law equation

$$\cos \alpha = \frac{1}{2abc}(-a^3 + ab^2 + ac^2) = \frac{b^2 + c^2 - a^2}{2bc},$$

and the corresponding expressions for $\cos \beta$ and $\cos \gamma$.

8.13 *Using the Gram–Schmidt procedure:*

(a) *construct an orthonormal set of vectors from the following:*

$$\mathsf{x}_1 = (0 \quad 0 \quad 1 \quad 1)^{\mathrm{T}}, \qquad \mathsf{x}_2 = (1 \quad 0 \quad -1 \quad 0)^{\mathrm{T}},$$
$$\mathsf{x}_3 = (1 \quad 2 \quad 0 \quad 2)^{\mathrm{T}}, \qquad \mathsf{x}_4 = (2 \quad 1 \quad 1 \quad 1)^{\mathrm{T}};$$

(b) *find an orthonormal basis, within a four-dimensional Euclidean space, for the subspace spanned by the three vectors*

$$(1 \quad 2 \quad 0 \quad 0)^{\mathrm{T}}, \quad (3 \quad -1 \quad 2 \quad 0)^{\mathrm{T}}, \quad (0 \quad 0 \quad 2 \quad 1)^{\mathrm{T}}.$$

The general procedure is to construct the orthonormal base set $\{\hat{\mathsf{z}}_i\}$ using the iteration procedure

$$\mathsf{z}_n = \mathsf{x}_n - \sum_{r=1}^{n-1} [\hat{\mathsf{z}}_r^{\dagger} \mathsf{x}_n] \hat{\mathsf{z}}_r \text{ with } \mathsf{z}_1 = \mathsf{x}_1.$$

The vector $\hat{\mathsf{z}}$ is the vector z after normalisation and the expression in square brackets is the (complex) inner product of $\hat{\mathsf{z}}_r$ and x_n.

(a) We start with $\hat{\mathsf{z}}_1 = 2^{-1/2} \mathsf{x}_1 = 2^{-1/2} [0 \quad 0 \quad 1 \quad 1]^{\mathrm{T}}$.

Next we calculate $(\hat{\mathsf{z}}_1)^{\dagger} \mathsf{x}_2$ as $-2^{-1/2}$ and then form z_2 as

$$\mathsf{z}_2 = \begin{bmatrix} 1 \\ 0 \\ -1 \\ 0 \end{bmatrix} - \frac{-1}{\sqrt{2}} \frac{1}{\sqrt{2}} \begin{bmatrix} 0 \\ 0 \\ 1 \\ 1 \end{bmatrix} = \begin{bmatrix} 1 \\ 0 \\ -\frac{1}{2} \\ \frac{1}{2} \end{bmatrix}.$$

The normalised vector $\hat{\mathsf{z}}_2$ is $6^{-1/2} (2 \quad 0 \quad -1)^{\mathrm{T}} 1$.

Proceeding in this way, but without detailed description, we obtain

$$\mathsf{z}_3 = \begin{bmatrix} 1 \\ 2 \\ 0 \\ 2 \end{bmatrix} - \frac{2}{\sqrt{2}} \frac{1}{\sqrt{2}} \begin{bmatrix} 0 \\ 0 \\ 1 \\ 1 \end{bmatrix} - \frac{4}{\sqrt{6}} \frac{1}{\sqrt{6}} \begin{bmatrix} 2 \\ 0 \\ -1 \\ 1 \end{bmatrix} = \begin{bmatrix} -\frac{1}{3} \\ 2 \\ -\frac{1}{3} \\ \frac{1}{3} \end{bmatrix}.$$

The normalised vector $\hat{\mathsf{z}}_3$ is $(39)^{-1/2} (-1 \quad 6 \quad -1)^{\mathrm{T}} 1$.

Finally,

$$\mathsf{z}_4 = \begin{bmatrix} 2 \\ 1 \\ 1 \\ 1 \end{bmatrix} - \frac{2}{\sqrt{2}} \frac{1}{\sqrt{2}} \begin{bmatrix} 0 \\ 0 \\ 1 \\ 1 \end{bmatrix} - \frac{4}{\sqrt{6}} \frac{1}{\sqrt{6}} \begin{bmatrix} 2 \\ 0 \\ -1 \\ 1 \end{bmatrix} - \frac{4}{\sqrt{39}} \frac{1}{\sqrt{39}} \begin{bmatrix} -1 \\ 6 \\ -1 \\ 1 \end{bmatrix}.$$

The normalised vector $\hat{z}_4$ is $(13)^{-1/2}(2 \quad 1 \quad 2)^T - 1$.

[Note that if the only requirement had been to find an orthonormal set of base vectors then the obvious $(1 \quad 0 \quad 0 \quad 0)^T$, $(0 \quad 1 \quad 0 \quad 0)^T$, etc. could have been chosen.]

(b) The procedure is as in part (a) except that we require only three orthonormal vectors. However, we *must* begin with the given vectors so as to ensure that the correct subspace is spanned.

We start with $\hat{z}_1 = 5^{-1/2}x_1 = 5^{-1/2}[1 \quad 2 \quad 0 \quad 0]^T$.

Next we calculate $(\hat{z}_1)^\dagger x_2$ as $-5^{-1/2}$ and then form z_2 as

$$z_2 = \begin{bmatrix} 3 \\ -1 \\ 2 \\ 0 \end{bmatrix} - \frac{1}{\sqrt{5}} \frac{1}{\sqrt{5}} \begin{bmatrix} 1 \\ 2 \\ 0 \\ 0 \end{bmatrix} = \begin{bmatrix} \frac{14}{5} \\ -\frac{7}{5} \\ 2 \\ 0 \end{bmatrix}.$$

The normalised vector $\hat{z}_2$ is $(345)^{-1/2}(14 \quad -7 \quad 10)^T \, 0$.

As the final base vector for the subspace we obtain

$$z_3 = \begin{bmatrix} 0 \\ 0 \\ 2 \\ 1 \end{bmatrix} - 0 \frac{1}{\sqrt{5}} \begin{bmatrix} 1 \\ 2 \\ 0 \\ 0 \end{bmatrix} - \frac{20}{\sqrt{345}} \frac{1}{\sqrt{345}} \begin{bmatrix} 14 \\ -7 \\ 10 \\ 0 \end{bmatrix} = \frac{1}{345} \begin{bmatrix} -280 \\ 140 \\ 490 \\ 345 \end{bmatrix}.$$

Thus, the normalised vector $\hat{z}_3$ is $(18285)^{-1/2}(-56 \quad 28 \quad 98)^T \, 69$. The fact that three orthonormal vectors can be found shows that the subspace is 3-dimensional and that the three original vectors are not linearly dependent.

8.15 *Determine which of the matrices below are mutually commuting, and, for those that are, demonstrate that they have a complete set of eigenvectors in common:*

$$A = \begin{pmatrix} 6 & -2 \\ -2 & 9 \end{pmatrix}, \qquad B = \begin{pmatrix} 1 & 8 \\ 8 & -11 \end{pmatrix},$$

$$C = \begin{pmatrix} -9 & -10 \\ -10 & 5 \end{pmatrix}, \qquad D = \begin{pmatrix} 14 & 2 \\ 2 & 11 \end{pmatrix}.$$

To establish the result we need to examine all pairs of products.

$$AB = \begin{pmatrix} 6 & -2 \\ -2 & 9 \end{pmatrix} \begin{pmatrix} 1 & 8 \\ 8 & -11 \end{pmatrix}$$

$$= \begin{pmatrix} -10 & 70 \\ 70 & -115 \end{pmatrix}$$

$$= \begin{pmatrix} 1 & 8 \\ 8 & -11 \end{pmatrix} \begin{pmatrix} 6 & -2 \\ -2 & 9 \end{pmatrix} = BA.$$

$$AC = \begin{pmatrix} 6 & -2 \\ -2 & 9 \end{pmatrix} \begin{pmatrix} -9 & -10 \\ -10 & 5 \end{pmatrix}$$

$$= \begin{pmatrix} -34 & -70 \\ -72 & 65 \end{pmatrix} \neq \begin{pmatrix} -34 & -72 \\ -70 & 65 \end{pmatrix}$$

$$= \begin{pmatrix} -9 & -10 \\ -10 & 5 \end{pmatrix} \begin{pmatrix} 6 & -2 \\ -2 & 9 \end{pmatrix} = CA.$$

Continuing in this way, we find:

$$AD = \begin{pmatrix} 80 & -10 \\ -10 & 95 \end{pmatrix} = DA.$$

$$BC = \begin{pmatrix} -89 & 30 \\ 38 & -135 \end{pmatrix} \neq \begin{pmatrix} -89 & 38 \\ 30 & -135 \end{pmatrix} = CB.$$

$$BD = \begin{pmatrix} 30 & 90 \\ 90 & -105 \end{pmatrix} = DB.$$

$$CD = \begin{pmatrix} -146 & -128 \\ -130 & 35 \end{pmatrix} \neq \begin{pmatrix} -146 & -130 \\ -128 & 35 \end{pmatrix} = DC.$$

These results show that whilst A, B and D are mutually commuting, none of them commutes with C.

We could use any of the three mutually commuting matrices to find the common set (actually a pair, as they are 2×2 matrices) of eigenvectors. We arbitrarily choose A. The eigenvalues of A satisfy

$$\begin{vmatrix} 6 - \lambda & -2 \\ -2 & 9 - \lambda \end{vmatrix} = 0,$$

$$\lambda^2 - 15\lambda + 50 = 0,$$

$$(\lambda - 5)(\lambda - 10) = 0.$$

For $\lambda = 5$, an eigenvector $(x, y)^T$ must satisfy $x - 2y = 0$, whilst, for $\lambda = 10$, $4x + 2y = 0$. Thus a pair of independent eigenvectors of A are $(2, 1)^T$ and $(1, -2)^T$. Direct substitution verifies that they are also eigenvectors of B and D with pairs of eigenvalues 5, -15 and 15, 10, respectively.

8.17 *Find three real orthogonal column matrices, each of which is a simultaneous eigenvector of*

$$A = \begin{pmatrix} 0 & 0 & 1 \\ 0 & 1 & 0 \\ 1 & 0 & 0 \end{pmatrix} \quad \text{and} \quad B = \begin{pmatrix} 0 & 1 & 1 \\ 1 & 0 & 1 \\ 1 & 1 & 0 \end{pmatrix}.$$

We first note that

$$AB = \begin{pmatrix} 1 & 1 & 0 \\ 1 & 0 & 1 \\ 0 & 1 & 1 \end{pmatrix} = BA.$$

The two matrices commute and so they *will* have a common set of eigenvectors. The eigenvalues of A are given by

$$\begin{vmatrix} -\lambda & 0 & 1 \\ 0 & 1-\lambda & 0 \\ 1 & 0 & -\lambda \end{vmatrix} = (1-\lambda)(\lambda^2 - 1) = 0,$$

i.e. $\lambda = 1$, $\lambda = 1$ and $\lambda = -1$, with corresponding eigenvectors $e^1 = (1, y_1, 1)^T$, $e^2 = (1, y_2, 1)^T$ and $e^3 = (1, 0, -1)^T$. For these to be mutually orthogonal requires that $y_1 y_2 = -2$.

The third vector, e^3, is clearly an eigenvector of B with eigenvalue $\mu_3 = -1$. For e^1 or e^2 to be an eigenvector of B with eigenvalue μ requires

$$\begin{pmatrix} 0-\mu & 1 & 1 \\ 1 & 0-\mu & 1 \\ 1 & 1 & 0-\mu \end{pmatrix} \begin{pmatrix} 1 \\ y \\ 1 \end{pmatrix} = \begin{pmatrix} 0 \\ 0 \\ 0 \end{pmatrix};$$

i.e.

$$-\mu + y + 1 = 0,$$

and

$$1 - \mu y + 1 = 0,$$

giving

$$-\frac{2}{y} + y + 1 = 0,$$

$$\Rightarrow \quad y^2 + y - 2 = 0,$$

$$\Rightarrow \quad y = 1 \quad \text{or} \quad -2.$$

Thus, $y_1 = 1$ with $\mu_1 = 2$, whilst $y_2 = -2$ with $\mu_2 = -1$.

The common eigenvectors are thus

$$e^1 = (1, 1, 1)^T, \quad e^2 = (1, -2, 1)^T, \quad e^3 = (1, 0, -1)^T.$$

We note, as a check, that $\sum_i \mu_i = 2 + (-1) + (-1) = 0 = \text{Tr } B$.

> **8.19** *Given that* A *is a real symmetric matrix with normalised eigenvectors* e^i, *obtain the coefficients* α_i *involved when column matrix* x, *which is the solution of*
>
> $$A x - \mu x = v, \qquad (*)$$
>
> *is expanded as* $x = \sum_i \alpha_i e^i$. *Here* μ *is a given constant and* v *is a given column matrix.*
>
> (a) *Solve* (*) *when*
>
> $$A = \begin{pmatrix} 2 & 1 & 0 \\ 1 & 2 & 0 \\ 0 & 0 & 3 \end{pmatrix},$$
>
> $\mu = 2$ *and* $v = (1 \quad 2 \quad 3)^T$.
>
> (b) *Would* (*) *have a solution if* (i) $\mu = 1$ *and* $v = (1 \quad 2 \quad 3)^T$, (ii) $v = (2 \quad 2 \quad 3)^T$? *Where it does, find it.*

Let $x = \sum_i \alpha_i e^i$, where $A e^i = \lambda_i e^i$. Then

$$A x - \mu x = v,$$
$$\sum_i A \alpha_i e^i - \sum_i \mu \alpha_i e^i = v,$$
$$\sum_i \left(\lambda_i \alpha_i e^i - \mu \alpha_i e^i \right) = v,$$
$$\alpha_j = \frac{(e^j)^\dagger v}{\lambda_j - \mu}.$$

To obtain the last line we have used the mutual orthogonality of the eigenvectors. We note, in passing, that if $\mu = \lambda_j$ for any j there is no solution unless $(e^j)^\dagger v = 0$.

(a) To obtain the eigenvalues of the given matrix A, consider

$$0 = |A - \lambda I| = (3 - \lambda)(4 - 4\lambda + \lambda^2 - 1) = (3 - \lambda)(3 - \lambda)(1 - \lambda).$$

The eigenvalues, and a possible set of corresponding normalised eigenvectors, are therefore,

$$\text{for} \quad \lambda = 3, \quad e^1 = (0,\, 0,\, 1)^T ;$$
$$\text{for} \quad \lambda = 3, \quad e^2 = 2^{-1/2}(1,\, 1,\, 0)^T ;$$
$$\text{for} \quad \lambda = 1, \quad e^3 = 2^{-1/2}(1,\, -1,\, 0)^T .$$

Since $\lambda = 3$ is a degenerate eigenvalue, there are infinitely many acceptable pairs of orthogonal eigenvectors corresponding to it; any pair of vectors of the form (a_i, a_i, b_i) with $2a_1 a_2 + b_1 b_2 = 0$ will suffice. The pair given is just about the simplest choice possible.

With $\mu = 2$ and $\mathbf{v} = (1, 2, 3)^{\mathrm{T}}$,

$$\alpha_1 = \frac{3}{3-2}, \quad \alpha_2 = \frac{3/\sqrt{2}}{3-2}, \quad \alpha_3 = \frac{-1/\sqrt{2}}{1-2}.$$

Thus the solution vector is

$$\mathbf{x} = 3 \begin{pmatrix} 0 \\ 0 \\ 1 \end{pmatrix} + \frac{3}{\sqrt{2}} \frac{1}{\sqrt{2}} \begin{pmatrix} 1 \\ 1 \\ 0 \end{pmatrix} + \frac{1}{\sqrt{2}} \frac{1}{\sqrt{2}} \begin{pmatrix} 1 \\ -1 \\ 0 \end{pmatrix} = \begin{pmatrix} 2 \\ 1 \\ 3 \end{pmatrix}.$$

(b) If $\mu = 1$ then it is equal to the third eigenvalue and a solution is only possible if $(\mathbf{e}^3)^{\dagger}\mathbf{v} = 0$.

For (i) $\mathbf{v} = (1, 2, 3)^{\mathrm{T}}$, $(\mathbf{e}^3)^{\dagger}\mathbf{v} = -1/\sqrt{2}$ and so no solution is possible.

For (ii) $\mathbf{v} = (2, 2, 3)^{\mathrm{T}}$, $(\mathbf{e}^3)^{\dagger}\mathbf{v} = 0$, and so a solution is possible. The other scalar products needed are $(\mathbf{e}^1)^{\dagger}\mathbf{v} = 3$ and $(\mathbf{e}^2)^{\dagger}\mathbf{v} = 2\sqrt{2}$. For this vector $\mathbf{v}$ the solution to the equation is

$$\mathbf{x} = \frac{3}{3-1} \begin{pmatrix} 0 \\ 0 \\ 1 \end{pmatrix} + \frac{2\sqrt{2}}{3-1} \frac{1}{\sqrt{2}} \begin{pmatrix} 1 \\ 1 \\ 0 \end{pmatrix} = \begin{pmatrix} 1 \\ 1 \\ \frac{3}{2} \end{pmatrix}.$$

[The solutions to both parts can be checked by resubstitution.]

8.21 *By finding the eigenvectors of the Hermitian matrix*

$$\mathbf{H} = \begin{pmatrix} 10 & 3i \\ -3i & 2 \end{pmatrix},$$

construct a unitary matrix $\mathbf{U}$ *such that* $\mathbf{U}^{\dagger}\mathbf{H}\mathbf{U} = \Lambda$, *where* Λ *is a real diagonal matrix.*

We start by finding the eigenvalues of $\mathbf{H}$ using

$$\begin{vmatrix} 10 - \lambda & 3i \\ -3i & 2 - \lambda \end{vmatrix} = 0,$$

$$20 - 12\lambda + \lambda^2 - 3 = 0,$$

$$\lambda = 1 \quad \text{or} \quad 11.$$

As expected for an hermitian matrix, the eigenvalues are real.

For $\lambda = 1$ and normalised eigenvector $(x, y)^{\mathrm{T}}$,

$$9x + 3iy = 0 \quad \Rightarrow \quad \mathbf{x}^1 = (10)^{-1/2} (1, 3i)^{\mathrm{T}}.$$

130

For $\lambda = 11$ and normalised eigenvector $(x, y)^{\mathrm{T}}$,

$$-x + 3iy = 0 \qquad \Rightarrow \qquad \mathbf{x}^2 = (10)^{-1/2} (3i, \ 1)^{\mathrm{T}}.$$

Again as expected, $(\mathbf{x}^1)^\dagger \mathbf{x}^2 = 0$, thus verifying the mutual orthogonality of the eigenvectors. It should be noted that the normalisation factor is determined by $(\mathbf{x}^i)^\dagger \mathbf{x}^i = 1$ (and *not* by $(\mathbf{x}^i)^{\mathrm{T}} \mathbf{x}^i = 1$).

We now use these normalised eigenvectors of H as the columns of the matrix U and check that it is unitary:

$$\mathsf{U} = \frac{1}{\sqrt{10}} \begin{pmatrix} 1 & 3i \\ 3i & 1 \end{pmatrix}, \quad \mathsf{U}^\dagger = \frac{1}{\sqrt{10}} \begin{pmatrix} 1 & -3i \\ -3i & 1 \end{pmatrix},$$

$$\mathsf{U}\mathsf{U}^\dagger = \frac{1}{10} \begin{pmatrix} 1 & 3i \\ 3i & 1 \end{pmatrix} \begin{pmatrix} 1 & -3i \\ -3i & 1 \end{pmatrix} = \frac{1}{10} \begin{pmatrix} 10 & 0 \\ 0 & 10 \end{pmatrix} = \mathsf{I}.$$

U has the further property that

$$\mathsf{U}^\dagger \mathsf{H} \mathsf{U} = \frac{1}{\sqrt{10}} \begin{pmatrix} 1 & -3i \\ -3i & 1 \end{pmatrix} \begin{pmatrix} 10 & 3i \\ -3i & 2 \end{pmatrix} \frac{1}{\sqrt{10}} \begin{pmatrix} 1 & 3i \\ 3i & 1 \end{pmatrix}$$

$$= \frac{1}{10} \begin{pmatrix} 1 & -3i \\ -3i & 1 \end{pmatrix} \begin{pmatrix} 1 & 33i \\ 3i & 11 \end{pmatrix}$$

$$= \frac{1}{10} \begin{pmatrix} 10 & 0 \\ 0 & 110 \end{pmatrix} = \begin{pmatrix} 1 & 0 \\ 0 & 11 \end{pmatrix} = \Lambda.$$

That the diagonal entries of Λ are the eigenvalues of H is in accord with the general theory of normal matrices.

8.23 *Given that the matrix*

$$\mathsf{A} = \begin{pmatrix} 2 & -1 & 0 \\ -1 & 2 & -1 \\ 0 & -1 & 2 \end{pmatrix}$$

has two eigenvectors of the form $(1 \quad y \quad 1)^{\mathrm{T}}$, *use the stationary property of the expression* $J(\mathbf{x}) = \mathbf{x}^{\mathrm{T}} \mathsf{A} \mathbf{x} / (\mathbf{x}^{\mathrm{T}} \mathbf{x})$ *to obtain the corresponding eigenvalues. Deduce the third eigenvalue.*

Since A is real and symmetric, each eigenvalue λ is real. Further, from the first component of $\mathsf{A}\mathbf{x} = \lambda\mathbf{x}$, we have that $2 - y = \lambda$, showing that y is also real. Considered as a function of a general vector of the form $(1, \ y, \ 1)^{\mathrm{T}}$, the quadratic form $x^T A x$ can be written explicitly as

$$x^T A x = (1 \ y \ 1) \begin{pmatrix} 2 & -1 & 0 \\ -1 & 2 & -1 \\ 0 & -1 & 2 \end{pmatrix} \begin{pmatrix} 1 \\ y \\ 1 \end{pmatrix}$$

$$= (1 \ y \ 1) \begin{pmatrix} 2-y \\ 2y-2 \\ 2-y \end{pmatrix}$$

$$= 2y^2 - 4y + 4.$$

The scalar product $x^T x$ has the value $2 + y^2$, and so we need to find the stationary values of

$$I = \frac{2y^2 - 4y + 4}{2 + y^2}.$$

These are given by

$$0 = \frac{dI}{dy} = \frac{(2+y^2)(4y-4) - (2y^2 - 4y + 4)2y}{(2+y^2)^2}$$

$$0 = 4y^2 - 8,$$

$$y = \pm\sqrt{2}.$$

The corresponding eigenvalues are the values of I at the stationary points, explicitly:

$$\text{for} \quad y = \sqrt{2}, \qquad \lambda_1 = \frac{2(2) - 4\sqrt{2} + 4}{2 + 2} = 2 - \sqrt{2};$$

$$\text{for} \quad y = -\sqrt{2}, \qquad \lambda_2 = \frac{2(2) + 4\sqrt{2} + 4}{2 + 2} = 2 + \sqrt{2}.$$

The final eigenvalue can be found using the fact that the sum of the eigenvalues is equal to the trace of the matrix; so

$$\lambda_3 = (2 + 2 + 2) - (2 - \sqrt{2}) - (2 + \sqrt{2}) = 2.$$

8.25 *The equation of a particular conic section is*

$$Q \equiv 8x_1^2 + 8x_2^2 - 6x_1 x_2 = 110.$$

Determine the type of conic section this represents, the orientation of its principal axes, and relevant lengths in the directions of these axes.

The eigenvalues of the matrix $\begin{pmatrix} 8 & -3 \\ -3 & 8 \end{pmatrix}$ associated with the quadratic form on the LHS (without any prior scaling) are given by

$$0 = \begin{vmatrix} 8 - \lambda & -3 \\ -3 & 8 - \lambda \end{vmatrix}$$
$$= \lambda^2 - 16\lambda + 55$$
$$= (\lambda - 5)(\lambda - 11).$$

Referred to the corresponding eigenvectors as axes, the conic section (an ellipse since both eigenvalues are positive) will take the form

$$5y_1^2 + 11y_2^2 = 110 \quad \text{or, in standard form,} \quad \frac{y_1^2}{22} + \frac{y_2^2}{10} = 1.$$

Thus the semi-axes are of lengths $\sqrt{22}$ and $\sqrt{10}$; the former is in the direction of the vector $(x_1, x_2)^{\mathrm{T}}$ given by $(8 - 5)x_1 - 3x_2 = 0$, i.e. it is the line $x_1 = x_2$. The other principal axis will be the line at right angles to this, namely the line $x_1 = -x_2$.

8.27 *Find the direction of the axis of symmetry of the quadratic surface*

$$7x^2 + 7y^2 + 7z^2 - 20yz - 20xz + 20xy = 3.$$

The straightforward, but longer, solution to this exercise is as follows.

Consider the characteristic polynomial of the matrix associated with the quadratic surface, namely,

$$f(\lambda) = \begin{vmatrix} 7 - \lambda & 10 & -10 \\ 10 & 7 - \lambda & -10 \\ -10 & -10 & 7 - \lambda \end{vmatrix}$$
$$= (7 - \lambda)(-51 - 14\lambda + \lambda^2) + 10(30 + 10\lambda) - 10(-30 - 10\lambda)$$
$$= -\lambda^3 + 21\lambda^2 + 153\lambda + 243.$$

If the quadratic surface has an axis of symmetry, it must have two equal major axes (perpendicular to it), and hence the characteristic equation must have a repeated root. This same root will therefore also be a root of $df/d\lambda = 0$, i.e. of

$$-3\lambda^2 + 42\lambda + 153 = 0,$$
$$\lambda^2 - 14\lambda - 51 = 0,$$
$$\lambda = 17 \quad \text{or} \quad -3.$$

Substitution shows that -3 is a root (and therefore a double root) of $f(\lambda) = 0$, but that 17 is not. The non-repeated root can be calculated as the trace of the matrix minus the repeated roots, i.e. $21 - (-3) - (-3) = 27$. It is the eigenvector that corresponds to this eigenvalue that gives the direction $(x, y, z)^T$ of the axis of symmetry. Its components must satisfy

$$(7 - 27)x + 10y - 10z = 0,$$
$$10x + (7 - 27)y - 10z = 0.$$

The axis of symmetry is therefore in the direction $(1, 1, -1)^T$.

A more subtle solution is obtained by noting that setting $\lambda = -3$ makes *all three* of the rows (or columns) of the determinant multiples of each other, i.e. it reduces the determinant to rank one. Thus -3 is a repeated root of the characteristic equation and the third root is $21 - 2(-3) = 27$. The rest of the analysis is as above.

We note in passing that, as two eigenvalues are negative and equal, the surface is the hyperboloid of revolution obtained by rotating a (two-branched) hyperbola about its axis of symmetry. Referred to this axis and two others forming a mutually orthogonal set, the equation of the quadratic surface takes the form $-3\chi^2 - 3\eta^2 + 27\zeta^2 = 3$ and so the tips of the two 'nose cones' ($\chi = \eta = 0$) are separated by $\frac{2}{3}$ of a unit.

8.29 *This exercise demonstrates the reverse of the usual procedure of diagonalising a matrix.*

(a) *Rearrange the result* $A' = S^{-1}AS$ *(which shows how to make a change of basis that diagonalises A) so as to express the original matrix A in terms of the unitary matrix S and the diagonal matrix A'. Hence show how to construct a matrix A that has given eigenvalues and given (orthogonal) column matrices as its eigenvectors.*

(b) *Find the matrix that has as eigenvectors* $(1 \quad 2 \quad 1)^T$, $(1 \quad -1 \quad 1)^T$ *and* $(1 \quad 0 \quad -1)^T$ *and corresponding eigenvalues* λ, μ *and* ν.

(c) *Try a particular case, say* $\lambda = 3$, $\mu = -2$ *and* $\nu = 1$, *and verify by explicit solution that the matrix so found does have these eigenvalues.*

(a) Since S is unitary, we can multiply the given result on the left by S and on the right by $S^\dagger$ to obtain

$$SA'S^\dagger = SS^{-1}ASS^\dagger = (I)\,A\,(I) = A.$$

More explicitly, in terms of the eigenvalues and normalised eigenvectors x^i of A,

$$\mathsf{A} = (\mathsf{x}^1 \quad \mathsf{x}^2 \quad \cdots \quad \mathsf{x}^n)\Lambda(\mathsf{x}^1 \quad \mathsf{x}^2 \quad \cdots \quad \mathsf{x}^n)^\dagger.$$

Here Λ is the diagonal matrix that has the eigenvalues of A as its diagonal elements.

Now, given normalised orthogonal column matrices and n specified values, we can use this result to construct a matrix that has the column matrices as eigenvectors and the values as eigenvalues.

(b) The normalised versions of the given column vectors are

$$\frac{1}{\sqrt{6}}(1,\,2,\,1)^\mathrm{T},\qquad \frac{1}{\sqrt{3}}(1,\,-1,\,1)^\mathrm{T},\qquad \frac{1}{\sqrt{2}}(1,\,0,\,-1)^\mathrm{T},$$

and the orthogonal matrix S can be constructed using these as its columns:

$$\mathsf{S} = \frac{1}{\sqrt{6}}\begin{pmatrix} 1 & \sqrt{2} & \sqrt{3} \\ 2 & -\sqrt{2} & 0 \\ 1 & \sqrt{2} & -\sqrt{3} \end{pmatrix}.$$

The required matrix A can now be formed as $\mathsf{S}\Lambda\mathsf{S}^\dagger$:

$$
\begin{aligned}
\mathsf{A} &= \frac{1}{6}\begin{pmatrix} 1 & \sqrt{2} & \sqrt{3} \\ 2 & -\sqrt{2} & 0 \\ 1 & \sqrt{2} & -\sqrt{3} \end{pmatrix}\begin{pmatrix} \lambda & 0 & 0 \\ 0 & \mu & 0 \\ 0 & 0 & \nu \end{pmatrix}\begin{pmatrix} 1 & 2 & 1 \\ \sqrt{2} & -\sqrt{2} & \sqrt{2} \\ \sqrt{3} & 0 & -\sqrt{3} \end{pmatrix} \\[2mm]
&= \frac{1}{6}\begin{pmatrix} 1 & \sqrt{2} & \sqrt{3} \\ 2 & -\sqrt{2} & 0 \\ 1 & \sqrt{2} & -\sqrt{3} \end{pmatrix}\begin{pmatrix} \lambda & 2\lambda & \lambda \\ \sqrt{2}\mu & -\sqrt{2}\mu & \sqrt{2}\mu \\ \sqrt{3}\nu & 0 & -\sqrt{3}\nu \end{pmatrix} \\[2mm]
&= \frac{1}{6}\begin{pmatrix} \lambda + 2\mu + 3\nu & 2\lambda - 2\mu & \lambda + 2\mu - 3\nu \\ 2\lambda - 2\mu & 4\lambda + 2\mu & 2\lambda - 2\mu \\ \lambda + 2\mu - 3\nu & 2\lambda - 2\mu & \lambda + 2\mu + 3\nu \end{pmatrix}.
\end{aligned}
$$

(c) Setting $\lambda = 3$, $\mu = -2$ and $\nu = 1$, as a particular case, gives A as

$$\mathsf{A} = \frac{1}{6}\begin{pmatrix} 2 & 10 & -4 \\ 10 & 8 & 10 \\ -4 & 10 & 2 \end{pmatrix}.$$

We complete the exercise by solving for the eigenvalues of A in the usual way. To avoid working with fractions, and any confusion with the value $\lambda = 3$ used when constructing A, we will find the eigenvalues of 6A and denote them by η.

$$0 = |6A - \eta I|$$

$$= \begin{vmatrix} 2-\eta & 10 & -4 \\ 10 & 8-\eta & 10 \\ -4 & 10 & 2-\eta \end{vmatrix}$$

$$= (2-\eta)(\eta^2 - 10\eta - 84) + 10(10\eta - 60) - 4(132 - 4\eta)$$

$$= -\eta^3 + 12\eta^2 + 180\eta - 1296$$

$$= -(\eta - 6)(\eta^2 - 6\eta - 216)$$

$$= -(\eta - 6)(\eta + 12)(\eta - 18).$$

Thus 6A has eigenvalues 6, -12 and 18; the values for A itself are 1, -2 and 3, as expected.

8.31 *One method of determining the nullity (and hence the rank) of an $M \times N$ matrix A is as follows.*

- *Write down an augmented transpose of A, by adding on the right an $N \times N$ unit matrix and thus producing an $N \times (M + N)$ array B.*
- *Subtract a suitable multiple of the first row of B from each of the other lower rows so as to make $B_{i1} = 0$ for $i > 1$.*
- *Subtract a suitable multiple of the second row (or the uppermost row that does not start with M zero values) from each of the other lower rows so as to make $B_{i2} = 0$ for $i > 2$.*
- *Continue in this way until all remaining rows have zeros in the first M places. The number of such rows is equal to the nullity of A, and the N rightmost entries of these rows are the components of vectors that span the null space. They can be made orthogonal if they are not so already.*

Use this method to show that the nullity of

$$A = \begin{pmatrix} -1 & 3 & 2 & 7 \\ 3 & 10 & -6 & 17 \\ -1 & -2 & 2 & -3 \\ 2 & 3 & -4 & 4 \\ 4 & 0 & -8 & -4 \end{pmatrix}$$

is 2 and that an orthogonal base for the null space of A is provided by any two column matrices of the form $(2 + \alpha_i \quad -2\alpha_i \quad 1 \quad \alpha_i)^T$, for which the α_i $(i = 1, 2)$ are real and satisfy $6\alpha_1\alpha_2 + 2(\alpha_1 + \alpha_2) + 5 = 0$.

We first construct B as

$$B = \begin{pmatrix} -1 & 3 & -1 & 2 & 4 & 1 & 0 & 0 & 0 \\ 3 & 10 & -2 & 3 & 0 & 0 & 1 & 0 & 0 \\ 2 & -6 & 2 & -4 & -8 & 0 & 0 & 1 & 0 \\ 7 & 17 & -3 & 4 & -4 & 0 & 0 & 0 & 1 \end{pmatrix}.$$

Now, following the bulleted steps in the question, we obtain, successively,

$$B_1 = \begin{pmatrix} -1 & 3 & -1 & 2 & 4 & 1 & 0 & 0 & 0 \\ 0 & 19 & -5 & 9 & 12 & 3 & 1 & 0 & 0 \\ 0 & 0 & 0 & 0 & 0 & 2 & 0 & 1 & 0 \\ 0 & 38 & -10 & 18 & 24 & 7 & 0 & 0 & 1 \end{pmatrix}$$

and

$$B_2 = \begin{pmatrix} -1 & 3 & -1 & 2 & 4 & 1 & 0 & 0 & 0 \\ 0 & 19 & -5 & 9 & 12 & 3 & 1 & 0 & 0 \\ 0 & 0 & 0 & 0 & 0 & 2 & 0 & 1 & 0 \\ 0 & 0 & 0 & 0 & 0 & 1 & -2 & 0 & 1 \end{pmatrix}.$$

Since there are two rows that have all zeros in the first five places, the nullity of A is 2, and hence its rank is $4 - 2 = 2$.

The same two rows show that the null space is spanned by the vectors $(2 \quad 0 \quad 1 \quad 0)^{\mathrm{T}}$ and $(1 \quad -2 \quad 0 \quad 1)^{\mathrm{T}}$ and, therefore, by any two linear combinations of them of the general form $(2 + \alpha_i \quad -2\alpha_i \quad 1 \quad \alpha_i)^{\mathrm{T}}$ for $i = 1, 2$, where α_i is any real number. If the basis is to be orthogonal then the scalar product of the two vectors must be zero, i.e.

$$(2 + \alpha_1)(2 + \alpha_2) + 4\alpha_1\alpha_2 + 1 + \alpha_1\alpha_2 = 0,$$
$$6\alpha_1\alpha_2 + 2(\alpha_1 + \alpha_2) + 5 = 0.$$

Thus α_1 may be chosen arbitrarily, but α_2 is then determined.

8.33 Solve the simultaneous equations

$$2x + 3y + z = 11,$$
$$x + y + z = 6,$$
$$5x - y + 10z = 34.$$

To eliminate z, (i) subtract the second equation from the first and (ii) subtract 10 times the second equation from the third.

$$x + 2y = 5,$$
$$-5x - 11y = -26.$$

137

To eliminate x add 5 times the first equation to the second

$$-y = -1.$$

Thus $y = 1$ and, by resubstitution, $x = 3$ and $z = 2$.

8.35 *Show that the following equations have solutions only if $\eta = 1$ or 2, and find them in these cases:*

$$x + y + z = 1, \qquad \text{(i)}$$
$$x + 2y + 4z = \eta, \qquad \text{(ii)}$$
$$x + 4y + 10z = \eta^2. \qquad \text{(iii)}$$

Expressing the equations in the form $\mathsf{A}\mathsf{x} = \mathsf{b}$, we first need to evaluate $|\mathsf{A}|$ as a preliminary to determining A^{-1}. However, we find that $|\mathsf{A}| = 1(20 - 16) + 1(4 - 10) + 1(4 - 2) = 0$. This result implies both that A is singular and has no inverse, and that the equations must be linearly dependent.

Either by observation or by solving for the combination coefficients, we see that for the LHS this linear dependence is expressed by

$$2 \times \text{(i)} + 1 \times \text{(iii)} - 3 \times \text{(ii)} = 0.$$

For a consistent solution, this must also be true for the RHSs, i.e.

$$2 + \eta^2 - 3\eta = 0.$$

This quadratic equation has solutions $\eta = 1$ and $\eta = 2$, which are therefore the only values of η for which the original equations have a solution. As the equations are linearly dependent, we may use any two to find these allowed solutions; for simplicity we use the first two in each case.

For $\eta = 1$,

$$x + y + z = 1, \quad x + 2y + 4z = 1 \Rightarrow \mathsf{x}^1 = (1 + 2\alpha, \ -3\alpha, \ \alpha)^{\mathsf{T}}.$$

For $\eta = 2$,

$$x + y + z = 1, \quad x + 2y + 4z = 2 \Rightarrow \mathsf{x}^2 = (2\alpha, \ 1 - 3\alpha, \ \alpha)^{\mathsf{T}}.$$

In both cases there is an infinity of solutions as α may take any finite value.

8.37 *Make an LU decomposition of the matrix*

$$A = \begin{pmatrix} 3 & 6 & 9 \\ 1 & 0 & 5 \\ 2 & -2 & 16 \end{pmatrix}$$

and hence solve $Ax = b$, *where* (i) $b = (21 \quad 9 \quad 28)^T$, (ii) $b = (21 \quad 7 \quad 22)^T$.

Using the notation

$$A = \begin{pmatrix} 1 & 0 & 0 \\ L_{21} & 1 & 0 \\ L_{31} & L_{32} & 1 \end{pmatrix} \begin{pmatrix} U_{11} & U_{12} & U_{13} \\ 0 & U_{22} & U_{23} \\ 0 & 0 & U_{33} \end{pmatrix},$$

and considering rows and columns alternately in the usual way for an *LU* decomposition, we require the following to be satisfied.

1st row: $U_{11} = 3$, $U_{12} = 6$, $U_{13} = 9$.
1st col: $L_{21}U_{11} = 1$, $L_{31}U_{11} = 2 \Rightarrow L_{21} = \frac{1}{3}$, $L_{31} = \frac{2}{3}$.
2nd row: $L_{21}U_{12} + U_{22} = 0$, $L_{21}U_{13} + U_{23} = 5 \Rightarrow U_{22} = -2$, $U_{23} = 2$.
2nd col: $L_{31}U_{12} + L_{32}U_{22} = -2 \Rightarrow L_{32} = 3$.
3rd row: $L_{31}U_{13} + L_{32}U_{23} + U_{33} = 16 \Rightarrow U_{33} = 4$.

Thus

$$L = \begin{pmatrix} 1 & 0 & 0 \\ \frac{1}{3} & 1 & 0 \\ \frac{2}{3} & 3 & 1 \end{pmatrix} \quad \text{and} \quad U = \begin{pmatrix} 3 & 6 & 9 \\ 0 & -2 & 2 \\ 0 & 0 & 4 \end{pmatrix}.$$

To solve $Ax = b$ with $A = LU$, we first determine y from $Ly = b$ and then solve $Ux = y$ for x.

(i) For $Ax = (21, 9, 28)^T$, we first solve

$$\begin{pmatrix} 1 & 0 & 0 \\ \frac{1}{3} & 1 & 0 \\ \frac{2}{3} & 3 & 1 \end{pmatrix} \begin{pmatrix} y_1 \\ y_2 \\ y_3 \end{pmatrix} = \begin{pmatrix} 21 \\ 9 \\ 28 \end{pmatrix}.$$

This can be done, almost by inspection, to give $y = (21, 2, 8)^T$.

We can now write $Ux = y$ explicitly as

$$\begin{pmatrix} 3 & 6 & 9 \\ 0 & -2 & 2 \\ 0 & 0 & 4 \end{pmatrix} \begin{pmatrix} x_1 \\ x_2 \\ x_3 \end{pmatrix} = \begin{pmatrix} 21 \\ 2 \\ 8 \end{pmatrix}$$

to give, equally easily, that the solution to the original matrix equation is $x = (-1, 1, 2)^T$.

(ii) To solve $Ax = (21,\ 7,\ 22)^T$ we use exactly the same forms for L and U, but the new values for the components of b, to obtain $y = (21,\ 0,\ 8)^T$ leading to the solution $x = (-3,\ 2,\ 2)^T$.

8.39 *Use the Cholesky separation method to determine whether the following matrices are positive definite. For each that is, determine the corresponding lower diagonal matrix* L :

$$A = \begin{pmatrix} 2 & 1 & 3 \\ 1 & 3 & -1 \\ 3 & -1 & 1 \end{pmatrix}, \qquad B = \begin{pmatrix} 5 & 0 & \sqrt{3} \\ 0 & 3 & 0 \\ \sqrt{3} & 0 & 3 \end{pmatrix}.$$

The matrix A is real and so we seek a real lower-diagonal matrix L such that $LL^T = A$. In order to avoid a lot of subscripts, we use lower-case letters as the non-zero elements of L:

$$\begin{pmatrix} a & 0 & 0 \\ b & c & 0 \\ d & e & f \end{pmatrix} \begin{pmatrix} a & b & d \\ 0 & c & e \\ 0 & 0 & f \end{pmatrix} = \begin{pmatrix} 2 & 1 & 3 \\ 1 & 3 & -1 \\ 3 & -1 & 1 \end{pmatrix}.$$

Firstly, from A_{11}, $a^2 = 2$. Since an overall negative sign multiplying the elements of L is irrelevant, we may choose $a = +\sqrt{2}$. Next, $ba = A_{12} = 1$, implying that $b = 1/\sqrt{2}$. Similarly, $d = 3/\sqrt{2}$.

From the second row of A we have

$$b^2 + c^2 = 3 \Rightarrow c = \sqrt{\tfrac{5}{2}},$$

$$bd + ce = -1 \Rightarrow e = \sqrt{\tfrac{2}{5}}(-1 - \tfrac{3}{2}) = -\sqrt{\tfrac{5}{2}}.$$

And, from the final row,

$$d^2 + e^2 + f^2 = 1 \Rightarrow f = (1 - \tfrac{9}{2} - \tfrac{5}{2})^{1/2} = \sqrt{-6}.$$

That f is imaginary shows that A is not a positive definite matrix.

The corresponding argument (keeping the same symbols but with different numerical values) for the matrix B is as follows.

Firstly, from A_{11}, $a^2 = 5$. Since an overall negative sign multiplying the elements of L is irrelevant, we may choose $a = +\sqrt{5}$. Next, $ba = B_{12} = 0$, implying that $b = 0$. Similarly, $d = \sqrt{3}/\sqrt{5}$.

From the second row of B we have

$$b^2 + c^2 = 3 \Rightarrow c = \sqrt{3},$$

$$bd + ce = 0 \Rightarrow e = \sqrt{\tfrac{1}{3}}(0 - 0) = 0.$$

And, from the final row,

$$d^2 + e^2 + f^2 = 3 \Rightarrow f = (3 - \tfrac{3}{5} - 0)^{1/2} = \sqrt{\tfrac{12}{5}}.$$

Thus all the elements of L have been calculated and found to be real and, in summary,

$$\mathsf{L} = \begin{pmatrix} \sqrt{5} & 0 & 0 \\ 0 & \sqrt{3} & 0 \\ \sqrt{\tfrac{3}{5}} & 0 & \sqrt{\tfrac{12}{5}} \end{pmatrix}.$$

That $\mathsf{L}\mathsf{L}^{\mathsf{T}} = \mathsf{B}$ can be confirmed by substitution.

8.41 *Find the SVD of*

$$\mathsf{A} = \begin{pmatrix} 0 & -1 \\ 1 & 1 \\ -1 & 0 \end{pmatrix},$$

showing that the singular values are $\sqrt{3}$ *and* 1.

With

$$\mathsf{A} = \begin{pmatrix} 0 & -1 \\ 1 & 1 \\ -1 & 0 \end{pmatrix} \quad \text{and} \quad \mathsf{A}^{\dagger} = \begin{pmatrix} 0 & 1 & -1 \\ -1 & 1 & 0 \end{pmatrix},$$

$$\mathsf{A}^{\dagger}\mathsf{A} = \begin{pmatrix} 2 & 1 \\ 1 & 2 \end{pmatrix},$$

which has eigenvalues given by $(2 - \lambda)(2 - \lambda) - 1 = 0$. The roots of this equation are $\lambda_1 = 3$ and $\lambda_2 = 1$, showing that the singular values s_i of A are $\sqrt{3}$ and $\sqrt{1}$.

The normalised eigenvectors $(x_1, x_2)^{\mathsf{T}}$ corresponding to these eigenvalues satisfy

$$(2 - 3)x_1 + x_2 = 0 \Rightarrow \mathsf{v}^1 = \frac{1}{\sqrt{2}} (1, \ 1)^{\mathsf{T}},$$

$$(2 - 1)x_1 + x_2 = 0 \Rightarrow \mathsf{v}^2 = \frac{1}{\sqrt{2}} (1, \ -1)^{\mathsf{T}}.$$

The next step is to calculate the (normalised) column vectors u^i from $(s_i)^{-1}\mathsf{A}\mathsf{v}^i = \mathsf{u}^i$:

$$\mathsf{u}^1 = \frac{1}{\sqrt{3}} \frac{1}{\sqrt{2}} \begin{pmatrix} 0 & -1 \\ 1 & 1 \\ -1 & 0 \end{pmatrix} \begin{pmatrix} 1 \\ 1 \end{pmatrix} = \frac{1}{\sqrt{6}} \begin{pmatrix} -1 \\ 2 \\ -1 \end{pmatrix},$$

and

$$u^2 = \frac{1}{\sqrt{1}}\frac{1}{\sqrt{2}} \begin{pmatrix} 0 & -1 \\ 1 & 1 \\ -1 & 0 \end{pmatrix} \begin{pmatrix} 1 \\ -1 \end{pmatrix} = \frac{1}{\sqrt{2}} \begin{pmatrix} 1 \\ 0 \\ -1 \end{pmatrix}.$$

For the third column vector we need one orthogonal to both u^1 and u^2; this can be obtained from their cross product and is $u^3 = (1/\sqrt{3})(1, 1, 1)^T$.

Finally, we can write A in SVD form:

$$A = USV^\dagger = \frac{1}{\sqrt{6}} \begin{pmatrix} -1 & \sqrt{3} & \sqrt{2} \\ 2 & 0 & \sqrt{2} \\ -1 & -\sqrt{3} & \sqrt{2} \end{pmatrix} \begin{pmatrix} \sqrt{3} & 0 \\ 0 & 1 \\ 0 & 0 \end{pmatrix} \frac{1}{\sqrt{2}} \begin{pmatrix} 1 & 1 \\ 1 & -1 \end{pmatrix},$$

where U and V are unitary. Both the unitarity and the decomposition can be checked by direct multiplication.

8.43 *Four experimental measurements of particular combinations of three physical variables, x, y and z, gave the following inconsistent results:*

$$13x + 22y - 13z = 4,$$
$$10x - 8y - 10z = 44,$$
$$10x - 8y - 10z = 47,$$
$$9x - 18y - 9z = 72.$$

Find the SVD best values for x, y and z. Denoting the equations by Ax = b, *identify the null space of* A *and hence obtain the general SVD solution.*

The method of finding the SVD follows that of exercise 8.41.

We start by computing

$$A^\dagger A = \begin{pmatrix} 13 & 10 & 10 & 9 \\ 22 & -8 & -8 & -18 \\ -13 & -10 & -10 & -9 \end{pmatrix} \begin{pmatrix} 13 & 22 & -13 \\ 10 & -8 & -10 \\ 10 & -8 & -10 \\ 9 & -18 & -9 \end{pmatrix}$$

$$= \begin{pmatrix} 450 & -36 & -450 \\ -36 & 936 & 36 \\ -450 & 36 & 450 \end{pmatrix}.$$

We next find its eigenvalues:

$$|A^\dagger A - \lambda| = \begin{vmatrix} 450 - \lambda & -36 & -450 \\ -36 & 936 - \lambda & 36 \\ -450 & 36 & 450 - \lambda \end{vmatrix}$$

$$= \begin{vmatrix} -\lambda & 0 & -\lambda \\ -36 & 936 - \lambda & 36 \\ -450 & 36 & 450 - \lambda \end{vmatrix}$$

$$= -\lambda(\lambda^2 - 1836\lambda + 839808)$$

$$= -\lambda(\lambda - 864)(\lambda - 972).$$

This shows that the singular values s_i are $\sqrt{972} = 18\sqrt{3}$, $\sqrt{864} = 12\sqrt{6}$ and 0.

The corresponding normalised eigenvectors $(x_1, x_2, x_3)^{\mathrm{T}}$, used to construct the orthogonal matrix V, satisfy

$$-522x_1 - 36x_2 - 450x_3 = 0,$$

$$-36x_1 - 36x_2 + 36x_3 = 0 \Rightarrow v^1 = \frac{1}{\sqrt{6}}(1, -2, -1)^{\mathrm{T}};$$

$$-414x_1 - 36x_2 - 450x_3 = 0,$$

$$-36x_1 + 72x_2 + 36x_3 = 0 \Rightarrow v^2 = \frac{1}{\sqrt{3}}(1, 1, -1)^{\mathrm{T}};$$

$$450x_1 - 36x_2 - 450x_3 = 0,$$

$$-36x_1 + 936x_2 + 36x_3 = 0 \Rightarrow v^3 = \frac{1}{\sqrt{2}}(1, 0, 1)^{\mathrm{T}}.$$

The singular value 0 implies that v^3 will be a vector in (and spanning) the null space of A, which therefore has rank 2 (rather than 3, as would be generally expected in this case).

For the non-zero singular values we now calculate the (normalised) column vectors u^i from $(s_i)^{-1}Av^i = u^i$:

$$u^1 = \frac{1}{18\sqrt{3}}\frac{1}{\sqrt{6}} \begin{pmatrix} 13 & 22 & -13 \\ 10 & -8 & -10 \\ 10 & -8 & -10 \\ 9 & -18 & -9 \end{pmatrix} \begin{pmatrix} 1 \\ -2 \\ -1 \end{pmatrix} = \frac{1}{3\sqrt{2}} \begin{bmatrix} -1 \\ 2 \\ 2 \\ 3 \end{bmatrix};$$

$$u^2 = \frac{1}{12\sqrt{6}}\frac{1}{\sqrt{3}} \begin{pmatrix} 13 & 22 & -13 \\ 10 & -8 & -10 \\ 10 & -8 & -10 \\ 9 & -18 & -9 \end{pmatrix} \begin{pmatrix} 1 \\ 1 \\ -1 \end{pmatrix} = \frac{1}{3\sqrt{2}} \begin{bmatrix} 4 \\ 1 \\ 1 \\ 0 \end{bmatrix}.$$

Although we will not need their components for the present exercise, we now find the third and fourth base vectors (to make U a unitary matrix). They must be solutions of $A^\dagger u^i = 0$; simple simultaneous equations show that, when normalised, two suitable vectors are

$$u^3 = \frac{1}{\sqrt{2}}(0,-1,1,0)^T \quad \text{and} \quad u^4 = \frac{1}{\sqrt{18}}(1,-2,-2,3)^T.$$

Thus, we are able to write $A = USV^\dagger$ explicitly as

$$\frac{1}{N}\begin{pmatrix} -1 & 4 & 0 & 1 \\ 2 & 1 & -3 & -2 \\ 2 & 1 & 3 & -2 \\ 3 & 0 & 0 & 3 \end{pmatrix}\begin{pmatrix} 18\sqrt{3} & 0 & 0 \\ 0 & 12\sqrt{6} & 0 \\ 0 & 0 & 0 \\ 0 & 0 & 0 \end{pmatrix}\begin{pmatrix} 1 & -2 & -1 \\ \sqrt{2} & \sqrt{2} & -\sqrt{2} \\ \sqrt{3} & 0 & \sqrt{3} \end{pmatrix},$$

where $N = \sqrt{18} \times \sqrt{6}$.

We now compute $R = V\bar{S}U^\dagger$ as (with N defined as before)

$$\frac{1}{N}\begin{pmatrix} 1 & \sqrt{2} & \sqrt{3} \\ -2 & \sqrt{2} & 0 \\ -1 & -\sqrt{2} & \sqrt{3} \end{pmatrix}\begin{pmatrix} \frac{1}{18\sqrt{3}} & 0 & 0 & 0 \\ 0 & \frac{1}{12\sqrt{6}} & 0 & 0 \\ 0 & 0 & 0 & 0 \end{pmatrix}\begin{pmatrix} -1 & 2 & 2 & 3 \\ 4 & 1 & 1 & 0 \\ 0 & -3 & 3 & 0 \\ 1 & -2 & -2 & 3 \end{pmatrix}$$

$$= \frac{1}{N}\begin{pmatrix} \frac{1}{18\sqrt{3}} & \frac{1}{12\sqrt{3}} & 0 & 0 \\ -\frac{1}{9\sqrt{3}} & \frac{1}{12\sqrt{3}} & 0 & 0 \\ -\frac{1}{18\sqrt{3}} & -\frac{1}{12\sqrt{3}} & 0 & 0 \end{pmatrix}\begin{pmatrix} -1 & 2 & 2 & 3 \\ 4 & 1 & 1 & 0 \\ 0 & -3 & 3 & 0 \\ 1 & -2 & -2 & 3 \end{pmatrix}$$

$$= \frac{1}{\sqrt{108}}\frac{1}{36\sqrt{3}}\begin{pmatrix} 10 & 7 & 7 & 6 \\ 16 & -5 & -5 & -12 \\ -10 & -7 & -7 & -6 \end{pmatrix}.$$

The best SVD solution is thus given by

$$Rb = \frac{1}{648}\begin{pmatrix} 10 & 7 & 7 & 6 \\ 16 & -5 & -5 & -12 \\ -10 & -7 & -7 & -6 \end{pmatrix}\begin{bmatrix} 4 \\ 44 \\ 47 \\ 72 \end{bmatrix} = \begin{pmatrix} 1.711 \\ -1.937 \\ -1.711 \end{pmatrix}.$$

As noted previously, the null space of A is spanned by the vector $x^3 = \frac{1}{\sqrt{2}}(1, 0, 1)^T$. The general SVD solution is therefore

$$(1.71 + \lambda, \, -1.94, \, -1.71 + \lambda)^T.$$

9

Normal modes

9.1 *Three coupled pendulums swing perpendicularly to the horizontal line containing their points of suspension, and the following equations of motion are satisfied:*

$$-m\ddot{x}_1 = cmx_1 + d(x_1 - x_2),$$
$$-M\ddot{x}_2 = cMx_2 + d(x_2 - x_1) + d(x_2 - x_3),$$
$$-m\ddot{x}_3 = cmx_3 + d(x_3 - x_2),$$

where x_1, x_2 and x_3 are measured from the equilibrium points; m, M and m are the masses of the pendulum bobs; and c and d are positive constants. Find the normal frequencies of the system and sketch the corresponding patterns of oscillation. What happens as $d \to 0$ or $d \to \infty$?

In a normal mode all three coordinates x_i oscillate with the same frequency and with fixed relative phases. When this is represented by solutions of the form $x_i = X_i \cos \omega t$, where the X_i are fixed constants, the equations become, in matrix and vector form,

$$\begin{pmatrix} cm + d - m\omega^2 & -d & 0 \\ -d & cM + 2d - M\omega^2 & -d \\ 0 & -d & cm + d - m\omega^2 \end{pmatrix} \begin{pmatrix} X_1 \\ X_2 \\ X_3 \end{pmatrix} = \mathbf{0}.$$

For there to be a non-trivial solution to these simultaneous homogeneous equations, we need

$$0 = \begin{vmatrix} (c-\omega^2)m+d & -d & 0 \\ -d & (c-\omega^2)M+2d & -d \\ 0 & -d & (c-\omega^2)m+d \end{vmatrix}$$

$$= \begin{vmatrix} (c-\omega^2)m+d & 0 & -(c-\omega^2)m-d \\ -d & (c-\omega^2)M+2d & -d \\ 0 & -d & (c-\omega^2)m+d \end{vmatrix}$$

$$= [(c-\omega^2)m+d]\{[(c-\omega^2)M+2d][(c-\omega^2)m+d]-d^2-d^2\}$$

$$= (cm-m\omega^2+d)(c-\omega^2)[Mm(c-\omega^2)+2dm+dM].$$

Thus, the normal (angular) frequencies are given by

$$\omega^2 = c, \quad \omega^2 = c + \frac{d}{m} \quad \text{and} \quad \omega^2 = c + \frac{2d}{M} + \frac{d}{m}.$$

If the solution column matrix is $X = (X_1, X_2, X_3)^\mathrm{T}$, then

(i) for $\omega^2 = c$, the components of X must satisfy

$$dX_1 - dX_2 = 0,$$
$$-dX_1 + 2dX_2 - dX_3 = 0, \quad \Rightarrow \quad X^1 = (1, 1, 1)^\mathrm{T};$$

(ii) for $\omega^2 = c + \dfrac{d}{m}$, we have

$$-dX_2 = 0,$$
$$-dX_1 + \left(-\frac{dM}{m} + 2d\right) X_2 - dX_3 = 0, \quad \Rightarrow \quad X^2 = (1, 0, -1)^\mathrm{T};$$

(iii) for $\omega^2 = c + \dfrac{2d}{M} + \dfrac{d}{m}$, the components must satisfy

$$\left[\left(-\frac{2d}{M} - \frac{d}{m}\right)m + d\right] X_1 - dX_2 = 0,$$

$$-dX_2 + \left[\left(-\frac{2d}{M} - \frac{d}{m}\right)m + d\right] X_3 = 0, \quad \Rightarrow \quad X^3 = \left(1, -\frac{2m}{M}, 1\right)^\mathrm{T}.$$

The corresponding patterns are shown in figure 9.1.

If $d \to 0$, the three oscillations decouple and each pendulum swings independently with angular frequency $\sqrt{c}$.

If $d \to \infty$, the three pendulums become rigidly coupled. The second and third modes have (theoretically) infinite frequency and therefore zero amplitude. The only sustainable mode is the one shown as case (b) in the figure; one in which all the pendulums swing as a single entity with angular frequency $\sqrt{c}$.

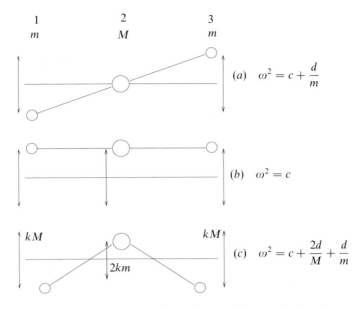

$$(a) \quad \omega^2 = c + \frac{d}{m}$$

$$(b) \quad \omega^2 = c$$

$$(c) \quad \omega^2 = c + \frac{2d}{M} + \frac{d}{m}$$

Figure 9.1 The normal modes, as viewed from above, of the coupled pendulums in exercise 9.1.

9.3 *Find the normal frequencies of a system consisting of three particles of masses* $m_1 = m$, $m_2 = \mu m$, $m_3 = m$ *connected in that order in a straight line by two equal light springs of force constant* k. *Describe the corresponding modes of oscillation.*

Now consider the particular case in which $\mu = 2$.

(a) *Show that the eigenvectors derived above have the expected orthogonality properties with respect to both the kinetic energy matrix* **A** *and the potential energy matrix* **B**.

(b) *For the situation in which the masses are released from rest with initial displacements (relative to their equilibrium positions) of* $x_1 = 2\epsilon$, $x_2 = -\epsilon$ *and* $x_3 = 0$, *determine their subsequent motions and maximum displacements.*

Let the coordinates of the particles, x_1, x_2, x_3, be measured from their equilibrium positions, at which the springs are neither extended nor compressed.

The kinetic energy of the system is simply

$$T = \tfrac{1}{2}m\left(\dot{x}_1^2 + \mu\dot{x}_2^2 + \dot{x}_3^2\right),$$

whilst the potential energy stored in the springs takes the form

$$V = \tfrac{1}{2}k\left[(x_2 - x_1)^2 + (x_3 - x_2)^2\right].$$

147

The kinetic- and potential-energy symmetric matrices are thus

$$A = \frac{m}{2} \begin{pmatrix} 1 & 0 & 0 \\ 0 & \mu & 0 \\ 0 & 0 & 1 \end{pmatrix}, \qquad B = \frac{k}{2} \begin{pmatrix} 1 & -1 & 0 \\ -1 & 2 & -1 \\ 0 & -1 & 1 \end{pmatrix}.$$

To find the normal frequencies we have to solve $|B - \omega^2 A| = 0$. Thus, writing $m\omega^2/k = \lambda$, we have

$$\begin{aligned} 0 &= \begin{vmatrix} 1-\lambda & -1 & 0 \\ -1 & 2-\mu\lambda & -1 \\ 0 & -1 & 1-\lambda \end{vmatrix} \\ &= (1-\lambda)(2 - \mu\lambda - 2\lambda + \mu\lambda^2 - 1) + (-1 + \lambda) \\ &= (1-\lambda)\lambda(-\mu - 2 + \mu\lambda), \end{aligned}$$

which leads to $\lambda = 0$, 1 or $1 + 2/\mu$.

The normalised eigenvectors corresponding to the first two eigenvalues can be found by inspection and are

$$x^1 = \frac{1}{\sqrt{3}} \begin{pmatrix} 1 \\ 1 \\ 1 \end{pmatrix}, \qquad x^2 = \frac{1}{\sqrt{2}} \begin{pmatrix} 1 \\ 0 \\ -1 \end{pmatrix}.$$

The components of the third eigenvector must satisfy

$$-\frac{2}{\mu} x_1 - x_2 = 0 \quad \text{and} \quad x_2 - \frac{2}{\mu} x_3 = 0.$$

The normalised third eigenvector is therefore

$$x^3 = \frac{1}{\sqrt{2 + (4/\mu^2)}} \left(1, -\frac{2}{\mu}, 1 \right)^{\mathrm{T}}.$$

The physical motions associated with these normal modes are as follows.

The first, with $\lambda = \omega = 0$ and all the x_i equal, merely describes bodily translation of the whole system, with no (i.e. zero-frequency) internal oscillations.

In the second solution, the central particle remains stationary, $x_2 = 0$, whilst the other two oscillate with equal amplitudes in antiphase with each other. This motion has frequency $\omega = (k/m)^{1/2}$, the same as that for the oscillations of a single mass m suspended from a single spring of force constant k.

The final and most complicated of the three normal modes has angular frequency $\omega = \{[(\mu + 2)/\mu](k/m)\}^{1/2}$, and involves a motion of the central particle which is in antiphase with that of the two outer ones and which has an amplitude $2/\mu$ times as great. In this motion the two springs are compressed and extended in turn. We also note that in the second and third normal modes the centre of mass of the system remains stationary.

Now setting $\mu = 2$, we have as the three normal (angular) frequencies 0, Ω and $\sqrt{2}\Omega$, where $\Omega^2 = k/m$. The corresponding (unnormalised) eigenvectors are

$$\mathsf{x}^1 = (1,\ 1,\ 1)^{\mathrm{T}}, \quad \mathsf{x}^2 = (1,\ 0,\ -1)^{\mathrm{T}}, \quad \mathsf{x}^3 = (1,\ -1,\ 1)^{\mathrm{T}}.$$

(a) The matrices A and B have the forms

$$\mathsf{A} = \begin{pmatrix} 1 & 0 & 0 \\ 0 & 2 & 0 \\ 0 & 0 & 1 \end{pmatrix}, \quad \mathsf{B} = \begin{pmatrix} 1 & -1 & 0 \\ -1 & 2 & -1 \\ 0 & -1 & 1 \end{pmatrix}.$$

To verify the standard orthogonality relations we need to show that the quadratic forms $(\mathsf{x}^i)^{\dagger}\mathsf{A}\mathsf{x}^j$ and $(\mathsf{x}^i)^{\dagger}\mathsf{B}\mathsf{x}^j$ have zero value for $i \neq j$. Direct evaluation of all the separate cases is as follows:

$$(\mathsf{x}^1)^{\dagger}\mathsf{A}\mathsf{x}^2 = 1 + 0 - 1 = 0,$$
$$(\mathsf{x}^1)^{\dagger}\mathsf{A}\mathsf{x}^3 = 1 - 2 + 1 = 0,$$
$$(\mathsf{x}^2)^{\dagger}\mathsf{A}\mathsf{x}^3 = 1 + 0 - 1 = 0,$$
$$(\mathsf{x}^1)^{\dagger}\mathsf{B}\mathsf{x}^2 = (\mathsf{x}^1)^{\dagger}(1,\ 0,\ -1)^{\mathrm{T}} = 1 + 0 - 1 = 0,$$
$$(\mathsf{x}^1)^{\dagger}\mathsf{B}\mathsf{x}^3 = (\mathsf{x}^1)^{\dagger}(2,\ -4,\ 2)^{\mathrm{T}} = 2 - 4 + 2 = 0,$$
$$(\mathsf{x}^2)^{\dagger}\mathsf{B}\mathsf{x}^3 = (\mathsf{x}^2)^{\dagger}(2,\ -4,\ 2)^{\mathrm{T}} = 2 + 0 - 2 = 0.$$

If $(\mathsf{x}^i)^{\dagger}\mathsf{A}\mathsf{x}^j$ has zero value then so does $(\mathsf{x}^j)^{\dagger}\mathsf{A}\mathsf{x}^i$ (and similarly for B). So there is no need to investigate the other six possibilities and the verification is complete.

(b) In order to determine the behaviour of the system we need to know which modes are present in the initial configuration. Each contributory mode will subsequently oscillate with its own frequency. In order to carry out this initial decomposition we write

$$(2\epsilon,\ -\epsilon,\ 0)^{\mathrm{T}} = a\,(1,\ 1,\ 1)^{\mathrm{T}} + b\,(1,\ 0,\ -1)^{\mathrm{T}} + c\,(1,\ -1,\ 1)^{\mathrm{T}},$$

from which it is clear that $a = 0$, $b = \epsilon$ and $c = \epsilon$. As each mode vibrates with its own frequency, the subsequent displacements are given by

$$x_1 = \epsilon(\cos\Omega t + \cos\sqrt{2}\Omega t),$$
$$x_2 = -\epsilon\cos\sqrt{2}\Omega t,$$
$$x_3 = \epsilon(-\cos\Omega t + \cos\sqrt{2}\Omega t).$$

Since Ω and $\sqrt{2}\Omega$ are not rationally related, at some times the two modes will, for all practical purposes (but not mathematically), be in phase and, at other times, be out of phase. Thus the maximum displacements will be $x_1(\max) = 2\epsilon$, $x_2(\max) = \epsilon$ and $x_3(\max) = 2\epsilon$.

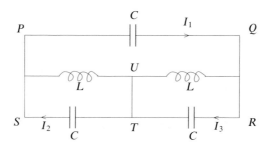

Figure 9.2   The circuit and notation for exercise 9.5.

9.5 *It is shown in physics and engineering textbooks that circuits containing capacitors and inductors can be analysed by replacing a capacitor of capacitance C by a 'complex impedance' $1/(i\omega C)$ and an inductor of inductance L by an impedance $i\omega L$, where ω is the angular frequency of the currents flowing and $i^2 = -1$.*

Use this approach and Kirchhoff's circuit laws to analyse the circuit shown in figure 9.2 and obtain three linear equations governing the currents I_1, I_2 and I_3. Show that the only possible frequencies of self-sustaining currents satisfy either (a) $\omega^2 LC = 1$ or (b) $3\omega^2 LC = 1$. Find the corresponding current patterns and, in each case, by identifying parts of the circuit in which no current flows, draw an equivalent circuit that contains only one capacitor and one inductor.

We apply Kirchhoff's laws to the three closed loops $PQUP$, $SUTS$ and $TURT$ and obtain, respectively,

$$\frac{1}{i\omega C}I_1 + i\omega L(I_1 - I_3) + i\omega L(I_1 - I_2) = 0,$$

$$i\omega L(I_2 - I_1) + \frac{1}{i\omega C}I_2 = 0,$$

$$i\omega L(I_3 - I_1) + \frac{1}{i\omega C}I_3 = 0.$$

For these simultaneous homogeneous linear equations to be consistent, it is necessary that

$$0 = \begin{vmatrix} \dfrac{1}{i\omega C} + 2i\omega L & -i\omega L & -i\omega L \\ -i\omega L & \dfrac{1}{i\omega C} + i\omega L & 0 \\ -i\omega L & 0 & \dfrac{1}{i\omega C} + i\omega L \end{vmatrix} = \begin{vmatrix} \lambda - 2 & 1 & 1 \\ 1 & \lambda - 1 & 0 \\ 1 & 0 & \lambda - 1 \end{vmatrix},$$

where, after dividing all entries by $-i\omega L$, we have written the combination

$(LC\omega^2)^{-1}$ as λ to save space. Expanding the determinant gives

$$\begin{aligned} 0 &= (\lambda - 2)(\lambda - 1)^2 - (\lambda - 1) - (\lambda - 1) \\ &= (\lambda - 1)(\lambda^2 - 3\lambda + 2 - 2) \\ &= \lambda(\lambda - 1)(\lambda - 3). \end{aligned}$$

Only the non-zero roots are of practical physical interest, and these are $\lambda = 1$ and $\lambda = 3$.

(a) The first of these eigenvalues has an eigenvector $\mathsf{I}^1 = (I_1, I_2, I_3)^{\mathrm{T}}$ that satisfies

$$-I_1 + I_2 + I_3 = 0,$$
$$I_1 = 0 \quad \Rightarrow \quad \mathsf{I}^1 = (0, 1, -1)^{\mathrm{T}}.$$

Thus there is no current in PQ and the capacitor in that link can be ignored. Equal currents circulate, in opposite directions, in the other two loops and, although the link TU carries both, there is no transfer between the two loops. Each loop is therefore equivalent to a capacitor of capacitance C in parallel with an inductor of inductance L.

(b) The second eigenvalue has an eigenvector $\mathsf{I}^2 = (I_1, I_2, I_3)^{\mathrm{T}}$ that satisfies

$$I_1 + I_2 + I_3 = 0,$$
$$I_1 + 2I_2 = 0 \quad \Rightarrow \quad \mathsf{I}^2 = (-2, 1, 1)^{\mathrm{T}}.$$

In this mode there is no current in TU and the circuit is equivalent to an inductor of inductance $L + L$ in parallel with a capacitor of capacitance $3C/2$; this latter capacitance is made up of C in parallel with the capacitance equivalent to two capacitors C in series, i.e. in parallel with $\frac{1}{2}C$. Thus, the equivalent single components are an inductance of $2L$ and a capacitance of $3C/2$.

9.7 *A double pendulum consists of two identical uniform rods, each of length ℓ and mass M, smoothly jointed together and suspended by attaching the free end of one rod to a fixed point. The system makes small oscillations in a vertical plane, with the angles made with the vertical by the upper and lower rods denoted by θ_1 and θ_2, respectively. The expressions for the kinetic energy T and the potential energy V of the system are (to second order in the θ_i)*

$$T \approx M l^2 \left(\tfrac{8}{3}\dot{\theta}_1^2 + 2\dot{\theta}_1\dot{\theta}_2 + \tfrac{2}{3}\dot{\theta}_2^2\right),$$
$$V \approx Mgl \left(\tfrac{3}{2}\theta_1^2 + \tfrac{1}{2}\theta_2^2\right).$$

Determine the normal frequencies of the system and find new variables ξ and η that will reduce these two expressions to diagonal form, i.e. to

$$a_1\dot{\xi}^2 + a_2\dot{\eta}^2 \qquad and \qquad b_1\xi^2 + b_2\eta^2.$$

To find the new variables we will use the following result. If the reader is not familiar with it, a standard textbook should be consulted.

If $Q_1 = u^TAu$ and $Q_2 = u^TBu$ are two real symmetric quadratic forms and u^n are those column matrices that satisfy

$$Bu^n = \lambda_n Au^n,$$

then the matrix P whose columns are the vectors u^n is such that the change of variables $u = Pv$ reduces both quadratic forms simultaneously to sums of squares, i.e. $Q_1 = v^TCv$ and $Q_2 = v^TDv$, with both C and D diagonal.

Further points to note are:

(i) that for the u^i as determined above, $(u^m)^TAu^n = 0$ if $m \neq n$ and similarly if A is replaced by B;

(ii) that P is not in general an orthogonal matrix, even if the vectors u^n are normalised.

(iii) In the special case that A is the identity matrix I: the above procedure is the same as diagonalising B; P *is* an orthogonal matrix if normalised vectors are used; mutual orthogonality of the eigenvectors takes on its usual form.

This exercise is a physical example to which the above mathematical result can be applied, the two real symmetric (actually positive-definite) matrices being the kinetic and potential energy matrices.

$$A = \begin{pmatrix} \frac{8}{3} & 1 \\ 1 & \frac{2}{3} \end{pmatrix}, \quad B = \begin{pmatrix} \frac{3}{2} & 0 \\ 0 & \frac{1}{2} \end{pmatrix} \quad \text{with} \quad \lambda_i = \frac{\omega_i^2 l}{g}.$$

We find the normal frequencies by solving

$$0 = |B - \lambda A|$$

$$= \begin{vmatrix} \frac{3}{2} - \frac{8}{3}\lambda & -\lambda \\ -\lambda & \frac{1}{2} - \frac{2}{3}\lambda \end{vmatrix}$$

$$= \frac{3}{4} - \frac{7}{3}\lambda + \frac{16}{9}\lambda^2 - \lambda^2$$

$$\Rightarrow \quad 0 = 28\lambda^2 - 84\lambda + 27.$$

Thus, $\lambda = 2.634$ or $\lambda = 0.3661$, and the normal frequencies are $(2.634g/l)^{1/2}$ and $(0.3661g/l)^{1/2}$.

The corresponding column vectors u^i have components that satisfy the following.

(i) For $\lambda = 0.3661$,

$$\left(\frac{3}{2} - \frac{8}{3}0.3661\right)\theta_1 - 0.3661\theta_2 = 0 \quad \Rightarrow \quad u^1 = (1, \ 1.431)^T.$$

(ii) For $\lambda = 2.634$,

$$\left(\frac{3}{2} - \frac{8}{3}2.634\right)\theta_1 - 2.634\theta_2 = 0 \quad \Rightarrow \quad u^2 = (1, \ -2.097)^T.$$

We can now construct P as

$$P = \begin{pmatrix} 1 & 1 \\ 1.431 & -2.097 \end{pmatrix}$$

and define new variables (ξ, η) by $(\theta_1, \theta_2)^\mathrm{T} = \mathsf{P}(\xi, \eta)^\mathrm{T}$. When the substitutions $\theta_1 = \xi + \eta$ and $\theta_2 = 1.431\xi - 2.097\eta \equiv \alpha\xi - \beta\eta$ are made into the expressions for T and V, they both take on diagonal forms. This can be checked by computing the coefficients of $\xi\eta$ in the two expressions. They are as follows.

$$\text{For } V: \quad 3 - \alpha\beta = 0, \qquad \text{and} \qquad \text{for } T: \quad \frac{16}{3} + 2(\alpha - \beta) - \frac{4}{3}\alpha\beta = 0.$$

As an example, the full expression for the potential energy becomes $V = Mg\ell\,(2.524\,\xi^2 + 3.699\,\eta^2)$.

9.9 *Three particles each of mass m are attached to a light horizontal string having fixed ends, the string being thus divided into four equal portions, each of length a and under a tension T. Show that for small transverse vibrations the amplitudes* x^i *of the normal modes satisfy* $\mathsf{B}\mathbf{x} = (ma\omega^2/T)\mathbf{x}$, *where* B *is the matrix*

$$\begin{pmatrix} 2 & -1 & 0 \\ -1 & 2 & -1 \\ 0 & -1 & 2 \end{pmatrix}.$$

Estimate the lowest and highest eigenfrequencies using trial vectors $(3, 4, 3)^\mathrm{T}$ *and* $(3, -4, 3)^\mathrm{T}$. *Use also the exact vectors* $\left(1, \sqrt{2}, 1\right)^\mathrm{T}$ *and* $\left(1, -\sqrt{2}, 1\right)^\mathrm{T}$ *and compare the results.*

For the ith mass, with displacement y_i, the force it experiences as a result of the tension in the string connecting it to the $(i+1)$th mass is the resolved component of that tension perpendicular to the equilibrium line, i.e. $f = \dfrac{y_{i+1} - y_i}{a}T$. Similarly the force due to the tension in the string connecting it to the $(i-1)$th mass is $f = \dfrac{y_{i-1} - y_i}{a}T$. Because the ends of the string are fixed the notional zeroth and fourth masses have $y_0 = y_4 = 0$.

The equations of motion are, therefore,

$$m\ddot{x}_1 = \frac{T}{a}[(0 - x_1) + (x_2 - x_1)],$$

$$m\ddot{x}_2 = \frac{T}{a}[(x_1 - x_2) + (x_3 - x_2)],$$

$$m\ddot{x}_3 = \frac{T}{a}[(x_2 - x_3) + (0 - x_3)].$$

If the displacements are written as $x_i = X_i \cos\omega t$ and $\mathbf{x} = (X_1, X_2, X_3)^\mathrm{T}$, then these equations become

$$-\frac{ma\omega^2}{T}X_1 = -2X_1 + X_2,$$

$$-\frac{ma\omega^2}{T}X_2 = X_1 - 2X_2 + X_3,$$

$$-\frac{ma\omega^2}{T}X_3 = X_2 - 2X_3.$$

This set of equations can be written as $\mathsf{B}\mathsf{x} = \dfrac{ma\omega^2}{T}\mathsf{x}$, with

$$\mathsf{B} = \begin{pmatrix} 2 & -1 & 0 \\ -1 & 2 & -1 \\ 0 & -1 & 2 \end{pmatrix}.$$

The Rayleigh–Ritz method shows that any estimate λ of $\dfrac{\mathsf{x}^{\mathsf{T}}\mathsf{B}\mathsf{x}}{\mathsf{x}^{\mathsf{T}}\mathsf{x}}$ always lies between the lowest and highest possible values of $ma\omega^2/T$.

Using the suggested *trial* vectors gives the following estimates for λ.

(i) For $\mathsf{x} = (3,\ 4,\ 3)^{\mathsf{T}}$

$$\lambda = [(3,\ 4,\ 3)\mathsf{B}\,(3,\ 4,\ 3)^{\mathsf{T}}]/34$$
$$= [(3,\ 4,\ 3)\,(2,\ 2,\ 2)^{\mathsf{T}}]/34$$
$$= 20/34 = 0.588.$$

(ii) For $\mathsf{x} = (3,\ -4,\ 3)^{\mathsf{T}}$

$$\lambda = [(3,\ -4,\ 3)\mathsf{B}\,(3,\ -4,\ 3)^{\mathsf{T}}]/34$$
$$= [(3,\ -4,\ 3)\,(10,\ -14,\ 10)^{\mathsf{T}}]/34$$
$$= 116/34 = 3.412.$$

Using, instead, the *exact* vectors yields the exact values of λ as follows.

(i) For the eigenvector corresponding to the lowest eigenvalue, $\mathsf{x} = (1,\ \sqrt{2},\ 1)^{\mathsf{T}}$,

$$\lambda = \left[(1,\ \sqrt{2},\ 1)\mathsf{B}(1,\ \sqrt{2},\ 1)^{\mathsf{T}}\right]/4$$
$$= \left[(1,\ \sqrt{2},\ 1)(2 - \sqrt{2},\ 2\sqrt{2} - 2,\ 2 - \sqrt{2})^{\mathsf{T}}\right]/4$$
$$= 2 - \sqrt{2} = 0.586.$$

(ii) For the eigenvector corresponding to the highest eigenvalue, $\mathsf{x} = (1,\ -\sqrt{2},\ 1)^{\mathsf{T}}$,

$$\lambda = \left[(1,\ -\sqrt{2},\ 1)\mathsf{B}(1,\ -\sqrt{2},\ 1)^{\mathsf{T}}\right]/4$$
$$= \left[(1,\ -\sqrt{2},\ 1)(2 + \sqrt{2},\ -2\sqrt{2} - 2,\ 2 + \sqrt{2})^{\mathsf{T}}\right]/4$$
$$= 2 + \sqrt{2} = 3.414.$$

As can be seen, the (crude) trial vectors give excellent approximations to the lowest and highest eigenfrequencies.

10

Vector calculus

In order to evaluate this integral, we need to group the terms in the integrand so that each is a part of the total derivative of a product of factors. Clearly, the first three terms are the derivative of $\mathbf{a}(\mathbf{b} \cdot \mathbf{a})$, i.e.

$$\frac{d}{dt}[\mathbf{a}(\mathbf{b} \cdot \mathbf{a})] = \dot{\mathbf{a}}(\mathbf{b} \cdot \mathbf{a}) + \mathbf{a}(\dot{\mathbf{b}} \cdot \mathbf{a}) + \mathbf{a}(\mathbf{b} \cdot \dot{\mathbf{a}}).$$

Similarly,
$$\frac{d}{dt}[\mathbf{b}(\mathbf{a} \cdot \mathbf{a})] = \dot{\mathbf{b}}(\mathbf{a} \cdot \mathbf{a}) + \mathbf{b}(\dot{\mathbf{a}} \cdot \mathbf{a}) + \mathbf{b}(\mathbf{a} \cdot \dot{\mathbf{a}}).$$

Hence,
$$I = \int \left\{ \frac{d}{dt}[\mathbf{a}(\mathbf{b} \cdot \mathbf{a})] - \frac{d}{dt}[\mathbf{b}(\mathbf{a} \cdot \mathbf{a})] \right\} \, dt$$
$$= \mathbf{a}(\mathbf{b} \cdot \mathbf{a}) - \mathbf{b}(\mathbf{a} \cdot \mathbf{a}) + \mathbf{h}$$
$$= \mathbf{a} \times (\mathbf{a} \times \mathbf{b}) + \mathbf{h},$$

where $\mathbf{h}$ is the (vector) constant of integration. To obtain the final line above, we used a special case of the expansion of a vector triple product.

10.3 *The general equation of motion of a (non-relativistic) particle of mass m and charge q when it is placed in a region where there is a magnetic field* **B** *and an electric field* **E** *is*

$$m\ddot{\mathbf{r}} = q(\mathbf{E} + \dot{\mathbf{r}} \times \mathbf{B});$$

here **r** *is the position of the particle at time t and* $\dot{\mathbf{r}} = d\mathbf{r}/dt$, *etc. Write this as three separate equations in terms of the Cartesian components of the vectors involved.*

For the simple case of crossed uniform fields $\mathbf{E} = E\mathbf{i}$, $\mathbf{B} = B\mathbf{j}$, *in which the particle starts from the origin at* $t = 0$ *with* $\dot{\mathbf{r}} = v_0\mathbf{k}$, *find the equations of motion and show the following:*

(a) *if* $v_0 = E/B$ *then the particle continues its initial motion;*
(b) *if* $v_0 = 0$ *then the particle follows the space curve given in terms of the parameter* ξ *by*

$$x = \frac{mE}{B^2q}(1 - \cos\xi), \qquad y = 0, \qquad z = \frac{mE}{B^2q}(\xi - \sin\xi).$$

Interpret this curve geometrically and relate ξ *to t. Show that the total distance travelled by the particle after time t is given by*

$$\frac{2E}{B} \int_0^t \left| \sin\frac{Bqt'}{2m} \right| dt'.$$

Expressed in Cartesian coordinates, the components of the vector equation read

$$m\ddot{x} = qE_x + q(\dot{y}B_z - \dot{z}B_y),$$
$$m\ddot{y} = qE_y + q(\dot{z}B_x - \dot{x}B_z),$$
$$m\ddot{z} = qE_z + q(\dot{x}B_y - \dot{y}B_x).$$

For $E_x = E$, $B_y = B$ and all other field components zero, the equations reduce to

$$m\ddot{x} = qE - qB\dot{z}, \qquad m\ddot{y} = 0, \qquad m\ddot{z} = qB\dot{x}.$$

The second of these, together with the initial conditions $y(0) = \dot{y}(0) = 0$, implies that $y(t) = 0$ for all t. The final equation can be integrated directly to give

$$m\dot{z} = qBx + mv_0, \qquad (*)$$

which can now be substituted into the first to give a differential equation for x:

$$m\ddot{x} = qE - qB\left(\frac{qB}{m}x + v_0\right),$$

$$\Rightarrow \quad \ddot{x} + \left(\frac{qB}{m}\right)^2 x = \frac{q}{m}(E - v_0 B).$$

(i) If $v_0 = E/B$ then the equation for x is that of simple harmonic motion and

$$x(t) = A \cos \omega t + B \sin \omega t,$$

where $\omega = qB/m$. However, in the present case, the initial conditions $x(0) = \dot{x}(0) = 0$ imply that $x(t) = 0$ for all t. Thus, there is no motion in either the x- or the y-direction and, as is then shown by $(*)$, the particle continues with its initial speed v_0 in the z-direction.

(ii) If $v_0 = 0$, the equation of motion is

$$\ddot{x} + \omega^2 x = \frac{qE}{m},$$

which again has sinusoidal solutions but has a non-zero RHS. The full solution consists of the same complementary function as in part (i) together with the simplest possible particular integral, namely $x = qE/m\omega^2$. It is therefore

$$x(t) = A \cos \omega t + B \sin \omega t + \frac{qE}{m\omega^2}.$$

The initial condition $x(0) = 0$ implies that $A = -qE/(m\omega^2)$, whilst $\dot{x}(0) = 0$ requires that $B = 0$. Thus,

$$x = \frac{qE}{m\omega^2}(1 - \cos \omega t),$$

$$\Rightarrow \quad \dot{z} = \frac{qB}{m}x = \omega \frac{qE}{m\omega^2}(1 - \cos \omega t) = \frac{qE}{m\omega}(1 - \cos \omega t).$$

Since $z(0) = 0$, straightforward integration gives

$$z = \frac{qE}{m\omega}\left(t - \frac{\sin \omega t}{\omega}\right) = \frac{qE}{m\omega^2}(\omega t - \sin \omega t).$$

Thus, since $qE/m\omega^2 = mE/B^2 q$, the path is of the given parametric form with $\xi = \omega t$. It is a cycloid in the plane $y = 0$; the x-coordinate varies in the restricted range $0 \le x \le 2qE/(m\omega^2)$, whilst the z-coordinate continually increases, though not at a uniform rate.

The element of path length is given by $ds^2 = dx^2 + dy^2 + dz^2$. In this case, writing $qE/(m\omega) = E/B$ as μ,

$$ds = \left[\left(\frac{dx}{dt}\right)^2 + \left(\frac{dz}{dt}\right)^2\right]^{1/2} dt$$

$$= \left[\mu^2 \sin^2 \omega t + \mu^2(1 - \cos \omega t)^2\right]^{1/2} dt$$

$$= \left[2\mu^2(1 - \cos \omega t)\right]^{1/2} dt = 2\mu|\sin \tfrac{1}{2}\omega t|\, dt.$$

Thus the total distance travelled after time t is given by

$$s = \int_0^t 2\mu|\sin \tfrac{1}{2}\omega t'|\, dt' = \frac{2E}{B}\int_0^t \left|\sin \frac{qBt'}{2m}\right| dt'.$$

10.5 *If two systems of coordinates with a common origin O are rotating with respect to each other, the measured accelerations differ in the two systems. Denoting by $\mathbf{r}$ and $\mathbf{r}'$ position vectors in frames $OXYZ$ and $OX'Y'Z'$, respectively, the connection between the two is*

$$\ddot{\mathbf{r}}' = \ddot{\mathbf{r}} + \dot{\boldsymbol{\omega}} \times \mathbf{r} + 2\boldsymbol{\omega} \times \dot{\mathbf{r}} + \boldsymbol{\omega} \times (\boldsymbol{\omega} \times \mathbf{r}),$$

where $\boldsymbol{\omega}$ is the angular velocity vector of the rotation of $OXYZ$ with respect to $OX'Y'Z'$ (taken as fixed). The third term on the RHS is known as the Coriolis acceleration, whilst the final term gives rise to a centrifugal force.

Consider the application of this result to the firing of a shell of mass m from a stationary ship on the steadily rotating earth, working to the first order in ω ($= 7.3 \times 10^{-5}\,\mathrm{rad\,s^{-1}}$). If the shell is fired with velocity $\mathbf{v}$ at time $t = 0$ and only reaches a height that is small compared with the radius of the earth, show that its acceleration, as recorded on the ship, is given approximately by

$$\ddot{\mathbf{r}} = \mathbf{g} - 2\boldsymbol{\omega} \times (\mathbf{v} + \mathbf{g}t),$$

where $m\mathbf{g}$ is the weight of the shell measured on the ship's deck.

The shell is fired at another stationary ship (a distance $\mathbf{s}$ away) and $\mathbf{v}$ is such that the shell would have hit its target had there been no Coriolis effect.

(a) *Show that without the Coriolis effect the time of flight of the shell would have been $\tau = -2\mathbf{g} \cdot \mathbf{v}/g^2$.*

(b) *Show further that when the shell actually hits the sea it is off-target by approximately*

$$\frac{2\tau}{g^2}[(\mathbf{g} \times \boldsymbol{\omega}) \cdot \mathbf{v}](\mathbf{g}\tau + \mathbf{v}) - (\boldsymbol{\omega} \times \mathbf{v})\tau^2 - \frac{1}{3}(\boldsymbol{\omega} \times \mathbf{g})\tau^3.$$

(c) *Estimate the order of magnitude Δ of this miss for a shell for which the initial speed v is $300\,\mathrm{m\,s^{-1}}$, firing close to its maximum range ($\mathbf{v}$ makes an angle of $\pi/4$ with the vertical) in a northerly direction, whilst the ship is stationed at latitude $45°$ North.*

As the Earth is rotating steadily $\dot{\boldsymbol{\omega}} = \mathbf{0}$, and for the mass at rest on the deck,

$$m\ddot{\mathbf{r}}' = m\mathbf{g} + \mathbf{0} + 2\boldsymbol{\omega} \times \dot{\mathbf{0}} + m\boldsymbol{\omega} \times (\boldsymbol{\omega} \times \mathbf{r}).$$

This, including the centrifugal effect, defines $\mathbf{g}$ which is assumed constant throughout the trajectory.

For the moving mass ($\ddot{\mathbf{r}}'$ is unchanged),

$$m\mathbf{g} + \boldsymbol{\omega} \times (\boldsymbol{\omega} \times \mathbf{r}) = m\ddot{\mathbf{r}} + 2m\boldsymbol{\omega} \times \dot{\mathbf{r}} + m\boldsymbol{\omega} \times (\boldsymbol{\omega} \times \mathbf{r}),$$

i.e.
$$\ddot{\mathbf{r}} = \mathbf{g} - 2\boldsymbol{\omega} \times \dot{\mathbf{r}}.$$

159

Now, $\omega \dot{r} \ll g$ and so to zeroth order in ω

$$\ddot{\mathbf{r}} = \mathbf{g} \quad \Rightarrow \quad \dot{\mathbf{r}} = \mathbf{g}t + \mathbf{v}.$$

Resubstituting this into the Coriolis term gives, to first order in ω,

$$\ddot{\mathbf{r}} = \mathbf{g} - 2\boldsymbol{\omega} \times (\mathbf{v} + \mathbf{g}t).$$

(a) With no Coriolis force,

$$\dot{\mathbf{r}} = \mathbf{g}t + \mathbf{v} \quad \text{and} \quad \mathbf{r} = \tfrac{1}{2}\mathbf{g}t^2 + \mathbf{v}t.$$

Let $\mathbf{s} = \tfrac{1}{2}\mathbf{g}\tau^2 + \mathbf{v}\tau$ and use the observation that $\mathbf{s} \cdot \mathbf{g} = 0$, giving

$$\tfrac{1}{2}g^2\tau^2 + \mathbf{v} \cdot \mathbf{g}\tau = 0 \quad \Rightarrow \quad \tau = -\frac{2\mathbf{v} \cdot \mathbf{g}}{g^2}.$$

(b) With Coriolis force,

$$\ddot{\mathbf{r}} = \mathbf{g} - 2(\boldsymbol{\omega} \times \mathbf{g})t - 2(\boldsymbol{\omega} \times \mathbf{v}),$$
$$\dot{\mathbf{r}} = \mathbf{g}t - (\boldsymbol{\omega} \times \mathbf{g})t^2 - 2(\boldsymbol{\omega} \times \mathbf{v})t + \mathbf{v},$$
$$\mathbf{r} = \tfrac{1}{2}\mathbf{g}t^2 - \tfrac{1}{3}(\boldsymbol{\omega} \times \mathbf{g})t^3 - (\boldsymbol{\omega} \times \mathbf{v})t^2 + \mathbf{v}t. \qquad (*)$$

If the shell hits the sea at time T in the position $\mathbf{r} = \mathbf{s} + \Delta$, then $(\mathbf{s} + \Delta) \cdot \mathbf{g} = 0$, i.e.

$$0 = (\mathbf{s} + \Delta) \cdot \mathbf{g} = \tfrac{1}{2}g^2\,T^2 - 0 - (\boldsymbol{\omega} \times \mathbf{v}) \cdot \mathbf{g}\,T^2 + \mathbf{v} \cdot \mathbf{g}\,T,$$
$$\Rightarrow \quad -\mathbf{v} \cdot \mathbf{g} = T(\tfrac{1}{2}g^2 - (\boldsymbol{\omega} \times \mathbf{v}) \cdot \mathbf{g}),$$
$$\Rightarrow \quad T = -\frac{\mathbf{v} \cdot \mathbf{g}}{\tfrac{1}{2}g^2}\left[1 - \frac{(\boldsymbol{\omega} \times \mathbf{v}) \cdot \mathbf{g}}{\tfrac{1}{2}g^2}\right]^{-1}$$
$$\approx \tau\left(1 + \frac{2(\boldsymbol{\omega} \times \mathbf{v}) \cdot \mathbf{g}}{g^2} + \cdots\right).$$

Working to first order in ω, we may put $T = \tau$ in those terms in $(*)$ that involve another factor ω, namely $\boldsymbol{\omega} \times \mathbf{v}$ and $\boldsymbol{\omega} \times \mathbf{g}$. We then find, to this order, that

$$\mathbf{s} + \Delta = \frac{1}{2}\mathbf{g}\left(\tau^2 + \frac{4(\boldsymbol{\omega} \times \mathbf{v}) \cdot \mathbf{g}}{g^2}\tau^2 + \cdots\right) - \frac{1}{3}(\boldsymbol{\omega} \times \mathbf{g})\tau^3$$
$$- (\boldsymbol{\omega} \times \mathbf{v})\tau^2 + \mathbf{v}\tau + 2\frac{(\boldsymbol{\omega} \times \mathbf{v}) \cdot \mathbf{g}}{g^2}\mathbf{v}\tau$$
$$= \mathbf{s} + \frac{(\boldsymbol{\omega} \times \mathbf{v}) \cdot \mathbf{g}}{g^2}(2\mathbf{g}\tau^2 + 2\mathbf{v}\tau) - \frac{1}{3}(\boldsymbol{\omega} \times \mathbf{g})\tau^3 - (\boldsymbol{\omega} \times \mathbf{v})\tau^2.$$

Hence, as stated in the question,

$$\Delta = \frac{2\tau}{g^2}[(\mathbf{g} \times \boldsymbol{\omega}) \cdot \mathbf{v}](\mathbf{g}\tau + \mathbf{v}) - (\boldsymbol{\omega} \times \mathbf{v})\tau^2 - \frac{1}{3}(\boldsymbol{\omega} \times \mathbf{g})\tau^3.$$

(c) With the ship at latitude $45°$ and firing the shell at close to $45°$ to the local horizontal, $\mathbf{v}$ and $\boldsymbol{\omega}$ are almost parallel and the $\boldsymbol{\omega} \times \mathbf{v}$ term can be set to zero. Further, with $\mathbf{v}$ in a northerly direction, $(\mathbf{g} \times \boldsymbol{\omega}) \cdot \mathbf{v} = 0$.

Thus we are left with only the cubic term in τ. In this,

$$\tau = \frac{2 \times 300 \cos(\pi/4)}{9.8} = 43.3 \text{ s},$$

and $\boldsymbol{\omega} \times \mathbf{g}$ is in a westerly direction (recall that $\boldsymbol{\omega}$ is directed northwards and $\mathbf{g}$ is directed downwards, towards the origin) and of magnitude $7 \, 10^{-5} \, 9.8 \sin(\pi/4) = 4.85 \, 10^{-4}$ m s^{-3}. Thus the miss is by approximately

$$-\tfrac{1}{3} \times 4.85 \, 10^{-4} \times (43.3)^3 = -13 \text{ m},$$

i.e. some $10 - 15$ m to the East of its intended target.

10.7 *For the twisted space curve* $y^3 + 27axz - 81a^2y = 0$, *given parametrically by*

$$x = au(3 - u^2), \qquad y = 3au^2, \qquad z = au(3 + u^2),$$

show that the following hold:

(a) $ds/du = 3\sqrt{2}a(1 + u^2)$, *where s is the distance along the curve measured from the origin;*

(b) *the length of the curve from the origin to the Cartesian point* $(2a, 3a, 4a)$ *is* $4\sqrt{2}a$;

(c) *the radius of curvature at the point with parameter u is* $3a(1 + u^2)^2$;

(d) *the torsion τ and curvature κ at a general point are equal;*

(e) *any of the Frenet–Serret formulae that you have not already used directly are satisfied.*

(a) We must first calculate

$$\frac{d\mathbf{r}}{du} = (3a - 3au^2, \, 6au, \, 3a + 3au^2),$$

from which it follows that

$$\frac{ds}{du} = \left(\frac{d\mathbf{r}}{du} \cdot \frac{d\mathbf{r}}{du} \right)^{1/2} = 3a(1 - 2u^2 + u^4 + 4u^2 + 1 + 2u^2 + u^4)^{1/2}$$

$$= 3\sqrt{2}a(1 + u^2).$$

(b) The point $(2a, 3a, 4a)$ is given by $u = 1$; the origin is $u = 0$. The length of the curve from the origin to the point is therefore given by

$$s = \int_0^1 3\sqrt{2}a(1 + u^2) \, du = 3\sqrt{2}a \left[u + \frac{u^3}{3} \right]_0^1 = 4\sqrt{2}a.$$

(c) Using

$$\hat{\mathbf{t}} = \frac{d\mathbf{r}}{ds} = \frac{d\mathbf{r}}{du}\frac{du}{ds} = \frac{3a}{3\sqrt{2a}(1+u^2)}(1-u^2,\ 2u,\ 1+u^2),$$

we find that

$$\frac{d\hat{\mathbf{t}}}{ds} = \frac{d\hat{\mathbf{t}}}{du}\frac{du}{ds}$$

$$= \frac{1}{3\sqrt{2a}(1+u^2)}\frac{1}{\sqrt{2}}\left[\frac{d}{du}\left(\frac{1-u^2}{1+u^2}\right),\ \frac{d}{du}\left(\frac{2u}{1+u^2}\right),\ \frac{d}{du}\left(\frac{1+u^2}{1+u^2}\right)\right]$$

$$= \frac{1}{6a(1+u^2)^3}(-4u,\ 2-2u^2,\ 0).$$

We now recall that $d\hat{\mathbf{t}}/ds = \kappa\hat{\mathbf{n}}$, where κ is the curvature and the principal normal $\hat{\mathbf{n}}$ is a unit vector in the same direction as $d\hat{\mathbf{t}}/ds$. Thus

$$\frac{1}{\rho} = \kappa = \left|\frac{d\hat{\mathbf{t}}}{ds}\right| = \frac{2(4u^2 + 1 - 2u^2 + u^4)^{1/2}}{6a(1+u^2)^3} = \frac{1}{3a(1+u^2)^2}.$$

(d) From part (c) we have the two results

$$\hat{\mathbf{t}} = \frac{1}{\sqrt{2}(1+u^2)}(1-u^2,\ 2u,\ 1+u^2),$$

$$\hat{\mathbf{n}} = \frac{1}{1+u^2}(-2u,\ 1-u^2,\ 0),$$

and so the binormal $\hat{\mathbf{b}}$ is given by

$$\hat{\mathbf{b}} = \hat{\mathbf{t}} \times \hat{\mathbf{n}}$$

$$= \frac{1}{\sqrt{2}(1+u^2)^2}\left[u^4 - 1,\ -2u(1+u^2),\ (1+u^2)^2\right]$$

$$= \frac{1}{\sqrt{2}}\left(\frac{u^2-1}{u^2+1},\ \frac{-2u}{u^2+1},\ 1\right).$$

From this it follows that

$$\frac{d\hat{\mathbf{b}}}{ds} = \frac{d\hat{\mathbf{b}}}{du}\frac{du}{ds}$$

$$= \frac{1}{3\sqrt{2a}(1+u^2)}\frac{1}{\sqrt{2}}\left(\frac{4u}{(1+u^2)^2},\ \frac{2(u^2-1)}{(1+u^2)^2},\ 0\right).$$

Comparing this with $-\tau\hat{\mathbf{n}}$, with $\hat{\mathbf{n}}$ as given above, shows that

$$\tau = \frac{2}{6a(1+u^2)^2}.$$

But

$$\kappa = \frac{1}{\rho} = \frac{1}{3a(1+u^2)^2},$$

162

thus establishing the result that τ equals κ for this curve.

(e) The remaining Frenet–Serret formula is

$$\frac{d\hat{\mathbf{n}}}{ds} = \tau\hat{\mathbf{b}} - \kappa\hat{\mathbf{t}}.$$

Consider the two sides of the equation separately:

$$\begin{aligned}
\text{LHS} &= \frac{d\hat{\mathbf{n}}}{ds} = \frac{d\hat{\mathbf{n}}}{du}\frac{du}{ds} \\
&= \frac{1}{3\sqrt{2}a(1+u^2)}\left[\frac{d}{du}\left(\frac{-2u}{1+u^2}\right), \frac{d}{du}\left(\frac{1-u^2}{1+u^2}\right), 0\right] \\
&= \frac{1}{3\sqrt{2}a(1+u^2)}\left[\frac{2u^2-2}{(1+u^2)^2}, \frac{-4u}{(1+u^2)^2}, 0\right] \\
&= \frac{1}{3\sqrt{2}a(1+u^2)^3}(2u^2-2, -4u, 0); \\
\text{RHS} &= \tau\hat{\mathbf{b}} - \kappa\hat{\mathbf{t}} = \kappa(\hat{\mathbf{b}} - \hat{\mathbf{t}}) \\
&= \frac{\kappa}{\sqrt{2}(1+u^2)}[u^2-1-(1-u^2), -2u-2u, 1+u^2-(1+u^2)] \\
&= \frac{1}{3\sqrt{2}a(1+u^2)^3}(2u^2-2, -4u, 0).
\end{aligned}$$

Thus, the two sides are equal and the unused formula is verified.

10.9 *In a magnetic field, field lines are curves to which the magnetic induction* **B** *is everywhere tangential. By evaluating* d**B**/ds, *where s is the distance measured along a field line, prove that the radius of curvature at any point on a line is given by*

$$\rho = \frac{B^3}{|\mathbf{B} \times (\mathbf{B} \cdot \nabla)\mathbf{B}|}.$$

We start with the three simple vector relationships

$$\frac{d\mathbf{r}}{ds} = \hat{\mathbf{t}}, \quad \frac{d\hat{\mathbf{t}}}{ds} = \frac{1}{\rho}\hat{\mathbf{n}} \quad \text{and} \quad \mathbf{B} = B\hat{\mathbf{t}},$$

and note that

$$d\mathbf{B} = \frac{\partial\mathbf{B}}{\partial x}dx + \frac{\partial\mathbf{B}}{\partial y}dy + \frac{\partial\mathbf{B}}{\partial z}dz = (d\mathbf{r}\cdot\nabla)\mathbf{B}.$$

Differentiating the third relationship with respect to s gives

$$\frac{d\mathbf{B}}{ds} = \frac{dB}{ds}\hat{\mathbf{t}} + B\frac{d\hat{\mathbf{t}}}{ds}.$$

We can replace the LHS of this equation with

$$\frac{d\mathbf{B}}{ds} = \left(\frac{d\mathbf{r} \cdot \nabla}{ds}\right)\mathbf{B} = (\hat{\mathbf{t}} \cdot \nabla)\mathbf{B} = \frac{\mathbf{B} \cdot \nabla}{B}\mathbf{B}$$

and obtain

$$\frac{\mathbf{B} \cdot \nabla}{B}\mathbf{B} = \frac{dB}{ds}\hat{\mathbf{t}} + \frac{B}{\rho}\hat{\mathbf{n}}.$$

Finally, we take the cross product of this equation with $\hat{\mathbf{t}}$ and obtain

$$\hat{\mathbf{t}} \times \frac{\mathbf{B} \cdot \nabla}{B}\mathbf{B} = 0 + \frac{B}{\rho}\hat{\mathbf{t}} \times \hat{\mathbf{n}},$$

$$\frac{\mathbf{B} \times (\mathbf{B} \cdot \nabla)\mathbf{B}}{B^2} = \frac{B}{\rho}\hat{\mathbf{b}},$$

$$\frac{|\mathbf{B} \times (\mathbf{B} \cdot \nabla)\mathbf{B}|}{B^2} = \frac{B}{\rho} \quad \Rightarrow \quad \rho = \frac{B^3}{|\mathbf{B} \times (\mathbf{B} \cdot \nabla)\mathbf{B}|}.$$

In the penultimate line we have given the unit vector $\hat{\mathbf{t}} \times \hat{\mathbf{n}}$ its usual symbol $\hat{\mathbf{b}}$ (for binormal), though the only property that is needed here is that it has unit length. To obtain the final line, we took the modulus of both sides of the equation on the previous one.

10.11 *Parameterising the hyperboloid*

$$\frac{x^2}{a^2} + \frac{y^2}{b^2} - \frac{z^2}{c^2} = 1$$

by $x = a\cos\theta\sec\phi$, $y = b\sin\theta\sec\phi$, $z = c\tan\phi$, *show that an area element on its surface is*

$$dS = \sec^2\phi \left[c^2\sec^2\phi \left(b^2\cos^2\theta + a^2\sin^2\theta\right) + a^2b^2\tan^2\phi\right]^{1/2} d\theta\, d\phi.$$

Use this formula to show that the area of the curved surface $x^2 + y^2 - z^2 = a^2$ *between the planes* $z = 0$ *and* $z = 2a$ *is*

$$\pi a^2 \left(6 + \frac{1}{\sqrt{2}}\sinh^{-1} 2\sqrt{2}\right).$$

With $x = a\cos\theta\sec\phi$, $y = b\sin\theta\sec\phi$ and $z = c\tan\phi$, the tangent vectors to the surface are given in Cartesian coordinates by

$$\frac{d\mathbf{r}}{d\theta} = (-a\sin\theta\sec\phi,\ b\cos\theta\sec\phi,\ 0),$$

$$\frac{d\mathbf{r}}{d\phi} = (a\cos\theta\sec\phi\tan\phi,\ b\sin\theta\sec\phi\tan\phi,\ c\sec^2\phi),$$

and the element of area by

$$dS = \left| \frac{d\mathbf{r}}{d\theta} \times \frac{d\mathbf{r}}{d\phi} \right| d\theta \, d\phi$$

$$= \left| (bc \cos \theta \sec^3 \phi, \ ac \sin \theta \sec^3 \phi, \ -ab \sec^2 \phi \tan \phi) \right| d\theta \, d\phi$$

$$= \sec^2 \phi \left[c^2 \sec^2 \phi \left(b^2 \cos^2 \theta + a^2 \sin^2 \theta \right) + a^2 b^2 \tan^2 \phi \right]^{1/2} d\theta \, d\phi.$$

We set $b = c = a$ and note that the plane $z = 2a$ corresponds to $\phi = \tan^{-1} 2$. The ranges of integration are therefore $0 \le \theta < 2\pi$ and $0 \le \phi \le \tan^{-1} 2$, whilst

$$dS = \sec^2 \phi (a^4 \sec^2 \phi + a^4 \tan^2 \phi)^{1/2} \, d\theta \, d\phi,$$

i.e. it is independent of θ.

To evaluate the integral of dS, we set $\tan \phi = \sinh \psi / \sqrt{2}$, with

$$\sec^2 \phi \, d\phi = \frac{1}{\sqrt{2}} \cosh \psi \, d\psi \quad \text{and} \quad \sec^2 \phi = 1 + \tfrac{1}{2} \sinh^2 \psi.$$

The upper limit for ψ will be given by $\Psi = \sinh^{-1} 2\sqrt{2}$; we note that $\cosh \Psi = 3$. Integrating over θ and making the above substitutions yields

$$S = 2\pi \int_0^{\Psi} \frac{1}{\sqrt{2}} \cosh \psi \, d\psi \, a^2 \left(1 + \frac{1}{2} \sinh^2 \psi + \frac{1}{2} \sinh^2 \psi \right)^{1/2}$$

$$= \sqrt{2} \pi a^2 \int_0^{\Psi} \cosh^2 \psi \, d\psi$$

$$= \frac{\sqrt{2} \pi a^2}{2} \int_0^{\Psi} (\cosh 2\psi + 1) \, d\psi$$

$$= \frac{\sqrt{2} \pi a^2}{2} \left[\frac{\sinh 2\psi}{2} + \psi \right]_0^{\Psi}$$

$$= \frac{\pi a^2}{\sqrt{2}} \left[\sinh \psi \cosh \psi + \psi \right]_0^{\Psi}$$

$$= \frac{\pi a^2}{\sqrt{2}} [(2\sqrt{2})(3) + \sinh^{-1} 2\sqrt{2}] = \pi a^2 \left(6 + \frac{1}{\sqrt{2}} \sinh^{-1} 2\sqrt{2} \right).$$

10.13 *Verify by direct calculation that*

$$\nabla \cdot (\mathbf{a} \times \mathbf{b}) = \mathbf{b} \cdot (\nabla \times \mathbf{a}) - \mathbf{a} \cdot (\nabla \times \mathbf{b}).$$

The proof of this standard result for the divergence of a vector product is most easily carried out in Cartesian coordinates though, of course, the result is valid in any three-dimensional coordinate system.

$$\text{LHS} = \nabla \cdot (\mathbf{a} \times \mathbf{b})$$

$$= \frac{\partial}{\partial x}(a_y b_z - a_z b_y) + \frac{\partial}{\partial y}(a_z b_x - a_x b_z) + \frac{\partial}{\partial z}(a_x b_y - a_y b_x)$$

$$= a_x \left(-\frac{\partial b_z}{\partial y} + \frac{\partial b_y}{\partial z} \right) + a_y \left(\frac{\partial b_z}{\partial x} - \frac{\partial b_x}{\partial z} \right) + a_z \left(-\frac{\partial b_y}{\partial x} + \frac{\partial b_x}{\partial y} \right)$$

$$+ b_x \left(\frac{\partial a_z}{\partial y} - \frac{\partial a_y}{\partial z} \right) + b_y \left(-\frac{\partial a_z}{\partial x} + \frac{\partial a_x}{\partial z} \right) + b_z \left(\frac{\partial a_y}{\partial x} - \frac{\partial a_x}{\partial y} \right)$$

$$= -\mathbf{a} \cdot (\nabla \times \mathbf{b}) + \mathbf{b} \cdot (\nabla \times \mathbf{a}) = \text{RHS}.$$

10.15 *Evaluate the Laplacian of the function*

$$\psi(x, y, z) = \frac{zx^2}{x^2 + y^2 + z^2}$$

(a) *directly in Cartesian coordinates, and* (b) *after changing to a spherical polar coordinate system. Verify that, as they must, the two methods give the same result.*

(a) In Cartesian coordinates we need to evaluate

$$\nabla^2 \psi = \frac{\partial^2 \psi}{\partial x^2} + \frac{\partial^2 \psi}{\partial y^2} + \frac{\partial^2 \psi}{\partial z^2}.$$

The required derivatives are

$$\frac{\partial \psi}{\partial x} = \frac{2xz(y^2 + z^2)}{(x^2 + y^2 + z^2)^2}, \qquad \frac{\partial^2 \psi}{\partial x^2} = \frac{(y^2 + z^2)(2zy^2 + 2z^3 - 6x^2 z)}{(x^2 + y^2 + z^2)^3},$$

$$\frac{\partial \psi}{\partial y} = \frac{-2x^2 yz}{(x^2 + y^2 + z^2)^2}, \qquad \frac{\partial^2 \psi}{\partial y^2} = -\frac{2zx^2(x^2 + z^2 - 3y^2)}{(x^2 + y^2 + z^2)^3},$$

$$\frac{\partial \psi}{\partial z} = \frac{x^2(x^2 + y^2 - z^2)}{(x^2 + y^2 + z^2)^2}, \qquad \frac{\partial^2 \psi}{\partial z^2} = -\frac{2zx^2(3x^2 + 3y^2 - z^2)}{(x^2 + y^2 + z^2)^3}.$$

Thus, writing $r^2 = x^2 + y^2 + z^2$,

$$\nabla^2 \psi = \frac{2z[(y^2 + z^2)(y^2 + z^2 - 3x^2) - 4x^4]}{(x^2 + y^2 + z^2)^3}$$

$$= \frac{2z[(r^2 - x^2)(r^2 - 4x^2) - 4x^4]}{r^6}$$

$$= \frac{2z(r^2 - 5x^2)}{r^4}.$$

(b) In spherical polar coordinates,

$$\psi(r,\theta,\phi) = \frac{r\cos\theta\, r^2 \sin^2\theta \cos^2\phi}{r^2} = r\cos\theta \sin^2\theta \cos^2\phi.$$

The three contributions to $\nabla^2\psi$ in spherical polars are

$$(\nabla^2\psi)_r = \frac{1}{r^2}\frac{\partial}{\partial r}\left(r^2\frac{\partial\psi}{\partial r}\right)$$

$$= \frac{2}{r}\cos\theta \sin^2\theta \cos^2\phi,$$

$$(\nabla^2\psi)_\theta = \frac{1}{r^2\sin\theta}\frac{\partial}{\partial\theta}\left(\sin\theta\frac{\partial\psi}{\partial\theta}\right)$$

$$= \frac{1}{r}\frac{\cos^2\phi}{\sin\theta}\frac{\partial}{\partial\theta}\left[\sin\theta\frac{\partial}{\partial\theta}(\cos\theta\sin^2\theta)\right]$$

$$= \frac{\cos^2\phi}{r}(4\cos^3\theta - 8\sin^2\theta\cos\theta),$$

$$(\nabla^2\psi)_\phi = \frac{1}{r^2\sin^2\theta}\frac{\partial^2\psi}{\partial\phi^2}$$

$$= \frac{\cos\theta}{r}(-2\cos^2\phi + 2\sin^2\phi).$$

Thus, the full Laplacian in spherical polar coordinates reads

$$\nabla^2\psi = \frac{\cos\theta}{r}(2\sin^2\theta\cos^2\phi + 4\cos^2\theta\cos^2\phi$$

$$- 8\sin^2\theta\cos^2\phi - 2\cos^2\phi + 2\sin^2\phi)$$

$$= \frac{\cos\theta}{r}(4\cos^2\phi - 10\sin^2\theta\cos^2\phi - 2\cos^2\phi + 2\sin^2\phi)$$

$$= \frac{\cos\theta}{r}(2 - 10\sin^2\theta\cos^2\phi)$$

$$= \frac{2r\cos\theta(r^2 - 5r^2\sin^2\theta\cos^2\phi)}{r^4}.$$

Rewriting this last expression in terms of Cartesian coordinates, one finally obtains

$$\nabla^2\psi = \frac{2z(r^2 - 5x^2)}{r^4},$$

which establishes the equivalence of the two approaches.

10.17 *The (Maxwell) relationship between a time-independent magnetic field* **B** *and the current density* **J** *(measured in SI units in* $A\,m^{-2}$*) producing it,*

$$\nabla \times \mathbf{B} = \mu_0 \mathbf{J},$$

can be applied to a long cylinder of conducting ionised gas which, in cylindrical polar coordinates, occupies the region $\rho < a$.

(a) *Show that a uniform current density* $(0, C, 0)$ *and a magnetic field* $(0, 0, B)$, *with B constant* $(= B_0)$ *for* $\rho > a$ *and* $B = B(\rho)$ *for* $\rho < a$, *are consistent with this equation. Given that* $B(0) = 0$ *and that* **B** *is continuous at* $\rho = a$, *obtain expressions for C and* $B(\rho)$ *in terms of* B_0 *and a.*

(b) *The magnetic field can be expressed as* $\mathbf{B} = \nabla \times \mathbf{A}$, *where* **A** *is known as the vector potential. Show that a suitable* **A** *can be found which has only one non-vanishing component,* $A_\phi(\rho)$, *and obtain explicit expressions for* $A_\phi(\rho)$ *for both* $\rho < a$ *and* $\rho > a$. *Like* **B**, *the vector potential is continuous at* $\rho = a$.

(c) *The gas pressure* $p(\rho)$ *satisfies the hydrostatic equation* $\nabla p = \mathbf{J} \times \mathbf{B}$ *and vanishes at the outer wall of the cylinder. Find a general expression for p.*

(a) In cylindrical polars with $\mathbf{B} = (0, 0, B(\rho))$, for $\rho \leq a$ we have

$$\mu_0(0,\ C,\ 0) = \nabla \times \mathbf{B} = \left(\frac{1}{\rho} \frac{\partial B}{\partial \phi},\ -\frac{\partial B}{\partial \rho},\ 0 \right).$$

As expected, $\partial B / \partial \phi = 0$. The azimuthal component of the equation gives

$$-\frac{\partial B}{\partial \rho} = \mu_0 C \quad \text{for} \quad \rho \leq a \quad \Rightarrow \quad B(\rho) = B(0) - \mu_0 C \rho.$$

Since **B** has to be differentiable at the origin of ρ and have no ϕ-dependence, $B(0)$ must be zero. This, together with $B = B_0$ for $\rho > a$ requires that $C = -B_0/(a\mu_0)$ and $B(\rho) = B_0 \rho / a$ for $0 \leq \rho \leq a$.

(b) With $\mathbf{B} = \nabla \times \mathbf{A}$, consider **A** of the form $\mathbf{A} = (0,\ A(\rho),\ 0)$. Then

$$(0,\ 0,\ B(\rho)) = \frac{1}{\rho} \left(\frac{\partial}{\partial z}(\rho A),\ 0,\ \frac{\partial}{\partial \rho}(\rho A) \right)$$

$$= \left(0,\ 0,\ \frac{1}{\rho} \frac{\partial}{\partial \rho}(\rho A) \right).$$

We now equate the only non-vanishing component on each side of the above equation, treating inside and outside the cylinder separately.

For $0 < \rho \leq a$,

$$\frac{1}{\rho}\frac{\partial}{\partial\rho}(\rho A) = \frac{B_0\rho}{a},$$

$$\rho A = \frac{B_0\rho^3}{3a} + D,$$

$$A(\rho) = \frac{B_0\rho^2}{3a} + \frac{D}{\rho}.$$

Since $A(0)$ must be finite (so that A is differentiable there), $D = 0$.

For $\rho > a$,

$$\frac{1}{\rho}\frac{\partial}{\partial\rho}(\rho A) = B_0,$$

$$\rho A = \frac{B_0\rho^2}{2} + E,$$

$$A(\rho) = \frac{1}{2}B_0\rho + \frac{E}{\rho}.$$

At $\rho = a$, the continuity of $\mathbf{A}$ requires

$$\frac{B_0 a^2}{3a} = \frac{1}{2}B_0 a + \frac{E}{a} \Rightarrow E = -\frac{B_0 a^2}{6}.$$

Thus, to summarise,

$$A(\rho) = \frac{B_0\rho^2}{3a} \quad \text{for} \quad 0 \leq \rho \leq a,$$

$$\text{and} \quad A(\rho) = B_0\left(\frac{\rho}{2} - \frac{a^2}{6\rho}\right) \quad \text{for} \quad \rho \geq a.$$

(c) For the gas pressure $p(\rho)$ in the region $0 < \rho \leq a$, we have $\nabla p = \mathbf{J} \times \mathbf{B}$. In component form,

$$\left(\frac{dp}{d\rho}, 0, 0\right) = \left(0, -\frac{B_0}{a\mu_0}, 0\right) \times \left(0, 0, \frac{B_0\rho}{a}\right),$$

with $p(a) = 0$.

$$\frac{dp}{d\rho} = -\frac{B_0^2\rho}{\mu_0 a^2} \quad \Rightarrow \quad p(\rho) = \frac{B_0^2}{2\mu_0}\left[1 - \left(\frac{\rho}{a}\right)^2\right].$$

10.19 *Maxwell's equations for electromagnetism in free space (i.e. in the absence of charges, currents and dielectric or magnetic media) can be written*

$$\text{(i)} \quad \nabla \cdot \mathbf{B} = 0, \qquad\qquad \text{(ii)} \quad \nabla \cdot \mathbf{E} = 0,$$

$$\text{(iii)} \quad \nabla \times \mathbf{E} + \frac{\partial \mathbf{B}}{\partial t} = \mathbf{0}, \quad \text{(iv)} \quad \nabla \times \mathbf{B} - \frac{1}{c^2}\frac{\partial \mathbf{E}}{\partial t} = \mathbf{0}.$$

A vector $\mathbf{A}$ *is defined by* $\mathbf{B} = \nabla \times \mathbf{A}$, *and a scalar* ϕ *by* $\mathbf{E} = -\nabla\phi - \partial\mathbf{A}/\partial t$. *Show that if the condition*

$$\text{(v)} \quad \nabla \cdot \mathbf{A} + \frac{1}{c^2}\frac{\partial \phi}{\partial t} = 0$$

is imposed (this is known as choosing the Lorentz gauge), then $\mathbf{A}$ *and* ϕ *satisfy wave equations as follows.*

$$\text{(vi)} \quad \nabla^2\phi - \frac{1}{c^2}\frac{\partial^2\phi}{\partial t^2} = 0,$$

$$\text{(vii)} \quad \nabla^2\mathbf{A} - \frac{1}{c^2}\frac{\partial^2\mathbf{A}}{\partial t^2} = \mathbf{0}.$$

The reader is invited to proceed as follows.

(a) *Verify that the expressions for* $\mathbf{B}$ *and* $\mathbf{E}$ *in terms of* $\mathbf{A}$ *and* ϕ *are consistent with* (i) *and* (iii).

(b) *Substitute for* $\mathbf{E}$ *in* (ii) *and use the derivative with respect to time of* (v) *to eliminate* $\mathbf{A}$ *from the resulting expression. Hence obtain* (vi).

(c) *Substitute for* $\mathbf{B}$ *and* $\mathbf{E}$ *in* (iv) *in terms of* $\mathbf{A}$ *and* ϕ. *Then use the gradient of* (v) *to simplify the resulting equation and so obtain* (vii).

(a) Substituting for $\mathbf{B}$ in (i),

$$\nabla \cdot \mathbf{B} = \nabla \cdot (\nabla \times \mathbf{A}) = 0, \quad \text{as it is for } any \text{ vector } \mathbf{A}.$$

Substituting for $\mathbf{E}$ and $\mathbf{B}$ in (iii),

$$\nabla \times \mathbf{E} + \frac{\partial \mathbf{B}}{\partial t} = -(\nabla \times \nabla\phi) - \nabla \times \frac{\partial \mathbf{A}}{\partial t} + \frac{\partial}{\partial t}(\nabla \times \mathbf{A}) = \mathbf{0}.$$

Here we have used the facts that $\nabla \times \nabla\phi = \mathbf{0}$ for any scalar, and that, since $\partial/\partial t$ and ∇ act on different variables, the order in which they are applied to $\mathbf{A}$ can be reversed. Thus (i) and (iii) are automatically satisfied if $\mathbf{E}$ and $\mathbf{B}$ are represented in terms of $\mathbf{A}$ and ϕ.

170

(b) Substituting for **E** in (ii) and taking the time derivative of (v),

$$0 = \nabla \cdot \mathbf{E} = -\nabla^2 \phi - \frac{\partial}{\partial t} (\nabla \cdot \mathbf{A}),$$

$$0 = \frac{\partial}{\partial t} (\nabla \cdot \mathbf{A}) + \frac{1}{c^2} \frac{\partial^2 \phi}{\partial t^2}.$$

Adding these equations gives

$$0 = -\nabla^2 \phi + \frac{1}{c^2} \frac{\partial^2 \phi}{\partial t^2}.$$

This is result (vi), the wave equation for ϕ.

(c) Substituting for **B** and **E** in (iv) and taking the gradient of (v),

$$\nabla \times (\nabla \times \mathbf{A}) - \frac{1}{c^2} \left(-\frac{\partial}{\partial t} \nabla \phi - \frac{\partial^2 \mathbf{A}}{\partial t^2} \right) = \mathbf{0},$$

$$\nabla (\nabla \cdot \mathbf{A}) - \nabla^2 \mathbf{A} + \frac{1}{c^2} \frac{\partial}{\partial t} (\nabla \phi) + \frac{1}{c^2} \frac{\partial^2 \mathbf{A}}{\partial t^2} = \mathbf{0}.$$

From (v), $\qquad\qquad\qquad \nabla (\nabla \cdot \mathbf{A}) + \frac{1}{c^2} \frac{\partial}{\partial t} (\nabla \phi) = \mathbf{0}.$

Subtracting these gives $\qquad\quad - \nabla^2 \mathbf{A} + \frac{1}{c^2} \frac{\partial^2 \mathbf{A}}{\partial t^2} = \mathbf{0}.$

In the second line we have used the vector identity

$$\nabla^2 \mathbf{F} = \nabla (\nabla \cdot \mathbf{F}) - \nabla \times (\nabla \times \mathbf{F})$$

to replace $\nabla \times (\nabla \times \mathbf{A})$. The final equation is result (vii).

10.21 *Paraboloidal coordinates u, v, ϕ are defined in terms of Cartesian coordinates by*

$$x = uv \cos \phi, \qquad y = uv \sin \phi, \qquad z = \tfrac{1}{2}(u^2 - v^2).$$

Identify the coordinate surfaces in the u, v, ϕ system. Verify that each coordinate surface ($u = $ constant, say) intersects every coordinate surface on which one of the other two coordinates (v, say) is constant. Show further that the system of coordinates is an orthogonal one and determine its scale factors. Prove that the u-component of $\nabla \times \mathbf{a}$ is given by

$$\frac{1}{(u^2 + v^2)^{1/2}} \left(\frac{a_\phi}{v} + \frac{\partial a_\phi}{\partial v} \right) - \frac{1}{uv} \frac{\partial a_v}{\partial \phi}.$$

To find a surface of constant u we eliminate v from the given relationships:

$$x^2 + y^2 = u^2 v^2 \quad \Rightarrow \quad 2z = u^2 - \frac{x^2 + y^2}{u^2}.$$

This is an inverted paraboloid of revolution about the z-axis. The range of z is $-\infty < z \leq \frac{1}{2}u^2$.

Similarly, the surface of constant v is given by

$$2z = \frac{x^2 + y^2}{v^2} - v^2.$$

This is also a paraboloid of revolution about the z-axis, but this time it is not inverted. The range of z is $-\frac{1}{2}v^2 \leq z < \infty$.

Since every constant-u paraboloid has some part of its surface in the region $z > 0$ and every constant-v paraboloid has some part of its surface in the region $z < 0$, it follows that every member of the first set intersects each member of the second, and vice-versa.

The surfaces of constant ϕ, $y = x \tan \phi$, are clearly (half-) planes containing the z-axis; each cuts the members of the other two sets in parabolic lines.

We now determine (the Cartesian components of) the tangential vectors and test their orthogonality:

$$\mathbf{e}_1 = \frac{\partial \mathbf{r}}{\partial u} = (v \cos \phi, \ v \sin \phi, \ u),$$

$$\mathbf{e}_2 = \frac{\partial \mathbf{r}}{\partial v} = (u \cos \phi, \ u \sin \phi, \ -v),$$

$$\mathbf{e}_3 = \frac{\partial \mathbf{r}}{\partial \phi} = (-uv \sin \phi, \ uv \cos \phi, \ 0),$$

$$\mathbf{e}_1 \cdot \mathbf{e}_2 = uv(\cos \phi \cos \phi + \sin \phi \sin \phi) - uv = 0,$$

$$\mathbf{e}_2 \cdot \mathbf{e}_3 = u^2 v(-\cos \phi \sin \phi + \sin \phi \cos \phi) = 0,$$

$$\mathbf{e}_1 \cdot \mathbf{e}_3 = uv^2(-\cos \phi \sin \phi + \sin \phi \cos \phi) = 0.$$

This shows that all pairs of tangential vectors are orthogonal and therefore that the coordinate system is an orthogonal one. Its scale factors are given by the magnitudes of these tangential vectors:

$$h_u^2 = |\mathbf{e}_1|^2 = (v \cos \phi)^2 + (v \sin \phi)^2 + u^2 = u^2 + v^2,$$

$$h_v^2 = |\mathbf{e}_2|^2 = (u \cos \phi)^2 + (u \sin \phi)^2 + v^2 = u^2 + v^2,$$

$$h_\phi^2 = |\mathbf{e}_3|^2 = (uv \sin \phi)^2 + (uv \cos \phi)^2 = u^2 v^2.$$

Thus

$$h_u = h_v = \sqrt{u^2 + v^2}, \qquad h_\phi = uv.$$

The u-component of $\nabla \times \mathbf{a}$ is given by

$$
\begin{aligned}
[\nabla \times \mathbf{a}]_u &= \frac{h_u}{h_u h_v h_\phi} \left[\frac{\partial}{\partial v}(h_\phi a_\phi) - \frac{\partial}{\partial \phi}(h_v a_v) \right] \\
&= \frac{1}{uv\sqrt{u^2 + v^2}} \left[\frac{\partial}{\partial v}(uv a_\phi) - \frac{\partial}{\partial \phi}(\sqrt{u^2 + v^2}\, a_v) \right] \\
&= \frac{1}{\sqrt{u^2 + v^2}} \left(\frac{a_\phi}{v} + \frac{\partial a_\phi}{\partial v} \right) - \frac{1}{uv} \frac{\partial a_v}{\partial \phi},
\end{aligned}
$$

as stated in the question.

10.23 *Hyperbolic coordinates u, v, ϕ are defined in terms of Cartesian coordinates by*

$$
x = \cosh u \cos v \cos \phi, \qquad y = \cosh u \cos v \sin \phi, \qquad z = \sinh u \sin v.
$$

Sketch the coordinate curves in the $\phi = 0$ plane, showing that far from the origin they become concentric circles and radial lines. In particular, identify the curves $u = 0$, $v = 0$, $v = \pi/2$ and $v = \pi$. Calculate the tangent vectors at a general point, show that they are mutually orthogonal and deduce that the appropriate scale factors are

$$
h_u = h_v = (\cosh^2 u - \cos^2 v)^{1/2}, \qquad h_\phi = \cosh u \cos v.
$$

Find the most general function $\psi(u)$ of u only that satisfies Laplace's equation $\nabla^2 \psi = 0$.

In the plane $\phi = 0$, i.e. $y = 0$, the curves $u = $ constant have x and z connected by

$$
\frac{x^2}{\cosh^2 u} + \frac{z^2}{\sinh^2 u} = 1.
$$

This general form is that of an ellipse, with foci at $(\pm 1, 0)$. With $u = 0$, it is the line joining the two foci (covered twice). As $u \to \infty$, and $\cosh u \approx \sinh u$ the form becomes that of a circle of very large radius.

The curves $v = $ constant are expressed by

$$
\frac{x^2}{\cos^2 v} - \frac{z^2}{\sin^2 v} = 1.
$$

These curves are hyperbolae that, for large x and z and fixed v, approximate $z = \pm x \tan v$, i.e. radial lines. The curve $v = 0$ is the part of the x-axis $1 \le x \le \infty$ (covered twice), whilst the curve $v = \pi$ is its reflection in the z-axis. The curve $v = \pi/2$ *is the z-axis.*

In Cartesian coordinates a general point and its derivatives with respect to u, v and ϕ are given by

$$\mathbf{r} = \cosh u \cos v \cos \phi \, \mathbf{i} + \cosh u \cos v \sin \phi \, \mathbf{j} + \sinh u \sin v \, \mathbf{k},$$

$$\mathbf{e}_1 = \frac{\partial \mathbf{r}}{\partial u} = \sinh u \cos v \cos \phi \, \mathbf{i} + \sinh u \cos v \sin \phi \, \mathbf{j} + \cosh u \sin v \, \mathbf{k},$$

$$\mathbf{e}_2 = \frac{\partial \mathbf{r}}{\partial v} = -\cosh u \sin v \cos \phi \, \mathbf{i} - \cosh u \sin v \sin \phi \, \mathbf{j} + \sinh u \cos v \, \mathbf{k},$$

$$\mathbf{e}_3 = \frac{\partial \mathbf{r}}{\partial \phi} = \cosh u \cos v(-\sin \phi \, \mathbf{i} + \cos \phi \, \mathbf{j}).$$

Now consider the scalar products:

$$\mathbf{e}_1 \cdot \mathbf{e}_2 = \sinh u \cos v \cosh u \sin v(-\cos^2 \phi - \sin^2 \phi + 1) = 0,$$

$$\mathbf{e}_1 \cdot \mathbf{e}_3 = \sinh u \cos^2 v \cosh u(-\sin \phi \cos \phi + \sin \phi \cos \phi) = 0,$$

$$\mathbf{e}_2 \cdot \mathbf{e}_3 = \cosh^2 u \sin v \cos v(\sin \phi \cos \phi - \sin \phi \cos \phi) = 0.$$

As each is zero, the system is an orthogonal one.

The scale factors are given by $|\mathbf{e}_i|$ and are thus found from:

$$\begin{aligned}
|\mathbf{e}_1|^2 &= \sinh^2 u \cos^2 v(\cos^2 \phi + \sin^2 \phi) + \cosh^2 u \sin^2 v \\
&= (\cosh^2 u - 1)\cos^2 v + \cosh^2 u(1 - \cos^2 v) \\
&= \cosh^2 u - \cos^2 v; \\
|\mathbf{e}_2|^2 &= \cosh^2 u \sin^2 v(\cos^2 \phi + \sin^2 \phi) + \sinh^2 u \cos^2 v \\
&= \cosh^2 u(1 - \cos^2 v) + (\cosh^2 u - 1)\cos^2 v \\
&= \cosh^2 u - \cos^2 v; \\
|\mathbf{e}_3|^2 &= \cosh^2 u \cos^2 v(\sin^2 \phi + \cos^2 \phi) = \cosh^2 u \cos^2 v.
\end{aligned}$$

The immediate deduction is that

$$h_u = h_v = (\cosh^2 u - \cos^2 v)^{1/2}, \qquad h_\phi = \cosh u \cos v.$$

An alternative form for h_u and h_v is $(\sinh^2 u + \sin^2 v)^{1/2}$.

If a solution of Laplace's equation is to be a function, $\psi(u)$, of u only, then all differentiation with respect to v and ϕ can be ignored. The expression for $\nabla^2 \psi$ reduces to

$$\begin{aligned}
\nabla^2 \psi &= \frac{1}{h_u h_v h_\phi} \left[\frac{\partial}{\partial u} \left(\frac{h_v h_\phi}{h_u} \frac{\partial \psi}{\partial u} \right) \right] \\
&= \frac{1}{\cosh u \cos v(\cosh^2 u - \cos^2 v)} \left[\frac{\partial}{\partial u} \left(\cosh u \cos v \frac{\partial \psi}{\partial u} \right) \right].
\end{aligned}$$

Laplace's equation itself is even simpler and reduces to

$$\frac{\partial}{\partial u} \left(\cosh u \frac{\partial \psi}{\partial u} \right) = 0.$$

This can be rewritten as

$$\frac{\partial \psi}{\partial u} = \frac{k}{\cosh u} = \frac{2k}{e^u + e^{-u}} = \frac{2ke^u}{e^{2u} + 1},$$

$$d\psi = \frac{Ae^u\, du}{1 + (e^u)^2} \quad \Rightarrow \quad \psi = B\tan^{-1} e^u + c.$$

This is the most general function of u only that satisfies Laplace's equation.

Line, surface and volume integrals

> **11.1** *The vector field* **F** *is defined by*
>
> $$\mathbf{F} = 2xz\mathbf{i} + 2yz^2\mathbf{j} + (x^2 + 2y^2z - 1)\mathbf{k}.$$
>
> *Calculate* $\nabla \times \mathbf{F}$ *and deduce that* **F** *can be written* $F = \nabla\phi$. *Determine the form of* ϕ.

With **F** as given, we calculate the curl of **F** to see whether or not it is the zero vector:

$$\nabla \times \mathbf{F} = (4yz - 4yz,\ 2x - 2x,\ 0 - 0) = \mathbf{0}.$$

The fact that it is implies that **F** can be written as $\nabla\phi$ for some scalar ϕ.

The form of $\phi(x, y, z)$ is found by integrating, in turn, the components of **F** until consistency is achieved, i.e. until a ϕ is found that has partial derivatives equal to the corresponding components of **F**:

$$2xz = F_x = \frac{\partial\phi}{\partial x} \quad \Rightarrow \quad \phi(x, y, z) = x^2z + g(y, z),$$

$$2yz^2 = F_y = \frac{\partial}{\partial y}[x^2z + g(y, z)] \quad \Rightarrow \quad g(y, z) = y^2z^2 + h(z),$$

$$x^2 + 2y^2z - 1 = F_z = \frac{\partial}{\partial z}[x^2z + y^2z^2 + h(z)]$$

$$\Rightarrow \quad h(z) = -z + k.$$

Hence, to within an unimportant constant, the form of ϕ is

$$\phi(x, y, z) = x^2z + y^2z^2 - z.$$

> **11.3** *A vector field* **F** *is given by* $\mathbf{F} = xy^2\mathbf{i} + 2\mathbf{j} + x\mathbf{k}$ *and L is a path parameterised by $x = ct$, $y = c/t$, $z = d$ for the range $1 \leq t \leq 2$. Evaluate the three integrals*
>
> $$\text{(a)} \quad \int_L \mathbf{F}\, dt, \qquad \text{(b)} \quad \int_L \mathbf{F}\, dy, \qquad \text{(c)} \quad \int_L \mathbf{F} \cdot d\mathbf{r}.$$

Although all three integrals are along the same path L, they are not necessarily of the same type. The vector or scalar nature of the integral is determined by that of the integrand when it is expressed in a form containing the infinitesimal dt.

(a) This is a vector integral and contains three separate integrations. We express each of the integrands in terms of t, according to the parameterisation of the integration path L, before integrating:

$$\int_L \mathbf{F}\, dt = \int_1^2 \left(\frac{c^3}{t}\mathbf{i} + 2\mathbf{j} + ct\,\mathbf{k} \right) dt$$

$$= \left[c^3 \ln t\, \mathbf{i} + 2t\,\mathbf{j} + \frac{1}{2}ct^2\,\mathbf{k} \right]_1^2$$

$$= c^3 \ln 2\, \mathbf{i} + 2\,\mathbf{j} + \frac{3}{2}c\,\mathbf{k}.$$

(b) This is a similar vector integral but here we must also replace the infinitesimal dy by the infinitesimal $-c\,dt/t^2$ before integrating:

$$\int_L \mathbf{F}\, dy = \int_1^2 \left(\frac{c^3}{t}\mathbf{i} + 2\mathbf{j} + ct\,\mathbf{k} \right)\left(\frac{-c}{t^2} \right) dt$$

$$= \left[\frac{c^4}{2t^2}\mathbf{i} + \frac{2c}{t}\mathbf{j} - c^2 \ln t\, \mathbf{k} \right]_1^2$$

$$= -\frac{3c^4}{8}\mathbf{i} - c\,\mathbf{j} - c^2 \ln 2\, \mathbf{k}.$$

(c) This is a scalar integral and before integrating we must take the scalar product of **F** with $d\mathbf{r} = dx\,\mathbf{i} + dy\,\mathbf{j} + dz\,\mathbf{k}$ to give a single integrand:

$$\int_L \mathbf{F} \cdot d\mathbf{r} = \int_1^2 \left(\frac{c^3}{t}\mathbf{i} + 2\mathbf{j} + ct\,\mathbf{k} \right) \cdot \left(c\,\mathbf{i} - \frac{c}{t^2}\mathbf{j} + 0\,\mathbf{k} \right) dt$$

$$= \int_1^2 \left(\frac{c^4}{t} - \frac{2c}{t^2} \right) dt$$

$$= \left[c^4 \ln t + \frac{2c}{t} \right]_1^2$$

$$= c^4 \ln 2 - c.$$

11.5 *Determine the point of intersection P, in the first quadrant, of the two ellipses*

$$\frac{x^2}{a^2} + \frac{y^2}{b^2} = 1 \quad \text{and} \quad \frac{x^2}{b^2} + \frac{y^2}{a^2} = 1.$$

Taking $b < a$, consider the contour L that bounds the area in the first quadrant that is common to the two ellipses. Show that the parts of L that lie along the coordinate axes contribute nothing to the line integral around L of $x\,dy - y\,dx$. Using a parameterisation of each ellipse of the general form $x = X\cos\phi$ and $y = Y\sin\phi$, evaluate the two remaining line integrals and hence find the total area common to the two ellipses.

Note: The line integral of $x\,dy - y\,dx$ around a general closed convex contour is equal to twice the area enclosed by that contour.

From the symmetry of the equations under the interchange of x and y, the point P must have $x = y$. Thus,

$$x^2 \left(\frac{1}{a^2} + \frac{1}{b^2} \right) = 1 \quad \Rightarrow \quad x = \frac{ab}{(a^2 + b^2)^{1/2}}.$$

Denoting as curve C_1 the part of

$$\frac{x^2}{a^2} + \frac{y^2}{b^2} = 1$$

that lies on the boundary of the common region, we parameterise it by $x = a\cos\theta_1$ and $y = b\sin\theta_1$. Curve C_1 starts from P and finishes on the y-axis. At P,

$$a\cos\theta_1 = x = \frac{ab}{(a^2 + b^2)^{1/2}} \quad \Rightarrow \quad \tan\theta_1 = \frac{a}{b}.$$

It follows that θ_1 lies in the range $\tan^{-1}(a/b) \leq \theta_1 \leq \pi/2$. Note that θ_1 is *not* the angle between the x-axis and the line joining the origin O to the corresponding point on the curve; for example, when the point is P itself then $\theta_1 = \tan^{-1} a/b$, whilst the line OP makes an angle of $\pi/4$ with the x-axis.

Similarly, referring to that part of

$$\frac{x^2}{b^2} + \frac{y^2}{a^2} = 1$$

that lies on the boundary of the common region as curve C_2, we parameterise it by $x = b\cos\theta_2$ and $y = a\sin\theta_2$ with $0 \leq \theta_2 \leq \tan^{-1}(b/a)$.

On the x-axis, both y and dy are zero and the integrand, $x\,dy - y\,dx$, vanishes.

Similarly, the integrand vanishes at all points on the y-axis. Hence,

$$
\begin{aligned}
I &= \oint_L (x\,dy - y\,dx) \\
&= \int_{C_2} (x\,dy - y\,dx) + \int_{C_1} (x\,dy - y\,dx) \\
&= \int_0^{\tan^{-1}(b/a)} [ab(\cos\theta_2 \cos\theta_2) - ab\sin\theta_2(-\sin\theta_2)]\,d\theta_2 \\
&\quad + \int_{\tan^{-1}(a/b)}^{\pi/2} [ab(\cos\theta_1 \cos\theta_1) - ab\sin\theta_1(-\sin\theta_1)]\,d\theta_1 \\
&= ab\tan^{-1}\frac{b}{a} + ab\left(\frac{\pi}{2} - \tan^{-1}\frac{a}{b}\right) \\
&= 2ab\tan^{-1}\frac{b}{a}.
\end{aligned}
$$

As noted in the question, the area enclosed by L is equal to $\frac{1}{2}$ of this value, i.e. the total common area in all four quadrants is

$$
4 \times \frac{1}{2} \times 2ab\tan^{-1}\frac{b}{a} = 4ab\tan^{-1}\frac{b}{a}.
$$

Note that if we let $b \to a$ then the two ellipses become identical circles and we recover the expected value of πa^2 for their common area.

11.7 *Evaluate the line integral*

$$
I = \oint_C \left[y(4x^2 + y^2)\,dx + x(2x^2 + 3y^2)\,dy \right]
$$

around the ellipse $x^2/a^2 + y^2/b^2 = 1$.

As it stands this integral is complicated and, in fact, it is the sum of two integrals. The form of the integrand, containing powers of x and y that can be differentiated easily, makes this problem one to which Green's theorem in a plane might usefully be applied. The theorem states that

$$
\oint_C (P\,dx + Q\,dy) = \int\int_R \left(\frac{\partial Q}{\partial x} - \frac{\partial P}{\partial y} \right) dx\,dy,
$$

where C is a closed contour enclosing the convex region R.

In the notation used above,

$$
P(x, y) = y(4x^2 + y^2) \quad \text{and} \quad Q(x, y) = x(2x^2 + 3y^2).
$$

It follows that

$$
\frac{\partial P}{\partial y} = 4x^2 + 3y^2 \quad \text{and} \quad \frac{\partial Q}{\partial x} = 6x^2 + 3y^2,
$$

179

leading to

$$\frac{\partial Q}{\partial x} - \frac{\partial P}{\partial y} = 2x^2.$$

This can now be substituted into Green's theorem and the y-integration carried out immediately as the integrand does not contain y. Hence,

$$
\begin{aligned}
I &= \int \int_R 2x^2 \, dx \, dy \\
&= \int_{-a}^{a} 2x^2 \, 2b \left(1 - \frac{x^2}{a^2} \right)^{1/2} dx \\
&= 4b \int_{\pi}^{0} a^2 \cos^2 \phi \, \sin \phi \, (-a \sin \phi \, d\phi), \text{ on setting } x = a \cos \phi, \\
&= -ba^3 \int_{\pi}^{0} \sin^2(2\phi) \, d\phi = \tfrac{1}{2}\pi b a^3.
\end{aligned}
$$

In the final line we have used the standard result for the integral of the square of a sinusoidal function.

11.9 *A single-turn coil C of arbitrary shape is placed in a magnetic field* **B** *and carries a current I. Show that the couple acting upon the coil can be written as*

$$\mathbf{M} = I \int_C (\mathbf{B} \cdot \mathbf{r}) \, d\mathbf{r} - I \int_C \mathbf{B}(\mathbf{r} \cdot d\mathbf{r}).$$

For a planar rectangular coil of sides 2a and 2b placed with its plane vertical and at an angle ϕ to a uniform horizontal field **B**, *show that* **M** *is, as expected,* $4abBI \cos \phi \, \mathbf{k}$.

For an arbitrarily shaped coil the total couple acting can only be found by considering that on an infinitesimal element and then integrating this over the whole coil. The force on an element $d\mathbf{r}$ of the coil is $d\mathbf{F} = I \, d\mathbf{r} \times \mathbf{B}$, and the moment of this force about the origin is $d\mathbf{M} = \mathbf{r} \times \mathbf{F}$. Thus the total moment is given by

$$
\begin{aligned}
\mathbf{M} &= \oint_C \mathbf{r} \times (I \, d\mathbf{r} \times \mathbf{B}) \\
&= I \oint_C (\mathbf{r} \cdot \mathbf{B}) \, d\mathbf{r} - I \oint_C \mathbf{B}(\mathbf{r} \cdot d\mathbf{r}).
\end{aligned}
$$

To obtain this second form we have used the vector identity

$$\mathbf{a} \times (\mathbf{b} \times \mathbf{c}) = (\mathbf{a} \cdot \mathbf{c})\mathbf{b} - (\mathbf{a} \cdot \mathbf{b})\mathbf{c}.$$

To determine the couple acting on the rectangular coil we work in Cartesian coordinates with the z-axis vertical and choose the orientation of axes in the horizontal plane such that the edge of the rectangle of length $2a$ is in the x-direction. Then

$$\mathbf{B} = B \cos \phi \, \mathbf{i} + B \sin \phi \, \mathbf{j}.$$

In the first term in $\mathbf{M}$,
(i) for the horizontal sides

$$\mathbf{r} = x \mathbf{i} \pm b \mathbf{k}, \quad d\mathbf{r} = dx \, \mathbf{i}, \quad \mathbf{r} \cdot \mathbf{B} = xB \cos \phi,$$

$$\int (\mathbf{r} \cdot \mathbf{B}) \, d\mathbf{r} = B \cos \phi \, \mathbf{i} \left(\int_{-a}^{a} x \, dx + \int_{a}^{-a} x \, dx \right) = \mathbf{0};$$

(ii) for the vertical sides

$$\mathbf{r} = \pm a \mathbf{i} + z \mathbf{k}, \quad d\mathbf{r} = dz \, \mathbf{k}, \quad \mathbf{r} \cdot \mathbf{B} = \pm aB \cos \phi,$$

$$\int (\mathbf{r} \cdot \mathbf{B}) \, d\mathbf{r} = B \cos \phi \, \mathbf{k} \left(\int_{-b}^{b} (+a) \, dz + \int_{b}^{-b} (-a) \, dz \right) = 4abB \cos \phi \, \mathbf{k}.$$

For the second term in $\mathbf{M}$, since the field is uniform it can be taken outside the integral as a (vector) constant. On the horizontal sides the remaining integral is

$$\int \mathbf{r} \cdot d\mathbf{r} = \pm \int_{-a}^{a} x \, dx = 0.$$

Similarly, the contribution from the vertical sides vanishes and the whole of the second term contributes nothing in this particular configuration.

The total moment is thus $4abB \cos \phi \, \mathbf{k}$, as expected.

11.11 *An axially symmetric solid body with its axis AB vertical is immersed in an incompressible fluid of density ρ_0. Use the following method to show that, whatever the shape of the body, for $\rho = \rho(z)$ in cylindrical polars the Archimedean upthrust is, as expected, $\rho_0 g V$, where V is the volume of the body.*

Express the vertical component of the resultant force $(-\int p \, d\mathbf{S}$, where p is the pressure) on the body in terms of an integral; note that $p = -\rho_0 g z$ and that for an annular surface element of width dl, $\mathbf{n} \cdot \mathbf{n}_z \, dl = -d\rho$. Integrate by parts and use the fact that $\rho(z_A) = \rho(z_B) = 0$.

We measure z negatively from the water's surface $z = 0$ so that the hydrostatic pressure is $p = -\rho_0 g z$. By symmetry, there is no net horizontal force acting on the body.

The upward force, F, is due to the net vertical component of the hydrostatic pressure acting upon the body's surface:

$$F = -\hat{\mathbf{n}}_z \cdot \int p \, d\mathbf{S}$$

$$= -\hat{\mathbf{n}}_z \cdot \int (-\rho_0 g z)(2\pi\rho \, \hat{\mathbf{n}} \, dl),$$

where $2\pi\rho \, dl$ is the area of the strip of surface lying between z and $z + dz$ and $\hat{\mathbf{n}}$ is the outward unit normal to that surface.

Now, from geometry, $\hat{\mathbf{n}}_z \cdot \hat{\mathbf{n}}$ is equal to minus the sine of the angle between dl and dz and so $\hat{\mathbf{n}}_z \cdot \hat{\mathbf{n}} \, dl$ is equal to $-d\rho$. Thus,

$$F = 2\pi\rho_0 g \int_{z_A}^{z_B} \rho z (-d\rho)$$

$$= -2\pi\rho_0 g \int_{z_A}^{z_B} \left(\rho \frac{\partial\rho}{\partial z} \right) z \, dz$$

$$= -2\pi\rho_0 g \left\{ \left[z \frac{\rho^2}{2} \right]_{z_A}^{z_B} - \int_{z_A}^{z_B} \frac{\rho^2}{2} \, dz \right\}.$$

But $\rho(z_A) = \rho(z_B) = 0$, and so the first contribution vanishes, leaving

$$F = \rho_0 g \int_{z_A}^{z_B} \pi\rho^2 \, dz = \rho_0 g V,$$

where V is the volume of the solid. This is the mathematical form of Archimedes' principle. Of course, the result is also valid for a closed body of arbitrary shape, $\rho = \rho(z, \phi)$, but a different method would be needed to prove it.

11.13 *A vector field* $\mathbf{a}$ *is given by* $-zxr^{-3}\mathbf{i} - zyr^{-3}\mathbf{j} + (x^2 + y^2)r^{-3}\mathbf{k}$, *where* $r^2 = x^2 + y^2 + z^2$. *Establish that the field is conservative* (a) *by showing that* $\nabla \times \mathbf{a} = \mathbf{0}$, *and* (b) *by constructing its potential function* ϕ.

We are told that

$$\mathbf{a} = -\frac{zx}{r^3}\mathbf{i} - \frac{zy}{r^3}\mathbf{j} + \frac{x^2 + y^2}{r^3}\mathbf{k},$$

with $r^2 = x^2 + y^2 + z^2$. We will need to differentiate r^{-3} with respect to x, y and z, using the chain rule, and so note that $\partial r / \partial x = x/r$, etc.

(a) Consider $\nabla \times \mathbf{a}$, term-by-term:

$$[\nabla \times \mathbf{a}]_x = \frac{\partial}{\partial y}\left(\frac{x^2 + y^2}{r^3}\right) - \frac{\partial}{\partial z}\left(\frac{-zy}{r^3}\right)$$

$$= \frac{-3(x^2 + y^2)y}{r^4 r} + \frac{2y}{r^3} + \frac{y}{r^3} - \frac{3(zy)z}{r^4 r}$$

$$= \frac{3y}{r^5}(-x^2 - y^2 + x^2 + y^2 + z^2 - z^2) = 0;$$

$$[\nabla \times \mathbf{a}]_y = \frac{\partial}{\partial z}\left(\frac{-zx}{r^3}\right) - \frac{\partial}{\partial x}\left(\frac{x^2 + y^2}{r^3}\right)$$

$$= \frac{3(zx)z}{r^4 r} - \frac{x}{r^3} - \frac{2x}{r^3} + \frac{3(x^2 + y^2)x}{r^4 r}$$

$$= \frac{3x}{r^5}(z^2 - x^2 - y^2 - z^2 + x^2 + y^2) = 0;$$

$$[\nabla \times \mathbf{a}]_z = \frac{\partial}{\partial x}\left(\frac{-zy}{r^3}\right) - \frac{\partial}{\partial y}\left(\frac{-zx}{r^3}\right)$$

$$= \frac{3(zy)x}{r^4 r} - \frac{3(zx)y}{r^4 r} = 0.$$

Thus all three components of $\nabla \times \mathbf{a}$ are zero, showing that $\mathbf{a}$ is a conservative field.

(b) To construct its potential function we proceed as follows:

$$\frac{\partial \phi}{\partial x} = \frac{-zx}{(x^2 + y^2 + z^2)^{3/2}} \Rightarrow \phi = \frac{z}{(x^2 + y^2 + z^2)^{1/2}} + f(y, z),$$

$$\frac{\partial \phi}{\partial y} = \frac{-zy}{(x^2 + y^2 + z^2)^{3/2}} = \frac{-zy}{(x^2 + y^2 + z^2)^{3/2}} + \frac{\partial f}{\partial y} \Rightarrow f(y, z) = g(z),$$

$$\frac{\partial \phi}{\partial z} = \frac{x^2 + y^2}{(x^2 + y^2 + z^2)^{3/2}}$$

$$= \frac{1}{(x^2 + y^2 + z^2)^{1/2}} + \frac{-z\,z}{(x^2 + y^2 + z^2)^{3/2}} + \frac{\partial g}{\partial z}$$

$$\Rightarrow g(z) = c.$$

Thus,

$$\phi(x, y, z) = c + \frac{z}{(x^2 + y^2 + z^2)^{1/2}} = c + \frac{z}{r}.$$

The very fact that we can construct a potential function $\phi = \phi(x, y, z)$ whose derivatives are the components of the vector field shows that the field is conservative.

11.15 *A force* $\mathbf{F}(\mathbf{r})$ *acts on a particle at* $\mathbf{r}$. *In which of the following cases can* $\mathbf{F}$ *be represented in terms of a potential? Where it can, find the potential.*

(a) $\mathbf{F} = F_0 \left[\mathbf{i} - \mathbf{j} - \dfrac{2(x-y)}{a^2} \mathbf{r} \right] \exp\left(-\dfrac{r^2}{a^2} \right)$;

(b) $\mathbf{F} = \dfrac{F_0}{a} \left[z\mathbf{k} + \dfrac{(x^2 + y^2 - a^2)}{a^2} \mathbf{r} \right] \exp\left(-\dfrac{r^2}{a^2} \right)$;

(c) $\mathbf{F} = F_0 \left[\mathbf{k} + \dfrac{a(\mathbf{r} \times \mathbf{k})}{r^2} \right]$.

(a) We first write the field entirely in terms of the Cartesian unit vectors using $\mathbf{r} = x\mathbf{i} + y\mathbf{j} + z\mathbf{k}$ and then attempt to construct a suitable potential function ϕ:

$$\mathbf{F} = F_0 \left[\mathbf{i} - \mathbf{j} - \frac{2(x-y)}{a^2}\mathbf{r} \right] \exp\left(-\frac{r^2}{a^2} \right)$$

$$= \frac{F_0}{a^2} \left[(a^2 - 2x^2 + 2xy)\mathbf{i} + (-a^2 - 2xy + 2y^2)\mathbf{j} \right.$$

$$\left. + \quad (-2xz + 2yz)\mathbf{k} \right] \exp\left(-\frac{r^2}{a^2} \right).$$

Since the partial derivative of $\exp(-r^2/a^2)$ with respect to any Cartesian coordinate u is $\exp(-r^2/a^2)(-2r/a^2)(u/r)$, the z-component of $\mathbf{F}$ appears to be the most straightforward to tackle first:

$$\frac{\partial \phi}{\partial z} = \frac{F_0}{a^2}(-2xz + 2yz)\exp\left(-\frac{r^2}{a^2} \right)$$

$$\Rightarrow \phi(x, y, z) = F_0(x - y)\exp\left(-\frac{r^2}{a^2} \right) + f(x, y)$$

$$\equiv \phi_1(x, y, z) + f(x, y).$$

Next we examine the derivatives of $\phi = \phi_1 + f$ with respect to x and y to see how closely they generate F_x and F_y:

$$\frac{\partial \phi_1}{\partial x} = F_0 \left[\exp\left(-\frac{r^2}{a^2} \right) + (x - y)\exp\left(-\frac{r^2}{a^2} \right)\left(\frac{-2x}{a^2} \right) \right]$$

$$= \frac{F_0}{a^2}(a^2 - 2x^2 + 2xy)\exp(-r^2/a^2) = F_x \quad \text{(as given)},$$

and $\dfrac{\partial \phi_1}{\partial y} = F_0 \left[-\exp\left(-\frac{r^2}{a^2} \right) + (x - y)\exp\left(-\frac{r^2}{a^2} \right)\left(\frac{-2y}{a^2} \right) \right]$

$$= \frac{F_0}{a^2}(-a^2 - 2xy + 2y^2)\exp(-r^2/a^2) = F_y \quad \text{(as given)}.$$

Thus, to within an arbitrary constant, $\phi_1(x, y, z) = F_0(x - y)\exp\left(-\dfrac{r^2}{a^2} \right)$ is a suitable potential function for the field, without the need for any additional function $f(x, y)$.

(b) We follow the same line of argument as in part (a). First, expressing $\mathbf{F}$ in terms of $\mathbf{i}$, $\mathbf{j}$ and $\mathbf{k}$,

$$\mathbf{F} = \frac{F_0}{a} \left[z\,\mathbf{k} + \frac{x^2 + y^2 - a^2}{a^2} \mathbf{r} \right] \exp\left(-\frac{r^2}{a^2}\right)$$

$$= \frac{F_0}{a^3} \left[x(x^2 + y^2 - a^2)\mathbf{i} + y(x^2 + y^2 - a^2)\mathbf{j} \right.$$

$$\left. + z(x^2 + y^2)\mathbf{k} \right] \exp\left(-\frac{r^2}{a^2}\right),$$

and then constructing a possible potential function ϕ. Again starting with the z-component:

$$\frac{\partial \phi}{\partial z} = \frac{F_0 z}{a^3}(x^2 + y^2) \exp\left(-\frac{r^2}{a^2}\right),$$

$$\Rightarrow \quad \phi(x, y, z) = -\frac{F_0}{2a}(x^2 + y^2) \exp\left(-\frac{r^2}{a^2}\right) + f(x, y)$$

$$\equiv \phi_1(x, y, z) + f(x, y),$$

Then, $\quad \dfrac{\partial \phi_1}{\partial x} = -\dfrac{F_0}{2a} \left[2x - \dfrac{2x(x^2 + y^2)}{a^2} \right] \exp\left(-\dfrac{r^2}{a^2}\right) = F_x \quad$ (as given),

and $\quad \dfrac{\partial \phi_1}{\partial y} = -\dfrac{F_0}{2a} \left[2y - \dfrac{2y(x^2 + y^2)}{a^2} \right] \exp\left(-\dfrac{r^2}{a^2}\right) = F_y \quad$ (as given).

Thus, $\phi_1(x, y, z) = \dfrac{F_0}{2a}(x^2 + y^2) \exp\left(-\dfrac{r^2}{a^2}\right)$, as it stands, is a suitable potential function for $\mathbf{F}(\mathbf{r})$ and establishes the conservative nature of the field.

(c) Again we express F in Cartesian components:

$$\mathbf{F} = F_0 \left[\mathbf{k} + \frac{a(\mathbf{r} \times \mathbf{k})}{r^2} \right] = \frac{ay}{r^2}\mathbf{i} - \frac{ax}{r^2}\mathbf{j} + \mathbf{k}.$$

That the z-component of $\mathbf{F}$ has no dependence on y whilst its y-component does depend upon z suggests that the x-component of $\nabla \times \mathbf{F}$ may not be zero. To test this out we compute

$$(\nabla \times \mathbf{F})_x = \frac{\partial(1)}{\partial y} - \frac{\partial}{\partial z}\left(\frac{-ax}{r^2}\right) = 0 - \frac{2axz}{r^4} \neq 0,$$

and find that it is not. To have even one component of $\nabla \times \mathbf{F}$ non-zero is sufficient to show that $\mathbf{F}$ is not conservative and that no potential function can be found. There is no point in searching further!

The same conclusion can be reached by considering the implication of $F_z = \mathbf{k}$, namely that any possible potential function has to have the form $\phi(x, y, z) =$

$z + f(x, y)$. However, $\partial\phi/\partial x$ is known to be $-ay/r^2 = -ay/(x^2 + y^2 + z^2)$. This yields a contradiction, as it requires $\partial f(x, y)/\partial x$ to depend on z, which is clearly impossible.

11.17 *The vector field* **f** *has components* $y\mathbf{i} - x\mathbf{j} + \mathbf{k}$ *and* γ *is a curve given parametrically by*

$$\mathbf{r} = (a - c + c\cos\theta)\mathbf{i} + (b + c\sin\theta)\mathbf{j} + c^2\theta\mathbf{k}, \quad 0 \le \theta \le 2\pi.$$

Describe the shape of the path γ *and show that the line integral* $\int_\gamma \mathbf{f} \cdot d\mathbf{r}$ *vanishes. Does this result imply that* **f** *is a conservative field?*

As θ increases from 0 to 2π, the x- and y-components of **r** vary sinusoidally and in quadrature about fixed values $a - c$ and b. Both variations have amplitude c and both return to their initial values when $\theta = 2\pi$. However, the z-component increases monotonically from 0 to a value of $2\pi c^2$. The curve γ is therefore one loop of a circular spiral of radius c and pitch $2\pi c^2$. Its axis is parallel to the z-axis and passes through the points $(a - c, b, z)$.

The line element $d\mathbf{r}$ has components $(-c\sin\theta\, d\theta, c\cos\theta\, d\theta, c^2\, d\theta)$ and so the line integral of f along γ is given by

$$\int_\gamma \mathbf{f} \cdot d\mathbf{r} = \int_0^{2\pi} \left[y(-c\sin\theta) - x(c\cos\theta) + c^2 \right] d\theta$$

$$= \int_0^{2\pi} \left[-c(b + c\sin\theta)\sin\theta - c(a - c + c\cos\theta)\cos\theta + c^2 \right] d\theta$$

$$= \int_0^{2\pi} \left(-bc\sin\theta - c^2\sin^2\theta - c(a - c)\cos\theta - c^2\cos^2\theta + c^2 \right) d\theta$$

$$= 0 - \pi c^2 - 0 - \pi c^2 + 2\pi c^2 = 0.$$

However, this does not imply that **f** is a conservative field since (i) γ is not a closed loop, and (ii) even if it were, the line integral has to vanish for *every* loop, not just for a particular one.

Further,

$$\nabla \times \mathbf{f} = (0 - 0,\ 0 - 0,\ -1 - 1) = (0,\ 0,\ -2) \ne \mathbf{0},$$

showing explicitly that **f** is not conservative.

186

11.19 *Evaluate the surface integral $\int \mathbf{r} \cdot d\mathbf{S}$, where $\mathbf{r}$ is the position vector, over that part of the surface $z = a^2 - x^2 - y^2$ for which $z \geq 0$, by each of the following methods.*

(a) *Parameterise the surface as $x = a \sin\theta \cos\phi$, $y = a \sin\theta \sin\phi$, $z = a^2 \cos^2\theta$, and show that*

$$\mathbf{r} \cdot d\mathbf{S} = a^4 (2 \sin^3\theta \cos\theta + \cos^3\theta \sin\theta)\, d\theta\, d\phi.$$

(b) *Apply the divergence theorem to the volume bounded by the surface and the plane $z = 0$.*

(a) With $x = a \sin\theta \cos\phi$, $y = a \sin\theta \sin\phi$, $z = a^2 \cos^2\theta$, we first check that this does parameterise the surface appropriately:

$$a^2 - x^2 - y^2 = a^2 - a^2 \sin^2\theta(\cos^2\phi + \sin^2\phi) = a^2(1 - \sin^2\theta) = a^2 \cos^2\theta = z.$$

We see that it does so for the relevant part of the surface, i.e. that which lies above the plane $z = 0$ with $0 \leq \theta \leq \pi/2$. It would not do so for the part with $z < 0$ for which $x^2 + y^2$ has to be greater than a^2; this is not catered for by the given parameterisation.

Having carried out this check, we calculate expressions for $d\mathbf{S}$ and hence $\mathbf{r} \cdot d\mathbf{S}$ in terms of θ and ϕ as follows:

$$\mathbf{r} = a \sin\theta \cos\phi\, \mathbf{i} + a \sin\theta \sin\phi\, \mathbf{j} + a^2 \cos^2\theta\, \mathbf{k},$$

and the tangent vectors at the point (θ, ϕ) on the surface are given by

$$\frac{\partial \mathbf{r}}{\partial \theta} = a \cos\theta \cos\phi\, \mathbf{i} + a \cos\theta \sin\phi\, \mathbf{j} - 2a^2 \cos\theta \sin\theta\, \mathbf{k},$$

$$\frac{\partial \mathbf{r}}{\partial \phi} = -a \sin\theta \sin\phi\, \mathbf{i} + a \sin\theta \cos\phi\, \mathbf{j}.$$

The corresponding vector element of surface area is thus

$$d\mathbf{S} = \frac{\partial \mathbf{r}}{\partial \theta} \times \frac{\partial \mathbf{r}}{\partial \phi}$$

$$= 2a^3 \cos\theta \sin^2\theta \cos\phi\, \mathbf{i} + 2a^3 \cos\theta \sin^2\theta \sin\phi\, \mathbf{j} + a^2 \cos\theta \sin\theta\, \mathbf{k},$$

giving $\mathbf{r} \cdot d\mathbf{S}$ as

$$\mathbf{r} \cdot d\mathbf{S} = 2a^4 \cos\theta \sin^3\theta \cos^2\phi + 2a^4 \cos\theta \sin^3\theta \sin^2\phi + a^4 \cos^3\theta \sin\theta$$

$$= 2a^4 \cos\theta \sin^3\theta + a^4 \cos^3\theta \sin\theta.$$

This is to be integrated over the ranges $0 \leq \phi < 2\pi$ and $0 \leq \theta \leq \pi/2$ as follows:

$$\int \mathbf{r} \cdot d\mathbf{S} = a^4 \int_0^{2\pi} d\phi \int_0^{\pi/2} (2\sin^3 \theta \cos \theta + \cos^3 \theta \sin \theta)\, d\theta$$

$$= 2\pi a^4 \left(2 \left[\frac{\sin^4 \theta}{4} \right]_0^{\pi/2} + \left[\frac{-\cos^4 \theta}{4} \right]_0^{\pi/2} \right)$$

$$= 2\pi a^4 \left(\frac{2}{4} + \frac{1}{4} \right) = \frac{3\pi a^4}{2}.$$

(b) The divergence of the vector field $\mathbf{r}$ is 3, a constant, and so the surface integral $\int \mathbf{r} \cdot d\mathbf{S}$ taken over the complete surface Σ (including the part that lies in the plane $z = 0$) is, by the divergence theorem, equal to three times the volume V of the region bounded by Σ. Now,

$$V = \int_0^{a^2} \pi \rho^2 \, dz = \int_0^{a^2} \pi (a^2 - z) \, dz = \pi (a^4 - \tfrac{1}{2}a^4) = \tfrac{1}{2}\pi a^4,$$

and so $\int_\Sigma \mathbf{r} \cdot d\mathbf{S} = 3\pi a^4/2$.

However, on the part of the surface lying in the plane $z = 0$, $\mathbf{r} = x\,\mathbf{i} + y\,\mathbf{j} + 0\,\mathbf{k}$, whilst $d\mathbf{S} = -dS\,\mathbf{k}$. Consequently the scalar product $\mathbf{r} \cdot d\mathbf{S} = 0$; in words, for any point on this face its position vector is orthogonal to the normal to the face. The surface integral over this face therefore contributes nothing to the total integral and the value obtained is that due to the curved surface alone, in agreement with the result in (a).

11.21 *Use the result*

$$\int_V \nabla \phi \, dV = \oint_S \phi \, d\mathbf{S},$$

together with an appropriately chosen scalar function ϕ, to prove that the position vector $\bar{\mathbf{r}}$ of the centre of mass of an arbitrarily shaped body of volume V and uniform density can be written

$$\bar{\mathbf{r}} = \frac{1}{V} \oint_S \tfrac{1}{2} r^2 \, d\mathbf{S}.$$

The position vector of the centre of mass is defined by

$$\bar{\mathbf{r}} \int_V \rho \, dV = \int_V \mathbf{r}\rho \, dV.$$

Now, we note that $\mathbf{r}$ can be written as $\nabla(\tfrac{1}{2}r^2)$. Thus, cancelling the constant ρ, we have

$$\bar{\mathbf{r}} V = \int_V \nabla(\tfrac{1}{2}r^2)\, dV n = \oint_S \tfrac{1}{2}r^2\, d\mathbf{S}$$
$$\Rightarrow \quad \bar{\mathbf{r}} = \frac{1}{V}\oint_S \tfrac{1}{2}r^2\, d\mathbf{S}.$$

This result provides an alternative method of finding the centre of mass $\bar{z}\mathbf{k}$ of the uniform hemisphere $r = a,\ 0 \le \theta \le \pi/2,\ 0 \le \phi < 2\pi$. The curved surface contributes $3a/4$ to $\bar{z}$ and the plane surface contributes $-3a/8$, giving $\bar{z} = 3a/8$.

11.23 *Demonstrate the validity of the divergence theorem:*

(a) *by calculating the flux of the vector*

$$\mathbf{F} = \frac{\alpha\mathbf{r}}{(r^2 + a^2)^{3/2}}$$

through the spherical surface $|\mathbf{r}| = \sqrt{3}a$;
(b) *by showing that*

$$\nabla \cdot \mathbf{F} = \frac{3\alpha a^2}{(r^2 + a^2)^{5/2}}$$

and evaluating the volume integral of $\nabla \cdot \mathbf{F}$ over the interior of the sphere $|\mathbf{r}| = \sqrt{3}a$. The substitution $r = a\tan\theta$ will prove useful in carrying out the integration.

(a) The field is radial with

$$\mathbf{F} = \frac{\alpha\,\mathbf{r}}{(r^2 + a^2)^{3/2}} = \frac{\alpha\,r}{(r^2 + a^2)^{3/2}}\hat{\mathbf{e}}_r.$$

The total flux is therefore given by

$$\Phi = \frac{4\pi r^2\,\alpha\,r}{(r^2 + a^2)^{3/2}}\bigg|_{r=a\sqrt{3}} = \frac{4\pi a^3\,\alpha\,3\sqrt{3}}{8a^3} = \frac{3\sqrt{3}\pi\alpha}{2}.$$

(b) From the divergrence theorem, the total flux over the surface of the sphere is equal to the volume integral of its divergence within the sphere. The divergence is given by

$$\nabla \cdot \mathbf{F} = \frac{1}{r^2}\frac{\partial}{\partial r}(r^2 F_r) = \frac{1}{r^2}\frac{\partial}{\partial r}\left(\frac{r^2\,\alpha\,r}{(r^2 + a^2)^{3/2}}\right)$$
$$= \frac{1}{r^2}\left[\frac{3\alpha r^2}{(r^2 + a^2)^{3/2}} - \frac{3\alpha r^4}{(r^2 + a^2)^{5/2}}\right]$$
$$= \frac{3\alpha a^2}{(r^2 + a^2)^{5/2}},$$

189

and on integrating over the sphere, we have

$$\int_V \nabla \cdot \mathbf{F} \, dV = \int_0^{\sqrt{3}a} \frac{3\alpha a^2}{(r^2 + a^2)^{5/2}} 4\pi r^2 \, dr, \text{ set } r = a \tan \theta, \, 0 \leq \theta \leq \tfrac{\pi}{3},$$

$$= 12\pi \alpha a^2 \int_0^{\pi/3} \frac{a^2 \tan^2 \theta \, a \sec^2 \theta}{a^5 \sec^5 \theta} \, d\theta$$

$$= 12\pi \alpha \int_0^{\pi/3} \sin^2 \theta \cos \theta \, d\theta$$

$$= 12\pi \alpha \left[\frac{\sin^3 \theta}{3} \right]_0^{\pi/3} = 12\pi \alpha \frac{\sqrt{3}}{8} = \frac{3\sqrt{3}\pi \alpha}{2}, \quad \text{as in (a).}$$

The equality of the results in parts (a) and (b) is in accordance with the divergence theorem.

11.25 *In a uniform conducting medium with unit relative permittivity, charge density ρ, current density $\mathbf{J}$, electric field $\mathbf{E}$ and magnetic field $\mathbf{B}$, Maxwell's electromagnetic equations take the form (with $\mu_0 \epsilon_0 = c^{-2}$)*

> (i) $\nabla \cdot \mathbf{B} = 0$, (ii) $\nabla \cdot \mathbf{E} = \rho/\epsilon_0$,

> (iii) $\nabla \times \mathbf{E} + \dot{\mathbf{B}} = 0$, (iv) $\nabla \times \mathbf{B} - (\dot{\mathbf{E}}/c^2) = \mu_0 \mathbf{J}$.

The density of stored energy in the medium is given by $\tfrac{1}{2}(\epsilon_0 E^2 + \mu_0^{-1} B^2)$. Show that the rate of change of the total stored energy in a volume V is equal to

$$- \int_V \mathbf{J} \cdot \mathbf{E} \, dV - \frac{1}{\mu_0} \oint_S (\mathbf{E} \times \mathbf{B}) \cdot d\mathbf{S},$$

where S is the surface bounding V.

[The first integral gives the ohmic heating loss, whilst the second gives the electromagnetic energy flux out of the bounding surface. The vector $\mu_0^{-1}(\mathbf{E} \times \mathbf{B})$ is known as the Poynting vector.]

The total stored energy is equal to the volume integral of the energy density. Let R be its rate of change. Then, differentiating under the integral sign, we have

$$R = \frac{d}{dt} \int_V \left(\frac{\epsilon_0}{2} E^2 + \frac{1}{2\mu_0} B^2 \right) dV$$

$$= \int_V \left(\epsilon_0 \mathbf{E} \cdot \dot{\mathbf{E}} + \frac{1}{\mu_0} \mathbf{B} \cdot \dot{\mathbf{B}} \right) dV.$$

Now using (iv) and (iii), we have

$$R = \int_V \left[\epsilon_0 \mathbf{E} \cdot (-\mu_0 c^2 \mathbf{J} + c^2 \nabla \times \mathbf{B}) - \frac{1}{\mu_0} \mathbf{B} \cdot (\nabla \times \mathbf{E}) \right] dV$$

$$= -\int_V \mathbf{E} \cdot \mathbf{J} \, dV + \int_V \left[\epsilon_0 c^2 \, \mathbf{E} \cdot (\nabla \times \mathbf{B}) - \frac{1}{\mu_0} \mathbf{B} \cdot (\nabla \times \mathbf{E}) \right] dV$$

$$= -\int_V \mathbf{E} \cdot \mathbf{J} \, dV - \frac{1}{\mu_0} \int_V \nabla \cdot (\mathbf{E} \times \mathbf{B}) \, dV$$

$$= -\int_V \mathbf{E} \cdot \mathbf{J} \, dV - \frac{1}{\mu_0} \oint_S (\mathbf{E} \times \mathbf{B}) \cdot d\mathbf{S}, \quad \text{by the divergence theorem.}$$

To obtain the penultimate line we used the vector identity

$$\nabla \cdot (\mathbf{a} \times \mathbf{b}) = \mathbf{b} \cdot (\nabla \times \mathbf{a}) - \mathbf{a} \cdot (\nabla \times \mathbf{b}).$$

11.27 *The vector field* **F** *is given by*

$$\mathbf{F} = (3x^2 yz + y^3 z + xe^{-x})\mathbf{i} + (3xy^2 z + x^3 z + ye^x)\mathbf{j} + (x^3 y + y^3 x + xy^2 z^2)\mathbf{k}.$$

Calculate (a) *directly, and* (b) *by using Stokes' theorem the value of the line integral* $\int_L \mathbf{F} \cdot d\mathbf{r}$, *where* L *is the (three-dimensional) closed contour* $OABCDEO$ *defined by the successive vertices* $(0,0,0)$, $(1,0,0)$, $(1,0,1)$, $(1,1,1)$, $(1,1,0)$, $(0,1,0)$, $(0,0,0)$.

(a) This calculation is a piece-wise evaluation of the line integral, made up of a series of scalar products of the length of a straight piece of the contour and the component of **F** parallel to it (integrated if that component varies along the particular straight section).

On OA, $y = z = 0$ and $F_x = xe^{-x}$;

$$I_1 = \int_0^1 xe^{-x} \, dx = \left[-xe^{-x} \right]_0^1 + \int_0^1 e^{-x} \, dx = 1 - 2e^{-1}.$$

On AB, $x = 1$ and $y = 0$ and $F_z = 0$; the integral I_2 is zero.

On BC, $x = 1$ and $z = 1$ and $F_y = 3y^2 + 1 + ey$;

$$I_3 = \int_0^1 (3y^2 + 1 + ey) \, dy = 1 + 1 + \tfrac{1}{2}e.$$

On CD, $x = 1$ and $y = 1$ and $F_z = 1 + 1 + z^2$;

$$I_4 = \int_1^0 (1 + 1 + z^2) \, dz = -1 - 1 - \tfrac{1}{3}.$$

On DE, $y = 1$ and $z = 0$ and $F_x = xe^{-x}$;

$$I_5 = \int_1^0 xe^{-x}\,dx = -1 + 2e^{-1}.$$

On EO, $x = z = 0$ and $F_y = ye^0$;

$$I_6 = \int_1^0 ye^0\,dy = -\tfrac{1}{2}.$$

Adding up these six contributions shows that the complete line integral has the value $\dfrac{e}{2} - \dfrac{5}{6}$.

(b) As a simple sketch shows, the given contour is three-dimensional. However, it is equivalent to two plane square contours, one $OADEO$ (denoted by S_1) lying in the plane $z = 0$ and the other $ABCDA$ (S_2) lying in the plane $x = 1$; the latter is traversed in the negative sense. The common segment AD does not form part of the original contour but, as it is traversed in opposite senses in the two constituent contours, it (correctly) contributes nothing to the line integral.

To use Stokes' theorem we first need to calculate

$$(\nabla \times \mathbf{F})_x = x^3 + 3y^2x + 2yxz^2 - 3xy^2 - x^3 = 2yxz^2,$$
$$(\nabla \times \mathbf{F})_y = 3x^2y + y^3 - 3x^2y - y^3 - y^2z^2 = -y^2z^2,$$
$$(\nabla \times \mathbf{F})_z = 3y^2z + 3x^2z + ye^x - 3x^2z - 3y^2z = ye^x.$$

Now, S_1 has its normal in the positive z-direction and so only the z-component of $\nabla \times \mathbf{F}$ is needed in the first surface integral of Stokes' theorem. Likewise only the x-component of $\nabla \times \mathbf{F}$ is needed in the second integral, but its value must be subtracted because of the sense in which its contour is traversed:

$$\int_{OABCDEO} (\nabla \times \mathbf{F}) \cdot d\mathbf{r} = \int_{S_1} (\nabla \times \mathbf{F})_z\,dx\,dy - \int_{S_2} (\nabla \times \mathbf{F})_x\,dy\,dz$$

$$= \int_0^1 \int_0^1 ye^x\,dx\,dy - \int_0^1 \int_0^1 2y \times 1 \times z^2\,dy\,dz$$

$$= \frac{1}{2}(e - 1) - 2\,\frac{1}{2}\,\frac{1}{3} = \frac{e}{2} - \frac{5}{6}.$$

As they must, the two methods give the same value.

12

Fourier series

12.1 *Prove the orthogonality relations that form the basis of the Fourier series representation of functions.*

All of the results are based on the values of the integrals

$$S(n) = \int_{x_0}^{x_0+L} \sin\left(\frac{2\pi n x}{L}\right) dx \quad \text{and} \quad C(n) = \int_{x_0}^{x_0+L} \cos\left(\frac{2\pi n x}{L}\right) dx$$

for integer values of n. Since in all cases with $n \geq 1$ the integrand goes through a whole number of complete cycles, the 'area under the curve' is zero. For the case $n = 0$, the integrand in $S(n)$ is zero and so therefore is $S(0)$; for $C(0)$ the integrand is unity and the value of $C(0)$ is L.

We now apply these observations to integrals whose integrands are the products of two sinusoidal functions with arguments that are multiples of a fundamental frequency. The integration interval is equal to the period of that fundamental frequency. To express the integrands in suitable forms, repeated use will be made of the expressions for the sums and differences of sinusoidal functions.

We consider first the product of a sine function and a cosine function:

$$I_1 = \int_{x_0}^{x_0+L} \sin\left(\frac{2\pi r x}{L}\right) \cos\left(\frac{2\pi p x}{L}\right)$$

$$= \int_{x_0}^{x_0+L} \frac{1}{2}\left[\sin\left(\frac{2\pi(r+p)x}{L}\right) + \sin\left(\frac{2\pi(r-p)x}{L}\right)\right] dx$$

$$= \frac{1}{2}[S(r+p) + S(r-p)] = 0, \text{ for all } r \text{ and } p.$$

Next, we consider the product of two cosines:

$$
\begin{aligned}
I_2 &= \int_{x_0}^{x_0+L} \cos\left(\frac{2\pi rx}{L}\right) \cos\left(\frac{2\pi px}{L}\right) \\
&= \int_{x_0}^{x_0+L} \frac{1}{2}\left[\cos\left(\frac{2\pi(r+p)x}{L}\right) + \cos\left(\frac{2\pi(r-p)x}{L}\right)\right] dx \\
&= \frac{1}{2}[C(r+p) + C(r-p)] = 0,
\end{aligned}
$$

unless $r = p > 0$ when $I_2 = \frac{1}{2}L$. If r and p are both zero, then the integrand is unity and $I_2 = L$.

Finally, for the product of two sine functions:

$$
\begin{aligned}
I_3 &= \int_{x_0}^{x_0+L} \sin\left(\frac{2\pi rx}{L}\right) \sin\left(\frac{2\pi px}{L}\right) \\
&= \int_{x_0}^{x_0+L} \frac{1}{2}\left[\cos\left(\frac{2\pi(r-p)x}{L}\right) - \cos\left(\frac{2\pi(r+p)x}{L}\right)\right] dx \\
&= \frac{1}{2}[C(r-p) - C(r+p)] = 0,
\end{aligned}
$$

unless $r = p > 0$ when $I_3 = \frac{1}{2}L$. If either of r and p is zero, then the integrand is zero and $I_3 = 0$.

In summary, all of the integrals have zero value except for those in which the integrand is the square of a single sinusoid. In these cases the integral has value $\frac{1}{2}L$ for all integers $r\,(=p)$ that are > 0. For $r\,(=p)$ equal to zero, the $\sin^2$ integral has value zero and the $\cos^2$ integral has value L.

12.3 *Which of the following functions of x could be represented by a Fourier series over the range indicated?*

(a) $\tanh^{-1}(x)$, $\qquad -\infty < x < \infty$;
(b) $\tan x$, $\qquad\qquad -\infty < x < \infty$;
(c) $|\sin x|^{-1/2}$, $\qquad -\infty < x < \infty$;
(d) $\cos^{-1}(\sin 2x)$, $\quad -\infty < x < \infty$;
(e) $x \sin(1/x)$, $\qquad\; -\pi^{-1} < x \le \pi^{-1}$, *cyclically repeated.*

The Dirichlet conditions that a function must satisfy before it can be represented by a Fourier series are:

(i) the function must be periodic;
(ii) it must be single-valued and continuous, except possibly at a finite number of finite discontinuities;

(iii) it must have only a finite number of maxima and minima within one period;

(iv) the integral over one period of $|f(x)|$ must converge.

We now test the given functions against these:

(a) $\tanh^{-1}(x)$ is not a periodic function, since it is only defined for $-1 \le x \le 1$ and changes (monotonically) from $-\infty$ to $+\infty$ as x varies over this restricted range. This function therefore fails condition (i) and *cannot* be represented as a Fourier series.

(b) $\tan x$ is a periodic function but its discontinuities are not finite, nor is its absolute modulus integrable. It therefore fails tests (ii) and (iv) and *cannot* be represented as a Fourier series.

(c) $|\sin x|^{-1/2}$ is a periodic function of period π and, although it becomes infinite at $x = n\pi$, there are no infinite discontinuities. Near $x = 0$, say, it behaves as $|x|^{-1/2}$ and its absolute modulus is therefore integrable. There is only one minimum in any one period. The function therefore satisfies all four Dirichlet conditions and *can* be represented as a Fourier series.

(d) $\cos^{-1}(\sin 2x)$ is clearly a multi-valued function and fails condition (ii); it *cannot* be represented as a Fourier series.

(e) $x \sin(1/x)$, for $-\pi^{-1} < x \le \pi^{-1}$ (cyclically repeated) is clearly cyclic (by definition), continuous, bounded, single-valued and integrable. However, since $\sin(1/x)$ oscillates with unlimited frequency near $x = 0$, there are an infinite number of maxima and minima in any region enclosing $x = 0$. Condition (iii) is therefore not satisfied and the function *cannot* be represented as a Fourier series.

12.5 *Find the Fourier series of the function $f(x) = x$ in the range $-\pi < x \le \pi$. Hence show that*

$$1 - \frac{1}{3} + \frac{1}{5} - \frac{1}{7} + \cdots = \frac{\pi}{4}.$$

This is an odd function in x and so a sine series with period 2π is appropriate. The coefficient of $\sin nx$ will be given by

$$
\begin{aligned}
b_n &= \frac{2}{2\pi} \int_{-\pi}^{\pi} x \sin nx \, dx \\
&= \frac{1}{\pi} \left\{ \left[-\frac{x \cos nx}{n} \right]_{-\pi}^{\pi} + \int_{-\pi}^{\pi} \frac{\cos nx}{n} \, dx \right\} \\
&= \frac{1}{\pi} \left[-\frac{\pi(-1)^n - (-\pi)(-1)^n}{n} + 0 \right] = \frac{2(-1)^{n+1}}{n}.
\end{aligned}
$$

Thus, $\qquad x = f(x) = 2 \sum_{n=1}^{\infty} \frac{(-1)^{n+1}}{n} \sin nx.$

We note in passing that although this series is convergent, as it must be, it has poor (i.e. n^{-1}) convergence; this can be put down to the periodic version of the function having a discontinuity (of 2π) at the end of each basic period.

To obtain the sum of a series from such a Fourier representation, we must make a judicious choice for the value of x – making such a choice is rather more of an art than a science! Here, setting $x = \pi/2$ gives

$$\frac{\pi}{2} = 2 \sum_{n=1}^{\infty} \frac{(-1)^{n+1} \sin(n\pi/2)}{n}$$

$$= 2 \sum_{n \text{ odd}} \frac{(-1)^{n+1}(-1)^{(n-1)/2}}{n},$$

$$\Rightarrow \quad \frac{\pi}{4} = \frac{1}{1} - \frac{1}{3} + \frac{1}{5} - \frac{1}{7} + \cdots.$$

12.7 *For the function*

$$f(x) = 1 - x, \qquad 0 \le x \le 1,$$

a Fourier sine series can be found by continuing it in the range $-1 < x \le 0$ as $f(x) = -1 - x$. The function thus has a discontinuity of 2 at $x = 0$. The series is

$$1 - x = f(x) = \frac{2}{\pi} \sum_{n=1}^{\infty} \frac{\sin n\pi x}{n}. \qquad (*)$$

In order to obtain a cosine series, the continuation has to be $f(x) = 1 + x$ in the range $-1 < x \le 0$. The function then has no discontinuity at $x = 0$ and the corresponding series is

$$1 - x = f(x) = \frac{1}{2} + \frac{4}{\pi^2} \sum_{n \text{ odd}} \frac{\cos n\pi x}{n^2}. \qquad (**)$$

For these continued functions and series, consider (i) *their derivatives and* (ii) *their integrals. Do they give meaningful equations? You will probably find it helpful to sketch all the functions involved.*

(i) *Derivatives*

(a) The sine series. With the continuation given, the derivative df/dx has the value -1 everywhere, except at the origin where the function is not defined (though $f(0) = 0$ seems the only possible choice), continuous or differentiable. Differentiating the given series (*) for $f(x)$ yields

$$\frac{df}{dx} = 2 \sum_{n=1}^{\infty} \cos n\pi x.$$

This series does not converge and the equation is not meaningful.

(b) The cosine series. With the stated continuation for $f(x)$ the derivative is $+1$ for $-1 < x \leq 0$ and is -1 for $0 \leq x \leq 1$. It is thus the negative of an odd (about $x = 0$) unit square-wave, whose Fourier series is

$$-\frac{4}{\pi} \sum_{n \text{ odd}} \frac{\sin n\pi x}{n}.$$

This is confirmed by differentiating (**) term by term to obtain the same result:

$$\frac{df}{dx} = \frac{4}{\pi^2} \sum_{n \text{ odd}} \frac{-n\pi \sin n\pi x}{n^2} = -\frac{4}{\pi} \sum_{n \text{ odd}} \frac{\sin n\pi x}{n}.$$

(ii) *Integrals*

Since integrals contain an arbitrary constant of integration, we will define $F(-1) = 0$, where $F(x)$ is the indefinite integral of $f(x)$.

(a) The sine series. For $-1 \leq x \leq 0$,

$$F_a(x) = F(-1) + \int_{-1}^{x} (-1 - x) \, dx = -x - \tfrac{1}{2}x^2 - \tfrac{1}{2}.$$

For $0 \leq x \leq 1$,

$$F_a(x) = F(0) + \int_{0}^{x} (1 - x) \, dx = -\tfrac{1}{2} + \left[x - \tfrac{1}{2}x^2 \right]_0^x = x - \tfrac{1}{2}x^2 - \tfrac{1}{2}.$$

This is a continuous function and, like all indefinite integrals, is 'smoother' than the function from which it is derived; this latter property will be reflected in the improved convergence of the derived series. Integrating term by term we find that its Fourier series is given by

$$F_a(x) = \frac{2}{\pi} \int_{-1}^{x} \sum_{n=1}^{\infty} \frac{\sin n\pi x'}{n} \, dx'$$

$$= \frac{2}{\pi} \sum_{n=1}^{\infty} \left[-\frac{\cos n\pi x'}{\pi n^2} \right]_{-1}^{x}$$

$$= \frac{2}{\pi^2} \sum_{n=1}^{\infty} \frac{(-1)^n - \cos n\pi x}{n^2}$$

$$= -\frac{1}{6} - \frac{2}{\pi^2} \sum_{n=1}^{\infty} \frac{\cos n\pi x}{n^2},$$

a series that has n^{-2} convergence. Here we have used the result that $\sum_{n=1}^{\infty}(-1)^n n^{-2} = -\pi^2/12$.

(b) The cosine series. The corresponding indefinite integral in this case is

$$F_b(x) = x + \tfrac{1}{2}x^2 + \tfrac{1}{2} \quad \text{for} \quad -1 \le x \le 0,$$
$$F_b(x) = x - \tfrac{1}{2}x^2 + \tfrac{1}{2} \quad \text{for} \quad 0 \le x \le 1,$$

and the corresponding integrated series, which has even better convergence (n^{-3}), is given by

$$\frac{1}{2}(x+1) + \frac{4}{\pi^3} \sum_{n \text{ odd}} \frac{\sin n\pi x}{n^3}.$$

However, to have a true Fourier series expression, we must substitute a Fourier series for the $x/2$ term that arises from integrating the constant ($\tfrac{1}{2}$) in (**). This series must be that for $x/2$ across the complete range $-1 \le x \le 1$, and so neither (*) nor (**) can be rearranged for the purpose. A straightforward calculation (see exercise 12.25 part (b), if necessary) yields the poorly convergent sine series

$$x = 2 \sum_{n=1}^{\infty} \frac{(-1)^{n+1}}{n\pi} \sin n\pi x,$$

and makes the final expression for $F_b(x)$

$$\frac{1}{2} + \sum_{n=1}^{\infty} \frac{(-1)^{n+1}}{n\pi} \sin n\pi x + \frac{4}{\pi^3} \sum_{n \text{ odd}} \frac{\sin n\pi x}{n^3}.$$

As will be apparent from a simple sketch, the first series in the above expression dominates; all of its terms are present and it has only n^{-1} convergence. The second series has alternate terms missing and its convergence $\sim n^{-3}$.

12.9 *Find the Fourier coefficients in the expansion of $f(x) = \exp x$ over the range $-1 < x < 1$. What value will the expansion have when $x = 2$?*

Since the Fourier series will have period 2, we can say immediately that at $x = 2$ the series will converge to the value it has at $x = 0$, namely 1.

As the function $f(x) = \exp x$ is neither even nor odd, its Fourier series will contain both sine and cosine terms. The cosine coefficients are given by

$$a_n = \frac{2}{2} \int_{-1}^{1} e^x \cos(n\pi x)\, dx$$

$$= [\cos(n\pi x)\, e^x]_{-1}^{1} + \int_{-1}^{1} n\pi \sin(n\pi x)\, e^x\, dx$$

$$= (-1)^n (e^1 - e^{-1}) + [n\pi \sin(n\pi x)\, e^x]_{-1}^{1}$$

$$\qquad\qquad - \int_{-1}^{1} n^2 \pi^2 \cos(n\pi x)\, e^x\, dx$$

$$= 2(-1)^n \sinh 1 - n^2 \pi^2 a_n,$$

$$\Rightarrow \quad a_n = \frac{2(-1)^n \sinh 1}{1 + n^2 \pi^2}.$$

Similarly, the sine coefficients are given by

$$b_n = \frac{2}{2} \int_{-1}^{1} e^x \sin(n\pi x)\, dx$$

$$= [\sin(n\pi x)\, e^x]_{-1}^{1} - \int_{-1}^{1} n\pi \cos(n\pi x)\, e^x\, dx$$

$$= 0 + [-n\pi \cos(n\pi x)\, e^x]_{-1}^{1} - \int_{-1}^{1} n^2 \pi^2 \sin(n\pi x)\, e^x\, dx$$

$$= 2(-1)^{n+1} n\pi \sinh 1 - n^2 \pi^2 b_n,$$

$$\Rightarrow \quad b_n = \frac{2(-1)^{n+1} n\pi \sinh 1}{1 + n^2 \pi^2}.$$

12.11 *Consider the function $f(x) = \exp(-x^2)$ in the range $0 \le x \le 1$. Show how it should be continued to give as its Fourier series a series (the actual form is not wanted) (a) with only cosine terms, (b) with only sine terms, (c) with period 1 and (d) with period 2.*

Would there be any difference between the values of the last two series at (i) $x = 0$, (ii) $x = 1$?

The function and its four continuations are shown as (a)–(d) in figure 12.1. Note that in the range $0 \le x \le 1$, all four graphs are identical.

Where a continued function has a discontinuity at the ends of its basic period, the series will yield a value at those end-points that is the average of the function's values on the two sides of the discontinuity. Thus for continuation (c) both (i) $x = 0$ and (ii) $x = 1$ are end-points, and the value of the series there will be

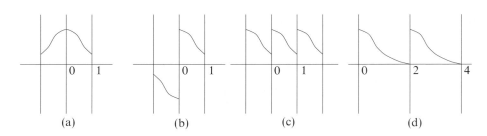

Figure 12.1   The solution to exercise 12.11, showing the continuations of $\exp(-x^2)$ in $0 \leq x \leq 1$ to give: (a) cosine terms only; (b) sine terms only; (c) period 1; (d) period 2.

$(1 + e^{-1})/2$. For continuation (d), $x = 0$ is an end-point, and the series will have value $\frac{1}{2}(1 + e^{-4})$. However, $x = 1$ is not a point of discontinuity, and the series will have the expected value of e^{-1}.

12.13 *Consider the representation as a Fourier series of the displacement of a string lying in the interval $0 \leq x \leq L$ and fixed at its ends, when it is pulled aside by y_0 at the point $x = L/4$. Sketch the continuations for the region outside the interval that will*

(a) *produce a series of period L,*
(b) *produce a series that is antisymmetric about $x = 0$, and*
(c) *produce a series that will contain only cosine terms.*
(d) *What are (i) the periods of the series in (b) and (c) and (ii) the value of the 'a$_0$-term' in (c)?*
(e) *Show that a typical term of the series obtained in (b) is*

$$\frac{32 y_0}{3 n^2 \pi^2} \sin \frac{n\pi}{4} \sin \frac{n\pi x}{L}.$$

Parts (a), (b) and (c) of figure 12.2 show the three required continuations. Condition (b) will result in a series containing only sine terms, whilst condition (c) requires the continued function to be symmetric about $x = 0$.

(d) (i) The period in both cases, (b) and (c), is clearly $2L$.

(ii) The average value of the displacement is found from 'the area under the triangular curve' to be $(\frac{1}{2} L y_0)/L = \frac{1}{2} y_0$, and this is the value of the 'a_0-term'.

(e) For the antisymmetric continuation there will be no cosine terms. The sine

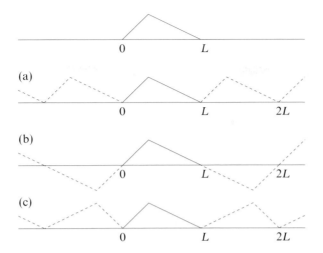

Figure 12.2 Plucked string with fixed ends: (a)–(c) show possible mathematical continuations; (b) is antisymmetric about 0 and (c) is symmetric.

term coefficients (for a period of $2L$) are given by

$$
b_n = 2\,\frac{2}{2L} \int_0^L f(x)\sin(nkx)\,dx, \quad \text{where } k = 2\pi/2L = \pi/L,
$$

$$
= \frac{2y_0}{L} \left[\int_0^{L/4} \frac{4x}{L} \sin(nkx)\,dx + \int_{L/4}^L \left(\frac{4}{3} - \frac{4x}{3L}\right) \sin(nkx)\,dx \right]
$$

$$
= \frac{8y_0}{3L^2} \left[\int_0^{L/4} 3x\sin(nkx)\,dx + \int_{L/4}^L (L-x)\sin(nkx)\,dx \right]
$$

$$
= \frac{8y_0}{3L^2} \left\{ \left[-\frac{3x\cos(nkx)}{nk} \right]_0^{L/4} + \int_0^{L/4} \frac{3\cos(nkx)}{nk}\,dx \right.
$$

$$
\left. + \left[-\frac{L\cos(nkx)}{nk} \right]_{L/4}^L + \left[\frac{x\cos(nkx)}{nk} \right]_{L/4}^L - \int_{L/4}^L \frac{\cos(nkx)}{nk}\,dx \right\}.
$$

Integrating by parts then yields

$$
b_n = \frac{8y_0}{3L^2} \left\{ -\frac{3L\cos(n\pi/4)}{4n(\pi/L)} - 0 + \left[\frac{3\sin(nkx)}{n^2k^2} \right]_0^{L/4} - \frac{L\cos(n\pi)}{n(\pi/L)} \right.
$$

$$
\left. + \frac{L\cos(n\pi/4)}{n(\pi/L)} + \frac{L\cos(n\pi)}{n(\pi/L)} - \frac{L\cos(n\pi/4)}{4n(\pi/L)} - \left[\frac{\sin(nkx)}{n^2k^2} \right]_{L/4}^L \right\}
$$

$$
= \frac{8y_0}{3L^2} \left[\frac{3L^2\sin(n\pi/4)}{n^2\pi^2} - \frac{L^2\sin(n\pi)}{n^2\pi^2} + \frac{L^2\sin(n\pi/4)}{n^2\pi^2} \right] = \frac{32y_0}{3n^2\pi^2} \sin\left(\frac{n\pi}{4}\right).
$$

201

A typical term is therefore

$$\frac{32y_0}{3n^2\pi^2} \sin\left(\frac{n\pi}{4}\right) \sin\left(\frac{n\pi x}{L}\right).$$

We note that every fourth term ($n = 4m$ with m an integer) will be missing.

12.15 *The Fourier series for the function $y(x) = |x|$ in the range $-\pi \le x < \pi$ is*

$$y(x) = \frac{\pi}{2} - \frac{4}{\pi} \sum_{m=0}^{\infty} \frac{\cos(2m+1)x}{(2m+1)^2}.$$

By integrating this equation term by term from 0 to x, find the function $g(x)$ whose Fourier series is

$$\frac{4}{\pi} \sum_{m=0}^{\infty} \frac{\sin(2m+1)x}{(2m+1)^3}.$$

Using these results, determine, as far as possible by inspection, the form of the functions of which the following are the Fourier series:

(a)

$$\cos\theta + \frac{1}{9}\cos 3\theta + \frac{1}{25}\cos 5\theta + \cdots ;$$

(b)

$$\sin\theta + \frac{1}{27}\sin 3\theta + \frac{1}{125}\sin 5\theta + \cdots ;$$

(c)

$$\frac{L^2}{3} - \frac{4L^2}{\pi^2} \left[\cos\frac{\pi x}{L} - \frac{1}{4}\cos\frac{2\pi x}{L} + \frac{1}{9}\cos\frac{3\pi x}{L} - \cdots \right].$$

[You may find it helpful to first set $x = 0$ in the quoted result and so obtain values for $S_o = \sum(2m+1)^{-2}$ and other sums derivable from it.]

First, define

$$S = \sum_{\text{all } n \ne 0} n^{-2}, \quad S_o = \sum_{\text{odd } n} n^{-2}, \quad S_e = \sum_{\text{even } n \ne 0} n^{-2}.$$

Clearly, $S_e = \frac{1}{4}S$.

Now set $x = 0$ in the quoted result to obtain

$$0 = \frac{\pi}{2} - \frac{4}{\pi} \sum_{m=0}^{\infty} \frac{1}{(2m+1)^2} = \frac{\pi}{2} - \frac{4}{\pi} S_o.$$

Thus, $S_o = \pi^2/8$. Further, $S = S_o + S_e = S_o + \frac{1}{4}S$; it follows that $S = \pi^2/6$ and, by subtraction, that $S_e = \pi^2/24$.

We now consider the integral of $y(x) = |x|$ from 0 to x.

(i) For $x < 0$, $\displaystyle\int_0^x |x|\, dx = \int_0^x (-x)\, dx = -\frac{1}{2}x^2$.

(ii) For $x > 0$, $\displaystyle\int_0^x |x|\, dx = \int_0^x x\, dx = \frac{1}{2}x^2$.

Integrating the series term by term gives

$$\frac{\pi x}{2} - \frac{4}{\pi} \sum_{m=0}^{\infty} \frac{\sin(2m+1)x}{(2m+1)^3}.$$

Equating these two results and isolating the series gives

$$\frac{4}{\pi} \sum_{m=0}^{\infty} \frac{\sin(2m+1)x}{(2m+1)^3} = \tfrac{1}{2}x(\pi - x) \text{ for } x \geq 0,$$

$$= \tfrac{1}{2}x(\pi + x) \text{ for } x \leq 0.$$

Questions (a)–(c) are to be solved largely through inspection and so detailed working is not (cannot be) given.

(a) Straightforward substitution of θ for x and rearrangement of the original Fourier series give $g_1(\theta) = \frac{1}{4}\pi(\frac{1}{2}\pi - |\theta|)$.

(b) Straightforward substitution of θ for x and rearrangement of the integrated Fourier series give $g_2(\theta) = \frac{1}{8}\pi\theta(\pi - |\theta|)$.

(c) This contains only cosine terms and is therefore an even function of x. Its average value (given by the a_0 term) is $\frac{1}{3}L^2$. Setting $x = 0$ gives

$$f(0) = \frac{L^2}{3} - \frac{4L^2}{\pi^2}\left(1 - \frac{1}{4} + \frac{1}{9} - \cdots\right)$$

$$= \frac{L^2}{3} - \frac{4L^2}{\pi^2}(S_o - S_e)$$

$$= \frac{L^2}{3} - \frac{4L^2}{\pi^2}\left(\frac{\pi^2}{8} - \frac{\pi^2}{24}\right) = 0.$$

Setting $x = L$ gives

$$f(L) = \frac{L^2}{3} - \frac{4L^2}{\pi^2}\left(-1 - \frac{1}{4} - \frac{1}{9} - \cdots\right)$$

$$= \frac{L^2}{3} - \frac{4L^2}{\pi^2}(-S) = L^2.$$

All of this evidence suggests that $f(x) = x^2$ (which it is).

12.17 *Find the (real) Fourier series of period 2 for $f(x) = \cosh x$ and $g(x) = x^2$ in the range $-1 \leq x \leq 1$. By integrating the series for $f(x)$ twice, prove that*

$$\sum_{n=1}^{\infty} \frac{(-1)^{n+1}}{n^2\pi^2(n^2\pi^2 + 1)} = \frac{1}{2}\left(\frac{1}{\sinh 1} - \frac{5}{6}\right).$$

Since both functions are even, we need consider only constants and cosine terms. The series for x^2 can be calculated directly or, more easily, by using the result of the final part of exercise 12.15 with L set equal to 1:

$$g(x) = x^2 = \frac{1}{3} + \frac{4}{\pi^2} \sum_{n=1}^{\infty} \frac{(-1)^n}{n^2} \cos \pi n x \text{ for } -1 \leq x \leq 1.$$

For $f(x) = \cosh x$,

$$a_0 = \frac{2}{2} 2 \int_0^1 \cosh x \, dx = 2\sinh(1),$$

$$a_n = \frac{2}{2} 2 \int_0^1 \cosh x \cos(n\pi x) \, dx$$

$$= 2\left[\frac{\cosh x \sin(n\pi x)}{n\pi}\right]_0^1 - 2\int_0^1 \frac{\sinh x \sin(n\pi x)}{n\pi} \, dx$$

$$= 0 + 2\left[\frac{\sinh x \cos(n\pi x)}{n^2\pi^2}\right]_0^1 - \frac{a_n}{n^2\pi^2}.$$

Rearranging this gives

$$a_n = \frac{(-1)^n 2\sinh(1)}{1 + n^2\pi^2}.$$

Thus,

$$\cosh x = \sinh(1)\left(1 + 2\sum_{n=1}^{\infty} \frac{(-1)^n}{1 + n^2\pi^2} \cos n\pi x\right).$$

We now integrate this expansion twice from 0 to x (anticipating that we will recover a hyperbolic cosine function plus some additional terms). Since $\sinh(0) = \sin(m\pi 0) = 0$, the first integration yields

$$\sinh x = \sinh(1)\left(x + 2\sum_{n=1}^{\infty} \frac{(-1)^n}{n\pi(1 + n^2\pi^2)} \sin n\pi x\right).$$

For the second integration we use $\cosh(0) = \cos(m\pi 0) = 1$ to obtain

$$\cosh(x) - 1 = \sinh(1)\left(\frac{1}{2}x^2 + 2\sum_{n=1}^{\infty} \frac{(-1)^{n+1}}{n^2\pi^2(1 + n^2\pi^2)}[\cos(n\pi x) - 1]\right).$$

However, this expansion must be the same as the original expansion for $\cosh(x)$ after a Fourier series has been substituted for the $\frac{1}{2}\sinh(1)x^2$ term. The coefficients of $\cos n\pi x$ in the two expressions must be equal; in particular, the equality of the constant terms (formally $\cos n\pi x$ with $n = 0$) requires that

$$\sinh(1) - 1 = \frac{1}{2}\sinh(1)\frac{1}{3} + 2\sinh(1)\sum_{n=1}^{\infty}\frac{(-1)^{n+2}}{n^2\pi^2(1 + n^2\pi^2)},$$

i.e.

$$\sum_{n=1}^{\infty}\frac{(-1)^{n+1}}{n^2\pi^2(n^2\pi^2 + 1)} = \frac{1}{2}\left(\frac{1}{\sinh 1} - \frac{5}{6}\right),$$

as stated in the question.

12.19 *Demonstrate explicitly for the odd (about $x = 0$) square-wave function that Parseval's theorem is valid. You will need to use the relationship*

$$\sum_{m=0}^{\infty}\frac{1}{(2m + 1)^2} = \frac{\pi^2}{8}.$$

Show that a filter that transmits frequencies only up to $8\pi/T$ will still transmit more than 90% of the power in a square-wave voltage signal of period T.

As stated in the solution to exercise 12.7, and in virtually every textbook, the odd square-wave function has only the odd harmonics present in its Fourier sine series representation. The coefficient of the $\sin(2m + 1)\pi x$ term is

$$b_{2m+1} = \frac{4}{(2m + 1)\pi}.$$

For a periodic function of period L whose complex Fourier coefficients are c_r, or whose cosine and sine coefficients are a_r and b_r, respectively, Parseval's theorem for one function states that

$$\frac{1}{L}\int_{x_0}^{x_0+L}|f(x)|^2 dx = \sum_{r=-\infty}^{\infty}|c_r|^2$$

$$= \left(\tfrac{1}{2}a_0\right)^2 + \tfrac{1}{2}\sum_{r=1}^{\infty}(a_r^2 + b_r^2),$$

and therefore requires in this particular case, in which all the a_r are zero and $L = 2$, that

$$\frac{1}{2}\sum_{m=0}^{\infty}\frac{16}{(2m + 1)^2\pi^2} = \frac{1}{2}\sum_{n=1}^{\infty}b_n^2 = \frac{1}{2}\int_{-1}^{1}|\pm 1|^2 dx = 1.$$

205

Since

$$\sum_{m=0}^{\infty} \frac{1}{(2m+1)^2} = \frac{\pi^2}{8},$$

this reduces to the identity

$$\frac{1}{2}\frac{16}{\pi^2}\frac{\pi^2}{8} = 1.$$

The power at any particular frequency in an electrical signal is proportional to the square of the amplitude at that frequency, i.e. to $|b_n|^2$ in the present case. If the filter passes only frequencies up to $8\pi/T = 4\omega$, then only the $n = 1$ and the $n = 3$ components will be passed. They contribute a fraction

$$\left(\frac{1}{1} + \frac{1}{9}\right) \div \frac{\pi^2}{8} = 0.901$$

of the total, i.e. more than 90%.

12.21 *Find the complex Fourier series for the periodic function of period 2π defined in the range $-\pi \le x \le \pi$ by $y(x) = \cosh x$. By setting $x = 0$ prove that*

$$\sum_{n=1}^{\infty} \frac{(-1)^n}{n^2+1} = \frac{1}{2}\left(\frac{\pi}{\sinh\pi} - 1\right).$$

We first note that, although $\cosh x$ is an even function of x, e^{-inx} is neither even nor odd. Consequently it will not be possible to convert the integral into one over the range $0 \le x \le \pi$. The complex Fourier coefficients c_n ($-\infty < n < \infty$) are therefore calculated as

$$\begin{aligned}
c_n &= \frac{1}{2\pi}\int_{-\pi}^{\pi} \cosh x\, e^{-inx}\, dx \\
&= \frac{1}{2\pi}\int_{-\pi}^{\pi} \frac{1}{2}\left(e^{-inx+x} + e^{-inx-x}\right)\, dx \\
&= \frac{1}{4\pi}\left[\frac{e^{(1-in)x}}{1-in}\right]_{-\pi}^{\pi} + \frac{1}{4\pi}\left[\frac{e^{(-1-in)x}}{-1-in}\right]_{-\pi}^{\pi} \\
&= \frac{1}{4\pi}\frac{(1+in)(-1)^n(2\sinh\pi) - (1-in)(-1)^n(-2\sinh\pi)}{1+n^2} \\
&= \frac{(-1)^n 4\sinh(\pi)}{4\pi(1+n^2)}.
\end{aligned}$$

Thus,

$$\cosh x = \sum_{n=-\infty}^{\infty} \frac{(-1)^n \sinh\pi}{\pi(1+n^2)}\, e^{inx}.$$

We now set $x = 0$ on both sides of the equation:

$$1 = \sum_{n=-\infty}^{\infty} \frac{(-1)^n \sinh \pi}{\pi(1 + n^2)},$$

$$\Rightarrow \quad \sum_{n=-\infty}^{\infty} \frac{(-1)^n}{1 + n^2} = \frac{\pi}{\sinh \pi}.$$

Separating out the $n = 0$ term, and noting that $(-1)^n = (-1)^{-n}$, now gives

$$1 + 2 \sum_{n=1}^{\infty} \frac{(-1)^n}{1 + n^2} = \frac{\pi}{\sinh \pi}$$

and hence the stated result.

12.23 *The complex Fourier series for the periodic function generated by* $f(t) = \sin t$ *for* $0 \leq t \leq \pi/2$, *and repeated in every subsequent interval of* $\pi/2$, *is*

$$\sin(t) = \frac{2}{\pi} \sum_{n=-\infty}^{\infty} \frac{4ni - 1}{16n^2 - 1} e^{i4nt}.$$

Apply Parseval's theorem to this series and so derive a value for the sum of the series

$$\frac{17}{(15)^2} + \frac{65}{(63)^2} + \frac{145}{(143)^2} + \cdots + \frac{16n^2 + 1}{(16n^2 - 1)^2} + \cdots.$$

Applying Parseval's theorem (see solution 12.19) in a straightforward manner to the given equation:

$$\frac{2}{\pi} \int_0^{\pi/2} \sin^2(t)\, dt = \frac{4}{\pi^2} \sum_{n=-\infty}^{\infty} \frac{4ni - 1}{16n^2 - 1} \frac{-4ni - 1}{16n^2 - 1},$$

$$\frac{2}{\pi} \frac{1}{2} \frac{\pi}{2} = \frac{4}{\pi^2} \sum_{n=-\infty}^{\infty} \frac{16n^2 + 1}{(16n^2 - 1)^2},$$

$$\frac{\pi^2}{8} = 1 + 2 \sum_{n=1}^{\infty} \frac{16n^2 + 1}{(16n^2 - 1)^2},$$

$$\Rightarrow \quad \sum_{n=1}^{\infty} \frac{16n^2 + 1}{(16n^2 - 1)^2} = \frac{\pi^2 - 8}{16}.$$

To obtain the second line we have used the standard result that the average value of the square of a sinusoid is $1/2$.

12.25 *Show that Parseval's theorem for two real functions whose Fourier expansions have cosine and sine coefficients a_n, b_n and α_n, β_n takes the form*

$$\frac{1}{L} \int_0^L f(x)g^*(x)\,dx = \frac{1}{4}a_0\alpha_0 + \frac{1}{2}\sum_{n=1}^{\infty}(a_n\alpha_n + b_n\beta_n).$$

(a) *Demonstrate that for $g(x) = \sin mx$ or $\cos mx$ this reduces to the definition of the Fourier coefficients.*

(b) *Explicitly verify the above result for the case in which $f(x) = x$ and $g(x)$ is the square-wave function, both in the interval $-1 \leq x \leq 1$.*

If c_n and γ_n are the complex Fourier coefficients for the real functions $f(x)$ and $g(x)$ that have real Fourier coefficients a_n, b_n and α_n, β_n, respectively, then

$$c_n = \tfrac{1}{2}(a_n - ib_n) \quad \text{and} \quad \gamma_n = \alpha_n - i\beta_n,$$
$$c_{-n} = \tfrac{1}{2}(a_n + ib_n) \quad \text{and} \quad \gamma_{-n} = \alpha_n + i\beta_n.$$

The two functions can be written as

$$f(x) = \sum_{n=-\infty}^{\infty} c_n \exp\left(\frac{2\pi inx}{L}\right),$$

$$g(x) = \sum_{n=-\infty}^{\infty} \gamma_n \exp\left(\frac{2\pi inx}{L}\right). \qquad (*)$$

Thus,

$$f(x)g^*(x) = \sum_{n=-\infty}^{\infty} c_n g^*(x) \exp\left(\frac{2\pi inx}{L}\right).$$

Integrating this equation with respect to x over the interval $(0, L)$ and dividing by L, we find

$$\frac{1}{L}\int_0^L f(x)g^*(x)\,dx = \sum_{n=-\infty}^{\infty} c_n \frac{1}{L}\int_0^L g^*(x)\exp\left(\frac{2\pi inx}{L}\right)dx$$

$$= \sum_{n=-\infty}^{\infty} c_n \left[\frac{1}{L}\int_0^L g(x)\exp\left(\frac{-2\pi inx}{L}\right)dx\right]^*$$

$$= \sum_{n=-\infty}^{\infty} c_n\gamma_n^*.$$

To obtain the last line we have used the inverse of relationship $(*)$.

Dividing up the sum over all n into a sum over positive n, a sum over negative n

and the $n = 0$ term, and then substituting for c_n and γ_n, gives

$$\frac{1}{L} \int_0^L f(x) g^*(x) \, dx = \frac{1}{4} \sum_{n=1}^{\infty} (a_n - ib_n)(\alpha_n + i\beta_n)$$

$$+ \frac{1}{4} \sum_{n=1}^{\infty} (a_n + ib_n)(\alpha_n - i\beta_n) + \frac{1}{4} a_0 \alpha_0$$

$$= \frac{1}{4} \sum_{n=1}^{\infty} (2 a_n \alpha_n + 2 b_n \beta_n) + \frac{1}{4} a_0 \alpha_0$$

$$= \frac{1}{2} \sum_{n=1}^{\infty} (a_n \alpha_n + b_n \beta_n) + \frac{1}{4} a_0 \alpha_0,$$

i.e. the stated result.

(a) For $g(x) = \sin mx$, $\beta_m = 1$ and all other α_n and β_n are zero. The above equation then reduces to

$$\frac{1}{L} \int_0^L f(x) \sin(mx) \, dx = \frac{1}{2} b_n,$$

which is the normal definition of b_n. Similarly, setting $g(x) = \cos mx$ leads to the normal definition of a_n.

(b) For the function $f(x) = x$ in the interval $-1 < x \le 1$, the sine coefficients are

$$b_n = \frac{2}{2} \int_{-1}^{1} x \sin n\pi x \, dx$$

$$= 2 \int_0^1 x \sin n\pi x \, dx$$

$$= 2 \left\{ \left[\frac{-x \cos n\pi x}{n\pi} \right]_0^1 + \int_0^1 \frac{\cos n\pi x}{n\pi} \, dx \right\}$$

$$= 2 \left\{ \frac{(-1)^{n+1}}{n\pi} + \left[\frac{\sin n\pi x}{n^2 \pi^2} \right]_0^1 \right\}$$

$$= \frac{2(-1)^{n+1}}{n\pi}.$$

As stated in exercise 12.19, for the (antisymmetric) square-wave function $\beta_n = 4/(n\pi)$ for odd n and $\beta_n = 0$ for even n.

Now the integral

$$\frac{1}{L} \int_0^L f(x) g^*(x) \, dx = \frac{1}{2} \left[\int_{-1}^{0} (-1) x \, dx + \int_0^1 (+1) x \, dx \right] = \frac{1}{2},$$

whilst

$$\frac{1}{2} \sum_{n=1}^{\infty} b_n \beta_n = \frac{1}{2} \sum_{n \text{ odd}} \frac{4}{n\pi} \frac{2(-1)^{n+1}}{n\pi} = \frac{4}{\pi^2} \sum_{n \text{ odd}} \frac{1}{n^2} = \frac{4}{\pi^2} \frac{\pi^2}{8} = \frac{1}{2}.$$

The value of the sum $\sum n^{-2}$ for odd n is taken from S_{o} in the solution to exercise 12.15. Thus, the two sides of the equation agree, verifying the validity of Parseval's theorem in this case.

13

Integral transforms

13.1 *Find the Fourier transform of the function* $f(t) = \exp(-|t|)$.

(a) *By applying Fourier's inversion theorem prove that*

$$\frac{\pi}{2} \exp(-|t|) = \int_0^\infty \frac{\cos \omega t}{1 + \omega^2}\, d\omega.$$

(b) *By making the substitution* $\omega = \tan \theta$, *demonstrate the validity of Parseval's theorem for this function.*

As the function $|t|$ is not representable by the same integrable function throughout the integration range, we must divide the range into two sections and use different explicit expressions for the integrand in each:

$$\tilde{f}(\omega) = \frac{1}{\sqrt{2\pi}} \int_{-\infty}^{\infty} e^{-|t|}\, e^{-i\omega t}\, dt$$

$$= \frac{1}{\sqrt{2\pi}} \int_0^{\infty} e^{-(1+i\omega)t}\, dt + \frac{1}{\sqrt{2\pi}} \int_{-\infty}^0 e^{(1-i\omega)t}\, dt$$

$$= \frac{1}{\sqrt{2\pi}} \left(\frac{1}{1 + i\omega} + \frac{1}{1 - i\omega} \right)$$

$$= \frac{1}{\sqrt{2\pi}} \frac{2}{1 + \omega^2}.$$

(a) Substituting this result into the inversion theorem gives

$$\exp^{-|t|} = \frac{1}{\sqrt{2\pi}} \int_{-\infty}^{\infty} \frac{2}{\sqrt{2\pi}(1 + \omega^2)}\, e^{i\omega t}\, d\omega.$$

Equating the real parts on the two sides of this equation and noting that the

211

resulting integrand is symmetric in ω, shows that

$$\exp^{-|t|} = \frac{2}{\pi} \int_0^\infty \frac{\cos \omega t}{(1 + \omega^2)} \, d\omega,$$

as given in the question.

(b) For Parseval's theorem, which states that

$$\int_{-\infty}^\infty |f(t)|^2 \, dt = \int_{-\infty}^\infty |\tilde{f}(\omega)|^2 \, d\omega,$$

we first evaluate

$$\int_{-\infty}^\infty |f(t)|^2 \, dt = \int_{-\infty}^0 e^{2t} \, dt + \int_0^\infty e^{-2t} \, dt$$

$$= 2 \int_0^\infty e^{-2t} \, dt$$

$$= 2 \left[\frac{e^{-2t}}{-2} \right]_0^\infty = 1.$$

The second integral, over ω, is

$$\int_{-\infty}^\infty |\tilde{f}(\omega)|^2 \, d\omega = 2 \int_0^\infty \frac{2}{\pi(1 + \omega^2)^2} \, d\omega, \quad \text{set } \omega \text{ equal to } \tan\theta,$$

$$= \frac{4}{\pi} \int_0^{\pi/2} \frac{1}{\sec^4\theta} \sec^2\theta \, d\theta$$

$$= \frac{4}{\pi} \int_0^{\pi/2} \cos^2\theta \, d\theta = \frac{4}{\pi} \frac{1}{2} \frac{\pi}{2} = 1,$$

i.e. the same as the first one, thus verifying the theorem for this function.

13.3 *Find the Fourier transform of $H(x - a)e^{-bx}$, where $H(x)$ is the Heaviside function.*

The Heaviside function $H(x)$ has value 0 for $x < 0$ and value 1 for $x \geq 0$. Write $H(x - a)e^{-bx} = h(x)$ with b assumed > 0. Then,

$$\tilde{h}(k) = \frac{1}{\sqrt{2\pi}} \int_{-\infty}^\infty H(x - a)e^{-bx} e^{-ikx} \, dx$$

$$= \frac{1}{\sqrt{2\pi}} \int_a^\infty e^{-bx - ikx} \, dx$$

$$= \frac{1}{\sqrt{2\pi}} \left[\frac{e^{-bx - ikx}}{-b - ik} \right]_a^\infty$$

$$= \frac{1}{\sqrt{2\pi}} \frac{e^{-ba} e^{-ika}}{b + ik} = e^{-ika} \frac{e^{-ba}}{\sqrt{2\pi}} \frac{b - ik}{b^2 + k^2}.$$

This same result could be obtained by setting $y = x - a$, finding the transform of $e^{-ba}e^{-by}$, and then using the translation property of Fourier transforms.

13.5 *By taking the Fourier transform of the equation*

$$\frac{d^2\phi}{dx^2} - K^2\phi = f(x),$$

show that its solution, $\phi(x)$, can be written as

$$\phi(x) = \frac{-1}{\sqrt{2\pi}} \int_{-\infty}^{\infty} \frac{e^{ikx}\widetilde{f}(k)}{k^2 + K^2} \, dk,$$

where $\widetilde{f}(k)$ is the Fourier transform of $f(x)$.

We take the Fourier transform of each term of

$$\frac{d^2\phi}{dx^2} - K^2\phi = f(x)$$

to give

$$\frac{1}{\sqrt{2\pi}} \int_{-\infty}^{\infty} \frac{d^2\phi}{dx^2} e^{-ikx} \, dx - K^2\tilde{\phi}(k) = \frac{1}{\sqrt{2\pi}} \int_{-\infty}^{\infty} f(x) e^{-ikx} \, dx.$$

Since ϕ must vanish at $\pm\infty$, the first term can be integrated twice by parts with no contributions at the end-points. This gives the full equation as

$$-k^2\tilde{\phi}(k) - K^2\tilde{\phi}(k) = \tilde{f}(k).$$

Now, by the Fourier inversion theorem,

$$\phi(x) = \frac{1}{\sqrt{2\pi}} \int_{-\infty}^{\infty} \tilde{\phi}(k) e^{ikx} \, dk$$

$$= -\frac{1}{\sqrt{2\pi}} \int_{-\infty}^{\infty} \frac{\tilde{f}(k) e^{ikx}}{k^2 + K^2} \, dk.$$

Note

The principal advantage of this Fourier approach to a set of one or more linear differential equations is that the differential operators act only on exponential functions whose exponents are linear in x. This means that the derivatives are no more than multiples of the original function and what were originally differential equations are turned into algebraic ones. As the differential equations are linear the algebraic equations can be solved explicitly for the transforms of their solutions, and the solutions themselves may then be found using the inversion theorem. The 'price' to be paid for this great simplification is that the inversion integral may not be tractable analytically, but, as a last resort, numerical integration can always be employed.

13.7 *Find the Fourier transform of the unit rectangular distribution*

$$f(t) = \begin{cases} 1 & |t| < 1 \\ 0 & otherwise. \end{cases}$$

Determine the convolution of f with itself and, without further integration, deduce its transform. Deduce that

$$\int_{-\infty}^{\infty} \frac{\sin^2 \omega}{\omega^2} \, d\omega = \pi,$$

$$\int_{-\infty}^{\infty} \frac{\sin^4 \omega}{\omega^4} \, d\omega = \frac{2\pi}{3}.$$

The function to be transformed is unity in the range $-1 \le t \le 1$ and so

$$\tilde{f}(\omega) = \frac{1}{\sqrt{2\pi}} \int_{-1}^{1} 1 \, e^{-i\omega t} \, dt = \frac{1}{\sqrt{2\pi}} \left[\frac{e^{-i\omega} - e^{i\omega}}{-i\omega} \right] = \frac{2\sin\omega}{\sqrt{2\pi}\omega}.$$

Denote by $p(t)$ the convolution of f with itself and, in the second line of the calculation below, change the integration variable from s to $u = t - s$:

$$p(t) \equiv \int_{-\infty}^{\infty} f(t-s)f(s) \, ds = \int_{-1}^{1} f(t-s) \, 1 \, ds$$

$$= \int_{t+1}^{t-1} f(u)(-du) = \int_{t-1}^{t+1} f(u) du.$$

It follows that

$$p(t) = \begin{cases} (t+1) - (-1) & 0 > t > -2 \\ 1 - (t-1) & 2 > t > 0 \end{cases} = \begin{cases} 2 - |t| & 0 < |t| < 2, \\ 0 & otherwise. \end{cases}$$

The transform of p is given directly by the convolution theorem [which states that if $h(t)$, given by $h = f * g$, is the convolution of f and g, then $\tilde{h} = \sqrt{2\pi} \, \tilde{f} \, \tilde{g}$] as

$$\tilde{p}(\omega) = \sqrt{2\pi} \frac{2\sin\omega}{\sqrt{2\pi}\omega} \frac{2\sin\omega}{\sqrt{2\pi}\omega} = \frac{4}{\sqrt{2\pi}} \frac{\sin^2 \omega}{\omega^2}.$$

Noting that the two integrals to be evaluated have as integrands the squares of functions that are essentially the known transforms of simple functions, we are led to apply Parseval's theorem to each. Applying the theorem to $f(t)$ and $p(t)$ yields

$$\int_{-\infty}^{\infty} \frac{4\sin^2 \omega}{2\pi\omega^2} \, d\omega = \int_{-\infty}^{\infty} |f(t)|^2 \, dt = 2 \quad \Rightarrow \quad \int_{-\infty}^{\infty} \frac{\sin^2 \omega}{\omega^2} = \pi,$$

and $\displaystyle\int_{-\infty}^{\infty} \frac{16}{2\pi} \frac{\sin^4 \omega}{\omega^4} d\omega = \int_{-2}^{0} (2+t)^2 \, dt + \int_{0}^{2} (2-t)^2 \, dt$

$$= \left[\frac{(2+t)^3}{3} \right]_{-2}^{0} - \left[\frac{(2-t)^3}{3} \right]_{0}^{2}$$

$$= \frac{8}{3} + \frac{8}{3},$$

$$\Rightarrow \quad \int_{-\infty}^{\infty} \frac{\sin^4 \omega}{\omega^4} d\omega = \frac{2\pi}{3}.$$

13.9 *By finding the complex Fourier series for its LHS show that either side of the equation*

$$\sum_{n=-\infty}^{\infty} \delta(t + nT) = \frac{1}{T} \sum_{n=-\infty}^{\infty} e^{-2\pi nit/T}$$

can represent a periodic train of impulses. By expressing the function $f(t+nX)$, in which X is a constant, in terms of the Fourier transform $\tilde{f}(\omega)$ *of $f(t)$, show that*

$$\sum_{n=-\infty}^{\infty} f(t + nX) = \frac{\sqrt{2\pi}}{X} \sum_{n=-\infty}^{\infty} \tilde{f}\left(\frac{2n\pi}{X} \right) e^{2\pi nit/X}.$$

This result is known as the Poisson summation formula.

Denote by $g(t)$ the periodic function $\sum_{n=-\infty}^{\infty} \delta(t+nT)$ with $2\pi/T = \omega$. Its complex Fourier coefficients are given by

$$c_n = \frac{1}{T} \int_0^T g(t) e^{-in\omega t} \, dt = \frac{1}{T} \int_0^T \delta(t) e^{-in\omega t} \, dt = \frac{1}{T}.$$

Thus, by the inversion theorem, its Fourier *series* representation is

$$g(t) = \sum_{n=-\infty}^{\infty} \frac{1}{T} e^{in\omega t} = \sum_{n=-\infty}^{\infty} \frac{1}{T} e^{-in\omega t} = \sum_{n=-\infty}^{\infty} \frac{1}{T} e^{-i2n\pi t/T},$$

showing that both this sum and the original one are representations of a periodic train of impulses.

In this result,

$$\sum_{n=-\infty}^{\infty} \delta(t + nT) = \frac{1}{T} \sum_{n=-\infty}^{\infty} e^{-2\pi nit/T},$$

we now make the changes of variable $t \to \omega$, $n \to -n$ and $T \to 2\pi/X$ and obtain

$$\sum_{n=-\infty}^{\infty} \delta\left(\omega - \frac{2\pi n}{X} \right) = \frac{X}{2\pi} \sum_{n=-\infty}^{\infty} e^{inX\omega}. \quad (*)$$

If we denote $f(t + nX)$ by $f_{nX}(t)$ then, by the translation theorem, we have $\tilde{f}_{nX}(\omega) = e^{inX\omega}\tilde{f}(\omega)$ and

$$f(t + nX) = \frac{1}{\sqrt{2\pi}} \int_{-\infty}^{\infty} \tilde{f}_{nX}(\omega) e^{i\omega t} \, d\omega$$

$$= \frac{1}{\sqrt{2\pi}} \int_{-\infty}^{\infty} e^{inX\omega} \tilde{f}(\omega) e^{i\omega t} \, d\omega,$$

$$\sum_{n=-\infty}^{\infty} f(t + nX) = \frac{1}{\sqrt{2\pi}} \int_{-\infty}^{\infty} \tilde{f}(\omega) e^{i\omega t} \sum_{n=-\infty}^{\infty} e^{inX\omega} \, d\omega, \quad \text{use } (*) \text{ above,}$$

$$= \frac{1}{\sqrt{2\pi}} \int_{-\infty}^{\infty} \tilde{f}(\omega) e^{i\omega t} \frac{2\pi}{X} \sum_{n=-\infty}^{\infty} \delta\left(\omega - \frac{2\pi n}{X}\right) \, d\omega$$

$$= \frac{\sqrt{2\pi}}{X} \sum_{n=-\infty}^{\infty} \tilde{f}\left(\frac{2\pi n}{X}\right) e^{i2\pi nt/X}.$$

In the final line we have made use of the properties of a δ-function when it appears as a factor in an integrand.

13.11 *For a function $f(t)$ that is non-zero only in the range $|t| < T/2$, the full frequency spectrum $\tilde{f}(\omega)$ can be constructed, in principle exactly, from values at discrete sample points $\omega = n(2\pi/T)$. Prove this as follows.*

(a) *Show that the coefficients of a complex Fourier series representation of $f(t)$ with period T can be written as*

$$c_n = \frac{\sqrt{2\pi}}{T} \tilde{f}\left(\frac{2\pi n}{T}\right).$$

(b) *Use this result to represent $f(t)$ as an infinite sum in the defining integral for $\tilde{f}(\omega)$, and hence show that*

$$\tilde{f}(\omega) = \sum_{n=-\infty}^{\infty} \tilde{f}\left(\frac{2\pi n}{T}\right) \text{sinc}\left(n\pi - \frac{\omega T}{2}\right),$$

where $\text{sinc}\, x$ is defined as $(\sin x)/x$.

(a) The complex coefficients for the Fourier series for $f(t)$ are given by

$$c_n = \frac{1}{T} \int_{-T/2}^{T/2} f(t) e^{-i2\pi nt/T} \, dt.$$

But, we also know that the Fourier transform of $f(t)$ is given by

$$\tilde{f}(\omega) = \frac{1}{\sqrt{2\pi}} \int_{-\infty}^{\infty} f(t) e^{-i\omega t} \, dt = \frac{1}{\sqrt{2\pi}} \int_{-T/2}^{T/2} f(t) e^{-i\omega t} \, dt.$$

Comparison of these two equations shows that $c_n = \frac{1}{T}\sqrt{2\pi}\tilde{f}\left(\frac{2\pi n}{T}\right)$.

(b) Using the Fourier series representation of $f(t)$, the frequency spectrum at a general frequency ω can now be constructed as

$$\tilde{f}(\omega) = \frac{1}{\sqrt{2\pi}} \int_{-T/2}^{T/2} f(t)\,e^{-i\omega t}\,dt$$

$$= \frac{1}{\sqrt{2\pi}} \int_{-T/2}^{T/2} \left[\sum_{n=-\infty}^{\infty} \frac{1}{T}\sqrt{2\pi}\tilde{f}\left(\frac{2\pi n}{T}\right) e^{i2\pi nt/T}\right] e^{-i\omega t}\,dt$$

$$= \frac{1}{T}\sum_{n=-\infty}^{\infty} \tilde{f}\left(\frac{2\pi n}{T}\right) \frac{2\sin\left(\frac{2\pi n}{2} - \frac{\omega T}{2}\right)}{\frac{2\pi n}{T} - \omega} = \sum_{n=-\infty}^{\infty} \tilde{f}\left(\frac{2\pi n}{T}\right) \operatorname{sinc}\left(n\pi - \frac{\omega T}{2}\right).$$

This final formula gives a prescription for calculating the frequency spectrum $\tilde{f}(\omega)$ of $f(t)$ for *any* ω, given the spectrum at the (admittedly infinite number of) discrete values $\omega = 2\pi n/T$. The sinc functions give the weights to be assigned to the known discrete values; of course, the weights vary as ω is varied, with, as expected, the largest weights for the nth contribution occurring when ω is close to $2\pi n/T$.

13.13 *Find the Fourier transform specified in part* (a) *and then use it to answer part* (b).

(a) *Find the Fourier transform of*

$$f(\gamma, p, t) = \begin{cases} e^{-\gamma t}\sin pt & t > 0 \\ 0 & t < 0, \end{cases}$$

where $\gamma\,(>0)$ and p are constant parameters.

(b) *The current $I(t)$ flowing through a certain system is related to the applied voltage $V(t)$ by the equation*

$$I(t) = \int_{-\infty}^{\infty} K(t-u)V(u)\,du,$$

where

$$K(\tau) = a_1 f(\gamma_1, p_1, \tau) + a_2 f(\gamma_2, p_2, \tau).$$

The function $f(\gamma, p, t)$ is as given in part (a) *and all the $a_i, \gamma_i\,(>0)$ and p_i are fixed parameters. By considering the Fourier transform of $I(t)$, find the relationship that must hold between a_1 and a_2 if the total net charge Q passed through the system (over a very long time) is to be zero for an arbitrary applied voltage.*

(a) Write the given sine function in terms of exponential functions. Its Fourier transform is then easily calculated as

$$\tilde{f}(\omega, \gamma, p) = \frac{1}{\sqrt{2\pi}} \int_0^\infty \frac{e^{(-\gamma - i\omega + ip)t} - e^{(-\gamma - i\omega - ip)t}}{2i} \, dt$$

$$= \frac{1}{\sqrt{2\pi}} \frac{1}{2i} \left(\frac{-1}{-\gamma - i\omega + ip} + \frac{1}{-\gamma - i\omega - ip} \right)$$

$$= \frac{1}{\sqrt{2\pi}} \frac{p}{(\gamma + i\omega)^2 + p^2}.$$

(b) Since the current is given by the convolution

$$I(t) = \int_{-\infty}^\infty K(t - u) V(u) \, du,$$

the convolution theorem implies that the Fourier transforms of I, K and V are related by $\tilde{I}(\omega) = \sqrt{2\pi}\, \tilde{K}(\omega)\, \tilde{V}(\omega)$ with, from part (a),

$$\tilde{K}(\omega) = \frac{1}{\sqrt{2\pi}} \left[\frac{a_1 p_1}{(\gamma_1 + i\omega)^2 + p_1^2} + \frac{a_2 p_2}{(\gamma_2 + i\omega)^2 + p_2^2} \right].$$

Now, by expressing $I(t')$ in its Fourier integral form, we can write

$$Q(\infty) = \int_{-\infty}^\infty I(t') \, dt' = \int_{-\infty}^\infty dt' \int_{-\infty}^\infty \frac{1}{\sqrt{2\pi}} \tilde{I}(\omega) e^{i\omega t'} \, d\omega.$$

But $\int_{-\infty}^\infty e^{i\omega t'} \, dt' = 2\pi\delta(\omega)$ and so

$$Q(\infty) = \int_{-\infty}^\infty \frac{1}{\sqrt{2\pi}} \tilde{I}(\omega)\, 2\pi\delta(\omega) \, d\omega$$

$$= \frac{2\pi}{\sqrt{2\pi}} \tilde{I}(0) = \sqrt{2\pi}\, \sqrt{2\pi}\, \tilde{K}(0)\, \tilde{V}(0)$$

$$= 2\pi \frac{1}{\sqrt{2\pi}} \left[\frac{a_1 p_1}{\gamma_1^2 + p_1^2} + \frac{a_2 p_2}{\gamma_2^2 + p_2^2} \right] \tilde{V}(0).$$

For $Q(\infty)$ to be zero for an arbitrary $V(t)$, we must have

$$\frac{a_1 p_1}{\gamma_1^2 + p_1^2} + \frac{a_2 p_2}{\gamma_2^2 + p_2^2} = 0,$$

and so this is the required relationship.

13.15 *Show that the Fourier transform of $tf(t)$ is $id\tilde{f}(\omega)/d\omega$.*

A linear amplifier produces an output that is the convolution of its input and its response function. The Fourier transform of the response function for a particular amplifier is

$$\tilde{K}(\omega) = \frac{i\omega}{\sqrt{2\pi}(\alpha + i\omega)^2}.$$

Determine the time variation of its output $g(t)$ when its input is the Heaviside step function.

This result is immediate, since differentiating the definition of a Fourier transform (under the integral sign) gives

$$i\frac{d\tilde{f}(\omega)}{d\omega} = \frac{i}{\sqrt{2\pi}}\frac{\partial}{\partial\omega}\left(\int_{-\infty}^{\infty} f(t)\, e^{-i\omega t}\, dt\right) = \frac{-i^2}{\sqrt{2\pi}}\int_{-\infty}^{\infty} tf(t)\, e^{-i\omega t}\, dt,$$

i.e. the transform of $tf(t)$.

Since the amplifier's output is the convolution of its input and response function, we will need the Fourier transforms of both to determine that of its output (using the convolution theorem). We already have that of its response function.

The input Heaviside step function $H(t)$ has a Fourier transform

$$\tilde{H}(\omega) = \frac{1}{\sqrt{2\pi}}\int_{-\infty}^{\infty} H(t)\, e^{-i\omega t}\, dt = \frac{1}{\sqrt{2\pi}}\int_{0}^{\infty} e^{-i\omega t}\, dt = \frac{1}{\sqrt{2\pi}}\frac{1}{i\omega}.$$

Thus, using the convolution theorem,

$$\begin{aligned}
\tilde{g}(\omega) &= \sqrt{2\pi}\,\frac{i\omega}{\sqrt{2\pi}(\alpha + i\omega)^2}\,\frac{1}{\sqrt{2\pi}}\frac{1}{i\omega} \\
&= \frac{1}{\sqrt{2\pi}}\frac{1}{(\alpha + i\omega)^2} \\
&= \frac{i}{\sqrt{2\pi}}\frac{\partial}{\partial\omega}\left(\frac{1}{\alpha + i\omega}\right) \\
&= i\frac{\partial}{\partial\omega}\left\{\mathscr{F}\left[e^{-\alpha t}H(t)\right]\right\} \\
&= \mathscr{F}\left[te^{-\alpha t}H(t)\right],
\end{aligned}$$

where we have used the 'library' result to recognise the transform of a decaying exponential in the penultimate line and the result proved above in the final step. The output of the amplifier is therefore of the form $g(t) = te^{-\alpha t}$ for $t > 0$ when its input takes the form of the Heaviside step function.

13.17 *In quantum mechanics, two equal-mass particles having momenta $\mathbf{p}_j = \hbar\mathbf{k}_j$ and energies $E_j = \hbar\omega_j$ and represented by plane wavefunctions $\phi_j = \exp[i(\mathbf{k}_j \cdot \mathbf{r}_j - \omega_j t)]$, $j = 1, 2$, interact through a potential $V = V(|\mathbf{r}_1 - \mathbf{r}_2|)$. In first-order perturbation theory the probability of scattering to a state with momenta and energies $\mathbf{p}'_j, E'_j$ is determined by the modulus squared of the quantity*

$$M = \iiint \psi_f^* V \psi_i \, d\mathbf{r}_1 \, d\mathbf{r}_2 \, dt.$$

The initial state ψ_i is $\phi_1\phi_2$ and the final state ψ_f is $\phi'_1\phi'_2$. It can be shown that M is proportional to the Fourier transform of V, i.e. to $\tilde{V}(\mathbf{k})$, where $2\hbar\mathbf{k} = (\mathbf{p}_2 - \mathbf{p}_1) - (\mathbf{p}'_2 - \mathbf{p}'_1)$.

For some ion–atom scattering processes, the spherically symmetric potential $V(\mathbf{r})$ may be approximated by $V = |\mathbf{r}_1 - \mathbf{r}_2|^{-1} \exp(-\mu|\mathbf{r}_1 - \mathbf{r}_2|)$. Show that the probability that the ion will scatter from, say, $\mathbf{p}_1$ to $\mathbf{p}'_1$ is proportional to $(\mu^2 + k^2)^{-2}$, where $k = |\mathbf{k}|$ and $\mathbf{k}$ is as given above.

We start by showing how to reduce the three-dimensional Fourier transform to a one-dimensional one whenever $V(\mathbf{r})$ is spherically symmetrical, i.e. $V(\mathbf{r}) = V(r)$. This result will be a general one and is not restricted to this particular example.

Choose spherical polar coordinates in which the vector $\mathbf{k}$ of the Fourier transform lies along the polar axis ($\theta = 0$); this can be done since $V(\mathbf{r})$ is spherically symmetric. We then have

$$d^3\mathbf{r} = r^2 \sin\theta \, dr \, d\theta \, d\phi \quad \text{and} \quad \mathbf{k} \cdot \mathbf{r} = kr\cos\theta,$$

where $k = |\mathbf{k}|$. The Fourier transform is given by

$$\begin{aligned}
\tilde{V}(\mathbf{k}) &= \frac{1}{(2\pi)^{3/2}} \int V(\mathbf{r}) e^{-i\mathbf{k}\cdot\mathbf{r}} \, d^3\mathbf{r} \\
&= \frac{1}{(2\pi)^{3/2}} \int_0^\infty dr \int_0^\pi d\theta \int_0^{2\pi} d\phi \, V(r) r^2 \sin\theta \, e^{-ikr\cos\theta} \\
&= \frac{1}{(2\pi)^{3/2}} \int_0^\infty dr \, 2\pi V(r) r^2 \int_0^\pi d\theta \, \sin\theta \, e^{-ikr\cos\theta}.
\end{aligned}$$

The integral over θ may be evaluated straightforwardly by noting that

$$\frac{d}{d\theta}(e^{-ikr\cos\theta}) = ikr\sin\theta \, e^{-ikr\cos\theta}.$$

This enables us to carry through the angular integration over θ and so reduce the multiple integral to a one-dimensional integral over the radial coordinate:

$$\tilde{V}(\mathbf{k}) = \frac{1}{(2\pi)^{3/2}} \int_0^\infty dr\, 2\pi V(r) r^2 \left[\frac{e^{-ikr\cos\theta}}{ikr} \right]_{\theta=0}^{\theta=\pi}$$

$$= \frac{1}{(2\pi)^{3/2}} \int_0^\infty 4\pi r^2 V(r) \left(\frac{\sin kr}{kr} \right) dr$$

$$= \frac{1}{(2\pi)^{3/2}k} \int_0^\infty 4\pi V(r) r \sin kr\, dr.$$

The ion–atom interaction potential in this particular example is $V(r) = r^{-1} \exp(-\mu r)$. As this is spherically symmetric, we may apply the result just derived to it. Substituting for $V(r)$ gives

$$M \propto \tilde{V}(\mathbf{k}) \propto \frac{1}{k} \int_0^\infty \frac{e^{-\mu r}}{r} r \sin kr\, dr$$

$$= \frac{1}{k} \,\mathrm{Im} \int_0^\infty e^{-\mu r + ikr}\, dr$$

$$= \frac{1}{k} \,\mathrm{Im} \left[\frac{-1}{-\mu + ik} \right]$$

$$= \frac{1}{k} \frac{k}{\mu^2 + k^2}.$$

Since the probability of the ion scattering from $\mathbf{p}_1$ to $\mathbf{p}'_1$ is proportional to the modulus squared of M, the probability is $\propto |M|^2 \propto (\mu^2 + k^2)^{-2}$.

13.19 *Calculate directly the auto-correlation function $a(z)$ for the product $f(t)$ of the exponential decay distribution and the Heaviside step function,*

$$f(t) = \frac{1}{\lambda} e^{-\lambda t} H(t).$$

Use the Fourier transform and energy spectrum of $f(t)$ to deduce that

$$\int_{-\infty}^\infty \frac{e^{i\omega z}}{\lambda^2 + \omega^2}\, d\omega = \frac{\pi}{\lambda} e^{-\lambda|z|}.$$

By definition,

$$a(z) = \int_{-\infty}^{\infty} \frac{1}{\lambda} e^{-\lambda t} H(t) \frac{1}{\lambda} e^{-\lambda(t+z)} H(t+z) \, dt$$

$$= \frac{e^{-\lambda z}}{\lambda^2} \int_{z_0}^{\infty} e^{-2\lambda t} \, dt,$$

where $z_0 = 0$ for $z > 0$ and $z_0 = |z|$ for $z < 0$; so

$$a(z) = \frac{e^{-\lambda z}}{\lambda^2} \left[\frac{e^{-2\lambda t}}{-2\lambda} \right]_{z_0}^{\infty}$$

$$= \frac{e^{-\lambda(z+2z_0)}}{2\lambda^3} = \frac{e^{-\lambda|z|}}{2\lambda^3}.$$

The Fourier transform of $f(t)$ is given by

$$\tilde{f}(\omega) = \frac{1}{\sqrt{2\pi}} \int_0^{\infty} \frac{1}{\lambda} e^{-\lambda t} e^{-i\omega t} \, dt = \frac{1}{\sqrt{2\pi}\lambda(\lambda + i\omega)}.$$

The special case of the Wiener–Kinchin theorem in which both functions are the same shows that the inverse Fourier transform of the energy spectrum, $\sqrt{2\pi}|\tilde{f}(\omega)|^2$, is equal to the auto-correlation function, i.e.

$$\frac{1}{\sqrt{2\pi}} \int_{-\infty}^{\infty} \sqrt{2\pi} \frac{e^{i\omega z}}{2\pi\lambda^2(\lambda^2 + \omega^2)} \, d\omega = \frac{e^{-\lambda|z|}}{2\lambda^3},$$

from which the stated result follows immediately.

13.21 *Find the Laplace transforms of $t^{-1/2}$ and $t^{1/2}$, by setting $x^2 = ts$ in the result*

$$\int_0^{\infty} \exp(-x^2) \, dx = \tfrac{1}{2}\sqrt{\pi}.$$

Setting $x^2 = st$, and hence $2x \, dx = s \, dt$ and $dx = s \, dt/(2\sqrt{st})$, we obtain

$$\int_0^{\infty} e^{-st} \frac{\sqrt{s}}{2} t^{-1/2} \, dt = \frac{\sqrt{\pi}}{2},$$

$$\Rightarrow \quad \mathcal{L}\left[t^{-1/2}\right] \equiv \int_0^{\infty} t^{-1/2} e^{-st} \, dt = \sqrt{\frac{\pi}{s}}.$$

Integrating the LHS of this result by parts yields

$$\left[e^{-st} 2t^{1/2} \right]_0^{\infty} - \int_0^{\infty} (-s) e^{-st} 2t^{1/2} \, dt = \sqrt{\frac{\pi}{s}}.$$

The first term vanishes at both limits, whilst the second is a multiple of the required Laplace transform of $t^{1/2}$. Hence,

$$\mathscr{L}\left[t^{1/2}\right] \equiv \int_0^\infty e^{-st} t^{1/2} \, dt = \frac{1}{2s}\sqrt{\frac{\pi}{s}}.$$

13.23 *Use the properties of Laplace transforms to prove the following without evaluating any Laplace integrals explicitly:*

(a) $\mathscr{L}\left[t^{5/2}\right] = \frac{15}{8}\sqrt{\pi}s^{-7/2}$;

(b) $\mathscr{L}\left[(\sinh at)/t\right] = \frac{1}{2}\ln\left[(s+a)/(s-a)\right], \quad s > |a|$;

(c) $\mathscr{L}\left[\sinh at \cos bt\right] = a(s^2 - a^2 + b^2)[(s-a)^2 + b^2]^{-1}[(s+a)^2 + b^2]^{-1}$.

(a) We use the general result for Laplace transforms that

$$\mathscr{L}\left[t^n f(t)\right] = (-1)^n \frac{d^n \bar{f}(s)}{ds^n}, \quad \text{for } n = 1, 2, 3, \ldots.$$

If we take $n = 2$, then $f(t)$ becomes $t^{1/2}$, for which we found the Laplace transform in exercise 13.21:

$$\mathscr{L}\left[t^{5/2}\right] = \mathscr{L}\left[t^2 \, t^{1/2}\right] = (-1)^2 \frac{d^2}{ds^2}\left(\frac{\sqrt{\pi}s^{-3/2}}{2}\right)$$

$$= \frac{\sqrt{\pi}}{2}\left(-\frac{3}{2}\right)\left(-\frac{5}{2}\right)s^{-7/2} = \frac{15\sqrt{\pi}}{8}s^{-7/2}.$$

(b) Here we apply a second general result for Laplace transforms which states that

$$\mathscr{L}\left[\frac{f(t)}{t}\right] = \int_s^\infty \bar{f}(u) \, du,$$

provided $\lim_{t\to 0}[f(t)/t]$ exists, which it does in this case.

$$\mathscr{L}\left[\frac{\sinh(at)}{t}\right] = \int_s^\infty \frac{a}{u^2 - a^2} \, du, \quad u > |a|,$$

$$= \frac{1}{2}\int_s^\infty \left(\frac{1}{u-a} - \frac{1}{u+a}\right) du$$

$$= \frac{1}{2}\ln\left(\frac{s+a}{s-a}\right), \quad s > |a|.$$

(c) The translation property of Laplace transforms can be used here to deal with

223

the sinh(at) factor, as it can be expressed in terms of exponential functions:

$$\mathscr{L}\left[\sinh(at)\cos(bt)\right] = \mathscr{L}\left[\tfrac{1}{2}e^{at}\cos(bt)\right] - \mathscr{L}\left[\tfrac{1}{2}e^{-at}\cos(bt)\right]$$

$$= \frac{1}{2}\frac{s-a}{(s-a)^2+b^2} - \frac{1}{2}\frac{s+a}{(s+a)^2+b^2}$$

$$= \frac{1}{2}\frac{(s^2-a^2)2a+2ab^2}{[(s-a)^2+b^2][(s+a)^2+b^2]}$$

$$= \frac{a(s^2-a^2+b^2)}{[(s-a)^2+b^2][(s+a)^2+b^2]}.$$

The result is valid for $s > |a|$.

13.25 *This exercise is concerned with the limiting behaviour of Laplace transforms.*

(a) *If $f(t) = A + g(t)$, where A is a constant and the indefinite integral of $g(t)$ is bounded as its upper limit tends to ∞, show that*

$$\lim_{s\to0} s\bar{f}(s) = A.$$

(a) *For $t > 0$, the function $y(t)$ obeys the differential equation*

$$\frac{d^2y}{dt^2} + a\frac{dy}{dt} + by = c\cos^2\omega t,$$

where a, b and c are positive constants. Find $\bar{y}(s)$ and show that $s\bar{y}(s) \to c/2b$ as $s \to 0$. Interpret the result in the t-domain.

(a) From the definition,

$$\bar{f}(s) = \int_0^\infty [A + g(t)]\, e^{-st}\, dt$$

$$= \left[\frac{A\,e^{-st}}{-s}\right]_0^\infty + \lim_{T\to\infty}\int_0^T g(t)\, e^{-st}\, dt,$$

$$s\bar{f}(s) = A + s\lim_{T\to\infty}\int_0^T g(t)\, e^{-st}\, dt.$$

Now, for $s \geq 0$,

$$\left|\lim_{T\to\infty}\int_0^T g(t)\, e^{-st}\, dt\right| \leq \left|\lim_{T\to\infty}\int_0^T g(t)\, dt\right| < B, \text{ say.}$$

Thus, taking the limit $s \to 0$,

$$\lim_{s\to0} s\bar{f}(s) = A \pm \lim_{s\to0} sB = A.$$

(b) We will need

$$\mathscr{L}\left[\cos^2 \omega t\right] = \mathscr{L}\left[\tfrac{1}{2}\cos 2\omega + \tfrac{1}{2}\right] = \frac{s}{2(s^2 + 4\omega^2)} + \frac{1}{2s}.$$

Taking the transform of the differential equation yields

$$-y'(0) - sy(0) + s^2\bar{y} + a[-y(0) + s\bar{y}] + b\bar{y} = c\left[\frac{s}{2(s^2 + 4\omega^2)} + \frac{1}{2s}\right].$$

This can be rearranged as

$$s\bar{y} = \frac{c\left(\dfrac{s^2}{2(s^2 + 4\omega^2)} + \dfrac{1}{2}\right) + sy'(0) + asy(0) + s^2 y(0)}{s^2 + as + b}.$$

In the limit $s \to 0$, this tends to $(c/2)/b = c/(2b)$, a value independent of that of a and the initial values of y and y'.

The $s = 0$ component of the transform corresponds to long-term values, when a steady state has been reached and rates of change are negligible. With the first two terms of the differential equation ignored, it reduces to $by = c\cos^2 \omega t$, and, as the average value of $\cos^2 \omega t$ is $\tfrac{1}{2}$, the solution is the more or less steady value of $y = \tfrac{1}{2}c/b$.

13.27 *The function $f_a(x)$ is defined as unity for $0 < x < a$ and zero otherwise. Find its Laplace transform $\bar{f}_a(s)$ and deduce that the transform of $xf_a(x)$ is*

$$\frac{1}{s^2}\left[1 - (1 + as)e^{-sa}\right].$$

Write $f_a(x)$ in terms of Heaviside functions and hence obtain an explicit expression for

$$g_a(x) = \int_0^x f_a(y)f_a(x - y)\, dy.$$

Use the expression to write $\bar{g}_a(s)$ in terms of the functions $\bar{f}_a(s)$ and $\bar{f}_{2a}(s)$, and their derivatives, and hence show that $\bar{g}_a(s)$ is equal to the square of $\bar{f}_a(s)$, in accordance with the convolution theorem.

From their definitions,

$$\bar{f}_a(s) = \int_0^a 1\, e^{-sx}\, dx = \frac{1}{s}(1 - e^{-sa}),$$

$$\int_0^a x f_a(x)\, e^{-sx}\, dx = -\frac{d\bar{f}_a}{ds} = \frac{1}{s^2}(1 - e^{-sa}) - \frac{a}{s}e^{-sa}$$

$$= \frac{1}{s^2}\left[1 - (1 + as)e^{-sa}\right]. \qquad (*)$$

225

In terms of Heaviside functions,

$$f(x) = H(x) - H(x - a),$$

and so the expression for $g_a(x) = \int_0^x f_a(y) f_a(x - y) \, dy$ is

$$\int_{-\infty}^{\infty} [H(y) - H(y - a)] [H(x - y) - H(x - y - a)] \, dy.$$

This can be expanded as the sum of four integrals, each of which contains the common factors $H(y)$ and $H(x - y)$, implying that, in all cases, unless x is positive and greater than y, the integral has zero value. The other factors in the four integrands are generated analogously to the terms of the expansion $(a - b)(c - d) = ac - ad - bc + bd$:

$$\int_{-\infty}^{\infty} H(y)H(x - y) \, dy$$

$$- \int_{-\infty}^{\infty} H(y)H(x - y - a) \, dy$$

$$- \int_{-\infty}^{\infty} H(y - a)H(x - y) \, dy$$

$$+ \int_{-\infty}^{\infty} H(y - a)H(x - y - a) \, dy.$$

In all four integrals the integrand is either 0 or 1 and the value of each integral is equal to the length of the y-interval in which the integrand is non-zero.

- The first integral requires $0 < y < x$ and therefore has value x for $x > 0$.
- The second integral requires $0 < y < x - a$ and therefore has value $x - a$ for $x > a$ and 0 for $x < a$.
- The third integral requires $a < y < x$ and therefore has value $x - a$ for $x > a$ and 0 for $x < a$.
- The final integral requires $a < y < x - a$ and therefore has value $x - 2a$ for $x > 2a$ and 0 for $x < 2a$.

Collecting these together:

$$
\begin{aligned}
x < 0 \qquad & g_a(x) = 0 - 0 - 0 + 0 = 0, \\
0 < x < a \qquad & g_a(x) = x - 0 - 0 + 0 = x, \\
a < x < 2a \qquad & g_a(x) = x - (x - a) - (x - a) + 0 = 2a - x, \\
2a < x \qquad & g_a(x) = x - (x - a) - (x - a) + (x - 2a) = 0.
\end{aligned}
$$

Consequently, the transform of $g_a(x)$ is given by

$$\bar{g}_a(s) = \int_0^a xe^{-sx}\,dx + \int_a^{2a} (2a - x)e^{-sx}\,dx$$

$$= -\int_0^{2a} xe^{-sx}\,dx + 2\int_0^a xe^{-sx}\,dx + 2a\int_a^{2a} e^{-sx}\,dx$$

$$= -\frac{1}{s^2}\left[1 - (1 + 2as)e^{-2sa}\right] + \frac{2}{s^2}\left[1 - (1 + as)e^{-sa}\right]$$

$$\quad + \frac{2a}{s}(e^{-sa} - e^{-2sa})$$

$$= \frac{1}{s^2}(1 - 2e^{-sa} + e^{-2sa})$$

$$= \frac{1}{s^2}(1 - e^{-as})^2 = [\bar{f}_a(s)]^2,$$

which is as expected. In order to adjust the integral limits in the second line, we both added and subtracted

$$\int_0^a (-x)e^{-sx}\,dx.$$

In the third line we used the result $(*)$ twice, once as it stands and once with a replaced by $2a$.

227

14

First-order ordinary differential equations

14.1 *A radioactive isotope decays in such a way that the number of atoms present at a given time, $N(t)$, obeys the equation*

$$\frac{dN}{dt} = -\lambda N.$$

If there are initially N_0 atoms present, find $N(t)$ at later times.

This is a straightforward separable equation with a well known solution:

$$\frac{dN}{dt} = -\lambda N.$$

Separating the variables,
$$\frac{dN}{N} = -\lambda \, dt.$$

Integrating,
$$\ln N(t) - \ln N(0) = -\lambda(t - 0).$$

Thus, since $N(0) = N_0$, we have that, at a later time,

$$N(t) = N_0 e^{-\lambda t}.$$

14.3 *Show that the following equations either are exact or can be made exact, and solve them:*

   (a) $y(2x^2y^2 + 1)y' + x(y^4 + 1) = 0$;
   (b) $2xy' + 3x + y = 0$;
   (c) $(\cos^2 x + y \sin 2x)y' + y^2 = 0$.

In general, given an equation expressed in the form $A \, dx + B \, dy = 0$, we consider the function

$$h(x, y) = \frac{1}{B} \left[\frac{\partial A}{\partial y} - \frac{\partial B}{\partial x} \right].$$

If this expression is zero, then the equation is exact and can be integrated as it stands to give a solution of the form $f(x, y) = c$. Even if $g(x, y)$ is non-zero, if it is a function of x alone then

$$\mu(x) = \exp \left\{ \int g(x) \, dx \right\}$$

provides an integrating factor (IF) that will make the equation exact. Similar considerations apply if $g(x, y)$ is a function of y alone. If g does actually depend on both x and y, then, in general, no further progress can be made using this method.

(a) Following the above procedure, we consider

$$h(x, y) = \frac{1}{2x^2 y^3 + y} \left[\frac{\partial}{\partial y} (xy^4 + x) - \frac{\partial}{\partial x} (2x^2 y^3 + y) \right] = \frac{4xy^3 - 4xy^3}{2x^2 y^3 + y} = 0.$$

It follows that the equation is exact and can be integrated as it stands:

$$c = f(x, y) = \int (2x^2 y^3 + y) \, dy + g(x)$$

$$= \frac{1}{2} x^2 y^4 + \frac{1}{2} y^2 + g(x), \quad \text{where}$$

$$xy^4 + x = \frac{\partial f}{\partial x} = xy^4 + 0 + g'(x), \quad \Rightarrow \quad g(x) = \frac{1}{2} x^2 + k,$$

$$\Rightarrow \quad c = f(x, y) = \frac{1}{2} (x^2 y^4 + y^2 + x^2).$$

The common factor of $\frac{1}{2}$ on the RHS can, of course, be absorbed into the constant on the LHS and has no particular significance.

(b) Again following the procedure, we consider

$$h(x, y) = \frac{1}{2x} \left[\frac{\partial}{\partial y} (3x + y) - \frac{\partial}{\partial x} (2x) \right] = -\frac{1}{2x}.$$

This is non-zero and implies that the equation is not exact. However, it is a function of x alone and so there is an IF given by

$$\mu(x) = \exp \left\{ \int -\frac{1}{2x} \, dx \right\} = \exp(-\frac{1}{2} \ln x) = \frac{1}{x^{1/2}}.$$

The exact equation is thus

$$2x^{1/2} \, dy + (3x^{1/2} + yx^{-1/2}) \, dx = 0,$$

229

and this can now be integrated:

$$c = f(x, y) = \int 2x^{1/2} \, dy + g(x)$$
$$= 2x^{1/2}y + g(x), \quad \text{where}$$
$$3x^{1/2} + yx^{-1/2} = \frac{\partial f}{\partial x} = x^{-1/2}y + g'(x), \quad \Rightarrow \quad g(x) = 2x^{3/2} + k,$$
$$\Rightarrow \quad c = f(x, y) = 2(x^{1/2}y + x^{3/2}).$$

Again, the overall numerical multiplicative factor on the RHS has no particular significance.

(c) Following the same general procedure,

$$h(x, y) = \frac{1}{\cos^2 x + y \sin 2x} \left[\frac{\partial}{\partial y}(y^2) - \frac{\partial}{\partial x}(\cos^2 x + y \sin 2x) \right]$$
$$= \frac{1}{\cos^2 x + y \sin 2x}(2y + \sin 2x - 2y \cos 2x)$$
$$= \frac{4y \sin^2 x + 2 \sin x \cos x}{\cos^2 x + y \sin 2x}$$
$$= \frac{2 \sin x(2y \sin x + \cos x)}{\cos x(\cos x + 2y \sin x)} = 2 \tan x.$$

This is non-zero and implies that the equation is not exact. However, it is a function of x alone and so there is an IF given by

$$\mu(x) = \exp \left\{ \int 2 \tan x \, dx \right\} = \exp(-2 \ln \cos x) = \frac{1}{\cos^2 x}.$$

The exact equation is thus

$$(1 + 2y \tan x) \, dy + y^2 \sec^2 x \, dx = 0,$$

and this can now be integrated:

$$c = f(x, y) = \int (1 + 2y \tan x) \, dy + g(x)$$
$$= y + y^2 \tan x + g(x), \quad \text{where}$$
$$y^2 \sec^2 x = \frac{\partial f}{\partial x} = 0 + y^2 \sec^2 x + g'(x), \quad \Rightarrow \quad g(x) = k,$$
$$\Rightarrow \quad c = f(x, y) = y + y^2 \tan x.$$

14.5 *By finding a suitable integrating factor, solve the following equations:*

(a) $(1 - x^2)y' + 2xy = (1 - x^2)^{3/2}$;
(b) $y' - y \cot x + \operatorname{cosec} x = 0$;
(c) $(x + y^3)y' = y$ *(treat y as the independent variable).*

(a) In standard form this is

$$y' + \frac{2xy}{1 - x^2} = (1 - x^2)^{1/2}.$$

The IF for this standard form is

$$\mu(x) = \exp\left\{ \int \frac{2x}{1 - x^2}\, dx \right\} = \exp[-\ln(1 - x^2)] = \frac{1}{1 - x^2},$$

i.e. $(1 - x^2)^{-2}$ for the original form. Applying it gives

$$\frac{y'}{1 - x^2} + \frac{2xy}{(1 - x^2)^2} = \frac{1}{(1 - x^2)^{1/2}},$$

$$\frac{d}{dx}\left(\frac{y}{1 - x^2} \right) = \frac{1}{(1 - x^2)^{1/2}},$$

$$\frac{y}{1 - x^2} = \sin^{-1} x + k,$$

$$\Rightarrow \quad y = (1 - x^2)(\sin^{-1} x + k).$$

(b) In standard form this is

$$y' - \frac{y \cos x}{\sin x} = -\frac{1}{\sin x}.$$

The IF for this standard form is given by

$$\mu(x) = \exp\left\{ -\int \frac{\cos x}{\sin x}\, dx \right\} = \exp[-\ln(\sin x)] = \frac{1}{\sin x}.$$

Applying it gives

$$\frac{y'}{\sin x} - \frac{y \cos x}{\sin^2 x} = -\frac{1}{\sin^2 x},$$

$$\frac{d}{dx}\left(\frac{y}{\sin x} \right) = -\operatorname{cosec}^2 x,$$

$$\frac{y}{\sin x} = \cot x + k,$$

$$\Rightarrow \quad y = \cos x + k \sin x.$$

(c) Rearranging this to make y the independent variable,

$$\frac{dx}{dy} - \frac{x}{y} = y^2.$$

231

By inspection (or by the standard method) the IF is y^{-1}, yielding

$$\frac{1}{y}\frac{dx}{dy} - \frac{x}{y^2} = \frac{y^2}{y}$$

$$\frac{d}{dy}\left(\frac{x}{y}\right) = y,$$

$$\frac{x}{y} = \frac{1}{2}y^2 + k,$$

$$\Rightarrow \quad x = \tfrac{1}{2}y^3 + ky.$$

14.7 *Find, in the form of an integral, the solution of the equation*

$$\alpha\frac{dy}{dt} + y = f(t)$$

for a general function $f(t)$. Find the specific solutions for

    (a) $f(t) = H(t)$,
    (b) $f(t) = \delta(t)$,
    (c) $f(t) = \beta^{-1}e^{-t/\beta}H(t)$ *with $\beta < \alpha$.*

For case (c), what happens if $\beta \to 0$?

The IF needed for the standard form is $\exp[\int \alpha^{-1}\,dt]$, i.e. $e^{t/\alpha}$. The equation then reads

$$e^{t/\alpha}\frac{dy}{dt} + \frac{y\,e^{t/\alpha}}{\alpha} = \frac{f(t)\,e^{t/\alpha}}{\alpha},$$

$$\frac{d}{dt}\left(y\,e^{t/\alpha}\right) = \frac{f(t)\,e^{t/\alpha}}{\alpha},$$

$$y(t) = e^{-t/\alpha}\int^t \frac{f(t')\,e^{t'/\alpha}}{\alpha}\,dt'.$$

We now apply this general result to the three specific cases.

(a) $f(t) = H(t)$, the Heaviside function. This is zero for $t < 0$ and so we can take the integral as running from 0 to t. The value of $H(t)$ for $t > 0$ is unity. Hence,

$$y(t) = e^{-t/\alpha}\int_0^t \frac{e^{t'/\alpha}}{\alpha}\,dt' = e^{-t/\alpha}[e^{t/\alpha} - 1] = 1 - e^{-t/\alpha}.$$

(b) With $f(t) = \delta(t)$, the integration will be trivial:

$$y(t) = e^{-t/\alpha}\int^t \frac{\delta(t')\,e^{t'/\alpha}}{\alpha}\,dt' = e^{-t/\alpha} \times \frac{1}{\alpha} = \frac{e^{-t/\alpha}}{\alpha}.$$

(c) For $f(t) = \beta^{-1} e^{-t/\beta} H(t)$, with $\beta < \alpha$, we have

$$
\begin{aligned}
y(t) &= e^{-t/\alpha} \int_0^t \frac{e^{t'/\alpha} e^{-t'/\beta}}{\alpha\beta} \, dt' \\
&= e^{-t/\alpha} \left[\frac{e^{(\alpha^{-1} - \beta^{-1})t'}}{\alpha\beta(\alpha^{-1} - \beta^{-1})} \right]_0^t \\
&= \frac{e^{-t/\beta}}{\beta - \alpha} - \frac{e^{-t/\alpha}}{\beta - \alpha} \\
&= \frac{e^{-t/\alpha} - e^{-t/\beta}}{\alpha - \beta}.
\end{aligned}
$$

As $\beta \to 0$, $f(t)$ becomes very strongly peaked near $t = 0$, but with the area under the peak remaining constant at unity. In the limit, the input $f(t)$ becomes a δ-function, the same as that in case (b). It can also be seen that in the same limit the solution $y(t)$ for case (c) tends to that for case (b), as is to be expected.

14.9 *A two-dimensional coordinate system that is useful for orbit problems is the tangential–polar coordinate system. In this system a curve is defined by r, the distance from a fixed point O to a general point P of the curve, and p, the perpendicular distance from O to the tangent to the curve at P. It can be shown that the instantaneous radius of curvature of the curve is given by $\rho = r \, dr/dp$.*

Using tangential–polar coordinates, consider a particle of mass m moving under the influence of a force f directed towards the origin O. By resolving forces along the instantaneous tangent and normal, prove that

$$
f = -mv \frac{dv}{dr} \qquad \text{and} \qquad mv^2 = fp \frac{dr}{dp}.
$$

Show further that $h = mpv$ is a constant of the motion and that the law of force can be deduced from

$$
f = \frac{h^2}{mp^3} \frac{dp}{dr}.
$$

Denote by ϕ the angle between the radius vector and the tangent to the orbit at any instant. Then, firstly, we note that $\cos\phi = dr/ds$, where s is the distance moved along the orbit curve and, secondly, that $p = r \sin\phi$.

Now we equate the tangential component of the central force $-f \cos\phi$ to the rate of change of the tangential momentum:

$$
-f \frac{dr}{ds} = -f \cos\phi = m \frac{dv}{dt} = m \frac{dv}{ds} \frac{ds}{dt} = mv \frac{dv}{ds}.
$$

Hence,

$$f = -mv \frac{dv}{ds} \frac{ds}{dr} = -mv \frac{dv}{dr}.$$

This is the first of the results.

Equating the normal component of the central force to that needed to keep the particle moving in an orbit with instantaneous radius of curvature $\rho = r\,dr/dp$ gives

$$\frac{mv^2}{\rho} = f \sin \phi = f \frac{p}{r} \quad \Rightarrow \quad mv^2 = f \frac{p}{r} r \frac{dr}{dp} = f p \frac{dr}{dp}.$$

Eliminating f from the two equations yields

$$mv^2 = -mvp \frac{dv}{dp} \quad \Rightarrow \quad mv + mp \frac{dv}{dp} = 0$$

$$\Rightarrow \quad h \equiv mpv \text{ is a constant of the motion.}$$

It follows that

$$f = \frac{mv^2}{p} \frac{dp}{dr} = \frac{h^2}{mp^3} \frac{dp}{dr},$$

from which the law of force can be deduced once p is given as a function of r.

14.11 *Solve*

$$(y - x)\frac{dy}{dx} + 2x + 3y = 0.$$

We first test whether the equation is exact, or can be made so with the help of an integrating factor. To do this, we write the equation as

$$(y - x)\,dy + (2x + 3y)\,dx = 0$$

and consider

$$h_x(x, y) = \frac{1}{y - x} \left[\frac{\partial}{\partial y}(2x + 3y) - \frac{\partial}{\partial x}(y - x) \right] = \frac{4}{y - x}.$$

This is not a function of x alone. Equally

$$h_y(x, y) = \frac{1}{2x + 3y} \left[-\frac{\partial}{\partial y}(2x + 3y) + \frac{\partial}{\partial x}(y - x) \right] = \frac{-4}{2x + 3y}$$

is not a function of y alone. We conclude that there is no straightforward IF and that another method has to be tried.

We note that the equation is homogeneous in x and y and so we set $y = vx$, with $\dfrac{\partial y}{\partial x} = v + x\dfrac{\partial v}{\partial x}$, and obtain

$$v + x\frac{\partial v}{\partial x} = -\frac{2 + 3v}{v - 1},$$
$$x\frac{\partial v}{\partial x} = \frac{-2 - 3v - v^2 + v}{v - 1} = -\frac{v^2 + 2v + 2}{v - 1},$$
$$\frac{dx}{x} = \frac{(1 - v)\,dv}{v^2 + 2v + 2}$$
$$= \frac{2}{(v + 1)^2 + 1} - \frac{v + 1}{(v + 1)^2 + 1},$$
$$\Rightarrow \quad \ln Ax = 2\tan^{-1}(v + 1) - \tfrac{1}{2}\ln[\,1 + (v + 1)^2\,],$$
$$\ln\left\{Bx^2[1 + (v + 1)^2]\right\} = 4\tan^{-1}(v + 1).$$

On setting $v = y/x$ this becomes

$$B[x^2 + (y + x)^2] = \exp\left[4\tan^{-1}\left(\frac{y + x}{x}\right)\right],$$

the final form of the solution.

14.13 *One of the properties of Laplace transforms is that the transform of the nth derivative of a function $f(t)$ is given by*

$$\mathscr{L}\left[\frac{d^n f}{dt^n}\right] = s^n \bar{f} - s^{n-1}f(0) - s^{n-2}\frac{df}{dt}(0) - \cdots - \frac{d^{n-1}f}{dt^{n-1}}(0), \quad \text{for } s > 0.$$

Using this and the result about the Laplace transform of $tf(t)$ obtained in exercise 13.25, show, for a function $y(t)$ that satisfies

$$t\frac{dy}{dt} + (t - 1)y = 0 \qquad (*)$$

with $y(0)$ finite, that $\bar{y}(s) = C(1 + s)^{-2}$ for some constant C.

Given that

$$y(t) = t + \sum_{n=2}^{\infty} a_n t^n,$$

determine C and show that $a_n = (-1)^{n-1}/(n - 1)!$. Compare this result with that obtained by integrating $(*)$ directly.

Using the stated property of derivatives with $n = 1$ and the result from the exercise, we Laplace transform the equation and obtain

$$-\frac{d}{ds}[s\bar{y}(s) - y(0)] - \frac{d\bar{y}(s)}{ds} - \bar{y}(s) = 0,$$

$$-s\frac{d\bar{y}}{ds} - \bar{y} + 0 - \frac{d\bar{y}}{ds} - \bar{y} = 0,$$

$$(1 + s)\frac{d\bar{y}}{ds} + 2\bar{y} = 0,$$

$$\frac{d\bar{y}}{\bar{y}} + \frac{2\,ds}{1 + s} = 0,$$

$$\Rightarrow \quad \ln\bar{y} + 2\ln(1 + s) = k,$$

$$\Rightarrow \quad \bar{y} = \frac{C}{(1 + s)^2}.$$

As a power series, $\bar{y}(s)$ takes the form

$$\bar{y}(s) = \frac{C}{s^2}\left(1 + \frac{1}{s}\right)^{-2}$$

$$= \frac{C}{s^2}\left(1 - \frac{2}{s} + \frac{(-2)(-3)}{2!}\frac{1}{s^2} + \cdots\right)$$

$$= \frac{C}{s^2} + \sum_{n=1}^{\infty}\frac{(-1)^n(n + 1)}{s^{n+2}}.$$

But, transforming the given solution,

$$y(t) = t + \sum_{m=2}^{\infty} a_m t^m,$$

yields

$$\bar{y} = \frac{1}{s^2} + \sum_{m=2}^{\infty} a_m \frac{m!}{s^{m+1}}.$$

Comparing coefficients in the two expressions for $\bar{y}$ shows that $C = 1$ and that $a_{m+1} = (-1)^m/m!$, i.e. $a_m = (-1)^{m-1}/(m-1)!$.

Direct integration of (∗) by separating the variables gives

$$0 = \frac{dy}{y} + \left(1 - \frac{1}{t}\right) dt,$$

$$\Rightarrow \quad A = \ln y + t - \ln t,$$

$$\Rightarrow \quad y = Bte^{-t}$$

$$= Bt + B\sum_{n=1}^{\infty}\frac{(-1)^n t^{n+1}}{n!}$$

$$= Bt + B\sum_{m=2}^{\infty}\frac{(-1)^{m-1} t^m}{(m-1)!}.$$

236

With B determined by the linear term as unity, the two solutions agree.

14.15 Solve
$$\frac{dy}{dx} = -\frac{x+y}{3x+3y-4}.$$

Since x and y only appear in the combination $x+y$ we set $v = x+y$ with $dv/dx = 1 + dy/dx$. The equation and its solution then become

$$\frac{dv}{dx} = 1 - \frac{v}{3v-4},$$

$$dx = \frac{3v-4}{2v-4}\, dv = \left(\frac{3}{2} + \frac{2}{2v-4}\right) dv,$$

$$\Rightarrow \quad x + k = \tfrac{3}{2}v + \ln(v-2) = \tfrac{3}{2}(x+y) + \ln(x+y-2),$$

$$\ln(x+y-2) = k - \tfrac{1}{2}(x+3y).$$

Although the initial equation might look as if it could be made exact with an integrating factor, applying the method descibed in exercise 14.3 shows that this not so; $B^{-1}[\partial A/\partial y - \partial B/\partial x]$ is neither zero nor a function of only one of the variables.

14.17 Solve
$$x(1 - 2x^2 y)\frac{dy}{dx} + y = 3x^2 y^2,$$

given that $y(1) = 1/2$.

Though this is clearly not a homogeneous equation, we test whether it might be an isobaric one by giving x a weight 1 and y a weight m and then seeing whether a suitable value for m can be found. From the presence of the term $1 - 2x^2 y$ it is clear that the only possible value of m is -2, since $2x^2 y$ must have the same weight as unity, namely weight 0. For this value of m the three terms in the equation have weights

$$1 + 0 + (-2) - 1, \quad -2, \quad 2 + 2(-2).$$

These are all the same (at -2) and so the equation is isobaric.

To find its solution we set $y = vx^m = vx^{-2}$ with

$$\frac{dy}{dx} = -\frac{2v}{x^3} + \frac{1}{x^2}\frac{dv}{dx}.$$

Substituting in the original equation produces

$$x(1 - 2v)\left(-\frac{2v}{x^3} + \frac{1}{x^2}\frac{dv}{dx}\right) + \frac{v}{x^2} = \frac{3x^2v^2}{x^4},$$

$$(1 - 2v)\left(-2v + x\frac{dv}{dx}\right) + v = 3v^2,$$

$$(1 - 2v)x\frac{dv}{dx} = v(1 - v),$$

$$\frac{1 - 2v}{v(1 - v)}dv = \frac{dx}{x},$$

$$\left(\frac{1}{v} - \frac{1}{1 - v}\right)dv = \frac{dx}{x},$$

$$\Rightarrow \quad \ln v + \ln(1 - v) = \ln x + A \quad \Rightarrow \quad v(1 - v) = Cx.$$

Expressing this in terms of the original variables by substituting $v = yx^2$ gives $yx^2(1 - yx^2) = Cx$, with $\frac{1}{2}(1 - \frac{1}{2}) = C$. Thus, after cancelling x from both sides, the solution is

$$4yx(1 - yx^2) = 1.$$

14.19 *Find the curve with the property that at each point on it the sum of the intercepts on the x- and y-axes of the tangent to the curve (taking account of sign) is equal to 1.*

At a point (X, Y) on the curve, the tangent to the curve is the straight line given by

$$y - Y = p(x - X),$$

where p is the slope of the tangent. This meets the axis $y = 0$ at $x = X - (Y/p)$ and the axis $x = 0$ at $y = Y - pX$. Thus, taking account of signs (i.e. some intercepts could be negative), the condition to be satisfied is

$$X - \frac{Y}{p} + Y - pX = 1.$$

Since (X, Y) lies on the required curve, the curve has an equation that satisfies

$$x - \frac{y}{p} + y - px = 1 \quad \Rightarrow \quad y = \frac{1 - x + px}{1 - p^{-1}} \quad (*).$$

Differentiating both sides of $(*)$ with respect to x, we now eliminate y by using the fact that its derivative with respect to x is p:

$$p = \frac{(1 - p^{-1})(-1 + p + xp') - (1 - x + px)p^{-2}p'}{(1 - p^{-1})^2},$$

$$p(p-1)^2 = (p^2 - p)(p-1) + p'[x(p^2 - p) - 1 + x - px].$$

The LHS and the first term on the RHS are equal, and so we have that either $p' = 0$ or

$$x(p^2 - 2p + 1) - 1 = 0,$$

$$\Rightarrow \quad x = \frac{1}{(p-1)^2},$$

$$\Rightarrow \quad p = 1 \pm \frac{1}{\sqrt{x}}.$$

From this and (*) it follows that

$$y = \frac{p[(1-x) + px]}{p-1} = \frac{\left(1 \pm \dfrac{1}{\sqrt{x}}\right)(1 - x + x \pm \sqrt{x})}{\pm \dfrac{1}{\sqrt{x}}}$$

$$= (\pm\sqrt{x} + 1)(1 \pm \sqrt{x}).$$

As expected, the solution is symmetric between x and y; this is demonstrated by the following rearrangement of the form just obtained:

$$y = (1 \pm \sqrt{x})^2,$$

$$\pm\sqrt{y} = 1 \pm \sqrt{x} \qquad (\pm \text{ signs not correlated}),$$

$$\pm\sqrt{y} - 1 = \pm\sqrt{x},$$

$$(1 \mp \sqrt{y})^2 = x.$$

Because of the square roots involved, a real curve exists only for x and y both positive, i.e. in the first quadrant. That curve is $\sqrt{x} + \sqrt{y} = 1$.

The singular solution $p' = 0$ (ignored earlier) corresponds to a set of curves, on each of which the slope is a constant. Any one such curve is a *straight* line joining the axial points $(\theta, 0)$ and $(0, 1 - \theta)$ for any arbitrary real θ; the tangent at any point on such a 'curve' is always the curve itself, whose intercepts, θ and $1 - \theta$, sum to unity.

> **14.21** Using the substitutions $u = x^2$ and $v = y^2$, reduce the equation
>
> $$xy \left(\frac{dy}{dx} \right)^2 - (x^2 + y^2 - 1)\frac{dy}{dx} + xy = 0$$
>
> to Clairaut's form. Hence show that the equation represents a family of conics and the four sides of a square.

Writing $dy/dx = p$ and $dv/du = q$, we have

$$\frac{du}{dx} = 2x, \quad \frac{dv}{dx} = 2yp, \quad q = \frac{dv}{du} = \frac{yp}{x}, \quad p = \frac{x}{y}q.$$

Making the substitutions yields

$$xy\frac{x^2}{y^2}q^2 - (u + v - 1)\frac{x}{y}q + xy = 0.$$

We now multiply by $\dfrac{y}{x}$ and substitute again:

$$uq^2 - (u + v - 1)q + v = 0,$$

$$v(1 - q) - uq + q + uq^2 = 0,$$

$$v = uq + \frac{q}{q - 1}, \quad \text{Clairaut's form } (*).$$

As the equation now has Clairaut's form it has two solutions.

(i) The first is

$$v = cu + \frac{c}{c - 1},$$

$$y^2 - cx^2 = \frac{c}{c - 1}.$$

- For $c > 1$, this is a hyperbola of the form $y^2 - \alpha^2 x^2 = \beta^2$.
- For $1 > c > 0$, it is a hyperbola of the form $x^2 - \alpha^2 y^2 = \beta^2$.
- For $c < 0$, the conic is an ellipse of the form $y^2 + \alpha^2 x^2 = \beta^2$.

In each case $\alpha > \beta > 0$.

(ii) The second (singular) solution is given by

$$\frac{d}{dq}\left(\frac{q}{q - 1} \right) + u = 0,$$

$$\frac{-1}{(q - 1)^2} + u = 0,$$

$$q = 1 \pm \frac{1}{\sqrt{u}}.$$

Substituting this into (∗) expressed in terms of x and y then gives

$$y^2 = x^2 \left(1 \pm \frac{1}{x}\right) + \frac{1 \pm \frac{1}{x}}{\pm \frac{1}{x}}$$

$$= x^2 \pm x \pm x + 1$$
$$= (x \pm 1)^2,$$
$$y = \pm(x \pm 1).$$

These lines are the four sides of the square that has corners at $(0, \pm 1)$ and $(\pm 1, 0)$.

14.23 *Find the general solutions of the following:*

$$\text{(a)} \ \frac{dy}{dx} + \frac{xy}{a^2 + x^2} = x, \qquad \text{(b)} \ \frac{dy}{dx} = \frac{4y^2}{x^2} - y^2.$$

(a) With dy/dx appearing in the first term and y in the second (and nowhere else), this is a linear first-order ODE and therefore has an IF given by

$$\mu(x) = \exp\left\{\int \frac{x}{a^2 + x^2}\right\} dx = \exp[\tfrac{1}{2}\ln(a^2 + x^2)] = (a^2 + x^2)^{1/2}.$$

When multiplied through by this, the equation becomes

$$\frac{d}{dx}[(a^2 + x^2)^{1/2}y] = x(a^2 + x^2)^{1/2},$$
$$\Rightarrow \quad (a^2 + x^2)^{1/2}y = \tfrac{2}{3}\tfrac{1}{2}(a^2 + x^2)^{3/2} + A,$$
$$\Rightarrow \quad y = \frac{a^2 + x^2}{3} + \frac{A}{(a^2 + x^2)^{1/2}}.$$

(b) The RHS can be written as the product of one function of x and another one of y; the equation is therefore separable:

$$\frac{dy}{y^2} = \left(\frac{4}{x^2} - 1\right) dx,$$
$$\Rightarrow \quad -\frac{1}{y} = -\frac{4}{x} - x + A,$$
$$\Rightarrow \quad y = \frac{x}{x^2 + Bx + 4},$$

where $B = -A$ and is the arbitrary integration constant.

14.25 *An electronic system has two inputs, to each of which a constant unit signal is applied, but starting at different times. The equations governing the system thus take the form*

$$\dot{x} + 2y = H(t),$$

$$\dot{y} - 2x = H(t - 3).$$

Initially (at $t = 0$), $x = 1$ and $y = 0$; find $x(t)$ at later times.

Since we have coupled equations, working with their Laplace transforms suggests itself. This will convert the equations into simultaneous algebraic equations – though there may be some difficulty in converting the solution back into t-space.

The transform of the Heaviside function is s^{-1}, and so the two transformed equations (incorporating the initial conditions and using the translation property of Laplace transforms) are

$$s\bar{x} - 1 + 2\bar{y} = \frac{1}{s},$$

$$s\bar{y} - 0 - 2\bar{x} = \frac{1}{s}e^{-3s}.$$

Since it is $x(t)$ that we require, we eliminate $\bar{y}$ to obtain

$$s^2\bar{x} - s + \frac{2}{s}e^{-3s} + 4\bar{x} = 1,$$

from which

$$\bar{x} = \frac{s^2 + s - 2e^{-3s}}{s(s^2 + 4)},$$

$$= \frac{s+1}{s^2 + 4} + \left[-\frac{1}{2s} + \frac{s}{2(s^2 + 4)} \right] e^{-3s}.$$

For the first term in square brackets, the coefficient in the partial fractions expansion was determined by considering the limit $s \to 0$; that for the second term was found by inspection.

Now, using a look-up table if necessary, we find that, in t-space, the function corresponding to the $\bar{x}$ found above is

$$x(t) = \tfrac{1}{2}\sin 2t + \cos 2t - \tfrac{1}{2}H(t - 3) + \tfrac{1}{2}H(t - 3)\cos 2(t - 3).$$

14.27 *Find the complete solution of*

$$\left(\frac{dy}{dx}\right)^2 - \frac{y}{x}\frac{dy}{dx} + \frac{A}{x} = 0,$$

where A is a positive constant.

At first sight this non-linear equation may appear to be homogeneous, but the term A/x rules this out. Since it is non-linear, we set $dy/dx = p$ and rearrange the equation to make y, which then appears only once, the subject:

$$p^2 - \frac{y}{x}p + \frac{A}{x} = 0,$$

$$xp - y + \frac{A}{p} = 0,$$

$$y = xp + \frac{A}{p}. \qquad (*)$$

This is now recognised as Clairaut's equation with $F(p) = A/p$. Its general solution is therefore given by

$$y = cx + \frac{A}{c} \qquad \text{for arbitrary } c.$$

It also has a singular solution (containing no arbitrary constants) given by

$$\frac{d}{dp}\left(\frac{A}{p}\right) + x = 0, \quad \Rightarrow \quad p = \sqrt{\frac{A}{x}} \quad \Rightarrow \quad y = x\sqrt{\frac{A}{x}} + \frac{A}{\sqrt{A/x}} = 2\sqrt{Ax}.$$

The final result was obtained by substituting for p in $(*)$.

14.29 *Find the solution $y = y(x)$ of*

$$x\frac{dy}{dx} + y - \frac{y^2}{x^{3/2}} = 0,$$

subject to $y(1) = 1$.

After being divided through by x, this equation is in the form of a Bernoulli equation with $n = 2$, i.e. it is of the form

$$\frac{dy}{dx} + P(x)y = Q(x)y^n.$$

Here, $P(x) = x^{-1}$ and $Q(x) = x^{-5/2}$. So we set $v = y^{1-2} = y^{-1}$ and obtain

$$\frac{dy}{dx} = \frac{d}{dx}\left(\frac{1}{v}\right) = -\frac{1}{v^2}\frac{dv}{dx}.$$

The equation then becomes

$$-\frac{1}{v^2}\frac{dv}{dx} + \frac{1}{vx} = \frac{1}{v^2 x^{5/2}},$$

$$\frac{dv}{dx} - \frac{v}{x} = -\frac{1}{x^{5/2}}, \quad \text{for which the IF is } 1/x,$$

$$\frac{d}{dx}\left(\frac{v}{x}\right) = -\frac{1}{x^{7/2}},$$

$$\frac{v}{x} = \frac{2}{5}\frac{1}{x^{5/2}} + \frac{3}{5}, \quad \text{using } y(1) = 1,$$

$$\frac{1}{y} = \frac{2}{5}\frac{1}{x^{3/2}} + \frac{3x}{5},$$

$$y = \frac{5x^{3/2}}{2 + 3x^{5/2}}.$$

The equation can also be treated as an isobaric one with $m = \frac{3}{2}$; the substitution $y = vx^{3/2}$ is made and the equation is reduced to the separable form

$$\frac{dv}{v(2v - 5)} = \frac{dx}{2x}.$$

After the LHS has been expressed in partial fractions, the integration can be carried out. The boundary condition, $v(1) = 1$, determines the constant of integration and after resubstituting $yx^{-3/2}$ for v, the same answer as obtained earlier is recovered, as it must be.

14.31 *Find the family of solutions of*

$$\frac{d^2y}{dx^2} + \left(\frac{dy}{dx}\right)^2 + \frac{dy}{dx} = 0$$

that satisfy $y(0) = 0$.

As the equation contains only derivatives, we write $dy/dx = p$ and $d^2y/dx^2 = dp/dx$; this will reduce the equation to one of first order:

$$\frac{dp}{dx} + p^2 + p = 0.$$

Separating the variables:

$$\frac{dp}{p(p + 1)} = -dx.$$

We now integrate and express the integrand in partial fractions:

$$\int \left(\frac{1}{p} - \frac{1}{p+1} \right) dp = -\int dx,$$

$$\ln(p) - \ln(p+1) = A - x,$$

$$\Rightarrow \quad \frac{p}{p+1} = Be^{-x},$$

$$\Rightarrow \quad p = \frac{e^{-x}}{C - e^{-x}}.$$

Now $p = dy/dx$ and so

$$\frac{dy}{dx} = \frac{e^{-x}}{C - e^{-x}},$$

$$y = \ln(C - e^{-x}) + D$$

$$= \ln(C - e^{-x}) - \ln(C - 1), \quad \text{since we require } y(0) = 0,$$

$$= \ln \frac{C - e^{-x}}{C - 1}.$$

This is as far as y can be determined since only one boundary condition is given for a second-order equation. As C is varied the solution generates a family of curves satisfying the original equation.

A variety of other forms of solution are possible and equally valid, the actual form obtained depending on where in the calculation the boundary condition is incorporated. They include

$$e^y = F(1 - e^{-x}) + 1, \quad y = \ln[G - (G - 1)e^{-x}], \quad y = \ln(e^{-K} + 1 - e^{-x}) + K.$$

15

Higher-order ordinary differential equations

15.1 *A simple harmonic oscillator, of mass m and natural frequency ω_0, experiences an oscillating driving force $f(t) = ma\cos\omega t$. Therefore, its equation of motion is*

$$\frac{d^2x}{dt^2} + \omega_0^2 x = a\cos\omega t,$$

where x is its position. Given that at $t = 0$ we have $x = dx/dt = 0$, find the function $x(t)$. Describe the solution if ω is approximately, but not exactly, equal to ω_0.

To find the full solution given the initial conditions, we need the complete general solution made up of a complementary function (CF) and a particular integral (PI). The CF is clearly of the form $A\cos\omega_0 t + B\sin\omega_0 t$ and, in view of the form of the RHS, we try $x(t) = C\cos\omega t + D\sin\omega t$ as a PI. Substituting this gives

$$-\omega^2 C\cos\omega t - \omega^2 D\sin\omega t + \omega_0^2 C\cos\omega t + \omega_0^2 D\sin\omega t = a\cos\omega t.$$

Equating coefficients of the independent functions $\cos\omega t$ and $\sin\omega t$ requires that

$$-\omega^2 C + \omega_0^2 C = a \quad \Rightarrow \quad C = \frac{a}{\omega_0^2 - \omega^2},$$

$$-\omega^2 D + \omega_0^2 D = 0 \quad \Rightarrow \quad D = 0.$$

Thus, the general solution is

$$x(t) = A\cos\omega_0 t + B\sin\omega_0 t + \frac{a}{\omega_0^2 - \omega^2}\cos\omega t.$$

The initial conditions impose the requirements

$$x(0) = 0 \quad \Rightarrow \quad 0 = A + \frac{a}{\omega_0^2 - \omega^2},$$

$$\text{and } \dot{x}(0) = 0 \quad \Rightarrow \quad 0 = \omega_0 B.$$

Incorporating the implications of these into the general solution gives

$$x(t) = \frac{a}{\omega_0^2 - \omega^2}(\cos \omega t - \cos \omega_0 t)$$

$$= \frac{2a \sin[\frac{1}{2}(\omega + \omega_0)t] \sin[\frac{1}{2}(\omega_0 - \omega)t]}{(\omega_0 + \omega)(\omega_0 - \omega)}.$$

For $\omega_0 - \omega = \epsilon$ with $|\epsilon|t \ll 1$,

$$x(t) \approx \frac{2a \sin \omega_0 t \frac{1}{2}\epsilon t}{2\omega_0 \epsilon} = \frac{at}{2\omega_0} \sin \omega_0 t.$$

Thus, for moderate t, $x(t)$ is a sine wave of linearly increasing amplitude.

Over a long time, $x(t)$ will vary between $\pm 2a/(\omega_0^2 - \omega^2)$ with sizeable intervals between the two extremes, i.e. it will show beats of amplitude $2a/(\omega_0^2 - \omega^2)$.

15.3 *The theory of bent beams shows that at any point in the beam the 'bending moment' is given by K/ρ, where K is a constant (that depends upon the beam material and cross-sectional shape) and ρ is the radius of curvature at that point. Consider a light beam of length L whose ends, $x = 0$ and $x = L$, are supported at the same vertical height and which has a weight W suspended from its centre. Verify that at any point x ($0 \le x \le L/2$ for definiteness) the net magnitude of the bending moment (bending moment = force $\times$ perpendicular distance) due to the weight and support reactions, evaluated on either side of x, is $Wx/2$.*

If the beam is only slightly bent, so that $(dy/dx)^2 \ll 1$, where $y = y(x)$ is the downward displacement of the beam at x, show that the beam profile satisfies the approximate equation

$$\frac{d^2 y}{dx^2} = -\frac{Wx}{2K}.$$

By integrating this equation twice and using physically imposed conditions on your solution at $x = 0$ and $x = L/2$, show that the downward displacement at the centre of the beam is $WL^3/(48K)$.

The upward reaction of the support at each end of the beam is $\frac{1}{2}W$.

At the position x the moment on the left is due to
     (i) the support at $x = 0$ providing a clockwise moment of $\frac{1}{2}Wx$.

The moment on the right is due to
     (ii) the support at $x = L$ providing an anticlockwise moment of $\frac{1}{2}W(L-x)$;
     (iii) the weight at $x = \frac{1}{2}L$ providing a clockwise moment of $W(\frac{1}{2}L - x)$.

The net clockwise moment on the right is therefore $W(\frac{1}{2}L - x) - \frac{1}{2}W(L - x) = -\frac{1}{2}Wx$, i.e. equal in magnitude, but opposite in sign, to that on the left.

The radius of curvature of the beam is $\rho = [1 + (-y')^2]^{3/2}/(-y'')$, but if $|y'| \ll 1$ this simplifies to $-1/y''$ and the equation of the beam profile satisfies

$$\frac{Wx}{2} = M = \frac{K}{\rho} = -K\frac{d^2 y}{dx^2}.$$

We now need to integrate this, taking into account the boundary conditions $y(0) = 0$ and, on symmetry grounds, $y'(\frac{1}{2}L) = 0$:

$$y' = -\frac{Wx^2}{4K} + A, \text{ with } y'(\tfrac{1}{2}L) = 0 \quad \Rightarrow \quad A = \frac{WL^2}{16K},$$

$$y' = \frac{W}{4K}\left(\frac{L^2}{4} - x^2\right),$$

$$y = \frac{W}{4K}\left(\frac{L^2 x}{4} - \frac{x^3}{3} + B\right), \text{ with } y(0) = 0 \quad \Rightarrow \quad B = 0.$$

The centre is lowered by

$$y(\tfrac{1}{2}L) = \frac{W}{4K}\left(\frac{L^2}{4}\frac{L}{2} - \frac{1}{3}\frac{L^3}{8}\right) = \frac{WL^3}{48K}.$$

Note that the derived analytic form for $y(x)$ is not applicable in the range $\frac{1}{2}L \leq x \leq L$; the beam profile is symmetrical about $x = \frac{1}{2}L$, but the expression $\frac{1}{4}L^2 x - \frac{1}{3}x^3$ is not invariant under the substitution $x \to L - x$.

15.5 *The function $f(t)$ satisfies the differential equation*

$$\frac{d^2 f}{dt^2} + 8\frac{df}{dt} + 12f = 12e^{-4t}.$$

For the following sets of boundary conditions determine whether it has solutions, and, if so, find them:

(a) $f(0) = 0$, $\quad f'(0) = 0$, $\quad f(\ln \sqrt{2}) = 0$;

(b) $f(0) = 0$, $\quad f'(0) = -2$, $\quad f(\ln \sqrt{2}) = 0$.

Three boundary conditions have been given, and, as this is a second-order linear equation for which only two independent conditions are needed, they may be inconsistent. The plan is to solve it using two of the conditions and then test whether the third one is compatible.

The auxiliary equation for obtaining the CF is

$$m^2 + 8m + 12 = 0 \quad \Rightarrow \quad m = -2 \text{ or } m = -6$$

$$\Rightarrow \quad f(t) = Ae^{-6t} + Be^{-2t}.$$

Since the form of the RHS, Ce^{-4t}, is not included in the CF, we can try it as the particular integral:

$$16C - 32C + 12C = 12 \quad \Rightarrow \quad C = -3.$$

The general solution is therefore

$$f(t) = Ae^{-6t} + Be^{-2t} - 3e^{-4t}.$$

(a) For boundary conditions $f(0) = 0$, $\quad f'(0) = 0$, $\quad f(\ln\sqrt{2}) = 0$:

$$f(0) = 0 \quad \Rightarrow \quad A + B - 3 = 0,$$
$$f'(0) = 0 \quad \Rightarrow \quad -6A - 2B + 12 = 0,$$
$$\Rightarrow \quad A = \tfrac{3}{2}, \quad B = \tfrac{3}{2}.$$
$$\text{Hence, } f(t) \quad = \quad \tfrac{3}{2}e^{-6t} + \tfrac{3}{2}e^{-2t} - 3e^{-4t}.$$

Recalling that $e^{-(\ln\sqrt{2})} = 1/\sqrt{2}$, we evaluate

$$f(\ln\sqrt{2}) = \frac{3}{2}\frac{1}{8} + \frac{3}{2}\frac{1}{2} - 3\frac{1}{4} = \frac{3}{16} \neq 0.$$

Thus the boundary conditions are inconsistent and there is no solution.

(b) For boundary conditions $f(0) = 0$, $\quad f'(0) = -2$, $\quad f(\ln\sqrt{2}) = 0$, we proceed as before:

$$f(0) = 0 \quad \Rightarrow \quad A + B - 3 = 0,$$
$$f'(0) = 0 \quad \Rightarrow \quad -6A - 2B + 12 = -2,$$
$$\Rightarrow \quad A = 2, \quad B = 1.$$
$$\text{Hence, } f(t) \quad = \quad 2e^{-6t} + e^{-2t} - 3e^{-4t}.$$

We again evaluate

$$f(\ln\sqrt{2}) = 2\frac{1}{8} + \frac{1}{2} - 3\frac{1}{4} = 0.$$

This time the boundary conditions are consistent and there is a unique solution as given above.

15.7 *A solution of the differential equation*

$$\frac{d^2y}{dx^2} + 2\frac{dy}{dx} + y = 4e^{-x}$$

takes the value 1 when $x = 0$ and the value e^{-1} when $x = 1$. What is its value when $x = 2$?

The auxiliary equation, $m^2 + 2m + 1 = 0$, has repeated roots $m = -1$, and so the general CF has the special form $y(x) = (A + Bx)e^{-x}$.

Turning to the PI, we note that the form of the RHS of the original equation is contained in the CF, and (to make matters worse) so is x times the RHS. We therefore need to take x^2 times the RHS as a trial PI:

$$y(x) = Cx^2 e^{-x}, \quad y' = C(2x - x^2)e^{-x}, \quad y'' = C(2 - 4x + x^2)e^{-x}.$$

Substituting these into the original equation shows that

$$2Ce^{-x} = 4e^{-x} \quad \Rightarrow \quad C = 2$$

and that the full general solution is given by

$$y(x) = (A + Bx)e^{-x} + 2x^2 e^{-x}.$$

We now determine the unknown constants using the information given about the solution. Since $y(0) = 1$, $A = 1$. Further, $y(1) = e^{-1}$ requires

$$e^{-1} = (1 + B)e^{-1} + 2e^{-1} \quad \Rightarrow \quad B = -2.$$

Finally, we conclude that $y(x) = (1 - 2x + 2x^2)e^{-x}$ and, therefore, that $y(2) = 5e^{-2}$.

15.9 *Find the general solutions of*

(a) $\dfrac{d^3 y}{dx^3} - 12\dfrac{dy}{dx} + 16y = 32x - 8$,

(b) $\dfrac{d}{dx}\left(\dfrac{1}{y}\dfrac{dy}{dx}\right) + (2a \coth 2ax)\left(\dfrac{1}{y}\dfrac{dy}{dx}\right) = 2a^2$,

where a is a constant.

(a) As this is a third-order equation, we expect three terms in the CF.

Since it is linear with constant coefficients, we can make use of the auxiliary equation, which is

$$m^3 - 12m + 16 = 0.$$

By inspection, $m = 2$ is one root; the other two can be found by factorisation:

$$m^3 - 12m + 16 = (m - 2)(m^2 + 2m - 8) = (m - 2)(m + 4)(m - 2) = 0.$$

Thus we have one repeated root ($m = 2$) and one other ($m = -4$) leading to a CF of the form

$$y(x) = (A + Bx)e^{2x} + Ce^{-4x}.$$

As the RHS contains no exponentials, we try $y(x) = Dx + E$ for the PI. We then need $16D = 32$ and $-12D + 16E = -8$, giving $D = 2$ and $E = 1$.

The general solution is therefore

$$y(x) = (A + Bx)e^{2x} + Ce^{-4x} + 2x + 1.$$

(b) The equation is already arranged in the form

$$\frac{dg(y)}{dx} + h(x)g(y) = j(x)$$

and so needs only an integrating factor to allow the first integration step to be made. For this equation the IF is

$$\exp\left\{\int 2a \coth 2ax \, dx\right\} = \exp(\ln \sinh 2ax) = \sinh 2ax.$$

After multiplication through by this factor, the equation can be written

$$\sinh 2ax \frac{d}{dx}\left(\frac{1}{y}\frac{dy}{dx}\right) + (2a\cosh 2ax)\left(\frac{1}{y}\frac{dy}{dx}\right) = 2a^2 \sinh 2ax,$$

$$\frac{d}{dx}\left(\sinh 2ax \frac{1}{y}\frac{dy}{dx}\right) = 2a^2 \sinh 2ax.$$

Integrating this gives

$$\sinh 2ax \frac{1}{y}\frac{dy}{dx} = \frac{2a^2}{2a}\cosh 2ax + A,$$

$$\Rightarrow \quad \frac{1}{y}\frac{dy}{dx} = a \coth 2ax + \frac{A}{\sinh 2ax}.$$

Integrating again, $\quad \ln y = \frac{1}{2}\ln(\sinh 2ax) + \int \frac{A}{\sinh 2ax}\, dx + B$

$$= \frac{1}{2}\ln(\sinh 2ax) + \frac{A}{2a}\ln(|\tanh ax|) + B,$$

$$\Rightarrow \quad y = C(\sinh 2ax)^{1/2}(|\tanh ax|)^D.$$

The indefinite integral of $(\sinh 2ax)^{-1}$ appearing in the fourth line can be verified by differentiating $y = \ln|\tanh ax|$ in the form $y = \frac{1}{2}\ln(\tanh^2 ax)$ and recalling that

$$\cosh ax \sinh ax = \frac{1}{2}\sinh 2ax.$$

15.11 *The quantities* $x(t)$, $y(t)$ *satisfy the simultaneous equations*

$$\ddot{x} + 2n\dot{x} + n^2 x = 0,$$

$$\ddot{y} + 2n\dot{y} + n^2 y = \mu\dot{x},$$

where $x(0) = y(0) = \dot{y}(0) = 0$ *and* $\dot{x}(0) = \lambda$. *Show that*

$$y(t) = \tfrac{1}{2}\mu\lambda t^2 \left(1 - \tfrac{1}{3}nt\right)\exp(-nt).$$

For these two coupled equations, in which an 'output' from the first acts as the 'driving input' for the second, we take Laplace transforms and incorporate the boundary conditions:

$$(s^2\bar{x} - 0 - \lambda) + 2n(s\bar{x} - 0) + n^2\bar{x} = 0,$$

$$(s^2\bar{y} - 0 - 0) + 2n(s\bar{y} - 0) + n^2\bar{y} = \mu(s\bar{x} - 0).$$

From the first transformed equation,

$$\bar{x} = \frac{\lambda}{s^2 + 2ns + n^2}.$$

Substituting this into the second transformed equation gives

$$\bar{y} = \frac{\mu s\bar{x}}{(s+n)^2} = \frac{\mu\lambda s}{(s+n)^2(s+n)^2}$$

$$= \frac{\mu\lambda}{(s+n)^3} - \frac{\mu\lambda n}{(s+n)^4},$$

$$\Rightarrow \quad y(t) = \mu\lambda\left(\frac{t^2}{2!}e^{-nt} - \frac{nt^3}{3!}e^{-nt}\right), \text{ from the look-up table,}$$

$$= \frac{1}{2}\mu\lambda t^2\left(1 - \frac{nt}{3}\right)e^{-nt},$$

i.e. as stated in the question.

15.13 *Two unstable isotopes* A *and* B *and a stable isotope* C *have the following decay rates per atom present:* $A \to B$, $3\,\mathrm{s}^{-1}$; $A \to C$, $1\,\mathrm{s}^{-1}$; $B \to C$, $2\,\mathrm{s}^{-1}$. *Initially a quantity* x_0 *of* A *is present but there are no atoms of the other two types. Using Laplace transforms, find the amount of* C *present at a later time* t.

Using the name symbol to represent the corresponding number of atoms and taking Laplace transforms, we have

$$\frac{dA}{dt} = -(3+1)A \quad \Rightarrow \quad s\bar{A} - x_0 = -4\bar{A}$$

$$\Rightarrow \quad \bar{A} = \frac{x_0}{s+4},$$

$$\frac{dB}{dt} = 3A - 2B \quad \Rightarrow \quad s\bar{B} = 3\bar{A} - 2\bar{B}$$

$$\Rightarrow \quad \bar{B} = \frac{3x_0}{(s+2)(s+4)},$$

$$\frac{dC}{dt} = A + 2B \quad \Rightarrow \quad s\bar{C} = \bar{A} + 2\bar{B}$$

$$\Rightarrow \quad \bar{C} = \frac{x_0(s+2) + 6x_0}{s(s+2)(s+4)}.$$

Using the 'cover-up' method for finding the coefficients of a partial fraction expansion without repeated factors, e.g. the coefficient of $(s+2)^{-1}$ is $[(-2 + 8)x_0]/[(-2)(-2+4)] = -6x_0/4$, we have

$$\bar{C} = \frac{x_0(s+8)}{s(s+2)(s+4)} = \frac{x_0}{s} - \frac{6x_0}{4(s+2)} + \frac{4x_0}{8(s+4)}$$

$$\Rightarrow \quad C(t) = x_0 \left(1 - \tfrac{3}{2}e^{-2t} + \tfrac{1}{2}e^{-4t} \right).$$

This is the required expression.

15.15 *The 'golden mean', which is said to describe the most aesthetically pleasing proportions for the sides of a rectangle (e.g. the ideal picture frame), is given by the limiting value of the ratio of successive terms of the Fibonacci series u_n, which is generated by*

$$u_{n+2} = u_{n+1} + u_n,$$

with $u_0 = 0$ and $u_1 = 1$. Find an expression for the general term of the series and verify that the golden mean is equal to the larger root of the recurrence relation's characteristic equation.

The recurrence relation is second order and its characteristic equation, obtained by setting $u_n = A\lambda^n$, is

$$\lambda^2 - \lambda - 1 = 0 \quad \Rightarrow \quad \lambda = \tfrac{1}{2}(1 \pm \sqrt{5}).$$

The general solution is therefore

$$u_n = A \left(\frac{1+\sqrt{5}}{2} \right)^n + B \left(\frac{1-\sqrt{5}}{2} \right)^n.$$

The initial values (boundary conditions) determine A and B:

$$u_0 = 0 \quad \Rightarrow \quad B = -A,$$

$$u_1 = 1 \quad \Rightarrow \quad A\left(\frac{1+\sqrt{5}}{2} - \frac{1-\sqrt{5}}{2}\right) = 1 \quad \Rightarrow \quad A = \frac{1}{\sqrt{5}},$$

$$\text{Hence, } u_n \quad = \quad \frac{1}{\sqrt{5}}\left[\left(\frac{1+\sqrt{5}}{2}\right)^n - \left(\frac{1-\sqrt{5}}{2}\right)^n\right].$$

If we write $(1 - \sqrt{5})/(1 + \sqrt{5}) = r < 1$, the ratio of successive terms in the series is

$$\frac{u_{n+1}}{u_n} = \frac{\frac{1}{2}[(1+\sqrt{5})^{n+1} - (1-\sqrt{5})^{n+1}]}{(1+\sqrt{5})^n - (1-\sqrt{5})^n}$$

$$= \frac{\frac{1}{2}[1+\sqrt{5} - (1-\sqrt{5})r^n]}{1 - r^n}$$

$$\to \frac{1+\sqrt{5}}{2} \text{ as } n \to \infty;$$

i.e. the limiting ratio is the same as the larger value of λ.

This result is a particular example of the more general one that the ratio of successive terms in a series generated by a recurrence relation tends to the largest (in absolute magnitude) of the roots of the characteristic equation. Here there are only two roots, but for an Nth-order relation there will be N roots.

15.17 *The first few terms of a series u_n, starting with u_0, are $1, 2, 2, 1, 6, -3$. The series is generated by a recurrence relation of the form*

$$u_n = Pu_{n-2} + Qu_{n-4},$$

where P and Q are constants. Find an expression for the general term of the series and show that, in fact, the series consists of two interleaved series given by

$$u_{2m} = \tfrac{2}{3} + \tfrac{1}{3}4^m,$$

$$u_{2m+1} = \tfrac{7}{3} - \tfrac{1}{3}4^m,$$

for $m = 0, 1, 2, \ldots$.

We first find P and Q using

$$n = 4 \qquad 6 = 2P + Q,$$

$$n = 5 \quad -3 = P + 2Q, \quad \Rightarrow \quad Q = -4 \text{ and } P = 5.$$

The recurrence relation is thus

$$u_n = 5u_{n-2} - 4u_{n-4}.$$

To solve this we try $u_n = A + B\lambda^n$ for arbitrary constants A and B and obtain

$$A + B\lambda^n = 5A + 5B\lambda^{n-2} - 4A - 4B\lambda^{n-4},$$
$$\Rightarrow \quad 0 = \lambda^4 - 5\lambda^2 + 4$$
$$= (\lambda^2 - 1)(\lambda^2 - 4) \quad \Rightarrow \quad \lambda = \pm 1, \pm 2.$$

The general solution is $\quad u_n = A + B(-1)^n + C2^n + D(-2)^n.$

We now need to solve the simultaneous equations for A, B, C and D provided by the values of $u_0, \ldots, u_3$:

$$1 = A + B + C + D,$$
$$2 = A - B + 2C - 2D,$$
$$2 = A + B + 4C + 4D,$$
$$1 = A - B + 8C - 8D.$$

These have the straightforward solution

$$A = \frac{3}{2}, \qquad B = -\frac{5}{6}, \qquad C = \frac{1}{12}, \qquad D = \frac{1}{4},$$

and so

$$u_n = \frac{3}{2} - \frac{5}{6}(-1)^n + \frac{1}{12}2^n + \frac{1}{4}(-2)^n.$$

When n is even and equal to $2m$,

$$u_{2m} = \frac{3}{2} - \frac{5}{6} + \frac{4^m}{12} + \frac{4^m}{4} = \frac{2}{3} + \frac{4^m}{3}.$$

When n is odd and equal to $2m + 1$,

$$u_{2m+1} = \frac{3}{2} + \frac{5}{6} + \frac{4^m}{6} - \frac{4^m}{2} = \frac{7}{3} - \frac{4^m}{3}.$$

In passing, we note that the fact that both P and Q, and all of the given values $u_0, \ldots, u_4$, are integers, and hence that all terms in the series are integers, provides an indirect proof that $4^m + 2$ is divisible by 3 (without remainder) for all non-negative integers m. This can be more easily proved by induction, as the reader may like to verify.

15.19 *Find the general expression for the u_n satisfying*

$$u_{n+1} = 2u_{n-2} - u_n$$

with $u_0 = u_1 = 0$ and $u_2 = 1$, and show that they can be written in the form

$$u_n = \frac{1}{5} - \frac{2^{n/2}}{\sqrt{5}} \cos\left(\frac{3\pi n}{4} - \phi\right),$$

where $\tan\phi = 2$.

The characteristic equation (which will be a cubic since the recurrence relation is third order) and its solution are given by

$$\lambda^{n+1} = 2\lambda^{n-2} - \lambda^n,$$

$$\lambda^3 + \lambda^2 - 2 = 0,$$

$$(\lambda - 1)(\lambda^2 + 2\lambda + 2) = 0 \quad \Rightarrow \quad \lambda = 1 \text{ or } \lambda = -1 \pm i.$$

Thus the general solution of the recurrence relation, which has the generic form $A\lambda_1^n + B\lambda_2^n + C\lambda_3^n$, is

$$u_n = A + B(-1 + i)^n + C(-1 - i)^n$$

$$= A + B\, 2^{n/2} e^{i3\pi n/4} + C\, 2^{n/2} e^{i5\pi n/4}.$$

To determine A, B and C we use

$$u_0 = 0, \qquad 0 = A + B + C,$$
$$u_1 = 0, \qquad 0 = A + B\, 2^{1/2} e^{i3\pi/4} + C\, 2^{1/2} e^{i5\pi/4}$$
$$= A + B(-1 + i) + C(-1 - i),$$
$$u_2 = 1, \qquad 1 = A + B\, 2 e^{i6\pi/4} + C\, 2 e^{i10\pi/4} = A + 2B(-i) + 2C(i).$$

Adding twice each of the first two equations to the last one gives $5A = 1$. Substituting this into the first and last equations then leads to

$$B + C = -\frac{1}{5} \qquad \text{and} \qquad -B + C = \frac{2}{5i},$$

from which it follows that

$$B = \frac{-1 + 2i}{10} = \frac{\sqrt{5}}{10} e^{i(\pi - \phi)}$$

$$\text{and} \qquad C = \frac{-1 - 2i}{10} = \frac{\sqrt{5}}{10} e^{i(\pi + \phi)},$$

where $\tan\phi = 2/1 = 2$.

Thus, collecting these results together, we have

$$u_n = \frac{1}{5} + \frac{2^{n/2}\sqrt{5}}{10}(e^{i3\pi n/4}e^{i(\pi-\phi)} + e^{i5\pi n/4}e^{i(\pi+\phi)})$$

$$= \frac{1}{5} - \frac{2^{n/2}\sqrt{5}}{10}(e^{i3\pi n/4}e^{-i\phi} + e^{-i3\pi n/4}e^{i\phi})$$

$$= \frac{1}{5} - \frac{2^{n/2}\sqrt{5}}{10}\left[2\cos\left(\frac{3\pi n}{4} - \phi\right)\right]$$

$$= \frac{1}{5} - \frac{2^{n/2}}{\sqrt{5}}\cos\left(\frac{3\pi n}{4} - \phi\right),$$

i.e. the form of solution given in the question.

15.21 *Find the general solution of*

$$x^2\frac{d^2y}{dx^2} - x\frac{dy}{dx} + y = x,$$

given that $y(1) = 1$ *and* $y(e) = 2e$.

This is Euler's equation and can be solved either by a change of variables, $x = e^t$, or by trying $y = x^\lambda$; we will adopt the second approach. Doing so in the homogeneous equation (RHS set to zero) gives

$$x^2\,\lambda(\lambda - 1)x^{\lambda-2} - x\,\lambda x^{\lambda-1} + x^\lambda = 0.$$

The CF is therefore obtained when λ satisfies

$$\lambda(\lambda - 1) - \lambda + 1 = 0 \quad \Rightarrow \quad (\lambda - 1)^2 = 0 \quad \Rightarrow \quad \lambda = 1 \text{ (repeated)}.$$

Thus, one solution is $y = x$; the other linearly independent solution implied by the repeated root is $x \ln x$ (see a textbook if this is not known).

There is now a further complication as the RHS of the original equation (x) is contained in the CF. We therefore need an extra factor of $\ln x$ in the trial PI, beyond those already in the CF. (This corresponds to the extra power of t needed in the PI if the transformation to a linear equation with constant coefficients is made via the $x = e^t$ change of variable.) As a consequence, the PI to be tried is $y = Cx(\ln x)^2$:

$$x^2\left[2C\frac{\ln x}{x} + \frac{2C}{x}\right] - x\left[Cx\frac{2\ln x}{x} + C(\ln x)^2\right] + Cx(\ln x)^2 = x.$$

This implies that $C = \frac{1}{2}$ and gives the general solution as

$$y(x) = Ax + Bx\ln x + \frac{1}{2}x(\ln x)^2.$$

It remains only to determine the unknown constants A and B; this is done using the two given values of $y(x)$. The boundary condition $y(1) = 1$ requires that $A = 1$, and $y(e) = 2e$ implies that $B = \frac{1}{2}$; the solution is now completely determined as

$$y(x) = x + \tfrac{1}{2}x \ln x(1 + \ln x).$$

15.23 *Prove that the general solution of*

$$(x - 2)\frac{d^2 y}{dx^2} + 3\frac{dy}{dx} + \frac{4y}{x^2} = 0$$

is given by

$$y(x) = \frac{1}{(x-2)^2}\left[k\left(\frac{2}{3x} - \frac{1}{2} \right) + cx^2 \right].$$

This equation is not of any plausible standard form, and the only solution method is to try to make it into an exact equation. If this is possible the order of the equation will be reduced by one.

We first multiply through by x^2 and then note that the resulting factor $3x^2$ in the second term can be written as $[x^2(x - 2)]' + 4x$, i.e. as the derivative of the function multiplying y'' together with another simple function. This latter can be combined with the undifferentiated term and allow the whole equation to be written as an exact equation:

$$\frac{d}{dx}\left[x^2(x-2)\frac{dy}{dx} \right] + 4x\frac{dy}{dx} + 4y = 0,$$

$$\frac{d}{dx}\left[x^2(x-2)\frac{dy}{dx} \right] + \frac{d(4xy)}{dx} = 0,$$

$$\Rightarrow \quad x^2(x-2)\frac{dy}{dx} + 4xy = k.$$

Either by inspection or by use of the standard formula, the IF is $(x - 2)/x^4$ and leads to

$$\frac{d}{dx}\left[\frac{(x-2)^2}{x^2}y \right] = \frac{k(x-2)}{x^4},$$

$$\Rightarrow \quad \frac{(x-2)^2}{x^2}y = k\left(-\frac{1}{2x^2} + \frac{2}{3x^3} \right) + c,$$

$$\Rightarrow \quad y = \frac{1}{(x-2)^2}\left(-\frac{k}{2} + \frac{2k}{3x} + cx^2 \right).$$

15.25 *Find the Green's function that satisfies*

$$\frac{d^2G(x,\xi)}{dx^2} - G(x,\xi) = \delta(x - \xi) \quad \text{with} \quad G(0,\xi) = G(1,\xi) = 0.$$

It is clear from inspection that the CF has solutions of the form $e^{\pm x}$. The other pair of solutions that may suggest themselves are $\sinh x$ and $\cosh x$, but these are merely independent linear combinations of the same two functions.

As both boundary conditions are given at finite values of x (rather than at $x \to \pm\infty$) and both are of the form $y(x) = 0$, it is more convenient to work with those particular linear combinations of e^x and e^{-x} that vanish at the boundary points. The only common linear combination of these two functions that vanishes at a finite value of x is a sinh function. To construct one that vanishes at $x = x_0$ the argument of the sinh function must be made to be $x - x_0$. For the present case the appropriate combinations are

$$\sinh x = \frac{1}{2}(e^x - e^{-x}) \quad \text{and} \quad \sinh(1-x) = \left(\frac{e}{2}\right)e^{-x} - \left(\frac{1}{2e}\right)e^x.$$

Thus, with $0 \le \xi \le 1$, we take

$$G(x,\xi) = \begin{cases} A(\xi)\sinh x & x < \xi, \\ B(\xi)\sinh(1-x) & x > \xi. \end{cases}$$

The continuity requirement on $G(x,\xi)$ at $x = \xi$ and the unit discontinuity requirement on its derivative at the same point give

$$A\sinh\xi - B\sinh(1-\xi) = 0$$

and

$$-B\cosh(1-\xi) - A\cosh\xi = 1,$$

leading to

$$A\sinh\xi\cosh(1-\xi) + A\cosh\xi\sinh(1-\xi) = -\sinh(1-\xi),$$
$$A[\sinh(\xi + 1 - \xi)] = -\sinh(1-\xi).$$

Hence,

$$A = -\frac{\sinh(1-\xi)}{\sinh 1} \quad \text{and} \quad B = -\frac{\sinh\xi}{\sinh 1},$$

giving the full Green's function as

$$G(x,\xi) = \begin{cases} -\dfrac{\sinh(1-\xi)}{\sinh 1}\sinh x & x < \xi, \\[2ex] -\dfrac{\sinh\xi}{\sinh 1}\sinh(1-x) & x > \xi. \end{cases}$$

15.27 *Show generally that if $y_1(x)$ and $y_2(x)$ are linearly independent solutions of*

$$\frac{d^2y}{dx^2} + p(x)\frac{dy}{dx} + q(x)y = 0,$$

with $y_1(0) = 0$ and $y_2(1) = 0$, then the Green's function $G(x, \xi)$ for the interval $0 \le x, \xi \le 1$ and with $G(0, \xi) = G(1, \xi) = 0$ can be written in the form

$$G(x, \xi) = \begin{cases} y_1(x)y_2(\xi)/W(\xi) & 0 < x < \xi, \\ y_2(x)y_1(\xi)/W(\xi) & \xi < x < 1, \end{cases}$$

where $W(x) = W[y_1(x), y_2(x)]$ is the Wronskian of $y_1(x)$ and $y_2(x)$.

As usual, we start by writing the general solution as a weighted sum of the linearly independent solutions, whilst leaving the possibility that the weights may be different for different x-ranges:

$$G(x, \xi) = \begin{cases} A(\xi)y_1(x) + B(\xi)y_2(x) & 0 < x < \xi, \\ C(\xi)y_1(x) + D(\xi)y_2(x) & \xi < x < 1. \end{cases}$$

Imposing the boundary conditions and using $y_1(0) = y_2(1) = 0$,

$$0 = G(0, \xi) = A(\xi)y_1(0) + B(\xi)y_2(0) \quad \Rightarrow \quad B(\xi) = 0,$$
$$0 = G(1, \xi) = C(\xi)y_1(1) + D(\xi)y_2(1) \quad \Rightarrow \quad C(\xi) = 0.$$

The continuity requirement on $G(x, \xi)$ at $x = \xi$ and the unit discontinuity requirement on its derivative at the same point give

$$A(\xi)y_1(\xi) - D(\xi)y_2(\xi) = 0,$$
$$A(\xi)y_1'(\xi) - D(\xi)y_2'(\xi) = -1,$$

leading to

$$A(\xi)[y_1y_2' - y_2y_1'] = y_2 \quad \Rightarrow \quad A(\xi) = \frac{y_2(\xi)}{W(\xi)},$$
$$D(\xi) = \frac{y_1(\xi)}{y_2(\xi)}A(\xi) = \frac{y_1(\xi)}{W(\xi)}.$$

Thus,

$$G(x, \xi) = \begin{cases} y_1(x)y_2(\xi)/W(\xi) & 0 < x < \xi, \\ y_2(x)y_1(\xi)/W(\xi) & \xi < x < 1. \end{cases}$$

This result is perfectly general for linear second-order equations of the type stated and can be a quick way to find the corresponding Green's function, provided the solutions that vanish at the end-points can be identified easily. Exercise 15.25 is a particular example of this general result.

15.29 *The equation of motion for a driven damped harmonic oscillator can be written*

$$\ddot{x} + 2\dot{x} + (1 + \kappa^2)x = f(t),$$

with $\kappa \neq 0$. If it starts from rest with $x(0) = 0$ and $\dot{x}(0) = 0$, find the corresponding Green's function $G(t, \tau)$ and verify that it can be written as a function of $t - \tau$ only. Find the explicit solution when the driving force is the unit step function, i.e. $f(t) = H(t)$. Confirm your solution by taking the Laplace transforms of both it and the original equation.

The auxiliary equation is

$$m^2 + 2m + (1 + \kappa^2) = 0 \quad \Rightarrow \quad m = -1 \pm i\kappa,$$

and the CF is $x(t) = Ae^{-t} \cos \kappa t + Be^{-t} \sin \kappa t$.

Let

$$G(t, \tau) = \begin{cases} A(\tau)e^{-t} \cos \kappa t + B(\tau)e^{-t} \sin \kappa t & 0 < t < \tau, \\ C(\tau)e^{-t} \cos \kappa t + D(\tau)e^{-t} \sin \kappa t & t > \tau. \end{cases}$$

The boundary condition $x(0) = 0$ implies that $A = 0$, and

$$\dot{x}(0) = 0 \quad \Rightarrow \quad B(-e^{-t} \sin \kappa t + \kappa e^{-t} \cos \kappa t) = 0 \quad \Rightarrow \quad B = 0.$$

Thus $G(t, \tau) = 0$ for $t < \tau$.

The continuity of G at $t = \tau$ gives

$$Ce^{-\tau} \cos \kappa\tau + De^{-\tau} \sin \kappa\tau = 0 \quad \Rightarrow \quad D = -\frac{C \cos \kappa\tau}{\sin \kappa\tau}.$$

The unit discontinuity in the derivative of G at $t = \tau$ requires (using $s = \sin \kappa\tau$ and $c = \cos \kappa\tau$ as shorthand)

$$Ce^{-\tau}(-c - \kappa s) + De^{-\tau}(-s + \kappa c) - 0 = 1,$$

$$C \left[-c - \kappa s - \frac{c}{s}(-s + \kappa c) \right] = e^{\tau},$$

$$C(-sc - \kappa s^2 + cs - \kappa c^2) = se^{\tau},$$

giving

$$C = -\frac{e^{\tau} \sin \kappa\tau}{\kappa} \quad \text{and} \quad D = \frac{e^{\tau} \cos \kappa\tau}{\kappa}.$$

Thus, for $t > \tau$,

$$G(t, \tau) = \frac{e^{\tau}}{\kappa}(-\sin \kappa\tau \cos \kappa t + \cos \kappa\tau \sin \kappa t)e^{-t}$$

$$= \frac{e^{-(t-\tau)}}{\kappa} \sin \kappa(t - \tau).$$

261

This form verifies that the Green's function is a function only of the difference $t - \tau$ and not of t and τ separately.

The explicit solution to the given equation when $f(t) = H(t)$ is thus

$$
\begin{aligned}
x(t) &= \int_0^\infty G(t, \tau) f(\tau) \, d\tau \\
&= \int_0^t G(t, \tau) H(\tau) \, d\tau, \text{ since } G(t, \tau) = 0 \text{ for } \tau > t, \\
&= \frac{1}{\kappa} \int_0^t e^{-(t-\tau)} \sin \kappa(t - \tau) \, d\tau \\
&= \frac{e^{-t}}{\kappa} \operatorname{Im} \int_0^t e^{\tau + i\kappa(t-\tau)} \, d\tau \\
&= \frac{e^{-t}}{\kappa} \operatorname{Im} \left[\frac{e^{i\kappa t} e^{\tau - i\kappa\tau}}{1 - i\kappa} \right]_{\tau=0}^{\tau=t} \\
&= \frac{e^{-t}}{\kappa} \operatorname{Im} \left[\frac{e^t - e^{i\kappa t}}{1 - i\kappa} \right].
\end{aligned}
$$

Now multiplying both numerator and denominator by $1 + i\kappa$ to make the latter real gives

$$
\begin{aligned}
x(t) &= \frac{e^{-t}}{\kappa(1 + \kappa^2)} \operatorname{Im} \left[(e^t - e^{i\kappa t})(1 + i\kappa) \right] \\
&= \frac{e^{-t}}{\kappa(1 + \kappa^2)} \left[\kappa(e^t - \cos \kappa t) - \sin \kappa t \right] \\
&= \frac{1}{1 + \kappa^2} \left(1 - e^{-t} \cos \kappa t - \frac{1}{\kappa} e^{-t} \sin \kappa t \right).
\end{aligned}
$$

The Laplace transform of this solution is given by

$$
\begin{aligned}
\bar{x} &= \frac{1}{1 + \kappa^2} \left(\frac{1}{s} - \frac{s + 1}{(s + 1)^2 + \kappa^2} - \frac{1}{\kappa} \frac{\kappa}{(s + 1)^2 + \kappa^2} \right) \\
&= \frac{(s + 1)^2 + \kappa^2 - s(s + 1) - s}{(1 + \kappa^2) s [(s + 1)^2 + \kappa^2]} \\
&= \frac{1}{s [(s + 1)^2 + \kappa^2]}.
\end{aligned}
$$

The Laplace transform of the original equation with the given initial conditions reads

$$
[s^2 \bar{x} - 0s - 0] + 2[s\bar{x} - 0] + (1 + \kappa^2)\bar{x} = \frac{1}{s},
$$

again showing that

$$
\bar{x} = \frac{1}{s[s^2 + 2s + 1 + \kappa^2]} = \frac{1}{s[(s + 1)^2 + \kappa^2]},
$$

and so confirming the solution reached using the Green's function approach.

15.31 *Find the Green's function $x = G(t, t_0)$ that solves*

$$\frac{d^2x}{dt^2} + \alpha\frac{dx}{dt} = \delta(t - t_0)$$

under the initial conditions $x = dx/dt = 0$ at $t = 0$. Hence solve

$$\frac{d^2x}{dt^2} + \alpha\frac{dx}{dt} = f(t),$$

where $f(t) = 0$ for $t < 0$. Evaluate your answer explicitly for $f(t) = Ae^{-\beta t}$ $(t > 0)$.

It is clear that one solution, $x(t)$, to the homogeneous equation has $\ddot{x} = -\alpha\dot{x}$ and is therefore $x(t) = Ae^{-\alpha t}$. The equation is of second order and therefore has a second solution; this is the trivial (but perfectly valid) x is a constant. The CF is thus $x(t) = Ae^{-\alpha t} + B$.

Let

$$G(t, t_0) = \begin{cases} Ae^{-\alpha t} + B, & 0 \le t \le t_0, \\ Ce^{-\alpha t} + D, & t > t_0. \end{cases}$$

Now, the initial conditions give

$$\begin{aligned} x(0) = 0 &\quad\Rightarrow\quad A + B = 0, \\ \dot{x}(0) = 0 &\quad\Rightarrow\quad -\alpha A = 0 \quad\Rightarrow\quad A = B = 0. \end{aligned}$$

Thus $G(t, t_0) = 0$ for $0 \le t \le t_0$.

The continuity/discontinuity conditions determine C and D through

$$Ce^{-\alpha t_0} + D - 0 = 0,$$

$$-\alpha Ce^{-\alpha t_0} - 0 = 1, \quad\Rightarrow\quad C = -\frac{e^{\alpha t_0}}{\alpha} \text{ and } D = \frac{1}{\alpha}.$$

It follows that $\qquad G(t, t_0) = \frac{1}{\alpha}[1 - e^{-\alpha(t - t_0)}]$ for $t > t_0$.

The general formalism now gives the solution of

$$\frac{d^2x}{dt^2} + \alpha\frac{dx}{dt} = f(t)$$

as

$$x(t) = \int_0^t \frac{1}{\alpha}[1 - e^{-\alpha(t - \tau)}]f(\tau)\,d\tau.$$

With $f(t) = Ae^{-\beta t}$ this becomes

$$x(t) = \int_0^t \frac{1}{\alpha}[1 - e^{-\alpha(t-\tau)}]Ae^{-\beta\tau}\, d\tau$$

$$= \frac{A}{\alpha} \int_0^t (e^{-\beta\tau} - e^{-\alpha t}e^{(\alpha-\beta)\tau})\, d\tau$$

$$= A\left[\frac{1 - e^{-\beta t}}{\alpha\beta} - \frac{e^{-\beta t} - e^{-\alpha t}}{\alpha(\alpha - \beta)}\right]$$

$$= A\left[\frac{\alpha - \beta - \alpha e^{-\beta t} + \beta e^{-\alpha t}}{\beta\alpha(\alpha - \beta)}\right]$$

$$= A\left[\frac{\alpha(1 - e^{-\beta t}) - \beta(1 - e^{-\alpha t})}{\beta\alpha(\alpha - \beta)}\right].$$

This is the required explicit solution.

15.33 *Solve*

$$2y\frac{d^3y}{dx^3} + 2\left(y + 3\frac{dy}{dx}\right)\frac{d^2y}{dx^2} + 2\left(\frac{dy}{dx}\right)^2 = \sin x.$$

The only realistic hope for this non-linear equation is to try to arrange it as an exact equation! We note that the second and fourth terms can be written as the derivative of a product, and that adding and subtracting $2y'y''$ will enable the first term to be written in a similar way. We therefore rewrite the equation as

$$\frac{d}{dx}\left(2y\frac{d^2y}{dx^2}\right) + \frac{d}{dx}\left(2y\frac{dy}{dx}\right) + (6 - 2)\frac{dy}{dx}\frac{d^2y}{dx^2} = \sin x,$$

$$\frac{d}{dx}\left(2y\frac{d^2y}{dx^2}\right) + \frac{d}{dx}\left(2y\frac{dy}{dx}\right) + \frac{d}{dx}\left[2\left(\frac{dy}{dx}\right)^2\right] = \sin x.$$

This second form is obtained by noting that the final term on the LHS of the first equation happens to be an exact differential. Thus the whole of the LHS is an exact differential and one stage of integration can be carried out:

$$2y\frac{d^2y}{dx^2} + 2y\frac{dy}{dx} + 2\left(\frac{dy}{dx}\right)^2 = -\cos x + A.$$

We now note that the first and third terms of this integrated equation can be combined as the derivative of a product, whilst the second term is the derivative of y^2. This allows a further step of integration:

$$\frac{d}{dx}\left(2y\frac{dy}{dx}\right) + 2y\frac{dy}{dx} = -\cos x + A,$$

$$\frac{d}{dx}\left(2y\frac{dy}{dx}\right) + \frac{d(y^2)}{dx} = -\cos x + A,$$

$$\Rightarrow \quad 2y\frac{dy}{dx} + y^2 = -\sin x + Ax + B,$$

$$\frac{d(y^2)}{dx} + y^2 = -\sin x + Ax + B.$$

At this stage an integrating factor is needed. However, as the LHS consists of the sum of the differentiated and undifferentiated forms of the same function, the required IF is simply e^x. After multiplying through by this, we obtain

$$\frac{d}{dx}\left(e^x y^2\right) = -e^x \sin x + Axe^x + Be^x,$$

$$\Rightarrow \quad y^2 = e^{-x}\left[C + \int^x (B + Au - \sin u)e^u \, du\right]$$

$$= Ce^{-x} + B + A(x-1) - \tfrac{1}{2}(\sin x - \cos x).$$

The last term in this final solution is obtained by considering

$$\int^x e^u \sin u \, du = \mathrm{Im}\, \int^x e^{(1+i)u} \, du$$

$$= \mathrm{Im}\, \left[\frac{e^{(1+i)u}}{1+i}\right]^x$$

$$= \mathrm{Im}\,[\tfrac{1}{2}(1-i)e^{(1+i)x}]$$

$$= \tfrac{1}{2}e^x(\sin x - \cos x).$$

15.35 *Express the equation*

$$\frac{d^2 y}{dx^2} + 4x\frac{dy}{dx} + (4x^2 + 6)y = e^{-x^2}\sin 2x$$

in canonical form and hence find its general solution.

In the standard shortened notation, we have

$$a_1(x) = 4x, \qquad a_0(x) = 4x^2 + 6, \qquad f(x) = e^{-x^2}\sin 2x.$$

Then, with $y(x)$ expressed as $y(x) = u(x)v(x)$, in order to have an equation with no v' term in it, we choose $u(x)$ as

$$u(x) = \exp\left\{-\tfrac{1}{2}\int^x a_1(z)\,dz\right\} = \exp\left\{-\tfrac{1}{2}\int^x 4z\,dz\right\} = e^{-x^2}.$$

The equation is then reduced to

$$\frac{d^2v}{dx^2} + g(x)v = h(x),$$

where

$$g(x) = a_0(x) - \tfrac{1}{4}[a_1(x)]^2 - \tfrac{1}{2}a_1'(x) = 4x^2 + 6 - 4x^2 - 2 = 4$$

and

$$h(x) = f(x)\exp\left\{\tfrac{1}{2}\int a_1(z)\,dz\right\} = (e^{-x^2}\sin 2x)\exp\left\{\tfrac{1}{2}\int 4z\,dz\right\}$$

$$= (e^{-x^2}\sin 2x)\,e^{x^2} = \sin 2x.$$

For this particular case the reduced equation is

$$v'' + 4v = \sin 2x.$$

This has CF $A\cos 2x + B\sin 2x$ but, because the RHS is contained in the CF, we need to try as a PI $y(x) = C(x)\cos 2x + D(x)\sin 2x$. Substituting this shows that C and D must satisfy

$$C''\cos 2x - 4C'\sin 2x + D''\sin 2x + 4D'\cos 2x = \sin 2x,$$

yielding the pair of simultaneous equations

$$C'' + 4D' = 0,$$
$$-4C' + D'' = 1.$$

Any solution will suffice, and the simplest is $C(x) = -\tfrac{1}{4}x$ with $D(x) = 0$.

We can now write the general solution and express it in terms of the original variables:

$$v(x) = (A - \tfrac{1}{4}x)\cos 2x + B\sin 2x,$$

$$y(x) = u(x)v(x) = [(A - \tfrac{1}{4}x)\cos 2x + B\sin 2x]e^{-x^2}.$$

15.37 *Consider the equation*

$$x^p y'' + \frac{n+3-2p}{n-1}x^{p-1}y' + \left(\frac{p-2}{n-1}\right)^2 x^{p-2}y = y^n,$$

in which $p \neq 2$ and $n > -1$ but $n \neq 1$. For the boundary conditions $y(1) = 0$ and $y'(1) = \lambda$, show that the solution is $y(x) = v(x)x^{(p-2)/(n-1)}$, where $v(x)$ is given by

$$\int_0^{v(x)} \frac{dz}{\left[\lambda^2 + 2z^{n+1}/(n+1)\right]^{1/2}} = \ln x.$$

To start, we test whether the equation is isobaric by giving y a weight m relative to x. The weights of the four terms are then

$$m - 2 + p, \quad m - 1 + p - 1, \quad m + p - 2, \quad mn.$$

These are all equal, provided m is chosen to satisfy $m + p - 2 = mn$, i.e. $m = (p - 2)/(n - 1)$. Thus the equation is isobaric.

Now set $y(x) = v(x)x^m$, noting that $y(1) = 0 \Rightarrow v(1) = 0$. As derivatives we have

$$y' = v'x^m + mvx^{m-1}, \quad y'' = v''x^m + 2mv'x^{m-1} + m(m - 1)vx^{m-2}.$$

We further note that, since $y'(1) = \lambda$ implies $v'(1) + mv(1) = \lambda$, we must have $v'(1) = \lambda$.

Substituting the derivatives into the equation, rewriting the constants in terms of m and dividing through by x^{p+m-2} gives

$$x^2v'' + 2mxv' + m(m - 1)v + (1 - 2m)(xv' + mv) + m^2v = v''x^0,$$
$$x^2v'' + xv' + [m(m - 1) + m - 2m^2 + m^2]v = v'',$$
$$x^2v'' + xv' = v''.$$

To solve this non-linear equation we set $x = e^t$ and $v(x) = u(t)$. The operator d/dx becomes $e^{-t}d/dt$. The initial conditions are that $u(0) = 0$ and

$$\frac{du}{dt} = \frac{dv}{dx}\frac{dx}{dt} = \lambda e^0 \text{ at } t = 0.$$

The equation itself transforms to

$$e^{2t}e^{-t}\frac{d}{dt}\left(e^{-t}\frac{du}{dt}\right) + e^t e^{-t}\frac{du}{dt} = u^n,$$
$$u'' - u' + u' = u^n,$$
$$u'' = u^n,$$
$$u'u'' = u'u^n,$$
$$\frac{1}{2}\left(\frac{du}{dt}\right)^2 = \frac{u^{n+1}}{n+1} + k.$$

Since $u'(0) = \lambda$ and $u(0) = 0$, it follows that $k = \frac{1}{2}\lambda^2$ and that

$$\frac{du}{dt} = \left(\frac{2u^{n+1}}{n+1} + \lambda^2\right)^{1/2}.$$

Integrating this gives

$$\int_0^{u(t)}\left(\frac{2z^{n+1}}{n+1} + \lambda^2\right)^{-1/2} dz = t - 0,$$

and, by changing back to the original variables,

$$\int_0^{v(x)} \left(\frac{2z^{n+1}}{n+1} + \lambda^2 \right)^{-1/2} dz = \ln x.$$

For any given x, this equation determines $v(x)$.

The solution $y(x)$ to the original equation is then given by

$$y(x) = v(x)x^{(p-2)/(n-1)}.$$

16

Series solutions of ordinary
differential equations

16.1 *Find two power series solutions about $z = 0$ of the differential equation*

$$(1 - z^2)y'' - 3zy' + \lambda y = 0.$$

Deduce that the value of λ for which the corresponding power series becomes an Nth-degree polynomial $U_N(z)$ is $N(N + 2)$. Construct $U_2(z)$ and $U_3(z)$.

If the equation is imagined divided through by $(1 - z^2)$ it is straightforward to see that, although $z = \pm 1$ are singular points of the equation, the point $z = 0$ is an *ordinary* point. We therefore expect two (uncomplicated!) series solutions with indicial values $\sigma = 0$ and $\sigma = 1$.

(a) $\sigma = 0$ and $y(z) = \sum_{n=0}^{\infty} a_n z^n$ with $a_0 \neq 0$.

Substituting and equating the coefficients of z^m,

$$(1 - z^2) \sum_{n=0}^{\infty} n(n - 1)a_n z^{n-2} - 3 \sum_{n=0}^{\infty} n a_n z^n + \lambda \sum_{n=0}^{\infty} a_n z^n = 0,$$

$$(m + 2)(m + 1)a_{m+2} - m(m - 1)a_m - 3m a_m + \lambda a_m = 0,$$

gives as the recurrence relation

$$a_{m+2} = \frac{m(m - 1) + 3m - \lambda}{(m + 2)(m + 1)} a_m = \frac{m(m + 2) - \lambda}{(m + 1)(m + 2)} a_m.$$

Since this recurrence relation connects alternate coefficients a_m, and $a_0 \neq 0$, only the coefficients with even indices are generated. All such coefficients with index higher than m will become zero, and the series will become an Nth-degree polynomial $U_N(z)$, if $\lambda = m(m + 2) = N(N + 2)$ for some (even) m appearing in the series; here, this means any positive even integer N.

269

To construct $U_2(z)$ we need to take $\lambda = 2(2 + 2) = 8$. The recurrence relation gives a_2 as

$$a_2 = \frac{0 - 8}{(0 + 1)(0 + 2)}a_0 = -4a_0 \quad \Rightarrow \quad U_2(z) = a_0(1 - 4z^2).$$

(b) $\sigma = 1$ and $y(z) = z \sum_{n=0}^{\infty} a_n z^n$ with $a_0 \neq 0$.

Substituting and equating the coefficients of z^{m+1},

$$(1 - z^2)\sum_{n=0}^{\infty}(n + 1)na_n z^{n-1} - 3\sum_{n=0}^{\infty}(n + 1)a_n z^{n+1} + \lambda \sum_{n=0}^{\infty} a_n z^{n+1} = 0,$$

$$(m + 3)(m + 2)a_{m+2} - (m + 1)ma_m - 3(m + 1)a_m + \lambda a_m = 0,$$

gives as the recurrence relation

$$a_{m+2} = \frac{m(m + 1) + 3(m + 1) - \lambda}{(m + 2)(m + 3)}a_m = \frac{(m + 1)(m + 3) - \lambda}{(m + 2)(m + 3)}a_m.$$

Again, all coefficients with index higher than m will become zero, and the series will become an Nth-degree polynomial $U_N(z)$, if $\lambda = (m + 1)(m + 3) = N(N + 2)$ for some (even) m appearing in the series; here, this means any positive odd integer N.

To construct $U_3(z)$ we need to take $\lambda = 3(3 + 2) = 15$. The recurrence relation gives a_2 as

$$a_2 = \frac{3 - 15}{(0 + 2)(0 + 3)}a_0 = -2a_0.$$

Thus,

$$U_3(z) = a_0(z - 2z^3).$$

16.3 *Find power series solutions in z of the differential equation*

$$zy'' - 2y' + 9z^5 y = 0.$$

Identify closed forms for the two series, calculate their Wronskian, and verify that they are linearly independent. Compare the Wronskian with that calculated from the differential equation.

Putting the equation in its standard form shows that $z = 0$ is a singular point of the equation but, as $-2z/z$ and $9z^7/z$ are finite as $z \to 0$, it is a regular singular point. We therefore substitute a Frobenius type solution,

$$y(z) = z^{\sigma}\sum_{n=0}^{\infty} a_n z^n \text{ with } a_0 \neq 0,$$

and obtain

$$\sum_{n=0}^{\infty}(n+\sigma)(n+\sigma-1)a_n z^{n+\sigma-1}$$

$$-2\sum_{n=0}^{\infty}(n+\sigma)a_n z^{n+\sigma-1} +9\sum_{n=0}^{\infty}a_n z^{n+\sigma+5} = 0.$$

Equating the coefficient of $z^{\sigma-1}$ to zero gives the indicial equation as

$$\sigma(\sigma-1)a_0 - 2\sigma a_0 = 0 \quad \Rightarrow \quad \sigma = 0,\ 3.$$

These differ by an integer and may or may not yield two independent solutions. The larger root, $\sigma = 3$, will give a solution; the smaller one, $\sigma = 0$, may not.

(a) $\sigma = 3$.

Equating the general coefficient of z^{m+2} to zero (with $\sigma = 3$) gives

$$(m+3)(m+2)a_m - 2(m+3)a_m + 9a_{m-6} = 0.$$

Hence the recurrence relation is

$$a_m = -\frac{9a_{m-6}}{m(m+3)},$$

$$\Rightarrow \quad a_{6p} = -\frac{9}{6p\,(6p+3)}a_{6p-6} = -\frac{a_{6p-6}}{2p\,(2p+1)} = \frac{(-1)^p a_0}{(2p+1)!}.$$

The first solution is therefore given by

$$y_1(x) = a_0 z^3 \sum_{n=0}^{\infty}\frac{(-1)^n}{(2n+1)!}z^{6n} = a_0\sum_{n=0}^{\infty}\frac{(-1)^n}{(2n+1)!}z^{3(2n+1)} = a_0\sin z^3.$$

(b) $\sigma = 0$.

Equating the general coefficient of z^{m-1} to zero (with $\sigma = 0$) gives

$$m(m-1)a_m - 2m a_m + 9a_{m-6} = 0.$$

Hence the recurrence relation is

$$a_m = -\frac{9a_{m-6}}{m(m-3)},$$

$$\Rightarrow \quad a_{6p} = -\frac{9}{6p\,(6p-3)}a_{6p-6} = -\frac{a_{6p-6}}{2p\,(2p-1)} = \frac{(-1)^p a_0}{(2p)!}.$$

A second solution is thus

$$y_2(x) = a_0\sum_{n=0}^{\infty}\frac{(-1)^n}{(2n)!}z^{6n} = a_0\cos z^3.$$

We see that $\sigma = 0$ does, in fact, produce a (different) series solution. This is because the recurrence relation relates a_n to a_{n+6} and does not involve a_{n+3};

the relevance here of considering the subscripted index '$m + 3$' is that '3' is the difference between the two indicial values.

We now calculate the Wronskian of the two solutions, $y_1 = a_0 \sin z^3$ and $y_2 = b_0 \cos z^3$:

$$
\begin{aligned}
W(y_1, y_2) &= y_1 y_2' - y_2 y_1' \\
&= a_0 \sin z^3 (-3b_0 z^2 \sin z^3) - b_0 \cos z^3 (3a_0 z^2 \cos z^3) \\
&= -3a_0 b_0 z^2 \neq 0.
\end{aligned}
$$

The fact that the Wronskian is non-zero shows that the two solutions are linearly independent.

We can also calculate the Wronskian from the original equation in its standard form,

$$
y'' - \frac{2}{z} y' + 9z^4 y = 0,
$$

as

$$
W = C \exp \left\{ - \int^z \frac{-2}{u} \, du \right\} = C \exp(2 \ln z) = Cz^2.
$$

This is in agreement with the Wronskian calculated from the solutions, as it must be.

16.5 *Investigate solutions of Legendre's equation at one of its singular points as follows.*

(a) *Verify that $z = 1$ is a regular singular point of Legendre's equation and that the indicial equation for a series solution in powers of $(z - 1)$ has a double root $\sigma = 0$.*

(b) *Obtain the corresponding recurrence relation and show that a polynomial solution is obtained if ℓ is a positive integer.*

(c) *Determine the radius of convergence R of the $\sigma = 0$ series and relate it to the positions of the singularities of Legendre's equation.*

(a) In standard form, Legendre's equation reads

$$
y'' - \frac{2z}{1 - z^2} y' + \frac{\ell(\ell + 1)}{1 - z^2} y = 0.
$$

This has a singularity at $z = 1$, but, since

$$
\frac{-2z(z - 1)}{1 - z^2} \to 1 \text{ and } \frac{\ell(\ell + 1)(z - 1)^2}{1 - z^2} \to 0 \text{ as } z \to 1,
$$

i.e. both limits are finite, the point is a regular singular point.

We next change the origin to the point $z = 1$ by writing $u = z - 1$ and $y(z) = f(u)$. The transformed equation is

$$f'' - \frac{2(u+1)}{-u(u+2)} f' + \frac{\ell(\ell+1)}{-u(u+2)} y = 0$$

or $\quad - u(u+2)f'' - 2(u+1)f' + \ell(\ell+1)f = 0.$

The point $u = 0$ is a regular singular point of this equation and so we set $f(u) = u^\sigma \sum_{n=0}^{\infty} a_n u^n$ and obtain

$$-u(u+2) \sum_{n=0}^{\infty} (\sigma+n)(\sigma+n-1)a_n u^{\sigma+n-2}$$

$$- 2(u+1) \sum_{n=0}^{\infty} (\sigma+n)a_n u^{\sigma+n-1} + \ell(\ell+1) \sum_{n=0}^{\infty} a_n u^{\sigma+n} = 0.$$

Equating to zero the coefficient of $u^{\sigma-1}$ gives

$$-2\sigma(\sigma-1)a_0 - 2\sigma a_0 = 0 \quad \Rightarrow \quad \sigma^2 = 0;$$

i.e. the indicial equation has a double root $\sigma = 0$.

(b) To obtain the recurrence relation we set the coefficient of u^m equal to zero for general m:

$$-m(m-1)a_m - 2(m+1)ma_{m+1} - 2ma_m - 2(m+1)a_{m+1} + \ell(\ell+1)a_m = 0.$$

Tidying this up gives

$$2(m+1)(m+1)a_{m+1} = [\ell(\ell+1) - m^2 + m - 2m]a_m,$$

$$\Rightarrow \quad a_{m+1} = \frac{\ell(\ell+1) - m(m+1)}{2(m+1)^2} a_m.$$

From this it is clear that, if ℓ is a positive integer, then $a_{\ell+1}$ and all further a_n are zero and that the solution is a polynomial (of degree ℓ).

(c) The limit of the ratio of successive terms in the series is given by

$$\left| \frac{a_{n+1} u^{n+1}}{a_n u^n} \right| = \left| \frac{u[\ell(\ell+1) - m(m+1)]}{2(m+1)^2} \right| \to \frac{|u|}{2} \quad \text{as } m \to \infty.$$

For convergence this limit needs to be < 1, i.e. $|u| < 2$. Thus the series converges in a circle of radius 2 centred on $u = 0$, i.e. on $z = 1$. The value 2 is to be expected, as it is the distance from $z = 1$ of the next nearest (actually the only other) singularity of the equation (at $z = -1$), excluding $z = 1$ itself.

16.7 *The first solution of Bessel's equation for $v = 0$ is*

$$J_0(z) = \sum_{n=0}^{\infty} \frac{(-1)^n}{n!\Gamma(n+1)} \left(\frac{z}{2}\right)^{2n}.$$

Use the derivative method to show that

$$J_0(z)\ln z - \sum_{n=1}^{\infty} \frac{(-1)^n}{(n!)^2} \left(\sum_{r=1}^{n} \frac{1}{r}\right) \left(\frac{z}{2}\right)^{2n}$$

is a second solution.

Bessel's equation with $v = 0$ reads

$$zy'' + y' + zy = 0.$$

The recurrence relations that gave rise to the first solution, $J_0(z)$, were $(\sigma+1)^2 a_1 = 0$ and $(\sigma + n)^2 a_n + a_{n-2} = 0$ for $n \geq 2$. Thus, in a general form as a function of σ, the solution is given by

$$y(\sigma, z) = a_0 z^{\sigma} \left\{ 1 - \frac{z^2}{(\sigma+2)^2} + \frac{z^4}{(\sigma+2)^2(\sigma+4)^2} - \cdots \right.$$
$$\left. + \frac{(-1)^n z^{2n}}{[(\sigma+2)(\sigma+4)\ldots(\sigma+2n)]^2} + \cdots \right\}.$$

Setting $\sigma = 0$ reproduces the first solution given above.

To obtain a second independent solution, we must differentiate the above expression with respect to σ, before setting σ equal to 0:

$$\frac{\partial y}{\partial \sigma} = \ln z \, J_0(z) + \sum_{n=1}^{\infty} \frac{da_{2n}(\sigma)}{d\sigma} z^{\sigma+2n} \text{ at } \sigma = 0.$$

Now

$$\frac{da_{2n}(\sigma)}{d\sigma}\bigg|_{\sigma=0} = \frac{d}{d\sigma} \left\{ \frac{(-1)^n}{[(\sigma+2)(\sigma+4)\ldots(\sigma+2n)]^2} \right\}_{\sigma=0}$$

$$= \frac{(-1)^n(-2)}{[\ldots]^3} \left(\frac{[\ldots]}{\sigma+2} + \frac{[\ldots]}{\sigma+4} + \cdots + \frac{[\ldots]}{\sigma+2n} \right)$$

$$= \frac{(-2)(-1)^n}{[\ldots]^2} \sum_{r=1}^{n} \frac{1}{\sigma+2r}$$

$$= \frac{-2(-1)^n}{2^{2n}(n!)^2} \sum_{r=1}^{n} \frac{1}{2r}, \quad \text{at } \sigma = 0.$$

Substituting this result, we obtain the second series as

$$J_0(z)\ln z - \sum_{n=1}^{\infty} \frac{(-1)^n}{(n!)^2}\left(\sum_{r=1}^{n}\frac{1}{r}\right)\left(\frac{z}{2}\right)^{2n}.$$

This is the form given in the question.

16.9 *Find series solutions of the equation $y'' - 2zy' - 2y = 0$. Identify one of the series as $y_1(z) = \exp z^2$ and verify this by direct substitution. By setting $y_2(z) = u(z)y_1(z)$ and solving the resulting equation for $u(z)$, find an explicit form for $y_2(z)$ and deduce that*

$$\int_0^x e^{-v^2}\,dv = e^{-x^2}\sum_{n=0}^{\infty}\frac{n!}{2(2n+1)!}(2x)^{2n+1}.$$

(a) The origin is an ordinary point of the equation and so power series solutions will be possible. Substituting $y(z) = \sum_{n=0}^{\infty}a_n z^n$ gives

$$\sum_{n=0}^{\infty}n(n-1)a_n z^{n-2} - 2\sum_{n=0}^{\infty}na_n z^n - 2\sum_{n=0}^{\infty}a_n z^n = 0.$$

Equating to zero the coefficient of z^{m-2} yields the recurrence relation

$$a_m = \frac{2m-2}{m(m-1)}a_{m-2} = \frac{2}{m}a_{m-2}.$$

The solution with $a_0 = 1$ and $a_1 = 0$ is therefore

$$y_1(z) = 1 + \frac{2z^2}{2} + \frac{2^2 z^4}{(2)(4)} + \cdots + \frac{2^n z^{2n}}{2^n\,n!} + \cdots$$

$$= \sum_{n=0}^{\infty}\frac{z^{2n}}{n!} = \exp z^2.$$

Putting this result into the original equation,

$$(4z^2 + 2)\exp z^2 - 2z\,2z\exp z^2 - 2\exp z^2 = 0,$$

shows directly that it is a valid solution.

The solution with $a_0 = 0$ and $a_1 = 1$ takes the form

$$y_2(z) = z + \frac{2z^3}{3} + \frac{2^2 z^5}{(3)(5)} + \cdots + \frac{2^n\,2^n\,n!\,z^{2n+1}}{(2n+1)!} + \cdots$$

$$= \sum_{n=0}^{\infty}\frac{n!\,(2z)^{2n+1}}{2(2n+1)!}.$$

275

We now set $y_2(z) = u(z)y_1(z)$ and substitute it into the original equation. As they must, the terms in which u is undifferentiated cancel and leave

$$u'' \exp z^2 + 2u'(2z \exp z^2) - 2zu' \exp z^2 = 0.$$

It follows that

$$\frac{u''}{u'} = -2z \quad \Rightarrow \quad u' = Ae^{-z^2} \quad \Rightarrow \quad u(x) = A \int^x e^{-v^2} dv.$$

Hence, setting the two derived forms for a second solution equal to each other, we have

$$\sum_{n=0}^{\infty} \frac{n!\,(2x)^{2n+1}}{2(2n+1)!} = y_2(x) = y_1(x)u(x) = e^{x^2} A \int^x e^{-v^2} dv.$$

For arbitrary small x, only the $n = 0$ term in the series is significant and takes the value $2x/2 = x$, whilst the integral is $A \int^x 1\, dv = Ax$. Thus $A = 1$ and the equality

$$\int_0^x e^{-v^2} dv = e^{-x^2} \sum_{n=0}^{\infty} \frac{n!\,(2x)^{2n+1}}{2(2n+1)!}$$

holds for all x.

16.11 *Find the general power series solution about $z = 0$ of the equation*

$$z\frac{d^2y}{dz^2} + (2z - 3)\frac{dy}{dz} + \frac{4}{z}y = 0.$$

The origin is clearly a singular point of this equation but, since $z(2z - 3)/z$ and $4z^2/z^2$ are finite as $z \to 0$, it is a regular singular point. The equation will therefore have at least one Frobenius-type solution of the form $y(z) = z^{\sigma} \sum_{n=0}^{\infty} a_n z^n$.

The indicial equation for the solution can be read off directly from $z^2 y'' + z(2z - 3)y' + 4y = 0$ as

$$\sigma(\sigma - 1) - 3\sigma + 4 = (\sigma - 2)^2 = 0 \quad \Rightarrow \quad \sigma = 2 \text{ (repeated root).}$$

The recurrence relation in terms of a general σ is needed and is provided by setting the coefficient of $z^{m+\sigma}$ equal to 0:

$$(m + \sigma)(m - 1 + \sigma)a_m + 2(m - 1 + \sigma)a_{m-1} - 3(m + \sigma)a_m + 4a_m = 0.$$

This relation can be simplified and then applied repeatedly to give a_m in terms of

a_0 and hence an explicit expression for $y(\sigma, z)$:

$$
\begin{aligned}
a_m &= \frac{-2(m-1+\sigma)}{(m+\sigma)^2 - (m+\sigma) - 3(m+\sigma) + 4} a_{m-1} \\
&= \frac{-2(m-1+\sigma)}{(m+\sigma-2)^2} a_{m-1} \qquad \text{for } m \geq 1 \\
&= (-2)^m \frac{(m-1+\sigma)(m-2+\sigma)\ldots\sigma}{(m-2+\sigma)^2(m-3+\sigma)^2\ldots(\sigma-1)^2} a_0 \\
&= (-2)^m \frac{(m-1+\sigma)}{(m-2+\sigma)(m-3+\sigma)\ldots\sigma(\sigma-1)^2} a_0.
\end{aligned}
$$

Because of the form of the recurrence relation, we write the $n = 0$ and $n = 1$ terms explicitly:

$$
\begin{aligned}
y(\sigma, z) = a_0 z^\sigma &- \frac{2\sigma}{(\sigma-1)^2} a_0 z^{\sigma+1} \\
&+ z^\sigma \sum_{n=2}^{\infty} \frac{(n-1+\sigma)(-2z)^n}{(n-2+\sigma)(n-3+\sigma)\ldots\sigma(\sigma-1)^2}.
\end{aligned}
$$

We also need the derivative of this with respect to σ. As always, the derivative consists of two terms, the first of which is $y(\sigma, z) \ln z$. The second, in this case, is

$$
\frac{2(\sigma+1)}{(\sigma-1)^3} a_0 z^{\sigma+1} + a_0 z^\sigma \sum_{n=2}^{\infty} \frac{(n-1+\sigma)(-2z)^n}{(n-2+\sigma)(n-3+\sigma)\ldots\sigma(\sigma-1)^2}
$$

$$
\times \left[\frac{1}{n-1+\sigma} - \frac{1}{n-2+\sigma} - \frac{1}{n-3+\sigma} - \cdots - \frac{1}{\sigma} - \frac{2}{\sigma-1} \right].
$$

The factor in square brackets is obtained by considering $a_n(\sigma)$ as the product of factors of the form $(\sigma + \alpha)^\beta$; differentiation of the product with respect to σ produces a sum of terms, each of which is the original product divided by $(\sigma + \alpha)$, for some α, and multiplied by the corresponding β. In the actual expression, β takes the values $+1$ (once), -1 (on $n-1$ occasions) and -2 (once).

To obtain two independent solutions, we finally set $\sigma = 2$ and $a_0 = 1$ obtaining

$$
y_1(z) = \sum_{n=0}^{\infty} \frac{(n+1)(-2)^n z^{n+2}}{n!},
$$

$$
\begin{aligned}
y_2(z) = y_1(z) \ln z &+ 6a_0 z^3 \\
&+ \sum_{n=2}^{\infty} \frac{(n+1)(-2)^n z^{n+2}}{n!} \left[\frac{1}{n+1} - \frac{1}{n} - \frac{1}{n-1} - \cdots - \frac{1}{2} - 2 \right].
\end{aligned}
$$

The general solution is any linear combination of $y_1(z)$ and $y_2(z)$.

16.13 *For the equation $y'' + z^{-3}y = 0$, show that the origin becomes a regular singular point if the independent variable is changed from z to $x = 1/z$. Hence find a series solution of the form $y_1(z) = \sum_0^\infty a_n z^{-n}$. By setting $y_2(z) = u(z)y_1(z)$ and expanding the resulting expression for du/dz in powers of z^{-1}, show that $y_2(z)$ is a second solution with asymptotic form*

$$y_2(z) = c\left[z + \ln z - \tfrac{1}{2} + O\left(\frac{\ln z}{z}\right)\right],$$

where c is an arbitrary constant.

With the equation in its original form, it is clear that, since $z^2/z^3 \to \infty$ as $z \to 0$, the origin is an irregular singular point. However, if we set $1/z = \xi$ and $y(z) = Y(\xi)$, with

$$\frac{d\xi}{dz} = -\frac{1}{z^2} = -\xi^2 \quad \Rightarrow \quad \frac{d}{dz} = -\xi^2 \frac{d}{d\xi},$$

then

$$-\xi^2 \frac{d}{d\xi}\left(-\xi^2 \frac{dY}{d\xi}\right) + \xi^3 Y = 0,$$

$$\xi^2 \frac{d^2Y}{d\xi^2} + 2\xi \frac{dY}{d\xi} + \xi Y = 0,$$

$$Y'' + \frac{2}{\xi}Y' + \frac{1}{\xi}Y = 0.$$

By inspection, $\xi = 0$ is a regular singular point of this equation, and its indicial equation is

$$\sigma(\sigma - 1) + 2\sigma = 0 \quad \Rightarrow \quad \sigma = 0, -1.$$

We start with the larger root, $\sigma = 0$, as this is 'guaranteed' to give a valid series solution and assume a solution of the form $Y(\xi) = \sum_{n=0}^\infty a_n \xi^n$, leading to

$$\sum_{n=0}^\infty n(n-1)a_n \xi^{n-1} + 2\sum_{n=0}^\infty na_n \xi^{n-1} + \sum_{n=0}^\infty a_n \xi^n = 0.$$

Equating to zero the coefficient of ξ^{m-1} gives the recurrence relation

$$a_m = \frac{-a_{m-1}}{m(m+1)} \quad \Rightarrow \quad a_m = \frac{(-1)^m}{(m+1)(m!)^2}a_0$$

and the series solution in inverse powers of z,

$$y_1(z) = a_0 \sum_{n=0}^\infty \frac{(-1)^n}{(n+1)(n!)^2 z^n}.$$

To find the second solution we set $y_2(z) = f(z)y_1(z)$. As usual (and as intended),

278

all terms with f undifferentiated vanish when this is substituted in the original equation. What is left is

$$0 = f''(z)y_1(z) + 2f'(z)y_1'(z),$$

which on rearrangement yields

$$\frac{f''}{f'} = -\frac{2y_1'}{y_1}.$$

This equation, although it contains a second derivative, is in fact only a first-order equation (for f'). It can be integrated directly to give

$$\ln f' = -2 \ln y_1 + c.$$

After exponentiation, this equation can be written as

$$\begin{aligned}
\frac{df}{dz} = \frac{A}{y_1^2(z)} &= \frac{A}{a_0^2} \left(1 - \frac{1}{2 \times 1^2 z} + \frac{1}{3 \times 2^2 z^2} - \cdots \right)^{-2} \\
&= \frac{A}{a_0^2} \left[1 + \frac{1}{z} + \mathrm{O}\left(\frac{1}{z^2}\right) \right],
\end{aligned}$$

where $A = e^c$.

Hence, on integrating a second time, one obtains

$$f(z) = \frac{A}{a_0^2} \left(z + \ln z + \mathrm{O}\left(\frac{1}{z}\right) \right),$$

which in turn implies

$$\begin{aligned}
y_2(z) &= \frac{A}{a_0^2} \left[z + \ln z + \mathrm{O}\left(\frac{1}{z}\right) \right] a_0 \left(1 - \frac{1}{2z} + \frac{1}{12z^2} - \cdots \right) \\
&= c \left[z + \ln z - \frac{1}{2} + \mathrm{O}\left(\frac{\ln z}{z}\right) \right].
\end{aligned}$$

This establishes the asymptotic form of the second solution.

16.15 *The origin is an ordinary point of the Chebyshev equation,*

$$(1 - z^2)y'' - zy' + m^2y = 0,$$

which therefore has series solutions of the form $z^\sigma \sum_0^\infty a_n z^n$ for $\sigma = 0$ and $\sigma = 1$.

(a) *Find the recurrence relationships for the a_n in the two cases and show that there exist polynomial solutions $T_m(z)$:*

    (i) *for $\sigma = 0$, when m is an even integer, the polynomial having $\frac{1}{2}(m+2)$ terms;*

    (ii) *for $\sigma = 1$, when m is an odd integer, the polynomial having $\frac{1}{2}(m+1)$ terms.*

(b) *$T_m(z)$ is normalised so as to have $T_m(1) = 1$. Find explicit forms for $T_m(z)$ for $m = 0, 1, 2, 3$.*

(c) *Show that the corresponding non-terminating series solutions $S_m(z)$ have as their first few terms*

$$S_0(z) = a_0 \left(z + \frac{1}{3!}z^3 + \frac{9}{5!}z^5 + \cdots \right),$$

$$S_1(z) = a_0 \left(1 - \frac{1}{2!}z^2 - \frac{3}{4!}z^4 - \cdots \right),$$

$$S_2(z) = a_0 \left(z - \frac{3}{3!}z^3 - \frac{15}{5!}z^5 - \cdots \right),$$

$$S_3(z) = a_0 \left(1 - \frac{9}{2!}z^2 + \frac{45}{4!}z^4 + \cdots \right).$$

(a)(i) If, for $\sigma = 0$, $y(z) = \sum_{n=0}^{\infty} a_n z^n$ with $a_0 \neq 0$, the condition for the coefficient of z^r in

$$(1 - z^2) \sum_{n=0}^{\infty} n(n-1)a_n z^{n-2} - z \sum_{n=0}^{\infty} n a_n z^{n-1} + m^2 \sum_{n=0}^{\infty} a_n z^n$$

to be zero is that

$$(r+2)(r+1)a_{r+2} - r(r-1)a_r - ra_r + m^2 a_r = 0,$$

$$\Rightarrow \quad a_{r+2} = \frac{r^2 - m^2}{(r+2)(r+1)} a_r.$$

This relation relates a_{r+2} to a_r and so to a_0 if r is even. For a_{r+2} to vanish, in this case, requires that $r = m$, which must therefore be an even integer. The non-vanishing coefficients will be $a_0, a_2, \ldots, a_m$, i.e. $\frac{1}{2}(m+2)$ of them in all.

(ii) If, for $\sigma = 1$, $y(z) = \sum_{n=0}^{\infty} a_n z^{n+1}$ with $a_0 \neq 0$, the condition for the coefficient of z^{r+1} in

$$(1 - z^2) \sum_{n=0}^{\infty} (n+1)na_n z^{n-1} - z \sum_{n=0}^{\infty} (n+1)a_n z^n + m^2 \sum_{n=0}^{\infty} a_n z^{n+1}$$

to be zero is that

$$(r+3)(r+2)a_{r+2} - (r+1)ra_r - (r+1)a_r + m^2 a_r = 0,$$
$$\Rightarrow \quad a_{r+2} = \frac{(r+1)^2 - m^2}{(r+3)(r+2)} a_r.$$

This relation relates a_{r+2} to a_r and so to a_0 if r is even. For a_{r+2} to vanish, in this case, requires that $r + 1 = m$, which must therefore be an odd integer. The non-vanishing coefficients will be, as before, $a_0, a_2, \ldots, a_{m-1}$, i.e. $\frac{1}{2}(m+1)$ of them in all.

(b) For $m = 0$, $T_0(z) = a_0$. With the given normalisation, $a_0 = 1$ and $T_0(z) = 1$.

For $m = 1$, $T_1(z) = a_0 z$. The required normalisation implies that $a_0 = 1$ and so $T_0(z) = z$.

For $m = 2$, we need the recurrence relation in (a)(i). This shows that

$$a_2 = \frac{0^2 - 2^2}{(2)(1)} a_0 = -2a_0 \quad \Rightarrow \quad T_2(z) = a_0(1 - 2z^2).$$

With the given normalisation, $a_0 = -1$ and $T_2(z) = 2z^2 - 1$.

For $m = 3$, we use the recurrence relation in (a)(ii) and obtain

$$a_2 = \frac{1^2 - 3^2}{(3)(2)} a_0 = -\frac{4}{3} a_0 \quad \Rightarrow \quad T_3(z) = a_0\left(z - \frac{4z^3}{3}\right).$$

For the required normalisation, we must have $a_0 = -\frac{1}{3}$ and consequently that $T_3(z) = 4z^3 - 3z$.

(c) The non-terminating series solutions $S_m(z)$ arise when $\sigma = 0$ but m is an odd integer and when $\sigma = 1$ with m an even integer. We take each in turn and apply the appropriate recurrence relation to generate the coefficients.

(i) $\sigma = 0$, $m = 1$, using the (a)(i) recurrence relation:

$$a_2 = \frac{0 - 1}{(2)(1)} a_0 = -\frac{1}{2!} a_0, \quad a_4 = \frac{4 - 1}{(4)(3)} a_2 = -\frac{3}{4!} a_0.$$

Hence,

$$S_1(z) = a_0 \left(1 - \frac{1}{2!}z^2 - \frac{3}{4!}z^4 - \cdots\right).$$

(ii) $\sigma = 0$, $m = 3$, using the (a)(i) recurrence relation:

$$a_2 = \frac{0 - 9}{(2)(1)} a_0 = -\frac{9}{2!} a_0, \quad a_4 = \frac{4 - 9}{(4)(3)} a_2 = \frac{45}{4!} a_0.$$

Hence,

$$S_3(z) = a_0 \left(1 - \frac{9}{2!}z^2 + \frac{45}{4!}z^4 + \cdots \right).$$

(iii) $\sigma = 1$, $m = 0$, using the (a)(ii) recurrence relation:

$$a_2 = \frac{1-0}{(3)(2)}a_0 = \frac{1}{3!}a_0, \quad a_4 = \frac{9-0}{(5)(4)}a_2 = \frac{9}{5!}a_0.$$

Hence,

$$S_0(z) = a_0 \left(z + \frac{1}{3!}z^3 + \frac{9}{5!}z^5 + \cdots \right).$$

(iv) $\sigma = 1$, $m = 2$, using the (a)(ii) recurrence relation:

$$a_2 = \frac{1-4}{(3)(2)}a_0 = -\frac{3}{3!}a_0, \quad a_4 = \frac{9-4}{(5)(4)}a_2 = -\frac{15}{5!}a_0.$$

Hence,

$$S_2(z) = a_0 \left(z - \frac{3}{3!}z^3 - \frac{15}{5!}z^5 - \cdots \right).$$

17

Eigenfunction methods for differential equations

17.1 *By considering $\langle h|h \rangle$, where $h = f + \lambda g$ with λ real, prove that, for two functions f and g,*

$$\langle f|f \rangle \langle g|g \rangle \geq \tfrac{1}{4}[\langle f|g \rangle + \langle g|f \rangle]^2.$$

The function $y(x)$ is real and positive for all x. Its Fourier cosine transform $\tilde{y}_c(k)$ is defined by

$$\tilde{y}_c(k) = \int_{-\infty}^{\infty} y(x) \cos(kx)\, dx,$$

and it is given that $\tilde{y}_c(0) = 1$. Prove that

$$\tilde{y}_c(2k) \geq 2[\tilde{y}_c(k)]^2 - 1.$$

For any $|h\rangle$ we have that $\langle h|h \rangle \geq 0$, with equality only if $|h\rangle = |0\rangle$. Hence, noting that λ is real, we have

$$0 \leq \langle h|h \rangle = \langle f + \lambda g|f + \lambda g \rangle = \langle f|f \rangle + \lambda \langle g|f \rangle + \lambda \langle f|g \rangle + \lambda^2 \langle g|g \rangle.$$

This equation, considered as a quadratic inequality in λ, states that the corresponding quadratic equation has no real roots. The condition for this ('$b^2 < 4ac$') is given by

$$[\langle g|f \rangle + \langle f|g \rangle]^2 \leq 4\langle f|f \rangle \langle g|g \rangle, \qquad (*)$$

from which the stated result follows immediately. Note that $\langle g|f \rangle + \langle f|g \rangle$ is real and its square is therefore non-negative.

The given datum is equivalent to

$$1 = \tilde{y}_c(0) = \int_{-\infty}^{\infty} y(x) \cos(0x)\, dx = \int_{-\infty}^{\infty} y(x)\, dx.$$

Now consider

$$\tilde{y}_c(2k) = \int_{-\infty}^{\infty} y(x)\cos(2kx)\,dx$$

$$= 2\int_{-\infty}^{\infty} y(x)\cos^2 kx - \int_{-\infty}^{\infty} y(x)\,dx,$$

$$\Rightarrow \quad \tilde{y}_c(2k) + 1 = 2\int_{-\infty}^{\infty} y(x)\cos^2 kx.$$

In order to use (*), we need to choose for $f(x)$ and $g(x)$ functions whose product will form the integrand defining $\tilde{y}_c(k)$. With this in mind, we take $f(x) = y^{1/2}(x)\cos kx$ and $g(x) = y^{1/2}(x)$; we may do this since $y(x) > 0$ for all x. Making these choices gives

$$\left(\int_{-\infty}^{\infty} y\cos kx\,dx + \int_{-\infty}^{\infty} y\cos kx\,dx\right)^2 \leq 4\int_{-\infty}^{\infty} y\cos^2 kx\,dx \int_{-\infty}^{\infty} y\,dx,$$

$$\left(\int_{-\infty}^{\infty} 2y\cos kx\,dx\right)^2 \leq 4\int_{-\infty}^{\infty} y\cos^2 kx\,dx \times 1,$$

$$4\tilde{y}_c^2(k) \leq 4\int_{-\infty}^{\infty} y\cos^2 kx\,dx.$$

Thus,

$$\tilde{y}_c(2k) + 1 = 2\int_{-\infty}^{\infty} y(x)\cos^2 kx \geq 2[\tilde{y}_c(k)]^2$$

and hence the stated result.

17.3 Consider the real eigenfunctions $y_n(x)$ of a Sturm–Liouville equation

$$(py')' + qy + \lambda\rho y = 0, \qquad a \leq x \leq b,$$

in which $p(x)$, $q(x)$ and $\rho(x)$ are continuously differentiable real functions and $p(x)$ does not change sign in $a \leq x \leq b$. Take $p(x)$ as positive throughout the interval, if necessary by changing the signs of all eigenvalues. For $a \leq x_1 \leq x_2 \leq b$, establish the identity

$$(\lambda_n - \lambda_m)\int_{x_1}^{x_2} \rho y_n y_m\,dx = \left[y_n p y_m' - y_m p y_n'\right]_{x_1}^{x_2}.$$

Deduce that if $\lambda_n > \lambda_m$ then $y_n(x)$ must change sign between two successive zeros of $y_m(x)$.

[*The reader may find it helpful to illustrate this result by sketching the first few eigenfunctions of the system $y'' + \lambda y = 0$, with $y(0) = y(\pi) = 0$, and the Legendre polynomials $P_n(z)$ for $n = 2, 3, 4, 5$.*]

The function $p(x)$ does not change sign in the interval $a \le x \le b$; we take it as positive, multiplying the equation all through by -1 if necessary. This means that the weight function ρ can still be taken as positive, but that we must consider all possible functions for $q(x)$ and eigenvalues λ of either sign.

We start with the eigenvalue equation for $y_n(x)$, multiply it through by $y_m(x)$ and then integrate from x_1 to x_2. From this result we subtract the same equation with the roles of n and m reversed, as follows. The integration limits are omitted until the explicit integration by parts is carried through:

$$\int y_m (p\, y_n')'\, dx + \int y_m q\, y_n\, dx + \int y_m \lambda_n \rho y_n\, dx = 0,$$

$$\int y_n (p\, y_m')'\, dx + \int y_n q\, y_m\, dx + \int y_n \lambda_m \rho y_m\, dx = 0,$$

$$\int \left[y_m (p\, y_n')' - y_n (p\, y_m')' \right] dx + (\lambda_n - \lambda_m) \int y_m \rho y_n\, dx = 0,$$

$$\left[y_m p\, y_n' \right]_{x_1}^{x_2} - \int y_m' p\, y_n'\, dx - \left[y_n p\, y_m' \right]_{x_1}^{x_2}$$

$$+ \int y_n' p\, y_m'\, dx + (\lambda_n - \lambda_m) \int y_m \rho y_n\, dx = 0.$$

Hence

$$(\lambda_n - \lambda_m) \int y_m \rho y_n\, dx = \left[y_n p\, y_m' - y_m p\, y_n' \right]_{x_1}^{x_2}. \qquad (*)$$

Now, in this general result, take x_1 and x_2 as successive zeros of $y_m(x)$, where m is determined by $\lambda_n > \lambda_m$ (after the signs have been changed, if that was necessary). Clearly the sign of $y_m(x)$ does not change in this interval; let it be α. It follows that the sign of $y_m'(x_1)$ is also α, whilst that of $y_m'(x_2)$ is $-\alpha$. In addition, the second term on the RHS of $(*)$ vanishes at both limits, as $y_m(x_1) = y_m(x_2) = 0$.

Let us now *suppose* that the sign of $y_n(x)$ does not change in this same interval and is always β. Then the sign of the expression on the LHS of $(*)$ is $(+1)(\alpha)(+1)\beta = \alpha\beta$. The first $(+1)$ appears because $\lambda_n > \lambda_m$.

The signs of the upper- and lower-limit contributions of the remaining term on the RHS of $(*)$ are $\beta(+1)(-\alpha)$ and $(-1)\beta(+1)\alpha$, respectively, the additional factor of (-1) in the second product arising from the fact that the contribution comes from a lower limit. The contributions at both limits have the same sign, $-\alpha\beta$, and so the sign of the total RHS must also be $-\alpha\beta$.

This contradicts, however, the sign of $+\alpha\beta$ found for the LHS. It follows that it was wrong to suppose that the sign of $y_n(x)$ does not change in the interval; in other words, a zero of $y_n(x)$ does appear between every pair of zeros of $y_m(x)$.

> **17.5** *Use the properties of Legendre polynomials to carry out the following exercises.*
>
> (a) *Find the solution of $(1 - x^2)y'' - 2xy' + by = f(x)$ that is valid in the range $-1 \leq x \leq 1$ and finite at $x = 0$, in terms of Legendre polynomials.*
>
> (b) *Find the explicit solution if $b = 14$ and $f(x) = 5x^3$. Verify it by direct substitution.*
>
> [*Explicit forms for the Legendre polynomials can be found in any textbook. In* Mathematical Methods for Physics and Engineering, 3rd edition, *they are given in Subsection 18.1.1.*]

(a) The LHS of the given equation is the same as that of Legendre's equation and so we substitute $y(x) = \sum_{n=0}^{\infty} a_n P_n(x)$ and use the fact that $(1 - x^2)P_n'' - 2xP_n' = -n(n + 1)P_n$. This results in

$$\sum_{n=0}^{\infty} a_n[b - n(n + 1)]P_n = f(x).$$

Now, using the mutual orthogonality and normalisation of the $P_n(x)$, we multiply both sides by $P_m(x)$ and integrate over x:

$$\sum_{n=0}^{\infty} a_n[b - n(n + 1)]\,\delta_{mn}\frac{2}{2m + 1} = \int_{-1}^{1} f(z)P_m(z)\,dz,$$

$$\Rightarrow \quad a_m = \frac{2m + 1}{2[b - m(m + 1)]}\int_{-1}^{1} f(z)P_m(z)\,dz.$$

This gives the coefficients in the solution $y(x)$.

(b) We now express $f(x)$ in terms of Legendre polynomials,

$$f(x) = 5x^3 = 2[\tfrac{1}{2}(5x^3 - 3x)] + 3[x] = 2P_3(x) + 3P_1(x),$$

and conclude that, because of the mutual orthogonality of the Legendre polynomials, only a_3 and a_1 in the series solution will be non-zero. To find them we need to evaluate

$$\int_{-1}^{1} f(z)P_3(z)\,dz = 2\frac{2}{2(3) + 1} = \frac{4}{7};$$

similarly, $\int_{-1}^{1} f(z)P_1(z)\,dz = 3 \times (2/3) = 2$.

Inserting these values gives

$$a_3 = \frac{7}{2(14 - 12)}\frac{4}{7} = 1 \text{ and } a_1 = \frac{3}{2(14 - 2)}2 = \frac{1}{4}.$$

Thus the solution is

$$y(x) = \frac{1}{4}P_1(x) + P_3(x) = \frac{1}{4}x + \frac{1}{2}(5x^3 - 3x) = \frac{5(2x^3 - x)}{4}.$$

Check:

$$(1 - x^2)\frac{60x}{4} - 2x\frac{30x^2 - 5}{4} + \frac{140x^3 - 70x}{4} = 5x^3,$$
$$\Rightarrow \quad 60x - 60x^3 - 60x^3 + 10x + 140x^3 - 70x = 20x^3,$$

which is satisfied.

17.7 *Consider the set of functions, $\{f(x)\}$, of the real variable x defined in the interval $-\infty < x < \infty$, that $\to 0$ at least as quickly as x^{-1}, as $x \to \pm\infty$. For unit weight function, determine whether each of the following linear operators is Hermitian when acting upon $\{f(x)\}$:*

$$\text{(a)} \ \frac{d}{dx} + x; \quad \text{(b)} \ -i\frac{d}{dx} + x^2; \quad \text{(c)} \ ix\frac{d}{dx}; \quad \text{(d)} \ i\frac{d^3}{dx^3}.$$

For an operator $\mathcal{L}$ to be Hermitian over the given range with respect to a unit weight function, the equation

$$\int_{-\infty}^{\infty} f^*(x)[\mathcal{L}g(x)]\,dx = \left\{\int_{-\infty}^{\infty} g^*(x)[\mathcal{L}f(x)]\,dx\right\}^* \qquad (*)$$

must be satisfied for general functions f and g.

(a) For $\mathcal{L} = \dfrac{d}{dx} + x$, the LHS of $(*)$ is

$$\int_{-\infty}^{\infty} f^*(x)\left(\frac{dg}{dx} + xg\right) dx = \left[f^*g\right]_{-\infty}^{\infty} - \int_{-\infty}^{\infty} \frac{df^*}{dx}g\,dx + \int_{-\infty}^{\infty} f^*xg\,dx$$

$$= 0 - \int_{-\infty}^{\infty} \frac{df^*}{dx}g\,dx + \int_{-\infty}^{\infty} f^*xg\,dx.$$

The RHS of $(*)$ is

$$\int_{-\infty}^{\infty} \left\{g^*(x)\left(\frac{df}{dx} + xf\right)dx\right\}^* = \left\{\int_{-\infty}^{\infty} g^*\frac{df}{dx}\,dx\right\}^* + \left\{\int_{-\infty}^{\infty} g^*xf\,dx\right\}^*$$

$$= \int_{-\infty}^{\infty} g\frac{df^*}{dx}\,dx + \int_{-\infty}^{\infty} gxf^*\,dx.$$

Since the sign of the first term differs in the two expressions, the LHS $\neq$ RHS and $\mathcal{L}$ is *not* Hermitian. It will also be apparent that purely multiplicative terms in the operator, such as x or x^2, will always be Hermitian; thus we can ignore the x^2 term in part (b).

287

(b) As explained above, we need only consider

$$\int_{-\infty}^{\infty} f^*(x) \left(-i\frac{dg}{dx}\right) dx = \left[-if^*g\right]_{-\infty}^{\infty} + i\int_{-\infty}^{\infty} \frac{df^*}{dx} g\, dx$$

$$= 0 + i\int_{-\infty}^{\infty} \frac{df^*}{dx} g\, dx$$

and

$$\left\{\int_{-\infty}^{\infty} g^*(x) \left(-i\frac{df}{dx}\right) dx\right\}^* = i\int_{-\infty}^{\infty} g\frac{df^*}{dx} dx.$$

These are equal, and so $\mathcal{L} = -i\dfrac{d}{dx}$ is Hermitian, as is $\mathcal{L} = -i\dfrac{d}{dx} + x^2$.

(c) For $\mathcal{L} = ix\dfrac{d}{dx}$, the LHS of (*) is

$$\int_{-\infty}^{\infty} f^*(x) \left(ix\frac{dg}{dx}\right) dx = \left[ixf^*g\right]_{-\infty}^{\infty} - i\int_{-\infty}^{\infty} x\frac{df^*}{dx} g\, dx - i\int_{-\infty}^{\infty} f^*g\, dx$$

$$= 0 - i\int_{-\infty}^{\infty} x\frac{df^*}{dx} g\, dx - i\int_{-\infty}^{\infty} f^*g\, dx.$$

The RHS of (*) is given by

$$\left\{\int_{-\infty}^{\infty} g^*(x)ix \left(\frac{df}{dx}\right) dx\right\}^* = -i\int_{-\infty}^{\infty} gx\frac{df^*}{dx} dx.$$

Since, in general, $-i\int_{-\infty}^{\infty} fg^*\, dx \neq 0$, the two sides are not equal; therefore $\mathcal{L}$ is not Hermitian.

(d) Since $\mathcal{L} = i\dfrac{d^3}{dx^3}$ is the cube of the operator $-i\dfrac{d}{dx}$, which was shown in part (b) to be Hermitian, it is expected that $\mathcal{L}$ is Hermitian. This can be verified directly as follows.

The LHS of (*) is given by

$$i\int_{-\infty}^{\infty} f^*\frac{d^3g}{dx^3} dx = \left[if^*\frac{d^2g}{dx^2}\right]_{-\infty}^{\infty} - i\int_{-\infty}^{\infty} \frac{df^*}{dx}\frac{d^2g}{dx^2} dx$$

$$= 0 - i\left[\frac{df^*}{dx}\frac{dg}{dx}\right]_{-\infty}^{\infty} + i\int_{-\infty}^{\infty} \frac{d^2f^*}{dx^2}\frac{dg}{dx} dx$$

$$= 0 + i\left[\frac{d^2f^*}{dx^2} g\right]_{-\infty}^{\infty} - i\int_{-\infty}^{\infty} \frac{d^3f^*}{dx^3} g\, dx$$

$$= 0 + \left\{\int_{-\infty}^{\infty} ig^*\frac{d^3f}{dx^3} dx\right\}^* = \text{RHS of (*).}$$

Thus $\mathcal{L}$ is confirmed as Hermitian.

17.9 *Find an eigenfunction expansion for the solution with boundary conditions* $y(0) = y(\pi) = 0$ *of the inhomogeneous equation*

$$\frac{d^2 y}{dx^2} + \kappa y = f(x),$$

where κ is a constant and

$$f(x) = \begin{cases} x & 0 \le x \le \pi/2, \\ \pi - x & \pi/2 < x \le \pi. \end{cases}$$

The eigenfunctions of the operator $\mathcal{L} = \dfrac{d^2}{dx^2} + \kappa$ are obviously

$$y_n(x) = A_n \sin nx + B_n \cos nx,$$

with corresponding eigenvalues $\lambda_n = n^2 - \kappa$.

The boundary conditions, $y(0) = y(\pi) = 0$, require that n is a positive integer and that $B_n = 0$, i.e.

$$y_n(x) = A_n \sin nx = \sqrt{\frac{2}{\pi}} \sin nx,$$

where A_n (for $n \ge 1$) has been chosen so that the eigenfunctions are normalised over the interval $x = 0$ to $x = \pi$. Since $\mathcal{L}$ is Hermitian on the range $0 \le x \le \pi$, the eigenfunctions are also mutually orthogonal, and so the $y_n(x)$ form an orthonormal set.

If the required solution is $y(x) = \sum_n a_n y_n(x)$, then direct substitution yields the result

$$\sum_{n=1}^{\infty} (\kappa - n^2) a_n y_n(x) = f(x).$$

Following the usual procedure for analysis using sets of orthonormal functions, this implies that

$$a_m = \frac{1}{\kappa - m^2} \int_0^{\pi} f(z) y_m(z) \, dz$$

and, consequently, that

$$y(x) = \sum_{n=1}^{\infty} \sqrt{\frac{2}{\pi}} \frac{\sin nx}{\kappa - n^2} \sqrt{\frac{2}{\pi}} \int_0^{\pi} f(z) \sin(nz) \, dz.$$

It only remains to evaluate

$$I_n = \int_0^\pi \sin(nx) f(x)\, dx$$

$$= \int_0^{\pi/2} x \sin nx\, dx + \int_{\pi/2}^\pi (\pi - x) \sin nx\, dx$$

$$= \left[\frac{-x \cos nx}{n} \right]_0^{\pi/2} + \int_0^{\pi/2} \frac{\cos nx}{n}\, dx$$

$$+ \left[\frac{-(\pi - x) \cos nx}{n} \right]_{\pi/2}^\pi + \int_{\pi/2}^\pi \frac{(-1) \cos nx}{n}\, dx$$

$$= -\frac{\pi}{2} \frac{\cos(n\pi/2)}{n} (1 - 1) + \left[\frac{\sin nx}{n^2} \right]_0^{\pi/2} - \left[\frac{\sin nx}{n^2} \right]_{\pi/2}^\pi$$

$$= 0 + \frac{(-1)^{(n-1)/2}}{n^2} (1 + 1) \text{ for odd } n \text{ and } = 0 \text{ for even } n.$$

Thus,

$$y(x) = \frac{4}{\pi} \sum_{n \text{ odd}} \frac{(-1)^{(n-1)/2}}{n^2(\kappa - n^2)} \sin nx$$

is the required solution.

17.11 *The differential operator $\mathcal{L}$ is defined by*

$$\mathcal{L}y = -\frac{d}{dx}\left(e^x \frac{dy}{dx} \right) - \tfrac{1}{4} e^x y.$$

Determine the eigenvalues λ_n of the problem

$$\mathcal{L}y_n = \lambda_n e^x y_n \qquad 0 < x < 1,$$

with boundary conditions

$$y(0) = 0, \qquad \frac{dy}{dx} + \tfrac{1}{2}y = 0 \quad at \quad x = 1.$$

(a) *Find the corresponding unnormalised y_n, and also a weight function $\rho(x)$ with respect to which the y_n are orthogonal. Hence, select a suitable normalisation for the y_n.*

(b) *By making an eigenfunction expansion, solve the equation*

$$\mathcal{L}y = -e^{x/2}, \qquad 0 < x < 1,$$

subject to the same boundary conditions as previously.

When written out explicitly, the eigenvalue equation is

$$-\frac{d}{dx}\left(e^x \frac{dy}{dx}\right) - \tfrac{1}{4}e^x y = \lambda e^x y, \qquad (*)$$

or, on differentiating out the product,

$$e^x y'' + e^x y' + (\lambda + \tfrac{1}{4})e^x y = 0.$$

The auxiliary equation is

$$m^2 + m + (\lambda + \tfrac{1}{4}) = 0 \quad \Rightarrow \quad m = -\tfrac{1}{2} \pm i\sqrt{\lambda}.$$

The general solution is thus given by

$$y(x) = Ae^{-x/2} \cos\sqrt{\lambda}x + Be^{-x/2} \sin\sqrt{\lambda}x,$$

with the condition $y(0) = 0$ implying that $A = 0$. The other boundary condition requires that, at $x = 1$,

$$-\tfrac{1}{2}Be^{-x/2} \sin\sqrt{\lambda}x + \sqrt{\lambda}Be^{-x/2} \cos\sqrt{\lambda}x + \tfrac{1}{2}Be^{-x/2}\sin\sqrt{\lambda}x = 0,$$

i.e. that $\cos\sqrt{\lambda} = 0$ and hence that $\lambda = (n + \tfrac{1}{2})^2\pi^2$ for non-negative integral n.

(a) The unnormalised eigenfunctions are

$$y_n(x) = B_n e^{-x/2} \sin(n + \tfrac{1}{2})\pi x$$

and $(*)$ is in Sturm–Liouville form. However, although $y_n(0) = 0$, the values at the upper limit in $\left[y'_m p\, y_n\right]_0^1$ are $y_n(1) = B_n e^{-1/2}(-1)^n$, $p(1) = e^1$ and $y'_m(1) = -\tfrac{1}{2}B_m e^{-1/2}(-1)^m$. Consequently, $\left[y'_m p\, y_n\right]_0^1 \neq 0$ and S–L theory cannot be applied. We therefore have to find a suitable weight function $\rho(x)$ by inspection. Given the general form of the eigenfunctions, ρ has to be taken as e^x, with the orthonormality integral taking the form

$$I_{nm} = \int_0^1 \rho(x)y_n(x)y_m^*(x)\,dx$$

$$= B_n B_m \int_0^1 e^x e^{-x/2} \sin[(n + \tfrac{1}{2})\pi x]e^{-x/2}\sin[(m + \tfrac{1}{2})\pi x]\,dx$$

$$= \begin{cases} 0 & \text{for } m \neq n, \\ \tfrac{1}{2}B_n B_m & \text{for } m = n. \end{cases}$$

It is clear that a suitable normalisation is $B_n = \sqrt{2}$ for all n.

(b) We write the solution as $y(x) = \sum_{n=0}^\infty a_n y_n(x)$, giving as the equation to be solved

$$-e^{x/2} = \mathcal{L}y = \mathcal{L} \sum_{n=0}^{\infty} a_n y_n(x)$$

$$= \sum_{n=0}^{\infty} a_n [\lambda_n \rho(x) y_n(x)]$$

$$= \sum_{n=0}^{\infty} a_n (n + \tfrac{1}{2})^2 \pi^2 e^x \sqrt{2} e^{-x/2} \sin[(n + \tfrac{1}{2})\pi x]$$

$$\Rightarrow \quad -1 = \sum_{n=0}^{\infty} a_n (n + \tfrac{1}{2})^2 \pi^2 \sqrt{2} \sin[(n + \tfrac{1}{2})\pi x].$$

After multiplying both sides of this equation by $\sin(m + \tfrac{1}{2})\pi x$ and integrating from 0 to 1, we obtain

$$a_m \int_0^1 \sin^2(m + \tfrac{1}{2})\pi x \, dx = \frac{-1}{(m + \tfrac{1}{2})^2 \pi^2 \sqrt{2}} \int_0^1 \sin(m + \tfrac{1}{2})\pi x \, dx,$$

$$\frac{a_m}{2} = \frac{-1}{(m + \tfrac{1}{2})^2 \pi^2 \sqrt{2}} \int_0^1 \sin(m + \tfrac{1}{2})\pi x \, dx$$

$$= \frac{1}{(m + \tfrac{1}{2})^2 \pi^2 \sqrt{2}} \left[\frac{\cos(m + \tfrac{1}{2})\pi x}{(m + \tfrac{1}{2})\pi} \right]_0^1,$$

$$a_m = -\frac{\sqrt{2}}{(m + \tfrac{1}{2})^3 \pi^3}.$$

Substituting this result into the assumed expansion, and recalling that $B_n = \sqrt{2}$, gives as the solution

$$y(x) = -\sum_{n=0}^{\infty} \frac{2}{(n + \tfrac{1}{2})^3 \pi^3} e^{-x/2} \sin(n + \tfrac{1}{2})\pi x.$$

17.13 By substituting $x = \exp t$, find the normalised eigenfunctions $y_n(x)$ and the eigenvalues λ_n of the operator $\mathcal{L}$ defined by

$$\mathcal{L}y = x^2 y'' + 2xy' + \tfrac{1}{4}y, \qquad 1 \le x \le e,$$

with $y(1) = y(e) = 0$. Find, as a series $\sum a_n y_n(x)$, the solution of $\mathcal{L}y = x^{-1/2}$.

Putting $x = e^t$ and $y(x) = u(t)$ with $u(0) = u(1) = 0$,

$$\frac{dx}{dt} = e^t \quad \Rightarrow \quad \frac{d}{dx} = e^{-t} \frac{d}{dt}$$

and the eigenvalue equation becomes

$$e^{2t}e^{-t}\frac{d}{dt}\left(e^{-t}\frac{du}{dt}\right) + 2e^te^{-t}\frac{du}{dt} + \frac{1}{4}u = \lambda u,$$

$$\frac{d^2u}{dt^2} - \frac{du}{dt} + 2\frac{du}{dt} + \left(\frac{1}{4} - \lambda\right) = 0.$$

The auxiliary equation to this constant-coefficient linear equation for u is

$$m^2 + m + (\tfrac{1}{4} - \lambda) = 0 \quad \Rightarrow \quad m = -\tfrac{1}{2} \pm \sqrt{\lambda},$$

leading to

$$u(t) = e^{-t/2}\left(Ae^{\sqrt{\lambda}t} + Be^{-\sqrt{\lambda}t}\right).$$

In view of the requirement that u vanishes at two different values of t (one of which is $t = 0$), we need $\lambda < 0$ and $u(t)$ to take the form

$$u(t) = Ae^{-t/2}\sin\sqrt{-\lambda}\,t \text{ with } \sqrt{-\lambda}\,1 = n\pi, \text{ i.e. } \lambda = -n^2\pi^2,$$

where n is an integer. Thus

$$u_n(t) = A_ne^{-t/2}\sin n\pi t \quad \text{or, in terms of } x, \quad y_n(x) = \frac{A_n}{\sqrt{x}}\sin(n\pi \ln x).$$

Normalisation requires that

$$1 = \int_1^e \frac{A_n^2}{x}\sin^2(n\pi \ln x)\,dx = \int_0^1 A_n^2 \sin^2(n\pi t)\,dt = \tfrac{1}{2}A_n^2 \quad \Rightarrow \quad A_n = \sqrt{2}.$$

To solve

$$\mathcal{L}y = x^2y'' + 2xy' + \tfrac{1}{4}y = \frac{1}{\sqrt{x}},$$

we set $y(x) = \sum_{n=0}^{\infty} a_n y_n(x)$. Then the equation becomes

$$\mathcal{L}y = \sum_{n=0}^{\infty} a_n(-n^2\pi^2)y_n(x) = \sum_{n=0}^{\infty} -n^2\pi^2 a_n \frac{\sqrt{2}}{\sqrt{x}}\sin(n\pi \ln x) = \frac{1}{\sqrt{x}}.$$

Multiplying through by $y_m(x)$ and integrating, as with ordinary Fourier series,

$$\int_1^e \frac{2a_n}{x}\sin(n\pi \ln x)\sin(m\pi \ln x)\,dx = -\frac{1}{n^2\pi^2}\int_1^e \frac{\sqrt{2}\sin(m\pi \ln x)}{x}\,dx.$$

The LHS of this equation is the normalisation integral just considered and has the value $a_m \delta_{mn}$. Thus

$$
\begin{aligned}
a_m &= -\frac{\sqrt{2}}{m^2 \pi^2} \int_1^e \frac{\sin(m\pi \ln x)}{x} \, dx \\
&= -\frac{\sqrt{2}}{m^2 \pi^2} \left[\frac{-\cos(m\pi \ln x)}{m\pi} \right]_1^e \\
&= -\frac{\sqrt{2}}{m^3 \pi^3} [1 - (-1)^m] \\
&= \begin{cases} -\dfrac{2\sqrt{2}}{m^3 \pi^3} & \text{for } m \text{ odd,} \\ 0 & \text{for } m \text{ even.} \end{cases}
\end{aligned}
$$

The explicit solution is therefore

$$
y(x) = -\frac{4}{\pi^3} \sum_{p=0}^{\infty} \frac{\sin[(2p+1)\pi \ln x]}{(2p+1)^3 \sqrt{x}}.
$$

17.15 *In the quantum mechanical study of the scattering of a particle by a potential, a Born-approximation solution can be obtained in terms of a function $y(\mathbf{r})$ that satisfies an equation of the form*

$$
(-\nabla^2 - K^2) y(\mathbf{r}) = F(\mathbf{r}).
$$

Assuming that $y_\mathbf{k}(\mathbf{r}) = (2\pi)^{-3/2} \exp(i\mathbf{k} \cdot \mathbf{r})$ is a suitably normalised eigenfunction of $-\nabla^2$ corresponding to eigenvalue k^2, find a suitable Green's function $G_K(\mathbf{r}, \mathbf{r}')$. By taking the direction of the vector $\mathbf{r} - \mathbf{r}'$ as the polar axis for a $\mathbf{k}$-space integration, show that $G_K(\mathbf{r}, \mathbf{r}')$ can be reduced to

$$
\frac{1}{4\pi^2 |\mathbf{r} - \mathbf{r}'|} \int_{-\infty}^{\infty} \frac{w \sin w}{w^2 - w_0^2} \, dw,
$$

where $w_0 = K|\mathbf{r} - \mathbf{r}'|$.

[This integral can be evaluated using contour integration and gives the Green's function explicitly as $(4\pi |\mathbf{r} - \mathbf{r}'|)^{-1} \exp(iK|\mathbf{r} - \mathbf{r}'|)$.]

Given that $y_\mathbf{k}(\mathbf{r}) = (2\pi)^{-3/2} \exp(i\mathbf{k} \cdot \mathbf{r})$ satisfies

$$
-\nabla^2 y_\mathbf{k}(\mathbf{r}) = k^2 y_\mathbf{k}(\mathbf{r}),
$$

it follows that

$$
(-\nabla^2 - K^2) y_\mathbf{k}(\mathbf{r}) = (k^2 - K^2) y_\mathbf{k}(\mathbf{r}).
$$

Thus the same functions are suitable eigenfunctions for the extended operator, but with different eigenvalues.

Its Green's function is therefore (from the general expression for Green's functions in terms of eigenfunctions)

$$
\begin{aligned}
G_K(\mathbf{r}, \mathbf{r}') &= \int \frac{1}{\lambda} y_{\mathbf{k}}(\mathbf{r}) y_{\mathbf{k}}^*(\mathbf{r}') \, d\mathbf{k} \\
&= \frac{1}{(2\pi)^3} \int \frac{\exp(i\mathbf{k} \cdot \mathbf{r}) \exp(-i\mathbf{k} \cdot \mathbf{r}')}{k^2 - K^2} \, d\mathbf{k}.
\end{aligned}
$$

We carry out the three-dimensional integration in $\mathbf{k}$-space using the direction $\mathbf{r} - \mathbf{r}'$ as the polar axis (and denote $\mathbf{r} - \mathbf{r}'$ by $\mathbf{R}$). The azimuthal integral is immediate. The remaining two-dimensional integration is as follows:

$$
\begin{aligned}
G_K(\mathbf{r}, \mathbf{r}') &= \frac{1}{(2\pi)^3} \int_0^\infty \int_0^\pi \frac{\exp(i\mathbf{k} \cdot \mathbf{R})}{k^2 - K^2} 2\pi k^2 \sin \theta_k \, d\theta_k \, dk \\
&= \frac{1}{(2\pi)^2} \int_0^\infty \int_0^\pi \frac{\exp(ikR \cos \theta_k)}{k^2 - K^2} k^2 \sin \theta_k \, d\theta_k \, dk \\
&= \frac{1}{(2\pi)^2} \int_0^\infty \frac{\exp(ikR) - \exp(-ikR)}{ikR(k^2 - K^2)} k^2 \, dk \\
&= \frac{1}{2\pi^2 R} \int_0^\infty \frac{k \sin kR}{k^2 - K^2} \, dk \\
&= \frac{1}{2\pi^2 R} \int_0^\infty \frac{w \sin w}{w^2 - w_0^2} \, dw, \quad \text{where } w = kR \text{ and } w_0 = kR, \\
&= \frac{1}{4\pi^2 R} \int_{-\infty}^\infty \frac{w \sin w}{w^2 - w_0^2} \, dw.
\end{aligned}
$$

Here, the final line is justified by noting that the integrand is an even function of the integration variable w.

18

Special functions

18.1 *Use the explicit expressions*

$$Y_0^0 = \sqrt{\tfrac{1}{4\pi}}, \qquad\qquad Y_1^0 = \sqrt{\tfrac{3}{4\pi}}\cos\theta,$$

$$Y_1^{\pm 1} = \mp\sqrt{\tfrac{3}{8\pi}}\sin\theta\exp(\pm i\phi), \qquad Y_2^0 = \sqrt{\tfrac{5}{16\pi}}(3\cos^2\theta - 1),$$

$$Y_2^{\pm 1} = \mp\sqrt{\tfrac{15}{8\pi}}\sin\theta\cos\theta\exp(\pm i\phi), \quad Y_2^{\pm 2} = \sqrt{\tfrac{15}{32\pi}}\sin^2\theta\exp(\pm 2i\phi),$$

to verify for $\ell = 0, 1, 2$ that

$$\sum_{m=-\ell}^{\ell} |Y_\ell^m(\theta,\phi)|^2 = \frac{2\ell+1}{4\pi}$$

and so is independent of the values of θ and ϕ. This is true for any ℓ, but a general proof is more involved. This result helps to reconcile intuition with the apparently arbitrary choice of polar axis in a general quantum mechanical system.

We first note that, since every term is the square of a modulus, factors of the form $\exp(\pm mi\phi)$ never appear in the sums. For each value of ℓ, let us denote the sum by S_ℓ. For $\ell = 0$ and $\ell = 1$, we have

$$S_0 = \sum_{m=0}^{0} |Y_0^m(\theta,\phi)|^2 = \frac{1}{4\pi},$$

$$S_1 = \sum_{m=-1}^{1} |Y_1^m(\theta,\phi)|^2 = \frac{3}{4\pi}\cos^2\theta + 2\frac{3}{8\pi}\sin^2\theta = \frac{3}{4\pi}.$$

296

For $\ell = 2$, the summation is more complicated but reads

$$S_2 = \sum_{m=-2}^{2} |Y_2^m(\theta, \phi)|^2$$

$$= \frac{5}{16\pi}(3\cos^2\theta - 1)^2 + 2\frac{15}{8\pi}\sin^2\theta\cos^2\theta + 2\frac{15}{32\pi}\sin^4\theta$$

$$= \frac{5}{16\pi}(9\cos^4\theta - 6\cos^2\theta + 1 + 12\sin^2\theta\cos^2\theta + 3\sin^4\theta)$$

$$= \frac{5}{16\pi}[6\cos^4\theta - 6\cos^2\theta + 1 + 6\sin^2\theta\cos^2\theta + 3(\cos^2\theta + \sin^2\theta)^2]$$

$$= \frac{5}{16\pi}[6\cos^2\theta(-\sin^2\theta) + 1 + 6\sin^2\theta\cos^2\theta + 3] = \frac{5}{4\pi}.$$

All three sums are independent of θ and ϕ, and are given by the general formula $(2\ell + 1)/4\pi$. It will, no doubt, be noted that $2\ell + 1$ is the number of terms in S_ℓ, i.e. the number of m values, and that 4π is the total solid angle subtended at the origin by all space.

18.3 *Use the generating function for the Legendre polynomials $P_n(x)$ to show that*

$$\int_0^1 P_{2n+1}(x)\,dx = (-1)^n \frac{(2n)!}{2^{2n+1}n!(n+1)!}$$

and that, except for the case $n = 0$,

$$\int_0^1 P_{2n}(x)\,dx = 0.$$

Denote $\int_0^1 P_n(x)\,dx$ by a_n. From the generating function for the Legendre polynomials, we have

$$\frac{1}{(1 - 2xh + h^2)^{1/2}} = \sum_{n=0}^{\infty} P_n(x)h^n.$$

Integrating this definition with respect to x gives

$$\int_0^1 \frac{dx}{(1 - 2xh + h^2)^{1/2}} = \sum_{n=0}^{\infty} \left(\int_0^1 P_n(x)\,dx \right) h^n,$$

$$\left[\frac{-(1 - 2xh + h^2)^{1/2}}{h} \right]_0^1 = \sum_{n=0}^{\infty} a_n h^n,$$

$$\frac{1}{h}[(1 + h^2)^{1/2} - 1 + h] = \sum_{n=0}^{\infty} a_n h^n.$$

Now expanding $(1 + h^2)^{1/2}$ using the binomial theorem yields

$$\sum_{n=0}^{\infty} a_n h^n = \frac{1}{h}\left[1 + \sum_{m=1}^{\infty} {}^{1/2}C_m h^{2m} - 1 + h\right] = 1 + \sum_{m=1}^{\infty} {}^{1/2}C_m h^{2m-1}.$$

Comparison of the coefficients of h^n on the two sides of the equation shows that all a_{2r} are zero except for $a_0 = 1$. For $n = 2r + 1$ we need $2m - 1 = n = 2r + 1$, i.e. $m = r + 1$, and the value of a_{2r+1} is ${}^{1/2}C_{r+1}$.

Now, the binomial coefficient ${}^{1/2}C_m$ can be written as

$$\begin{aligned}
{}^{1/2}C_m &= \frac{\frac{1}{2}(\frac{1}{2} - 1)(\frac{1}{2} - 2)\cdots(\frac{1}{2} - m + 1)}{m!}, \\
&= \frac{1(1 - 2)(1 - 4)\cdots(1 - 2m + 2)}{2^m\, m!} \\
&= (-1)^{m-1}\frac{(1)(1)(3)\cdots(2m - 3)}{2^m\, m!} \\
&= (-1)^{m-1}\frac{(2m - 2)!}{2^m\, m!\, 2^{m-1}(m - 1)!} \\
&= (-1)^{m-1}\frac{(2m - 2)!}{2^{2m-1}\, m!\,(m - 1)!}.
\end{aligned}$$

Thus, setting $m = r + 1$ gives the value of the integral a_{2r+1} as

$$a_{2r+1} = {}^{1/2}C_{r+1} = (-1)^r\frac{(2r)!}{2^{2r+1}(r + 1)!\,r!},$$

as stated in the question.

18.5   *The Hermite polynomials $H_n(x)$ may be defined by*

$$\Phi(x, h) = \exp(2xh - h^2) = \sum_{n=0}^{\infty} \frac{1}{n!}H_n(x)h^n.$$

Show that

$$\frac{\partial^2 \Phi}{\partial x^2} - 2x\frac{\partial \Phi}{\partial x} + 2h\frac{\partial \Phi}{\partial h} = 0,$$

and hence that the $H_n(x)$ satisfy the Hermite equation,

$$y'' - 2xy' + 2ny = 0,$$

where n is an integer ≥ 0.

Use Φ to prove that

(a) $H_n'(x) = 2nH_{n-1}(x)$,
(b) $H_{n+1}(x) - 2xH_n(x) + 2nH_{n-1}(x) = 0$.

With

$$\Phi(x, h) = \exp(2xh - h^2) = \sum_{n=0}^{\infty} \frac{1}{n!} H_n(x) h^n,$$

we have

$$\frac{\partial \Phi}{\partial x} = 2h\Phi, \quad \frac{\partial \Phi}{\partial h} = (2x - 2h)\Phi, \quad \frac{\partial^2 \Phi}{\partial x^2} = 4h^2 \Phi.$$

It then follows that

$$\frac{\partial^2 \Phi}{\partial x^2} - 2x \frac{\partial \Phi}{\partial x} + 2h \frac{\partial \Phi}{\partial h} = (4h^2 - 4hx + 4hx - 4h^2)\Phi = 0.$$

Substituting the series form into this result gives

$$\sum_{n=0}^{\infty} \left(\frac{1}{n!} H_n'' - \frac{2x}{n!} H_n' + \frac{2n}{n!} \right) h^n = 0,$$

$$\Rightarrow \quad H_n'' - 2x H_n' + 2n H_n = 0.$$

This is the equation satisfied by $H_n(x)$, as stated in the question.

(a) From the first relationship derived above, we have that

$$\frac{\partial \Phi}{\partial x} = 2h\Phi,$$

$$\sum_{n=0}^{\infty} \frac{1}{n!} H_n'(x) h^n = 2h \sum_{n=0}^{\infty} \frac{1}{n!} H_n(x) h^n,$$

$$\Rightarrow \quad \frac{1}{m!} H_m' = \frac{2}{(m-1)!} H_{m-1}, \text{ from the coefficients of } h^m.$$

Hence, $\qquad H_n'(x) = 2n H_{n-1}(x).$

(b) Differentiating result (a) and then applying it again yields

$$H_n'' = 2n H_{n-1}' = 2n\, 2(n-1) H_{n-2}.$$

Using this in the differential equation satisfied by the H_n, we obtain

$$4n(n-1)H_{n-2} - 2x\, 2n H_{n-1} + 2n H_n = 0.$$

This gives

$$H_{n+1}(x) - 2x H_n(x) + 2n H_{n-1}(x) = 0$$

after dividing through by $2n$ and changing $n \to n+1$.

18.7 *For the associated Laguerre polynomials, carry through the following exercises.*

(a) *Prove the Rodrigues' formula*

$$L_n^m(x) = \frac{e^x x^{-m}}{n!} \frac{d^n}{dx^n}(x^{n+m}e^{-x}),$$

taking the polynomials to be defined by

$$L_n^m(x) = \sum_{k=0}^{n}(-1)^k \frac{(n+m)!}{k!(n-k)!(k+m)!}x^k.$$

(b) *Prove the recurrence relations*

$$(n+1)L_{n+1}^m(x) = (2n+m+1-x)L_n^m(x) - (n+m)L_{n-1}^m(x),$$

$$x(L_n^m)'(x) = nL_n^m(x) - (n+m)L_{n-1}^m(x),$$

but this time taking the polynomial as defined by

$$L_n^m(x) = (-1)^m \frac{d^m}{dx^m}L_{n+m}(x)$$

or the generating function.

(a) It is most convenient to evaluate the nth derivative directly, using Leibnitz' theorem. This gives

$$L_n^m(x) = \frac{e^x x^{-m}}{n!} \sum_{r=0}^{n} \frac{n!}{r!(n-r)!} \frac{d^r}{dx^r}(x^{n+m}) \frac{d^{n-r}}{dx^{n-r}}(e^{-x})$$

$$= e^x x^{-m} \sum_{r=0}^{n} \frac{1}{r!(n-r)!} \frac{(n+m)!}{(n+m-r)!}x^{n+m-r}(-1)^{n-r}e^{-x}$$

$$= \sum_{r=0}^{n} \frac{(-1)^{n-r}}{r!(n-r)!} \frac{(n+m)!}{(n+m-r)!}x^{n-r}.$$

Relabelling the summation using the new index $k = n - r$, we immediately obtain

$$L_n^m(x) = \sum_{k=0}^{n}(-1)^k \frac{(n+m)!}{k!(n-k)!(k+m)!}x^k,$$

which is as given in the question.

(b) The first recurrence relation can be proved using the generating function for

300

the associated Laguerre functions:

$$G(x,h) = \frac{e^{-xh/(1-h)}}{(1-h)^{m+1}} = \sum_{n=0}^{\infty} L_n^m(x)h^n.$$

Differentiating the second equality with respect to h, we obtain

$$\frac{(m+1)(1-h)-x}{(1-h)^{m+3}}e^{-xh/(1-h)} = \sum nL_n^m h^{n-1}.$$

Using the generating function for a second time, we may rewrite this as

$$[(m+1)(1-h)-x]\sum L_n^m h^n = (1-h)^2\sum nL_n^m h^{n-1}.$$

Equating the coefficients of h^n now yields

$$(m+1)L_n^m - (m+1)L_{n-1}^m - xL_n^m = (n+1)L_{n+1}^m - 2nL_n^m + (n-1)L_{n-1}^m,$$

which can be rearranged and simplified to give the first recurrence relation.

The second result is most easily proved by differentiating one of the standard recurrence relations satisfied by the ordinary Laguerre polynomials, but with n replaced by $n+m$. This standard equality reads

$$xL_{n+m}'(x) = (n+m)L_{n+m}(x) - (n+m)L_{n-1+m}(x).$$

We convert this into an equation for the associated polynomials,

$$L_n^m(x) = (-1)^m\frac{d^m}{dx^m}L_{n+m}(x),$$

by differentiating it m times with respect to x and multiplying through by $(-1)^m$. The result is

$$x(L_n^m)' + mL_n^m = (n+m)L_n^m - (n+m)L_{n-1}^m,$$

which immediately simplifies to give the second recurrence relation satisfied by the associated Laguerre polynomials.

18.9 *By initially writing $y(x)$ as $x^{1/2}f(x)$ and then making subsequent changes of variable, reduce Stokes' equation,*

$$\frac{d^2y}{dx^2} + \lambda xy = 0,$$

to Bessel's equation. Hence show that a solution that is finite at $x=0$ is a multiple of $x^{1/2}J_{1/3}(\frac{2}{3}\sqrt{\lambda x^3})$.

With $y(x) = x^{1/2}f(x)$,

$$y' = \frac{f}{2x^{1/2}} + x^{1/2}f' \text{ and } y'' = -\frac{f}{4x^{3/2}} + \frac{f'}{x^{1/2}} + x^{1/2}f''$$

and the equation becomes

$$-\frac{f}{4x^{3/2}} + \frac{f'}{x^{1/2}} + x^{1/2}f'' + \lambda x^{3/2}f = 0,$$
$$x^2 f'' + xf' + (\lambda x^3 - \tfrac{1}{4})f = 0.$$

Now, guided by the known form of Bessel's equation, change the independent variable to $u = x^{3/2}$ with $f(x) = g(u)$ and

$$\frac{du}{dx} = \frac{3}{2}x^{1/2} \quad \Rightarrow \quad \frac{d}{dx} = \frac{3}{2}u^{1/3}\frac{d}{du}.$$

This gives

$$u^{4/3}\frac{3}{2}u^{1/3}\frac{d}{du}\left(\frac{3}{2}u^{1/3}\frac{dg}{du}\right) + u^{2/3}\frac{3}{2}u^{1/3}\frac{dg}{du} + \left(\lambda u^2 - \frac{1}{4}\right)g = 0,$$

$$\frac{3}{2}u^{5/3}\left(\frac{3}{2}u^{1/3}\frac{d^2g}{du^2} + \frac{1}{2}u^{-2/3}\frac{dg}{du}\right) + \frac{3}{2}u\frac{dg}{du} + \left(\lambda u^2 - \frac{1}{4}\right)g = 0,$$

$$\frac{9}{4}u^2\frac{d^2g}{du^2} + \frac{9}{4}u\frac{dg}{du} + \left(\lambda u^2 - \frac{1}{4}\right)g = 0,$$

$$u^2\frac{d^2g}{du^2} + u\frac{dg}{du} + \left(\frac{4}{9}\lambda u^2 - \frac{1}{9}\right)g = 0.$$

This is close to Bessel's equation but still needs a scaling of the variables. So, set $\frac{2}{3}\sqrt{\lambda}u \equiv \mu u = v$ and $g(u) = h(v)$, obtaining

$$\frac{v^2}{\mu^2}\mu^2\frac{d^2h}{dv^2} + \frac{v}{\mu}\mu\frac{dh}{dv} + \left(v^2 - \frac{1}{9}\right)h = 0.$$

This *is* Bessel's equation and has a general solution

$$h(v) = c_1 J_{1/3}(v) + c_2 J_{-1/3}(v),$$
$$\Rightarrow \quad g(u) = c_1 J_{1/3}(\tfrac{2\sqrt{\lambda}}{3}u) + c_2 J_{-1/3}(\tfrac{2\sqrt{\lambda}}{3}u),$$
$$\Rightarrow \quad f(x) = c_1 J_{1/3}(\tfrac{2\sqrt{\lambda}}{3}x^{3/2}) + c_2 J_{-1/3}(\tfrac{2\sqrt{\lambda}}{3}x^{3/2}).$$

For a solution that is finite at $x = 0$, only the Bessel function with a positive subscript can be accepted. Therefore the required solution is

$$y(x) = c_1 x^{1/2} J_{1/3}(\tfrac{2\sqrt{\lambda}}{3}x^{3/2}).$$

18.11 *Identify the series for the following hypergeometric functions, writing them in terms of better-known functions.*

(a) $F(a, b, b; z)$,
(b) $F(1, 1, 2; -x)$,
(c) $F(\frac{1}{2}, 1, \frac{3}{2}; -x^2)$,
(d) $F(\frac{1}{2}, \frac{1}{2}, \frac{3}{2}; x^2)$,
(e) $F(-a, a, \frac{1}{2}; \sin^2 x)$; *this is a much more difficult exercise.*

The hypergeometric equation is

$$z(1 - z)y'' + [c - (a + b + 1)z]y' - aby = 0.$$

The $(n + 1)$th term of its series solution, the hypergeometric function $F(a, b, c; z)$, is given by

$$\frac{a(a + 1) \cdots (a + n - 1)\, b(b + 1) \cdots (b + n - 1)}{c(c + 1) \cdots (c + n - 1)} \frac{z^n}{n!}$$

for $n \geq 1$ and unity for $n = 0$.

(a) $F(a, b, b; z)$. In each term the equal factors arising from the second and third parameters cancel, as one is in the numerator and the other in the denominator. Thus,

$$F(a, b, b; z) = 1 + az + \frac{a(a + 1)}{2!} z^2 + \frac{a(a + 1)(a + 2)}{3!} z^3 + \cdots$$
$$= (1 - z)^{-a}.$$

(b) $F(1, 1, 2; -x)$. The $n + 1$th term is

$$\frac{(n!)(n!)}{(n + 1)!(n!)} (-x)^n = \frac{(-1)^n x^n}{n + 1}$$

making the series

$$\sum_{n=0}^{\infty} \frac{(-1)^n x^n}{n + 1} = 1 - \frac{x}{2} + \frac{x^2}{3} - \frac{x^3}{4} + \cdots = \frac{1}{x} \ln(1 + x).$$

(c) $F(\frac{1}{2}, 1, \frac{3}{2}; -x^2)$. Directly from the series:

$$F(\tfrac{1}{2}, 1, \tfrac{3}{2}; -x^2) = 1 + \frac{(\frac{1}{2})(1)}{1!(\frac{3}{2})} (-x^2) + \frac{(\frac{1}{2})(\frac{3}{2})(1)(2)}{2!(\frac{3}{2})(\frac{5}{2})} (-x^2)^2 + \cdots$$
$$= 1 - \frac{x^2}{3} + \frac{x^4}{5} - \frac{x^6}{7} + \cdots.$$

The coefficients are those of $\tan^{-1} x$, though the powers of x are all too small by one. Thus $F(\frac{1}{2}, 1, \frac{3}{2}; -x^2) = x^{-1} \tan^{-1} x$.

(d) $F(\frac{1}{2}, \frac{1}{2}, \frac{3}{2}; x^2)$. Again, directly from the series:

$$F(\tfrac{1}{2}, \tfrac{1}{2}, \tfrac{3}{2}; x^2) = 1 + \frac{(\frac{1}{2})^2}{1!(\frac{3}{2})}(x^2) + \frac{(\frac{1}{2})^2(\frac{3}{2})^2}{2!(\frac{3}{2})(\frac{5}{2})}(x^2)^2 + \cdots$$

$$= 1 + \tfrac{1}{6}x^2 + \tfrac{3}{40}5x^4 + \frac{15}{336}x^6 + \cdots.$$

From the larger standard tables of Maclaurin series it can be seen that, although the successive coefficients are those of $\sin^{-1} x$, the powers of x are all too small by one. Thus $F(\frac{1}{2}, \frac{1}{2}, \frac{3}{2}; -x^2) = x^{-1}\sin^{-1} x$.

(e) $F(-a, a, \frac{1}{2}; \sin^2 x)$. Since we will obtain a series involving terms such as $\sin^{2m} x$, the series may be difficult to identify. The series is

$$1 + \frac{(-a)(a)}{(\frac{1}{2})}\sin^2 x + \frac{(-a)(-a+1)(a)(a+1)}{2!(\frac{1}{2})(\frac{3}{2})}\sin^4 x + \cdots. \qquad (*)$$

Clearly, this contains only even powers of x, though just the first two terms alone constitute an infinite power series in x. However, a term containing x^{2m} can only arise from the first $m + 1$ terms of $(*)$ and a few trials may be helpful.

If $F(-a, a, \frac{1}{2}; \sin^2 x) = \sum_{n=0}^{\infty} b_n x^{2n}$, then $b_0 = 1$ and $b_1 = -2a^2$ since the corresponding powers of x can only arise from the first and second terms of $(*)$, respectively. The coefficient of x^4 is determined by the second and third terms of $(*)$ and is given by

$$b_2 = -2a^2\left(-\frac{2}{3!}\right) + \frac{2a^2(a^2-1)}{3}(1) = \frac{2a^4}{3}.$$

The coefficient of x^6, namely b_3, has contributions from the second, third and fourth terms of $(*)$ and is given by

$$-2a^2\left[\left(\frac{1}{3!}\right)^2 + \frac{2}{5!}\right] + \frac{2a^2(a^2-1)}{3}\left(\frac{-4}{3!}\right) + \frac{4a^2(a^2-1)(4-a^2)}{45}\quad (1)$$

$$= -2a^2\left(\frac{20+12}{720}\right) - \frac{8a^4}{18} + \frac{8a^2}{18} + \frac{4}{45}(-4a^2 + 5a^4 - a^6)$$

$$= \left(-\frac{64}{720} + \frac{8}{18} - \frac{16}{45}\right)a^2 + \left(-\frac{8}{18} + \frac{20}{45}\right)a^4 + \left(-\frac{4}{45}\right)a^6$$

$$= -\frac{4}{45}a^6.$$

Thus, in powers up to x^6,

$$F(-a, a, \tfrac{1}{2}; \sin^2 x) = 1 - 2a^2x^2 + \tfrac{2}{3}a^4x^4 - \tfrac{4}{45}a^6x^6$$

$$= 1 - \frac{(2ax)^2}{2!} + \frac{(2ax)^4}{4!} - \frac{(2ax)^6}{6!}.$$

Though not totally conclusive, this sequence of coefficients strongly suggests that $F(-a, a, \frac{1}{2}; \sin^2 x) = \cos 2ax$. Note that a does not need to be an integer.

This tentative conclusion can be tested by transforming the original hypergeometric equation as follows. With $z = \sin^2 x$, we have that $dz/dx = 2 \sin x \cos x = \sin 2x$, implying that $d/dz = (\sin 2x)^{-1} d/dx$. The equation becomes

$$\sin^2 x (1 - \sin^2 x) \frac{1}{\sin 2x} \frac{d}{dx} \left(\frac{1}{\sin 2x} \frac{dy}{dx} \right)$$
$$+ \left[\frac{1}{2} - (-a + a + 1) \sin^2 x \right] \frac{1}{\sin 2x} \frac{dy}{dx} + a^2 y = 0.$$

This can be simplified as follows:

$$\frac{1}{4} \sin 2x \left(\frac{1}{\sin 2x} \frac{d^2 y}{dx^2} - \frac{2 \cos 2x}{\sin^2 2x} \frac{dy}{dx} \right) + \frac{1 - 2 \sin^2 x}{2 \sin 2x} \frac{dy}{dx} + a^2 y = 0,$$
$$\frac{1}{4} \frac{d^2 y}{dx^2} - \frac{\cos 2x}{2 \sin 2x} \frac{dy}{dx} + \frac{\cos 2x}{2 \sin 2x} \frac{dy}{dx} + a^2 y = 0,$$
$$\frac{d^2 y}{dx^2} + 4a^2 y = 0.$$

For a solution with $y(0) = 1$, this implies that $y(x) = \cos 2ax$, thus confirming our provisional conclusion.

18.13 *Find a change of variable that will allow the integral*

$$I = \int_1^\infty \frac{\sqrt{u - 1}}{(u + 1)^2} \, du$$

to be expressed in terms of the beta function and so evaluate it.

The beta function is normally expressed in terms of an integral, over the range 0 to 1, of an integrand of the form $v^m (1 - v)^n$, with $m, n > -1$. We therefore need a change of variable $u = f(x)$ such that $u + 1$ is an inverse power of x; this being so, we also need $f(0) = \infty$ and $f(1) = 1$.

Consider

$$u + 1 = \frac{A}{x}, \quad \text{i.e.} \quad f(x) = \frac{A}{x} - 1.$$

This satisfies the first two requirements, and also satisfies the third one if we choose $A = 2$.

So, substitute $u = \frac{2}{x} - 1 = \frac{2 - x}{x}$, with $u - 1 = \frac{2(1 - x)}{x}$ and $du = -\frac{2}{x^2}$. The integral then becomes

$$
\begin{aligned}
I &= \int_1^0 \frac{2^{1/2}(1-x)^{1/2}\,x^2\,(-2)}{x^{1/2}\,2^2\,x^2}\,dx \\
&= \frac{1}{\sqrt{2}} \int_0^1 (1-x)^{1/2} x^{-1/2}\,dx \\
&= \frac{1}{\sqrt{2}}\,B(\tfrac{1}{2},\tfrac{3}{2}) = \frac{\Gamma(\tfrac{1}{2})\,\Gamma(\tfrac{3}{2})}{\sqrt{2}\,\Gamma(\tfrac{1}{2}+\tfrac{3}{2})} \\
&= \frac{\sqrt{\pi}\,\tfrac{1}{2}\sqrt{\pi}}{\sqrt{2}\,1} = \frac{\pi}{2\sqrt{2}}.
\end{aligned}
$$

18.15 *The complex function $z!$ is defined by*

$$
z! = \int_0^\infty u^z e^{-u}\,du \qquad \textit{for } \operatorname{Re} z > -1.
$$

For $\operatorname{Re} z \le -1$ it is defined by

$$
z! = \frac{(z+n)!}{(z+n)(z+n-1)\cdots(z+1)},
$$

where n is any (positive) integer $> -\operatorname{Re} z$. Being the ratio of two polynomials, $z!$ is analytic everywhere in the finite complex plane except at the poles that occur when z is a negative integer.

(a) *Show that the definition of $z!$ for $\operatorname{Re} z \le -1$ is independent of the value of n chosen.*

(b) *Prove that the residue of $z!$ at the pole $z = -m$, where m is an integer > 0, is $(-1)^{m-1}/(m-1)!$.*

(a) Let m and n be two choices of integer with $m > n > -\operatorname{Re} z$. Denote the corresponding definitions of $z!$ by $(z!)_m$ and $(z!)_n$ and consider the ratio of these two functions:

$$
\begin{aligned}
\frac{(z!)_m}{(z!)_n} &= \frac{(z+m)!}{(z+m)(z+m-1)\cdots(z+1)}\,\frac{(z+n)(z+n-1)\cdots(z+1)}{(z+n)!} \\
&= \frac{(z+m)!}{(z+m)(z+m-1)\cdots(z+n+1) \times (z+n)!} \\
&= \frac{(z+m)!}{(z+m)!} = 1.
\end{aligned}
$$

Thus the two functions are identical for all z, i.e the definition of $z!$ is independent of the choice of n, provided that $n > -\operatorname{Re} z$.

(b) From the given definition of $z!$ it is clear that its pole at $z = -m$ is a simple one. The residue R at the pole is therefore given by

$$R = \lim_{z \to -m} (z + m)z!$$

$$= \lim_{z \to -m} \frac{(z + m)(z + n)!}{(z + n)(z + n - 1)\cdots(z + 1)} \quad \text{(integer n is chosen $> m$)}$$

$$= \lim_{z \to -m} \frac{(z + n)!}{(z + n)(z + n - 1)\cdots(z + m + 1)(z + m - 1)\cdots(z + 1)}$$

$$= \frac{(-m + n)!}{(-m + n)\cdots(-m + m + 1)(-m + m - 1)\cdots(-m + 1)}$$

$$= \frac{1}{[-1][-2]\cdots[-(m-1)]}$$

$$= (-1)^{m-1}\frac{1}{(m-1)!},$$

as stated in the question.

18.17 *The integral*

$$I = \int_{-\infty}^{\infty} \frac{e^{-k^2}}{k^2 + a^2}\, dk, \quad (*)$$

in which $a > 0$, occurs in some statistical mechanics problems. By first considering the integral

$$J = \int_0^{\infty} e^{iu(k + ia)}\, du$$

and a suitable variation of it, show that $I = (\pi/a)\exp(a^2)\operatorname{erfc}(a)$, where $\operatorname{erfc}(x)$ is the complementary error function.

The fact that $a > 0$ will ensure that the improper integral J is well defined. It is

$$J = \int_0^{\infty} e^{iu(k + ia)}\, du = \left[\frac{e^{iu(k + ia)}}{i(k + ia)}\right]_0^{\infty} = \frac{i}{k + ia}.$$

We note that this result contains one of the factors that would appear as a denominator in one term of a partial fraction expansion of the integrand in (*). Another term would contain a factor $(k - ia)^{-1}$, and this can be generated by

$$J' = \int_0^{\infty} e^{-iu(k - ia)}\, du = \left[\frac{e^{-iu(k - ia)}}{-i(k - ia)}\right]_0^{\infty} = \frac{-i}{k - ia}.$$

Now, actually expressing the integrand in partial fractions, using the integral expressions J and J' for the factors, and then reversing the order of integration gives

$$I = \frac{1}{2a} \int_{-\infty}^{\infty} \left(\frac{ie^{-k^2}}{k+ia} - \frac{ie^{-k^2}}{k-ia} \right) dk$$

$$= \frac{1}{2a} \int_{-\infty}^{\infty} e^{-k^2} dk \int_0^{\infty} e^{iu(k+ia)} du + \frac{1}{2a} \int_{-\infty}^{\infty} e^{-k^2} dk \int_0^{\infty} e^{-iu(k-ia)} du,$$

$$\Rightarrow \quad 2aI = \int_0^{\infty} du \int_{-\infty}^{\infty} e^{-k^2+iuk-ua} dk + \int_0^{\infty} du \int_{-\infty}^{\infty} e^{-k^2-iuk-ua} dk$$

$$= \int_0^{\infty} du \int_{-\infty}^{\infty} e^{-(k-iu/2)^2-u^2/4-ua} dk$$

$$+ \int_0^{\infty} du \int_{-\infty}^{\infty} e^{-(k+iu/2)^2-u^2/4-ua} dk$$

$$= 2\sqrt{\pi} \int_0^{\infty} e^{-u^2/4-ua} du,$$

using the standard Gaussian result. We now complete the square in the exponent and set $2v = u + 2$, obtaining

$$2aI = 2\sqrt{\pi} \int_0^{\infty} e^{-(u+2a)^2/4+a^2} du,$$

$$= 2\sqrt{\pi} \int_a^{\infty} e^{-v^2} e^{a^2} 2dv.$$

From this it follows that

$$I = \frac{\sqrt{\pi}}{a} 2e^{a^2} \frac{\sqrt{\pi}}{2} \operatorname{erfc}(a) = \frac{\pi}{a} e^{a^2} \operatorname{erfc}(a),$$

as stated in the question.

18.19 *For the functions $M(a,c;z)$ that are the solutions of the confluent hypergeometric equation:*

(a) *use their series representation to prove that*

$$c \frac{d}{dz} M(a,c;z) = a M(a+1,c+1;z);$$

(b) *use an integral representation to prove that*

$$M(a,c;z) = e^z M(c-a,c;-z).$$

(a) Directly differentiating the explicit series term by term gives

$$\frac{d}{dz} M(a,c;z) = \frac{d}{dz} \left[1 + \frac{a}{c} z + \frac{a(a+1)}{2!\,c(c+1)} z^2 + \cdots \right]$$

$$= \frac{a}{c} + \frac{2a(a+1)}{2!\,c(c+1)} z + \frac{3a(a+1)(a+2)}{3!\,c(c+1)(c+2)} z^2 + \cdots$$

$$= \frac{a}{c} \left[1 + \frac{a+1}{c+1} z + \frac{(a+1)(a+2)}{2!\,(c+1)(c+2)} z^2 + \cdots \right]$$

$$= \frac{a}{c} M(a+1, c+1; z).$$

The quoted result follows immediately.

(b) This will be achieved most simply if we choose a representation in which the parameters can be rearranged without having to perform any actual integration. We therefore take the representation

$$M(a,c;z) = \frac{\Gamma(c)}{\Gamma(c-a)\Gamma(a)} \int_0^1 e^{zt} \, t^{a-1} \, (1-t)^{c-a-1} \, dt$$

and change the variable of integration to $s = 1 - t$ whilst regrouping the parameters (without changing their values, of course). This gives

$$M(a,c;z)$$

$$= \frac{\Gamma(c)}{\Gamma(a)\Gamma(c-a)} \int_1^0 e^z \, e^{-zs} \, (1-s)^{a-1} \, s^{c-a-1} \, (-ds)$$

$$= \frac{\Gamma(c)\,e^z}{\Gamma[c-(c-a)]\,\Gamma[(c-a)]} \int_0^1 e^{-zs} \, (1-s)^{c-(c-a)-1} \, s^{(c-a)-1} \, ds$$

$$= e^z \, M(c-a, c, -z),$$

thus establishing the identity, in which $a \to c - a$ and $z \to -z$ whilst c remains unchanged.

18.21 *Find the differential equation satisfied by the function $y(x)$ defined by*

$$y(x) = Ax^{-n} \int_0^x e^{-t} t^{n-1} \, dt \equiv Ax^{-n} \gamma(n, x),$$

and, by comparing it with the confluent hypergeometric function, express y as a multiple of the solution $M(a,c;z)$ of that equation. Determine the value of A that makes y equal to M.

As the comparison is to be made with the hypergeometric equation, which is a second-order differential equation, we must calculate the first two derivatives of

309

$y(x)$. Further, as it is a homogeneous equation, we may omit the multiplicative constant A for the time being:

$$y(x) = x^{-n}\gamma(n, x),$$
$$y'(x) = -nx^{-n-1}\gamma(n, x) + x^{-n}e^{-x}x^{n-1}$$
$$= -nx^{-1}y + x^{-1}e^{-x},$$
$$y''(x) = nx^{-2}y - nx^{-1}y' - x^{-2}e^{-x} - x^{-1}e^{-x}$$
$$= nx^{-2}y - nx^{-1}y' - (x^{-1} + 1)(y' + nx^{-1}y),$$
$$x^2y'' = (-nx - x - x^2)y' + (n - n - nx)y.$$

The second line uses the standard result for differentiating an indefinite integral with respect to its upper limit. In the fifth line we substituted for $x^{-1}e^{-x}$ from the expression obtained for $y'(x)$ in the third line. Thus the equation to be compared with the confluent hypergeometric equation is

$$xy'' + (n + 1 + x)y' + ny = 0.$$

This has to be compared with

$$zw'' + (c - z)w' - aw = 0.$$

Now xy'' and xy' terms have the same signs (both positive), whereas the zw'' and zw' terms have opposite signs. To deal with this, we must set $z = -x$ in the confluent hypergeometric equation; renaming $w(z) = y(x)$ gives $w' = -y'$ and $w'' = y''$. The equation then becomes (after an additional overall sign change)

$$xy'' + (c + x)y' + ay = 0.$$

The obvious assignments, to go with $z \to -x$, are now $a \to n$ and $c \to n + 1$. We therefore conclude that $y(x)$ is a multiple of $M(n, n + 1; -x)$.

To determine the constant A in the given form of $y(x)$ we expand both its definition and $M(n, n + 1; -x)$ in powers of x. Strictly, only the first term is necessary, but the second acts as a check.

From the hypergeometric series,

$$M(n, n + 1; -x) = 1 + \frac{n(-x)}{n + 1} + \cdots.$$

From the definition of $y(x)$,

$$y(x) = Ax^{-n} \int_0^x e^{-t} t^{n-1} \, dt$$

$$= Ax^{-n} \int_0^x \left[t^{n-1} - t(t^{n-1}) + \frac{t^2}{2!}(t^{n-1}) + \cdots \right]$$

$$= Ax^{-n} \left[\frac{t^n}{n} - \frac{t^{n+1}}{n+1} + \cdots \right]_0^x$$

$$= Ax^{-n} \left[\frac{x^n}{n} - \frac{x^{n+1}}{n+1} + \cdots \right]$$

$$= \frac{A}{n} - \frac{Ax}{n+1} + \cdots .$$

This reproduces the first two terms of $M(n, n+1; -x)$ if $A = n$, yielding, finally, that

$$y(x) = nx^{-n} \gamma(n, x) = M(n, n+1; -x).$$

18.23 *Prove two of the properties of the incomplete gamma function $P(a, x^2)$ as follows.*

(a) *By considering its form for a suitable value of a, show that the error function can be expressed as a particular case of the incomplete gamma function.*

(b) *The Fresnel integrals, of importance in the study of the diffraction of light, are given by*

$$C(x) = \int_0^x \cos\left(\frac{\pi}{2} t^2\right) dt, \qquad S(x) = \int_0^x \sin\left(\frac{\pi}{2} t^2\right) dt.$$

Show that they can be expressed in terms of the error function by

$$C(x) + iS(x) = A \operatorname{erf}\left[\frac{\sqrt{\pi}}{2}(1 - i)x \right],$$

where A is a (complex) constant, which you should determine. Hence express $C(x) + iS(x)$ in terms of the incomplete gamma function.

(a) From the definition of the incomplete gamma function, we have

$$P(a, x^2) = \frac{1}{\Gamma(a)} \int_0^{x^2} e^{-t} t^{a-1} \, dt.$$

Guided by the x^2 in the upper limit, we now change the integration variable to

$y = +\sqrt{t}$, with $2y\,dy = dt$, and obtain

$$P(a, x^2) = \frac{1}{\Gamma(a)} \int_0^x e^{-y^2} y^{2(a-1)}\, 2y\, dy.$$

To make the RHS into an error function we need to remove the y-term; to do this we choose a such that $2(a-1)+1 = 0$, i.e. $a = \frac{1}{2}$. With this choice, $\Gamma(a) = \sqrt{\pi}$ and

$$P\left(\tfrac{1}{2}, x^2\right) = \frac{2}{\sqrt{\pi}} \int_0^x e^{-y^2}\, dy,$$

i.e. a correctly normalised error function.

(b) Consider the given expression:

$$z = A \operatorname{erf}\left[\frac{\sqrt{\pi}}{2}(1-i)x\right] = \frac{2A}{\sqrt{\pi}} \int_0^{\sqrt{\pi}(1-i)x/2} e^{-u^2}\, du.$$

Changing the variable of integration to s, given by $u = \frac{1}{2}\sqrt{\pi}(1-i)s$, and recalling that $(1-i)^2 = -2i$, we obtain

$$z = \frac{2A}{\sqrt{\pi}} \int_0^x e^{-s^2\pi(-2i)/4} \frac{\sqrt{\pi}}{2}(1-i)\, ds$$

$$= A(1-i) \int_0^x e^{i\pi s^2/2}\, ds$$

$$= A(1-i) \int_0^x \left[\cos\left(\frac{\pi s^2}{2}\right) + i \sin\left(\frac{\pi s^2}{2}\right)\right] ds$$

$$= A(1-i)\,[C(x) + iS(x)].$$

For the correct normalisation we need $A(1-i) = 1$, implying that $A = (1+i)/2$.

Now, from part (a), the error function can be expressed in terms of the incomplete gamma function $P(a, x)$ by

$$\operatorname{erf}(x) = P(\tfrac{1}{2}, x^2).$$

Here the argument of the error function is $\frac{1}{2}\sqrt{\pi}(1-i)x$, whose square is $-\frac{1}{2}\pi i x^2$, and so

$$C(x) + iS(x) = \frac{1+i}{2} P\left(\frac{1}{2}, -\frac{i\pi}{2}x^2\right).$$

19

Quantum operators

19.1 *Show that the commutator of two operators that correspond to two physical observables cannot itself correspond to another physical observable.*

Let the two operators be A and B, both of which must be Hermitian since they correspond to physical variables, and consider the Hermitian conjugate of their commutator:

$$[A, B]^{\dagger} = (AB)^{\dagger} - (BA)^{\dagger} = B^{\dagger}A^{\dagger} - A^{\dagger}B^{\dagger} = BA - AB = -[A, B].$$

Thus, the commutator is anti-Hermitian or zero and therefore cannot represent a non-trivial physical variable (as its eigenvalues are imaginary).

19.3 *In quantum mechanics, the time dependence of the state function $|\psi\rangle$ of a system is given, as a further postulate, by the equation*

$$i\hbar \frac{\partial}{\partial t}|\psi\rangle = H|\psi\rangle,$$

where H is the Hamiltonian of the system. Use this to find the time dependence of the expectation value $\langle A \rangle$ of an operator A that itself has no explicit time dependence. Hence show that operators that commute with the Hamiltonian correspond to the classical 'constants of the motion'.

For a particle of mass m moving in a one-dimensional potential $V(x)$, prove Ehrenfest's theorem:

$$\frac{d\langle p_x \rangle}{dt} = -\left\langle \frac{dV}{dx} \right\rangle \qquad and \qquad \frac{d\langle x \rangle}{dt} = \frac{\langle p_x \rangle}{m}.$$

The expectation value of A at any time is $\langle \psi(x,t)|A|\psi(x,t)\rangle$, where we have explicitly indicated that the state function varies with time. Now

$$\frac{d}{dt}\langle \psi\,|\,A\,|\,\psi\rangle = \left(\frac{\partial}{\partial t}\langle \psi|\right)A\,|\,\psi\rangle + \langle \psi\,|\,\frac{\partial A}{\partial t}\,|\,\psi\rangle + \langle \psi\,|\,A\left(\frac{\partial}{\partial t}|\,\psi\rangle\right).$$

Since A has no explicit time dependence, $\partial A/\partial t = 0$ and the second term drops out.

The given (quantum) equation of motion and its Hermitian conjugate are

$$i\hbar\frac{\partial}{\partial t}|\psi\rangle = H|\psi\rangle \quad \text{and} \quad \frac{\partial}{\partial t}\langle \psi\,| = -\frac{1}{i\hbar}\langle \psi\,|H^\dagger = -\frac{1}{i\hbar}\langle \psi\,|H,$$

since H is Hermitian. Thus,

$$\frac{d}{dt}\langle \psi\,|\,A\,|\,\psi\rangle = -\frac{1}{i\hbar}\langle \psi\,|\,HA\,|\,\psi\rangle + \frac{1}{i\hbar}\langle \psi\,|\,AH\,|\,\psi\rangle$$

$$= -\frac{1}{i\hbar}\langle \psi\,|\,[H,A]\,|\,\psi\rangle$$

$$= \frac{i}{\hbar}\langle \psi\,|\,[H,A]\,|\,\psi\rangle.$$

This shows that the rate of change of the expectation value of A is proportional to the expectation value of the commutator of A and the Hamiltonian. If A and H commute, the RHS is zero, the expectation value of A has a zero rate of change, and $\langle \psi\,|\,A\,|\,\psi\rangle$ is a constant of the motion.

For the particle moving in the one-dimensional potential $V(x)$,

$$H = \frac{p_x^2}{2m} + V(x).$$

(i) For $\langle p_x\rangle$,

$$[H, p_x]\,|\,\psi\rangle = [V, p_x]\,|\,\psi\rangle, \text{ since } p_x \text{ clearly commutes with } p_x^2,$$

$$= -i\hbar V\frac{\partial}{\partial x}|\,\psi\rangle + i\hbar\frac{\partial}{\partial x}\left(V|\,\psi\rangle\right)$$

$$= -i\hbar V\frac{\partial}{\partial x}|\,\psi\rangle + i\hbar V\frac{\partial}{\partial x}|\,\psi\rangle + i\hbar\frac{\partial V}{\partial x}|\,\psi\rangle$$

$$= i\hbar\frac{\partial V}{\partial x}|\,\psi\rangle,$$

implying that

$$\frac{d}{dt}\langle p_x\rangle = \frac{d}{dt}\langle \psi\,|\,p_x\,|\,\psi\rangle = \frac{i}{\hbar}\langle \psi\,|\,[H, p_x]\,|\,\psi\rangle$$

$$= \frac{i}{\hbar}\langle \psi\,|\,i\hbar\frac{\partial V}{\partial x}\,|\,\psi\rangle = -\left\langle\frac{\partial V}{\partial x}\right\rangle.$$

(ii) For $\langle x\rangle$ we will need the general commutator property $[AB, C] = A[B, C] + $

$[A, C] B$ to evaluate $\left[p_x^2, x\right]$:

$$[H, x] | \psi \rangle = \frac{1}{2m} \left[p_x^2, x\right] | \psi \rangle, \quad \text{since } x \text{ clearly commutes with } V(x),$$

$$= \frac{1}{2m} \left\{ p_x \left[p_x, x\right] | \psi \rangle + \left[p_x, x\right] p_x | \psi \rangle \right\}, \quad \text{as above},$$

$$= \frac{1}{2m} \left\{ -i\hbar p_x | \psi \rangle - i\hbar p_x | \psi \rangle \right\} = -\frac{i\hbar}{m} p_x | \psi \rangle,$$

implying that

$$\frac{d}{dt} \langle x \rangle = \frac{d}{dt} \langle \psi | x | \psi \rangle = \frac{i}{\hbar} \langle \psi | [H, x] | \psi \rangle$$

$$= \frac{i}{\hbar} \langle \psi | - \frac{i\hbar}{m} p_x | \psi \rangle = \frac{1}{m} \langle \psi | p_x | \psi \rangle = \frac{1}{m} \langle p_x \rangle.$$

Ehrenfest's theorem should be compared to the classical statements 'momentum equals mass times velocity', 'the force is given by minus the gradient of the potential' and 'force is equal to the rate of change of momentum'.

19.5 *Find closed-form expressions for* $\cos C$ *and* $\sin C$, *where* C *is the matrix*

$$C = \begin{pmatrix} 1 & 1 \\ 1 & -1 \end{pmatrix}.$$

Demonstrate that the 'expected' relationships

$$\cos^2 C + \sin^2 C = I \quad and \quad \sin 2C = 2 \sin C \cos C$$

are valid.

Consider the square of C:

$$C^2 = \begin{pmatrix} 1 & 1 \\ 1 & -1 \end{pmatrix} \begin{pmatrix} 1 & 1 \\ 1 & -1 \end{pmatrix} = \begin{pmatrix} 2 & 0 \\ 0 & 2 \end{pmatrix} = 2I.$$

Now

$$\cos C = \sum_{n=0}^{\infty} \frac{(-1)^n}{(2n)!} C^{2n} = \sum_{n=0}^{\infty} \frac{(-1)^n}{(2n)!} 2^n I^n = (\cos \sqrt{2}) I$$

and

$$\sin C = \sum_{n=0}^{\infty} \frac{(-1)^n}{(2n+1)!} C^{2n+1} = \sum_{n=0}^{\infty} \frac{(-1)^n}{(2n+1)!} 2^n I^n C = \frac{1}{\sqrt{2}} (\sin \sqrt{2}) C.$$

To test the analogue of '$\cos^2 \theta + \sin^2 \theta = 1$':

$$\cos^2 C + \sin^2 C = (\cos^2 \sqrt{2}) I + \tfrac{1}{2} (\sin^2 \sqrt{2}) C^2$$

$$= (\cos^2 \sqrt{2}) I + \tfrac{1}{2} (\sin^2 \sqrt{2}) 2 I = I,$$

as expected.

To test the analogue of 'sin $2\theta = 2 \sin \theta \cos \theta$', we note that $(2C)^{2n} = 2^{2n}(2I)^n = (2^2 \, 2I)^n = 8^n I^n$ and obtain

$$\sin 2C = \sin \begin{pmatrix} 2 & 2 \\ 2 & -2 \end{pmatrix}$$

$$= \sum_{n=0}^{\infty} \frac{(-1)^n}{(2n+1)!} \, 8^n \, I^n \begin{pmatrix} 2 & 2 \\ 2 & -2 \end{pmatrix}$$

$$= \frac{1}{\sqrt{8}} (\sin \sqrt{8}) \begin{pmatrix} 2 & 2 \\ 2 & -2 \end{pmatrix}$$

$$= \frac{\sin 2\sqrt{2}}{\sqrt{2}} \begin{pmatrix} 1 & 1 \\ 1 & -1 \end{pmatrix}.$$

But we also have that

$$2 \sin C \cos C = 2 \frac{1}{\sqrt{2}} (\sin \sqrt{2}) \begin{pmatrix} 1 & 1 \\ 1 & -1 \end{pmatrix} (\cos \sqrt{2}) \begin{pmatrix} 1 & 0 \\ 0 & 1 \end{pmatrix}$$

$$= \frac{2 \sin \sqrt{2} \cos \sqrt{2}}{\sqrt{2}} \begin{pmatrix} 1 & 1 \\ 1 & -1 \end{pmatrix},$$

$$= \frac{\sin 2\sqrt{2}}{\sqrt{2}} \begin{pmatrix} 1 & 1 \\ 1 & -1 \end{pmatrix},$$

thus confirming the relationship (at least in this case).

19.7 *Expressed in terms of the annihilation and creation operators A and $A^\dagger$, a system has an unperturbed Hamiltonian $H_0 = \hbar\omega A^\dagger A$. The system is disturbed by the addition of a perturbing Hamiltonian $H_1 = g\hbar\omega(A + A^\dagger)$, where g is real. Show that the effect of the perturbation is to move the whole energy spectrum of the system down by $g^2\hbar\omega$.*

The total Hamiltonian H for the system is $H_0 + H_1$, where

$$H_0 = \hbar\omega A^\dagger A \qquad \text{and} \qquad H_1 = g\hbar\omega(A + A^\dagger).$$

We note that both terms are Hermitian ($H_0^\dagger = H_0$, $H_1^\dagger = H_1$) and that the energy spectrum of the system is given by the eigenvalues μ_i for which

$$(H_0 + H_1)\,|\,\psi_i\rangle = \mu_i\,|\,\psi_i\rangle$$

has solutions.

Now,
$$\begin{aligned} H &= H_0 + H_1 \\ &= \hbar\omega[A^\dagger A + g(A + A^\dagger)] \\ &= \hbar\omega[(A^\dagger + gI)(A + gI) - g^2 I\,]. \end{aligned}$$

We define B by $B = A + gI$, with $B^\dagger = A^\dagger + gI$, and consider

$$[B, B^\dagger] = [A + gI, A^\dagger + gI] = [A, A^\dagger],$$

since $[C, I] = 0$ for any C and clearly $[I, I] = 0$.

Thus, H is expressible as

$$H = \hbar\omega B^\dagger B - g^2 \hbar\omega I$$

with $[B, B^\dagger] = [A, A^\dagger]$. That is, H has the same structure with respect to B as H_0 has with respect to A (apart from an additional term proportional to the identity operator) and B and $B^\dagger$ have the same commutation relation as A and $A^\dagger$. This implies that H has the same spectrum of eigenvalues λ_i as H_0, except for a (downward) shift of $-g^2 \hbar\omega$, i.e. $\mu_i = \lambda_i - g^2 \hbar\omega$ for each value of i. Thus the whole spectrum is lowered by this amount.

19.9 *By considering the function*

$$F(\lambda) = \exp(\lambda A)B\exp(-\lambda A),$$

where A and B are linear operators and λ a parameter, and finding its derivatives with respect to λ, prove that

$$e^A B e^{-A} = B + [A, B] + \frac{1}{2!}[A, [A, B]] + \frac{1}{3!}[A, [A, [A, B]]] + \cdots.$$

Use this result to express

$$\exp\left(\frac{iL_x\theta}{\hbar}\right) L_y \exp\left(\frac{-iL_x\theta}{\hbar}\right)$$

as a linear combination of the angular momentum operators L_x, L_y and L_z.

Starting from the definition of $F(\lambda)$, we calculate its first few derivatives with respect to λ, remembering that operator A commutes with any function of A but not necessarily with any function of B:

$$F(\lambda) = \exp(\lambda A)B\exp(-\lambda A),$$

$$\frac{dF(\lambda)}{d\lambda} = A\exp(\lambda A)B\exp(-\lambda A) - \exp(\lambda A)B\exp(-\lambda A)A$$

$$= AF(\lambda) - F(\lambda)A = [A, F(\lambda)],$$

$$\frac{d^2F(\lambda)}{d\lambda^2} = A\frac{dF}{d\lambda} - \frac{dF}{d\lambda}A = \left[A, \frac{dF(\lambda)}{d\lambda}\right] = [A, [A, F(\lambda)]],$$

$$\frac{d^3F(\lambda)}{d\lambda^3} = [A, [A, [A, F(\lambda)]]], \text{ and so on for higher derivatives.}$$

Now we use a Taylor series in λ, based on the values of the derivatives at $\lambda = 0$, to evaluate $F(1)$. At $\lambda = 0$, $F(\lambda) = B$, and we obtain

$$e^A B e^{-A} = F(1)$$

$$= F(0) + \frac{dF(0)}{d\lambda} + \frac{1}{2!} \frac{d^2 F(0)}{d\lambda^2} + \frac{1}{3!} \frac{d^3 F(0)}{d\lambda^3} + \cdots$$

$$= B + [A, B] + \frac{1}{2!} [A, [A, B]] + \frac{1}{3!} [A, [A, [A, B]]] + \cdots .$$

To apply this result to

$$\Theta \equiv \exp\left(\frac{iL_x \theta}{\hbar}\right) L_y \exp\left(\frac{-iL_x \theta}{\hbar}\right),$$

we need to take A as $iL_x\theta/\hbar$ and B as L_y. The corresponding commutator is given by

$$\left[\frac{iL_x\theta}{\hbar}, L_y\right] = \frac{i\theta}{\hbar} [L_x, L_y] = -\theta L_z.$$

Because multiple commutators are involved, we will also need

$$\left[\frac{iL_x\theta}{\hbar}, L_z\right] = \frac{i\theta}{\hbar} [L_x, L_z] = \theta L_y.$$

Substituting in the derived series, we obtain

$$\Theta = L_y + (-\theta L_z) + \frac{1}{2!} \left[\frac{i\theta L_x}{\hbar}, -\theta L_z\right] + \cdots$$

$$= L_y - \theta L_z + \frac{1}{2!} (-\theta^2 L_y) + \frac{1}{3!} \left[\frac{i\theta L_x}{\hbar}, -\theta^2 L_y\right] + \cdots$$

$$= L_y - \theta L_z - \frac{1}{2!} \theta^2 L_y + \frac{1}{3!} (\theta^3 L_z) + \frac{1}{4!} \left[\frac{i\theta L_x}{\hbar}, \theta^3 L_z\right] + \cdots$$

$$= L_y - \theta L_z - \frac{1}{2!} \theta^2 L_y + \frac{1}{3!} \theta^3 L_z + \frac{1}{4!} (\theta^4 L_y) + \frac{1}{5!} \left[\frac{i\theta L_x}{\hbar}, \theta^4 L_y\right] +$$

$$= \left(1 - \frac{\theta^2}{2!} + \frac{\theta^4}{4!} - \cdots\right) L_y - \left(\theta - \frac{\theta^3}{3!} + \frac{\theta^5}{5!} - \cdots\right) L_z$$

$$= \cos\theta\, L_y - \sin\theta\, L_z.$$

At each stage in the above calculation, the value of the commutator in the nth term of the series has been used to reduce the $(n+1)$th term from a multiple commutator, with n levels of nesting, to a single commutator.

20

Partial differential equations: general and particular solutions

20.1 *Determine whether the following can be written as functions of $p = x^2 + 2y$ only, and hence whether they are solutions of*

$$\frac{\partial u}{\partial x} = x \frac{\partial u}{\partial y}.$$

(a) $x^2(x^2 - 4) + 4y(x^2 - 2) + 4(y^2 - 1)$;

(b) $x^4 + 2x^2 y + y^2$;

(c) $[x^4 + 4x^2 y + 4y^2 + 4]/[2x^4 + x^2(8y + 1) + 8y^2 + 2y]$.

As a first step, we verify that any function of $p = x^2 + 2y$ will satisfy the given equation. Using the chain rule, we have

$$\frac{\partial u}{\partial p} \frac{\partial p}{\partial x} = x \frac{\partial u}{\partial p} \frac{\partial p}{\partial y},$$

$$\Rightarrow \quad \frac{\partial u}{\partial p} 2x = x \frac{\partial u}{\partial p} 2.$$

This is satisfied for *any* function $u(p)$, thus completing the verification.

To test the given functions we substitute for $y = \frac{1}{2}(p - x^2)$ or for $x^2 = p - 2y$ in each of the $f(x, y)$ and then examine whether the resulting forms are independent of x or y, respectively.

(a) $f(x, y) = x^2(x^2 - 4) + 4y(x^2 - 2) + 4(y^2 - 1)$

$\qquad = x^2(x^2 - 4) + 2(p - x^2)(x^2 - 2) + p^2 - 2p x^2 + x^4 - 4$

$\qquad = x^4(1 - 2 + 1) + x^2(-4 + 2p + 4 - 2p) - 4p + p^2 - 4$

$\qquad = p^2 - 4p - 4 = g(p).$

This is a function of p only, and therefore the original $f(x, y)$ is a solution of the PDE.

Though not necessary for answering the question, we will repeat the verification, but this time by substituting for x rather than for y:

$$\begin{aligned} f(x, y) &= x^2(x^2 - 4) + 4y(x^2 - 2) + 4(y^2 - 1) \\ &= (p - 2y)(p - 2y - 4) + 4y(p - 2y - 2) + 4(y^2 - 1) \\ &= p^2 - 4py + 4y^2 - 4p + 8y + 4yp - 8y^2 - 8y + 4y^2 - 4 \\ &= p^2 - 4p - 4 = g(p); \end{aligned}$$

i.e. it is the same as before, as it must be, and again this shows that $f(x, y)$ is a solution of the PDE.

(b) $\begin{aligned}[t] f(x, y) &= x^4 + 2x^2 y + y^2 \\ &= (p - 2y)^2 + 2y(p - 2y) + y^2 \\ &= p^2 - 4p y + 4y^2 + 2p y - 4y^2 + y^2 \\ &= (p - y)^2 \neq g(p). \end{aligned}$

As this is a function of both p and y, it is not a solution of the PDE.

(c) $\begin{aligned}[t] f(x, y) &= \frac{x^4 + 4x^2 y + 4y^2 + 4}{2x^4 + x^2(8y + 1) + 8y^2 + 2y} \\ &= \frac{p^2 - 4p y + 4y^2 + 4yp - 8y^2 + 4y^2 + 4}{2p^2 - 8p y + 8y^2 + 8yp + p - 16y^2 - 2y + 8y^2 + 2y} \\ &= \frac{p^2 + 4}{2p^2 + p} = g(p). \end{aligned}$

This is a function of p only and therefore $f(x, y)$ is a solution of the PDE.

20.3 *Solve the following partial differential equations for $u(x, y)$ with the boundary conditions given:*

(a) $x \dfrac{\partial u}{\partial x} + xy = u, \quad u = 2y$ *on the line* $x = 1$;

(b) $1 + x \dfrac{\partial u}{\partial y} = xu, \quad u(x, 0) = x.$

(a) This can be solved as an ODE for u as a function of x, though the 'constant of integration' will be a function of y. In standard form, the equation reads

$$\frac{\partial u}{\partial x} - \frac{u}{x} = -y.$$

By inspection (or formal calculation) the IF for this is x^{-1} and the equation can be rearranged as

$$\frac{\partial}{\partial x}\left(\frac{u}{x}\right) = -\frac{y}{x},$$

$$\Rightarrow \quad \frac{u}{x} = -y\ln x + f(y),$$

$$u = 2y \text{ on } x = 1 \Rightarrow f(y) = 2y,$$

$$\text{and so } u(x, y) = xy(2 - \ln x).$$

(b) This equation can be written in standard form, with u as a function of y:

$$\frac{\partial u}{\partial y} - u = -\frac{1}{x},$$

for which the IF is clearly e^{-y}, leading to

$$\frac{\partial}{\partial y}\left(e^{-y}u\right) = -\frac{e^{-y}}{x},$$

$$\Rightarrow \quad e^{-y}u = \frac{e^{-y}}{x} + f(x),$$

$$u(x, 0) = x \Rightarrow f(x) = x - \frac{1}{x}.$$

Substituting this result, and multiplying through by e^y, gives $u(x, y)$ as

$$u(x, y) = \frac{1}{x} + \left(x - \frac{1}{x}\right)e^y.$$

20.5 *Find solutions of*

$$\frac{1}{x}\frac{\partial u}{\partial x} + \frac{1}{y}\frac{\partial u}{\partial y} = 0$$

for which (a) $u(0, y) = y$ *and* (b) $u(1, 1) = 1$.

As usual, we find $p(x, y)$ from

$$\frac{dx}{x^{-1}} = \frac{dy}{y^{-1}} \quad \Rightarrow \quad x^2 - y^2 = p.$$

(a) On $x = 0$, $p = -y^2$ and

$$u(0, y) = y = (-p)^{1/2} \quad \Rightarrow \quad u(x, y) = [-(x^2 - y^2)]^{1/2} = (y^2 - x^2)^{1/2}.$$

(b) At $(1, 1)$, $p = 0$ and

$$u(1, 1) = 1 \quad \Rightarrow \quad u(x, y) = 1 + g(x^2 - y^2),$$

321

where g is any function that has $g(0) = 0$.

We note that in part (a) the solution is uniquely determined because the boundary values are given along a line, whereas in part (b), where the value is fixed at only an isolated point, the solution is indeterminate to the extent of a loosely determined function. This is the normal situation, though it is modified if the line in (a) happens to be a characteristic of the PDE.

20.7 *Solve*

$$\sin x \frac{\partial u}{\partial x} + \cos x \frac{\partial u}{\partial y} = \cos x \qquad (*)$$

subject to (a) $u(\pi/2, y) = 0$ *and (b)* $u(\pi/2, y) = y(y + 1)$.

As usual, the CF is found from

$$\frac{dx}{\sin x} = \frac{dy}{\cos x} \quad \Rightarrow \quad y - \ln \sin x = p.$$

Since the RHS of (*) is a factor in one of the terms on the LHS, a trivial PI is any function of y only whose derivative (with respect to y) is unity, of which the simplest is $u(x, y) = y$. The general solution is therefore

$$u(x, y) = f(y - \ln \sin x) + y.$$

The actual form of the arbitrary function $f(p)$ is determined by the form that $u(x, y)$ takes on the boundary, here the line $x = \pi/2$.

(a) With $u(\pi/2, y) = 0$:

$$0 = f(y - 0) + y \quad \Rightarrow \quad f(p) = -p$$
$$\Rightarrow \quad u(x, y) = \ln \sin x - y + y = \ln \sin x.$$

(b) With $u(\pi/2, y) = y(y + 1)$:

$$y(y + 1) = f(y - 0) + y \quad \Rightarrow \quad f(p) = p^2$$
$$\Rightarrow \quad u(x, y) = (y - \ln \sin x)^2 + y.$$

20.9 *If* $u(x, y)$ *satisfies*

$$\frac{\partial^2 u}{\partial x^2} - 3 \frac{\partial^2 u}{\partial x \partial y} + 2 \frac{\partial^2 u}{\partial y^2} = 0$$

and $u = -x^2$ *and* $\partial u/\partial y = 0$ *for* $y = 0$ *and all* x, *find the value of* $u(0, 1)$.

If we are to find solutions to this homogeneous second-order PDE of the form $u(x, y) = f(x + \lambda y)$, then λ must satisfy

$$1 - 3\lambda + 2\lambda^2 = 0 \quad \Rightarrow \quad \lambda = \tfrac{1}{2}, \ 1.$$

Thus $u(x, y) = g(x + \tfrac{1}{2}y) + f(x + y) \equiv g(p_1) + f(p_2)$.

On $y = 0$, $p_1 = p_2 = x$ and

$$-x^2 = u(x, 0) = g(x) + f(x), \qquad (*)$$

$$0 = \frac{\partial u}{\partial y}(x, 0) = \tfrac{1}{2}g'(x) + f'(x).$$

From $(*)$, $\qquad - 2x = g'(x) + f'(x).$

Subtracting, $\qquad 2x = -\tfrac{1}{2}g'(x).$

Integrating, $\qquad g(x) = -2x^2 + k \quad \Rightarrow \quad f(x) = x^2 - k, \quad$ from $(*)$.

Hence, $\qquad u(x, y) = -2(x + \tfrac{1}{2}y)^2 + k + (x + y)^2 - k$

$$= -x^2 + \tfrac{1}{2}y^2.$$

At the particular point $(0, 1)$ we have $u(0, 1) = -0^2 + \tfrac{1}{2}(1)^2 = \tfrac{1}{2}$.

20.11 *In those cases in which it is possible to do so, evaluate $u(2, 2)$, where $u(x, y)$ is the solution of*

$$2y\frac{\partial u}{\partial x} - x\frac{\partial u}{\partial y} = xy(2y^2 - x^2)$$

that satisfies the (separate) boundary conditions given below.

(a) $u(x, 1) = x^2$ for all x.
(b) $u(x, 1) = x^2$ for $x \geq 0$.
(c) $u(x, 1) = x^2$ for $0 \leq x \leq 3$.
(d) $u(x, 0) = x$ for $x \geq 0$.
(e) $u(x, 0) = x$ for all x.
(f) $u(1, \sqrt{10}) = 5$.
(g) $u(\sqrt{10}, 1) = 5$.

To find the CF, $u(x, y) = f(p)$, we set

$$\frac{dx}{2y} = -\frac{dy}{x} \quad \Rightarrow \quad x^2 + 2y^2 = p.$$

This result also defines the characteristic curves, which are right ellipses centred on the origin with semi-axes of lengths $\sqrt{p}$ and $\sqrt{p/2}$. The point $(2, 2)$ lies on the characteristic with $p = 2^2 + 2(2^2) = 12$; we will only be able to determine the value of $u(2, 2)$ if this curve cuts the boundary on which u is specified.

For a PI we try $u(x, y) = Ax^n y^m$:

$$2Anx^{n-1}y^{m+1} - Amx^{n+1}y^{m-1} = 2xy^3 - x^3y,$$

which has a solution, $n = m = 2$ with $A = \frac{1}{2}$. Thus the general solution is

$$u(x, y) = f(x^2 + 2y^2) + \tfrac{1}{2}x^2y^2.$$

(a) With $u(x, 1) = x^2$ for all x:

$$\begin{aligned}
x^2 = u(x, 1) &= f(x^2 + 2) + \tfrac{1}{2}x^2 \\
\Rightarrow \qquad f(p) &= \tfrac{1}{2}(p - 2) \\
\Rightarrow \qquad u(x, y) &= \tfrac{1}{2}(x^2 + 2y^2 - 2) + \tfrac{1}{2}x^2y^2 \\
&= \tfrac{1}{2}(x^2 + x^2y^2 + 2y^2 - 2), \\
u(2, 2) &= \tfrac{1}{2}(4 + 16 + 8 - 2) = 13.
\end{aligned}$$

The line $y = 1$ cuts each characteristic in zero (for $p < 2$) or two (for $p > 2$) distinct points. Here $p = 12$ (> 2) and the characteristic (ellipse) that passes through $(2, 2)$ cuts the boundary (the line $y = 1$) in two places. In general, a double intersection can lead to inconsistencies and hence to no solution. However, it causes no difficulty with the given boundary conditions since the required values of x^2 at $x = \pm\sqrt{12 - 2(1^2)}$ are equal and u is a even function of x.

(b) With $u(x, 1) = x^2$ for $x \geq 0$.
Since every characteristic ellipse (with $p > 2$) cuts the line $y = 1$ once (and only once in $x > 0$), this gives the same result as in part (a).

(c) With $u(x, 1) = x^2$ for $0 \leq x \leq 3$.
The ellipses that cut the line $y = 1$ with $0 \leq x \leq 3$ have p-values lying between $0^2 + 2(1)^2 = 2$ and $3^2 + 2(1)^2 = 11$. Thus the $p = 12$ curve does not do so and $u(2, 2)$ is undetermined.

(d) With $u(x, 0) = x$ for $x \geq 0$:

$$\begin{aligned}
x = u(x, 0) &= f(x^2 + 0) + 0 \\
\Rightarrow \qquad f(p) &= p^{1/2} \\
\Rightarrow \qquad u(x, y) &= (x^2 + 2y^2)^{1/2} + \tfrac{1}{2}x^2y^2 \\
\Rightarrow \qquad u(2, 2) &= \sqrt{12} + 8.
\end{aligned}$$

The characteristic (ellipse) $p = 12$ cuts the positive x-axis (i.e. $y = 0$) in one and only one place ($x = +\sqrt{12}$) and so the solution is well defined and the above evaluation valid.

(e) With $u(x, 0) = x$ for all x.
This is as in part (d) except that now a characteristic ellipse cuts the defining boundary in two places, $x = \pm\sqrt{p}$, and requires both $u(\sqrt{p}, 0) = \sqrt{p}$ and

$u(-\sqrt{p},0) = -\sqrt{p}$. Since $\sqrt{z}$ is not differentiable at $z = 0$, this is not possible and no solution exists.

(f) With $u(1,\sqrt{10}) = 5$.
At the point $(1,\sqrt{10})$ the value of p is $1 + 2(10) = 21$. As the 'boundary' consists of just this one point, it is only at the points that lie on the characteristic $p = 21$ that the value of $u(x,y)$ can be known. Since for the point $(2,2)$ the value of p is 12, the value of $u(2,2)$ cannot be determined.

(g) With $u(\sqrt{10},1) = 5$.
At the point $(\sqrt{10},1)$ the value of p is $10 + 2(1) = 12$. Since for $(2,2)$ it is also 12 the value of $u(2,2)$ is determined and is given by $f(12) + \frac{1}{2}(4)(4) = 5 + 8 = 13$.

20.13 *The solution to the equation*

$$6\frac{\partial^2 u}{\partial x^2} - 5\frac{\partial^2 u}{\partial x \partial y} + \frac{\partial^2 u}{\partial y^2} = 14$$

that satisfies $u = 2x + 1$ and $\partial u/\partial y = 4 - 6x$, both on the line $y = 0$, is

$$u(x,y) = -8y^2 - 6xy + 2x + 4y + 1.$$

By changing the independent variables in the equation to

$$\xi = x + 2y \quad and \quad \eta = x + 3y,$$

show that it must be possible to write $14(x^2 + 5xy + 6y^2)$ in the form

$$f_1(x + 2y) + f_2(x + 3y) - (x^2 + y^2),$$

and determine the forms of $f_1(z)$ and $f_2(z)$.

Let $u(x,y) = v(\xi,\eta)$, with $\xi = x + 2y$ and $\eta = x + 3y$. We must first express the differential operators $\partial/\partial x$ and $\partial/\partial y$ in terms of differentiation with respect to ξ and η; to do this we use the chain rule:

$$\frac{\partial}{\partial x} = \frac{\partial \xi}{\partial x}\frac{\partial}{\partial \xi} + \frac{\partial \eta}{\partial x}\frac{\partial}{\partial \eta} = \frac{\partial}{\partial \xi} + \frac{\partial}{\partial \eta}; \quad \text{similarly} \quad \frac{\partial}{\partial y} = 2\frac{\partial}{\partial \xi} + 3\frac{\partial}{\partial \eta}.$$

The equation becomes

$$6\left(\frac{\partial}{\partial \xi} + \frac{\partial}{\partial \eta}\right)\left(\frac{\partial v}{\partial \xi} + \frac{\partial v}{\partial \eta}\right) - 5\left(\frac{\partial}{\partial \xi} + \frac{\partial}{\partial \eta}\right)\left(2\frac{\partial v}{\partial \xi} + 3\frac{\partial v}{\partial \eta}\right)$$
$$+ \left(2\frac{\partial}{\partial \xi} + 3\frac{\partial}{\partial \eta}\right)\left(2\frac{\partial v}{\partial \xi} + 3\frac{\partial v}{\partial \eta}\right) = 14,$$

$$(6 - 10 + 4)\frac{\partial^2 v}{\partial \xi^2} + (12 - 25 + 12)\frac{\partial^2 v}{\partial \xi \partial \eta} + (6 - 15 + 9)\frac{\partial^2 v}{\partial \eta^2} = 14.$$

Collecting similar terms together, we find

$$\frac{\partial^2 v}{\partial \xi \partial \eta} = -14.$$

This equation has a CF of the form $f(\xi) + g(\eta)$ and a PI of $-14\xi\eta$. Thus its general solution is

$$v(\xi, \eta) = f(\xi) + g(\eta) - 14\xi\eta.$$

This must be the same as the given answer, i.e.

$$-8y^2 - 6xy + 2x + 4y + 1 = f(x + 2y) + g(x + 3y) - 14(x + 2y)(x + 3y)$$

for some functions f and g. Thus

$$
\begin{aligned}
w(x, y) &= 14(x^2 + 5xy + 6y^2) \\
&= 14(x + 2y)(x + 3y) \\
&= f(x + 2y) + g(x + 3y) + 8y^2 + 6xy - 2x - 4y - 1 \\
&= f(x + 2y) + g(x + 3y) - (x^2 + y^2) + h(x, y),
\end{aligned}
$$

where
$$
\begin{aligned}
h(x, y) &= x^2 + 9y^2 + 6xy - 2x - 4y - 1 \\
&= (x + 3y)^2 - 2(x + 2y) - 1 \\
&= F(x + 2y) + G(x + 3y).
\end{aligned}
$$

It follows that

$$w(x, y) = f_1(x + 2y) + f_2(x + 3y) - (x^2 + y^2),$$

where $f_1(z) = f(z) - 2z - 1$ and $f_2(z) = g(z) + z^2$.

After rearrangement this reads

$$15x^2 + 70xy + 85y^2 = f_1(x + 2y) + f_2(x + 3y). \qquad (**)$$

Taking second derivatives with respect to x and y separately,

$$
\begin{aligned}
30 &= f_1'' + f_2'', \\
170 &= 4f_1'' + 9f_2'', \\
\Rightarrow \quad 50 &= 5f_2'' \quad \Rightarrow \quad f_2(z) = 5z^2 + \alpha z + \beta \\
\text{and} \quad 100 &= 5f_1'' \quad \Rightarrow \quad f_1(z) = 10z^2 + \gamma z + \delta.
\end{aligned}
$$

Equating the coefficients of xy, x and y and the constants in $(**)$ gives $70 = 40 + 30$, $0 = \alpha + \gamma$, $0 = 3\alpha + 2\gamma$, $0 = \beta + \delta$. These equations have the solution $\alpha = \gamma = 0$ and $\beta = k = -\delta$. Thus

$$f_1(z) = 10z^2 - k \qquad \text{and} \qquad f_2(z) = 5z^2 + k.$$

Clearly, k can take any value without affecting the final form given in the question.

20.15 *Find the most general solution of $\partial^2 u/\partial x^2 + \partial^2 u/\partial y^2 = x^2 y^2$.*

The complementary function for this equation is the solution to the two-dimensional Laplace equation and [either as a general known result or from substituting the trial form $h(x + \lambda y)$ which leads to $\lambda^2 = -1$ and hence to $\lambda = \pm i$] has the form $f(x + iy) + g(x - iy)$ for arbitrary functions f and g.

It therefore remains only to find a suitable PI. As f and g are not specified, there are infinitely many possibilities and which one we finish up with will depend upon the details of the approach adopted. When a solution has been obtained it should be checked by substitution.

As no PI is obvious by inspection, we make a change of variables with the object of obtaining one by means of an explicit integration. To do this, we use as new variables the arguments of the arbitrary functions appearing in the CF.

Setting $\xi = x + iy$ and $\eta = x - iy$, with $u(x, y) = v(\xi, \eta)$, gives

$$\left(\frac{\partial}{\partial \xi} + \frac{\partial}{\partial \eta} \right) \left(\frac{\partial v}{\partial \xi} + \frac{\partial v}{\partial \eta} \right)$$

$$+ \left(i\frac{\partial}{\partial \xi} - i\frac{\partial}{\partial \eta} \right) \left(i\frac{\partial v}{\partial \xi} - i\frac{\partial v}{\partial \eta} \right) = \left(\frac{\xi + \eta}{2} \right)^2 \left(\frac{\xi - \eta}{2i} \right)^2,$$

$$(1 - 1)\frac{\partial^2 v}{\partial \xi^2} + (2 + 2)\frac{\partial^2 v}{\partial \xi \partial \eta} + (1 - 1)\frac{\partial^2 v}{\partial \eta^2} = -\frac{1}{16}(\xi^2 - \eta^2)^2,$$

$$\frac{\partial^2 v}{\partial \xi \partial \eta} = -\frac{1}{64}(\xi^2 - \eta^2)^2.$$

When we integrate this we can set all constants of integration and all arbitrary functions equal to zero as *any* solution will suffice:

$$\frac{\partial^2 v}{\partial \xi \partial \eta} = -\frac{1}{64}(\xi^4 - 2\xi^2\eta^2 + \eta^4),$$

$$\frac{\partial v}{\partial \eta} = -\frac{1}{64} \left(\frac{\xi^5}{5} - \frac{2\xi^3\eta^2}{3} + \xi\eta^4 \right),$$

$$v = -\frac{1}{64} \left(\frac{\xi^5\eta}{5} - \frac{2\xi^3\eta^3}{9} + \frac{\xi\eta^5}{5} \right).$$

Re-expressing this solution as a function of x and y (noting that $\xi\eta = x^2 + y^2$)

gives

$$u(x, y) = \frac{1}{(64)(45)}[10\xi^3\eta^3 - 9\xi\eta(\xi^4 + \eta^4)]$$

$$= \frac{1}{(64)(45)}[10(x^2 + y^2)^3 - 18(x^2 + y^2)(x^4 - 6x^2y^2 + y^4)]$$

$$= \frac{x^2 + y^2}{(64)(45)}(10x^4 + 20x^2y^2 + 10y^4 - 18x^4 + 108x^2y^2 - 18y^4)$$

$$= \frac{x^2 + y^2}{(64)(45)}(128x^2y^2 - 8x^4 - 8y^4)$$

$$= \frac{1}{360}(15x^4y^2 - x^6 + 15x^2y^4 - y^6).$$

Check

Applying $\dfrac{\partial^2}{\partial x^2} + \dfrac{\partial^2}{\partial y^2}$ to the final expression yields

$$\frac{1}{360}[15(12)x^2y^2 - 30x^4 + 30y^4 + 30x^4 + 15(12)x^2y^2 - 30y^4] = x^2y^2,$$

as it should.

20.17 *The non-relativistic Schrödinger equation,*

$$-\frac{\hbar^2}{2m}\nabla^2 u + V(\mathbf{r})u = i\hbar\frac{\partial u}{\partial t},$$

is similar to the diffusion equation in having different orders of derivatives in its various terms; this precludes solutions that are arbitrary functions of particular linear combinations of variables. However, since exponential functions do not change their forms under differentiation, solutions in the form of exponential functions of combinations of the variables may still be possible.

Consider the Schrödinger equation for the case of a constant potential, i.e. for a free particle, and show that it has solutions of the form $A\exp(lx + my + nz + \lambda t)$, where the only requirement is that

$$-\frac{\hbar^2}{2m}(l^2 + m^2 + n^2) = i\hbar\lambda.$$

In particular, identify the equation and wavefunction obtained by taking λ as $-iE/\hbar$, and l, m and n as $ip_x/\hbar, ip_y/\hbar$ and $ip_z/\hbar$, respectively, where E is the energy and $\mathbf{p}$ the momentum of the particle; these identifications are essentially the content of the de Broglie and Einstein relationships.

For a free particle we may omit the potential term $V(\mathbf{r})$ from the Schrödinger equation, which then reads (in Cartesian coordinates)

$$-\frac{\hbar^2}{2m}\left(\frac{\partial^2 u}{\partial x^2} + \frac{\partial^2 u}{\partial y^2} + \frac{\partial^2 u}{\partial z^2}\right) = i\hbar\frac{du}{dt}.$$

We try $u(x, y, z, t) = A\exp(lx + my + nz + \lambda t)$, i.e. the product of four exponential functions, and obtain

$$-\frac{\hbar^2}{2m}(l^2 + m^2 + n^2)u = i\hbar\lambda u.$$

This equation is clearly satisfied provided

$$-\frac{\hbar^2}{2m}(l^2 + m^2 + n^2) = i\hbar\lambda.$$

With λ as $-iE/\hbar$, and l, m and n as $ip_x/\hbar, ip_y/\hbar$ and $ip_z/\hbar$, respectively, where E is the energy and $\mathbf{p}$ is the momentum of the particle, we have

$$-\frac{\hbar^2}{2m}\left(-\frac{p_x^2}{\hbar^2} - \frac{p_y^2}{\hbar^2} - \frac{p_z^2}{\hbar^2}\right) = E,$$

which can be written more compactly as $E = p^2/2m$, the classical non-relativistic relationship between the (kinetic) energy and momentum of a free particle.

The wavefunction obtained is

$$u(\mathbf{r}, t) = A\exp\left[\frac{i}{\hbar}(p_x x + p_y y + p_z z - Et)\right]$$

$$= A\exp\left[\frac{i}{\hbar}(\mathbf{p} \cdot \mathbf{r} - Et)\right],$$

i.e. a classical plane wave of wave number $\mathbf{k} = \mathbf{p}/\hbar$ and angular frequency $\omega = E/\hbar$ travelling in the direction $\mathbf{p}/p$.

20.19 *An incompressible fluid of density ρ and negligible viscosity flows with velocity v along a thin, straight, perfectly light and flexible tube, of cross-section A which is held under tension T. Assume that small transverse displacements u of the tube are governed by*

$$\frac{\partial^2 u}{\partial t^2} + 2v\frac{\partial^2 u}{\partial x\partial t} + \left(v^2 - \frac{T}{\rho A}\right)\frac{\partial^2 u}{\partial x^2} = 0.$$

(a) *Show that the general solution consists of a superposition of two waveforms travelling with different speeds.*

(b) *The tube initially has a small transverse displacement $u = a\cos kx$ and is suddenly released from rest. Find its subsequent motion.*

(a) This is a second-order equation and will (in general) have two solutions of the form $u(x, t) = f(x + \lambda t)$, where both λ satisfy

$$\lambda^2 + 2v\lambda + \left(v^2 - \frac{T}{\rho A}\right) = 0 \quad \Rightarrow \quad \lambda = -v \pm \sqrt{v^2 - v^2 + \frac{T}{\rho A}} \equiv -v \pm \alpha,$$

and gives (minus) the speed of the corresponding profile. Thus the general displacement consists of a superposition of waveforms travelling with speeds $v \mp \alpha$.

(b) Now $u(x, 0) = a \cos kx$ and $\dot{u}(x, 0) = 0$, where the dot denotes differentiation with respect to time t. Let the general solution be given by

$$u(x, t) = f[x - (v + \alpha)t] + g[x - (v - \alpha)t],$$

with $\qquad a \cos kx = f(x) + g(x)$

and $\qquad 0 = -(v + \alpha)f'(x) - (v - \alpha)g'(x).$

We differentiate the first of these with respect to x and then eliminate the function $f'(x)$:

$$-ka \sin kx = f'(x) + g'(x),$$
$$-ka(v + \alpha) \sin kx = (v + \alpha - v + \alpha)g'(x),$$
$$g'(x) = -\frac{ka(v + \alpha)}{2\alpha} \sin kx,$$
$$\Rightarrow \quad g(x) = \frac{v + \alpha}{2\alpha} a \cos kx + c,$$
$$\Rightarrow \quad f(x) = \frac{\alpha - v}{2\alpha} a \cos kx - c.$$

Now that the forms of the initially arbitrary functions $f(x)$ and $g(x)$ have been determined, it follows that, for a *general* time t,

$$u(x, t) = \frac{\alpha - v}{2\alpha} a \cos[kx - k(v + \alpha)t] + \frac{\alpha + v}{2\alpha} a \cos[kx - k(v - \alpha)t]$$
$$= \frac{a}{2} 2 \cos(kx - kvt) \cos k\alpha t + \frac{va}{2\alpha} 2 \sin(kx - kvt) \sin(-k\alpha t)$$
$$= a \cos[k(x - vt)] \cos k\alpha t - \frac{va}{\alpha} \sin[k(x - vt)] \sin k\alpha t.$$

20.21 *In an electrical cable of resistance R and capacitance C, each per unit length, voltage signals obey the equation $\partial^2 V/\partial x^2 = RC \partial V/\partial t$. This (diffusion-type) equation has solutions of the form*

$$f(\zeta) = \frac{2}{\sqrt{\pi}} \int_0^{\zeta} \exp(-v^2)\,dv, \quad \text{where } \zeta = \frac{x(RC)^{1/2}}{2t^{1/2}}.$$

It also has solutions of the form $V = Ax + D$.

(a) *Find a combination of these that represents the situation after a steady volt-age V_0 is applied at $x = 0$ at time $t = 0$.*
(b) *Obtain a solution describing the propagation of the voltage signal resulting from the application of the signal $V = V_0$ for $0 < t < T$, $V = 0$ otherwise, to the end $x = 0$ of an infinite cable.*
(c) *Show that for $t \gg T$ the maximum signal occurs at a value of x proportional to $t^{1/2}$ and has a magnitude proportional to t^{-1}.*

(a) Consider the given function

$$f(\zeta) = \frac{2}{\sqrt{\pi}} \int_0^{\zeta} \exp(-v^2)\,dv, \quad \text{where } \zeta = \frac{x(RC)^{1/2}}{2t^{1/2}}.$$

The requirements to be satisfied by the correct combination of this function and $V(x,t) = Ax + D$ are (i) that, at $t = 0$, V is zero for all x, except $x = 0$ where it is V_0, and (ii) that, as $t \to \infty$, V is V_0 for all x.

(i) At $t = 0$, $\zeta = \infty$ and $f(\zeta) = 1$ for all $x \neq 0$.
(ii) As $t \to \infty$, $\zeta \to 0$ and $f(\zeta) \to 0$ for all finite x.

The required combination is therefore $D = V_0$ and $-V_0 f(\zeta)$, i.e.

$$V(x,t) = V_0 \left[1 - \frac{2}{\sqrt{\pi}} \int_0^{\frac{1}{2}x(CR/t)^{1/2}} \exp(-v^2)\,dv \right].$$

(b) The equation is linear and so we may superpose solutions. The response to the input $V = V_0$ for $0 < t < T$ can be considered as that to V_0 applied at $t = 0$ and continued, together with $-V_0$ applied at $t = T$ and continued. The solution is therefore the difference between two solutions of the form found in part (a):

$$V(x,t) = \frac{2V_0}{\sqrt{\pi}} \int_{\frac{1}{2}x(CR/t)^{1/2}}^{\frac{1}{2}x[CR/(t-T)]^{1/2}} \exp\left(-v^2\right)\,dv.$$

(c) To find the maximum signal we set $\partial V/\partial x$ equal to zero. Remembering that we are differentiating with respect to the limits of an integral (whose integrand

331

does not contain x explicitly), we obtain

$$\frac{1}{2}\left(\frac{CR}{t-T}\right)^{1/2}\exp\left[-\frac{x^2CR}{4(t-T)}\right]-\frac{1}{2}\left(\frac{CR}{t}\right)^{1/2}\exp\left[-\frac{x^2CR}{4t}\right]=0.$$

This requires

$$\left(\frac{t-T}{t}\right)^{1/2}=\exp\left[-\frac{x^2CR}{4(t-T)}+\frac{x^2CR}{4t}\right]$$
$$=\exp\left[\frac{x^2CR(-t+t-T)}{4t(t-T)}\right].$$

For $t\gg T$, we expand both sides:

$$1-\frac{1}{2}\frac{T}{t}+\cdots=1-\frac{Tx^2CR}{4t^2}+\cdots,$$

$$\Rightarrow\quad x^2\approx\frac{2t}{CR}\quad\Rightarrow\quad v=\frac{1}{2}\sqrt{\frac{2t}{CR}}\left(\frac{CR}{t}\right)^{1/2}=\frac{1}{\sqrt{2}}.$$

The corresponding value of V is approximately equal to the value of the integrand, evaluated at this value of v, multiplied by the difference between the two limits of the integral. Thus

$$V_{\text{max}}\approx\frac{2V_0}{\sqrt{\pi}}\exp(-v^2)\frac{x\sqrt{CR}}{2}\left[\frac{1}{(t-T)^{1/2}}-\frac{1}{t^{1/2}}\right]$$
$$\approx\frac{2V_0}{\sqrt{\pi}}e^{-1/2}\frac{x\sqrt{CR}}{2}\frac{1}{2}\frac{T}{t^{3/2}}$$
$$=\frac{V_0Te^{-1/2}}{\sqrt{2\pi}\,t}.$$

In summary, for $t\gg T$ the maximum signal occurs at a value of x proportional to $t^{1/2}$ and has a magnitude proportional to t^{-1}.

20.23 *Consider each of the following situations in a qualitative way and determine the equation type, the nature of the boundary curve and the type of boundary conditions involved:*

(a) *a conducting bar given an initial temperature distribution and then thermally isolated;*

(b) *two long conducting concentric cylinders, on each of which the voltage distribution is specified;*

(c) *two long conducting concentric cylinders, on each of which the charge distribution is specified;*

(d) *a semi-infinite string, the end of which is made to move in a prescribed way.*

We use the notation

$$A\frac{\partial^2 u}{\partial x^2} + B\frac{\partial^2 u}{\partial x \partial y} + C\frac{\partial^2 u}{\partial y^2} + D\frac{\partial u}{\partial x} + E\frac{\partial u}{\partial y} + Fu = R(x,y)$$

to express the most general type of PDE, and the following table

Equation type	Boundary	Conditions
hyperbolic	open	Cauchy
parabolic	open	Dirichlet or Neumann
elliptic	closed	Dirichlet or Neumann

to determine the appropriate boundary type and hence conditions.

(a) The diffusion equation $\kappa\dfrac{\partial^2 T}{\partial x^2} = \dfrac{\partial T}{\partial t}$ has $A = \kappa$, $B = 0$ and $C = 0$; thus $B^2 = 4AC$ and the equation is parabolic. This needs an open boundary. In the present case, the initial heat distribution (at the $t = 0$ boundary) is a Dirichlet condition and the insulation (no temperature gradient at the external surfaces) is a Neumann condition.

(b) The governing equation in two-dimensional Cartesians (not the natural choice for this situation, but this does not matter for the present purpose) is the Laplace equation, $\dfrac{\partial^2 \phi}{\partial x^2} + \dfrac{\partial^2 \phi}{\partial y^2} = 0$, which has $A = 1$, $B = 0$ and $C = 1$ and therefore $B^2 < 4AC$. The equation is therefore elliptic and requires a closed boundary. Since ϕ is specified on the cylinders, the boundary conditions are Dirichlet in this particular situation.

(c) This is the same as part (b) except that the specified charge distribution σ determines $\partial\phi/\partial n$, through $\partial\phi/\partial n = \sigma/\epsilon_0$, and imposes Neumann boundary conditions.

(d) For the wave equation $\dfrac{\partial^2 u}{\partial x^2} - \dfrac{1}{c^2}\dfrac{\partial^2 u}{\partial t^2} = 0$, we have $A = 1$, $B = 0$ and $C = -c^{-2}$, thus making $B^2 > 4AC$ and the equation hyperbolic. We thus require an open boundary and Cauchy conditions, with the displacement of the end of the string having to be specified at all times — this is equivalent to the displacement and the velocity of the end of the string being specified at all times.

20.25 *The Klein–Gordon equation (which is satisfied by the quantum-mechanical wavefunction $\Phi(\mathbf{r})$ of a relativistic spinless particle of non-zero mass m) is*

$$\nabla^2\Phi - m^2\Phi = 0.$$

Show that the solution for the scalar field $\Phi(\mathbf{r})$ in any volume V bounded by a surface S is unique if either Dirichlet or Neumann boundary conditions are specified on S.

Suppose that, for a given set of boundary conditions ($\Phi = f$ or $\partial\Phi/\partial n = g$ on S), there are two solutions to the Klein–Gordon equation, Φ_1 and Φ_2. Then consider $\Phi_3 = \Phi_1 - \Phi_2$, which satisfies

$$\nabla^2\Phi_3 = \nabla^2\Phi_1 - \nabla^2\Phi_2 = m^2\Phi_1 - m^2\Phi_2 = m^2\Phi_3$$

and

$$\text{either } \Phi_3 = f - f = 0, \text{ or } \frac{\partial\Phi_3}{\partial n} = g - g = 0 \text{ on } S.$$

Now apply Green's first theorem with the scalar functions equal to Φ_3 and Φ_3^*:

$$\int^S \Phi_3^* \frac{\partial\Phi_3}{\partial n} \, dS = \int_V [\Phi_3^* \nabla^2\Phi_3 + (\nabla\Phi_3^*) \cdot (\nabla\Phi_3)] \, dV,$$

$$\Rightarrow \quad 0 = \int_V (m^2|\Phi_3|^2 + |\nabla\Phi_3|^2) \, dV,$$

whichever set of boundary conditions applies. Since both terms in the integrand on the RHS are non-negative, each must be equal to zero. In particular, $|\Phi_3| = 0$ implies that $\Phi_3 = 0$ everywhere, i.e $\Phi_1 = \Phi_2$ everywhere; the solution is unique.

21

Partial differential equations: separation of variables and other methods

In each case we write $u(x, y) = X(x)Y(y)$, separate the variables into groups that each depend on only one variable, and then assert that each must be equal to a constant, with the several constants satisfying an arithmetic identity.

(a)
$$\frac{\partial u}{\partial x} - x\frac{\partial u}{\partial y} = 0,$$

$$X'Y - xXY' = 0,$$

$$\frac{X'}{xX} = \frac{Y'}{Y} = k \quad \Rightarrow \quad \ln X = \tfrac{1}{2}kx^2 + c_1, \quad \ln Y = ky + c_2,$$

$$\Rightarrow \quad X = Ae^{kx^2/2}, \quad Y = Be^{ky},$$

$$\Rightarrow \quad u(x, y) = Ce^{\lambda(x^2+2y)}, \text{ where } k = 2\lambda.$$

(b)
$$x\frac{\partial u}{\partial x} - 2y\frac{\partial u}{\partial y} = 0,$$

$$xX'Y - 2yXY' = 0,$$

$$\frac{xX'}{X} = \frac{2yY'}{Y} = k \quad \Rightarrow \quad \ln X = k \ln x + c_1,$$

$$\ln Y = \tfrac{1}{2}k \ln y + c_2,$$

$$\Rightarrow \quad X = Ax^k, \quad Y = By^{k/2},$$

$$\Rightarrow \quad u(x, y) = C(x^2 y)^{\lambda}, \text{ where } k = 2\lambda.$$

> **21.3**   *The wave equation describing the transverse vibrations of a stretched membrane under tension T and having a uniform surface density ρ is*
>
> $$T \left(\frac{\partial^2 u}{\partial x^2} + \frac{\partial^2 u}{\partial y^2} \right) = \rho \frac{\partial^2 u}{\partial t^2}.$$
>
> *Find a separable solution appropriate to a membrane stretched on a frame of length a and width b, showing that the natural angular frequencies of such a membrane are given by*
>
> $$\omega^2 = \frac{\pi^2 T}{\rho} \left(\frac{n^2}{a^2} + \frac{m^2}{b^2} \right),$$
>
> *where n and m are any positive integers.*

We seek solutions $u(x, y, t)$ that are periodic in time and have $u(0, y, t) = u(a, y, t) = u(x, 0, t) = u(x, b, t) = 0$. Write $u(x, y, t) = X(x)Y(y)S(t)$ and substitute, obtaining

$$T(X''YS + XY''S) = \rho XYS'',$$

which, when divided through by XYS, gives

$$\frac{X''}{X} + \frac{Y''}{Y} = \frac{\rho}{T} \frac{S''}{S} = -\frac{\omega^2 \rho}{T}.$$

The second equality, obtained by applying the separation of variables principle with separation constant $-\omega^2 \rho / T$, gives $S(t)$ as a sinusoidal function of t of frequency ω, i.e. $A \cos(\omega t) + B \sin(\omega t)$.

We then have, on applying the separation of variables principle a second time, that

$$\frac{X''}{X} = \lambda \text{ and } \frac{Y''}{Y} = \mu, \text{ where } \lambda + \mu = -\frac{\omega^2 \rho}{T}. \qquad (*)$$

These equations must also have sinusoidal solutions. This is because, since $u(0, y, t) = u(a, y, t) = u(x, 0, t) = u(x, b, t) = 0$, each solution has to have zeros at two different values of its argument. We are thus led to

$$X = A \sin(p\, x) \text{ and } Y = B \sin(q\, x), \text{ where } p^2 = -\lambda \text{ and } q^2 = -\mu.$$

Further, since $u(a, y, t) = u(x, b, t) = 0$, we must have $p = n\pi / a$ and $q = m\pi / b$, where n and m are integers. Putting these values back into $(*)$ gives

$$-p^2 - q^2 = -\frac{\omega^2 \rho}{T} \quad \Rightarrow \quad \pi^2 \left(\frac{n^2}{a^2} + \frac{m^2}{b^2} \right) = \frac{\omega^2 \rho}{T}.$$

Hence the quoted result.

21.5 *Denoting the three terms of ∇^2 in spherical polars by $\nabla_r^2, \nabla_\theta^2, \nabla_\phi^2$ in an obvious way, evaluate $\nabla_r^2 u$, etc. for the two functions given below and verify that, in each case, although the individual terms are not necessarily zero their sum $\nabla^2 u$ is zero. Identify the corresponding values of ℓ and m.*

(a) $u(r, \theta, \phi) = \left(Ar^2 + \dfrac{B}{r^3} \right) \dfrac{3 \cos^2 \theta - 1}{2}.$

(b) $u(r, \theta, \phi) = \left(Ar + \dfrac{B}{r^2} \right) \sin \theta \exp i\phi.$

In both cases we write $u(r, \theta, \phi)$ as $R(r)\Theta(\theta)\Phi(\phi)$ with

$$\nabla_r^2 = \frac{1}{r^2} \frac{\partial}{\partial r} \left(r^2 \frac{\partial}{\partial r} \right), \quad \nabla_\theta^2 = \frac{1}{r^2 \sin \theta} \frac{\partial}{\partial \theta} \left(\sin \theta \frac{\partial}{\partial \theta} \right), \quad \nabla_\phi^2 = \frac{1}{r^2 \sin^2 \theta} \frac{\partial^2}{\partial \phi^2}.$$

(a) $u(r, \theta, \phi) = \left(Ar^2 + \dfrac{B}{r^3} \right) \dfrac{3 \cos^2 \theta - 1}{2}.$

$$\nabla_r^2 u = \frac{1}{r^2} \frac{\partial}{\partial r} \left(2Ar^3 - \frac{3B}{r^2} \right) \Theta = \left(6A + \frac{6B}{r^5} \right) \Theta = \frac{6u}{r^2},$$

$$\nabla_\theta^2 u = \frac{R}{r^2} \frac{1}{\sin \theta} \frac{\partial}{\partial \theta}(-3 \sin^2 \theta \cos \theta) = \frac{R}{r^2} \left(\frac{-6 \sin \theta \cos^2 \theta + 3 \sin^3 \theta}{\sin \theta} \right)$$

$$= \frac{R}{r^2}(-9 \cos^2 \theta + 3) = -\frac{6u}{r^2},$$

$$\nabla_\phi^2 u = 0.$$

Thus, although $\nabla_r^2 u$ and $\nabla_\theta^2 u$ are not individually zero, their sum is. From $\nabla_r^2 u = \ell(\ell + 1)u = 6u$, we deduce that $\ell = 2$ (or -3) and from $\nabla_\phi^2 u = 0$ that $m = 0$.

(b) $u(r, \theta, \phi) = \left(Ar + \dfrac{B}{r^2} \right) \sin \theta \, e^{i\phi}.$

$$\nabla_r^2 u = \frac{1}{r^2} \frac{\partial}{\partial r} \left(Ar^2 - \frac{2B}{r} \right) \Theta\Phi = \left(\frac{2A}{r} + \frac{2B}{r^4} \right) \Theta\Phi = \frac{2u}{r^2},$$

$$\nabla_\theta^2 u = \frac{R\Phi}{r^2} \frac{1}{\sin \theta} \frac{\partial}{\partial \theta}(\sin \theta \cos \theta) = \frac{R\Phi}{r^2} \left(\frac{-\sin^2 \theta + \cos^2 \theta}{\sin \theta} \right)$$

$$= -\frac{u}{r^2} + \frac{\cos^2 \theta}{\sin^2 \theta} \frac{u}{r^2},$$

$$\nabla_\phi^2 u = \frac{R\Theta}{r^2 \sin^2 \theta} \frac{\partial^2}{\partial \phi^2}(e^{i\phi}) = -\frac{u}{r^2 \sin^2 \theta}.$$

Hence,

$$\nabla^2 u = \frac{2u}{r^2} - \frac{u}{r^2} + \frac{\cos^2\theta}{\sin^2\theta}\frac{u}{r^2} - \frac{u}{r^2\sin^2\theta} = \frac{u}{r^2}\left(1 + \frac{\cos^2\theta - 1}{\sin^2\theta}\right) = 0.$$

Here each individual term is non-zero, but their sum *is* zero. Further, $\ell(\ell+1) = 2$ and so $\ell = 1$ (or -2), and from $\nabla_\phi^2 u = -u/(r^2 \sin\theta)$ it follows that $m^2 = 1$. In fact, from the normal definition of spherical harmonics, $m = +1$.

21.7 *If the stream function $\psi(r,\theta)$ for the flow of a very viscous fluid past a sphere is written as $\psi(r,\theta) = f(r)\sin^2\theta$, then $f(r)$ satisfies the equation*

$$f^{(4)} - \frac{4f''}{r^2} + \frac{8f'}{r^3} - \frac{8f}{r^4} = 0.$$

At the surface of the sphere $r = a$ the velocity field $\mathbf{u} = \mathbf{0}$, whilst far from the sphere $\psi \simeq (Ur^2\sin^2\theta)/2$.

Show that $f(r)$ can be expressed as a superposition of powers of r, and determine which powers give acceptable solutions. Hence show that

$$\psi(r,\theta) = \frac{U}{4}\left(2r^2 - 3ar + \frac{a^3}{r}\right)\sin^2\theta.$$

For solutions of

$$f^{(4)} - \frac{4f''}{r^2} + \frac{8f'}{r^3} - \frac{8f}{r^4} = 0$$

that are powers of r, i.e. have the form Ar^n, n must satisfy the quartic equation

$$n(n-1)(n-2)(n-3) - 4n(n-1) + 8n - 8 = 0,$$
$$(n-1)[n(n-2)(n-3) - 4n + 8] = 0,$$
$$(n-1)(n-2)[n(n-3) - 4] = 0,$$
$$(n-1)(n-2)(n-4)(n+1) = 0.$$

Thus the possible powers are 1, 2, 4 and -1.

Since $\psi \to \frac{1}{2}Ur^2\sin^2\theta$ as $r \to \infty$, the solution can contain no higher (positive) power of r than the second. Thus there is no $n = 4$ term and the solution has the form

$$\psi(r,\theta) = \left(\frac{Ur^2}{2} + Ar + \frac{B}{r}\right)\sin^2\theta.$$

On the surface of the sphere $r = a$ both velocity components, u_r and u_θ, are zero. These components are given in terms of the stream functions, as shown below;

note that u_r is found by differentiating with respect to θ and u_θ by differentiating with respect to r.

$$u_r = 0 \quad \Rightarrow \quad \frac{1}{a^2 \sin\theta} \frac{\partial \psi}{\partial \theta} = 0 \quad \Rightarrow \quad \frac{Ua^2}{2} + Aa + \frac{B}{a} = 0,$$

$$u_\theta = 0 \quad \Rightarrow \quad \frac{-1}{a \sin\theta} \frac{\partial \psi}{\partial r} = 0 \quad \Rightarrow \quad Ua + A - \frac{B}{a^2} = 0,$$

$$\Rightarrow \quad A = -\tfrac{3}{4}Ua \text{ and } B = \tfrac{1}{4}Ua^3.$$

The full solution is thus

$$\psi(r,\theta) = \frac{U}{4}\left(2r^2 - 3ar + \frac{a^3}{r}\right)\sin^2\theta.$$

21.9 *A circular disc of radius a is heated in such a way that its perimeter $\rho = a$ has a steady temperature distribution $A + B\cos^2\phi$, where ρ and ϕ are plane polar coordinates and A and B are constants. Find the temperature $T(\rho,\phi)$ everywhere in the region $\rho < a$.*

This is a steady state problem, for which the (heat) diffusion equation becomes the Laplace equation. The most general single-valued solution to the Lapace equation in plane polar coordinates is given by

$$T(\rho,\phi) = C\ln\rho + D + \sum_{n=1}^{\infty}(A_n \cos n\phi + B_n \sin n\phi)(C_n\rho^n + D_n\rho^{-n}).$$

The region $\rho < a$ contains the point $\rho = 0$; since $\ln\rho$ and all ρ^{-n} become infinite at that point, $C = D_n = 0$ for all n.

On $\rho = a$

$$T(a,\phi) = A + B\cos^2\phi = A + \tfrac{1}{2}B(\cos 2\phi + 1).$$

Equating the coefficients of $\cos n\phi$, including $n = 0$, gives $A + \tfrac{1}{2}B = D$, $A_2 C_2 a^2 = \tfrac{1}{2}B$ and $A_n C_n a^n = 0$ for all $n \neq 2$; further, all $B_n = 0$. The solution everywhere (not just on the perimeter) is therefore

$$T(\rho,\phi) = A + \frac{B}{2} + \frac{B\rho^2}{2a^2}\cos 2\phi.$$

It should be noted that 'equating coefficients' to determine unknown constants is justified by the fact that the sinusoidal functions in the sum are mutually orthogonal over the range $0 \leq \phi < 2\pi$.

21.11 *The free transverse vibrations of a thick rod satisfy the equation*

$$a^4 \frac{\partial^4 u}{\partial x^4} + \frac{\partial^2 u}{\partial t^2} = 0.$$

Obtain a solution in separated-variable form and, for a rod clamped at one end, $x = 0$, and free at the other, $x = L$, show that the angular frequency of vibration ω satisfies

$$\cosh\left(\frac{\omega^{1/2} L}{a}\right) = -\sec\left(\frac{\omega^{1/2} L}{a}\right).$$

[At a clamped end both u and $\partial u/\partial x$ vanish, whilst at a free end, where there is no bending moment, $\partial^2 u/\partial x^2$ and $\partial^3 u/\partial x^3$ are both zero.]

The general solution is written as the product $u(x,t) = X(x)T(t)$, which, on substitution, produces the separated equation

$$a^4 \frac{X^{(4)}}{X} = -\frac{T''}{T} = \omega^2.$$

Here the separation constant has been chosen so as to give oscillatory behaviour (in the time variable). The spatial equation then becomes

$$X^{(4)} - \mu^4 X = 0, \quad \text{where } \mu = \omega^{1/2}/a.$$

The required auxiliary equation is $\lambda^4 - \mu^4 = 0$, leading to the general solution

$$X(x) = A \sin \mu x + B \cos \mu x + C \sinh \mu x + D \cosh \mu x.$$

The constants A, B, C and D are to be determined by requiring $X(0) = X'(0) = 0$ and $X''(L) = X'''(L) = 0$.

At the clamped end,

$$X(0) = 0 \quad \Rightarrow \quad D = -B,$$
$$X' \quad = \quad \mu(A \cos \mu x - B \sin \mu x + C \cosh \mu x - B \sinh \mu x),$$
$$X'(0) = 0 \quad \Rightarrow \quad C = -A.$$

At the free end,

$$X'' = \mu^2(-A \sin \mu x - B \cos \mu x - A \sinh \mu x - B \cosh \mu x),$$
$$X''' = \mu^3(-A \cos \mu x + B \sin \mu x - A \cosh \mu x - B \sinh \mu x),$$
$$X''(L) = 0 \quad \Rightarrow \quad A(\sin \mu L + \sinh \mu L) + B(\cos \mu L + \cosh \mu L) = 0,$$
$$X'''(L) = 0 \quad \Rightarrow \quad A(-\cos \mu L - \cosh \mu L) + B(\sin \mu L - \sinh \mu L) = 0.$$

340

Cross-multiplying then gives

$$-\sin^2 \mu L + \sinh^2 \mu L = \cos^2 \mu L + 2\cos \mu L \cosh \mu L + \cosh^2 \mu L,$$

$$0 = 1 + 2\cos \mu L \cosh \mu L + 1,$$

$$-1 = \cos \mu L \cosh \mu L,$$

$$\cosh \left(\frac{\omega^{1/2} L}{a} \right) = -\sec \left(\frac{\omega^{1/2} L}{a} \right).$$

Because sinusoidal and hyperbolic functions can all be written in terms of exponential functions, this problem could also be approached by assuming solutions that are (exponential) functions of linear combinations of x and t (as in Chapter 20). However, in practice, eliminating the t-dependent terms leads to involved algebra.

21.13 *A string of length L, fixed at its two ends, is plucked at its mid-point by an amount A and then released. Prove that the subsequent displacement is given by*

$$u(x,t) = \sum_{n=0}^{\infty} \frac{8A}{\pi^2 (2n+1)^2} \sin \left[\frac{(2n+1)\pi x}{L} \right] \cos \left[\frac{(2n+1)\pi ct}{L} \right],$$

where, in the usual notation, $c^2 = T/\rho$.

Find the total kinetic energy of the string when it passes through its unplucked position, by calculating it in each mode (each n) and summing, using the result

$$\sum_{0}^{\infty} \frac{1}{(2n+1)^2} = \frac{\pi^2}{8}.$$

Confirm that the total energy is equal to the work done in plucking the string initially.

We start with the wave equation:

$$\frac{\partial^2 u}{\partial x^2} - \frac{1}{c^2} \frac{\partial^2 u}{\partial t^2} = 0$$

and assume a separated-variable solution $u(x,t) = X(x)S(t)$. This leads to

$$\frac{X''}{X} = \frac{1}{c^2} \frac{S''}{S} = -k^2.$$

The solution to the spatial equation is given by

$$X(x) = B \cos kx + C \sin kx.$$

Taking the string as anchored at $x = 0$ and $x = L$, we must have $B = 0$ and k constrained by $\sin kL = 0 \Rightarrow k = n\pi/L$ with n an integer.

The solution to the corresponding temporal equation is

$$S(t) = D \cos kct + E \sin kct.$$

Since there is no initial motion, i.e. $\dot{S}(0) = 0$, it follows that $E = 0$.
For any particular value of k, the constants C and D can be amalgamated. The general solution is given by a superposition of the allowed functions, i.e.

$$u(x, t) = \sum_{n=1}^{\infty} C_n \sin \frac{n\pi x}{L} \cos \frac{n\pi ct}{L}.$$

We now have to determine the C_n by making $u(x, 0)$ match the given initial configuration, which is

$$u(x, 0) = \begin{cases} \dfrac{2Ax}{L} & \text{for } 0 \le x \le \dfrac{L}{2}, \\ \dfrac{2A(L - x)}{L} & \dfrac{L}{2} < x \le L. \end{cases}$$

This is now a Fourier series calculation yielding

$$\frac{C_n L}{2} = \int_0^{L/2} \frac{2Ax}{L} \sin \frac{n\pi x}{L} \, dx + \int_{L/2}^{L} \frac{2A(L - x)}{L} \sin \frac{n\pi x}{L} \, dx$$

$$= \frac{2A}{L} J_1 + 2A J_2 - \frac{2A}{L} J_3,$$

with

$$J_1 = \left[-\frac{xL}{n\pi} \cos \frac{n\pi x}{L} \right]_0^{L/2} + \int_0^{L/2} \frac{L}{n\pi} \cos \frac{n\pi x}{L} \, dx$$

$$= -\frac{L^2}{2\pi n} \cos \frac{n\pi}{2} + \frac{L^2}{n^2\pi^2} \sin \frac{n\pi}{2},$$

$$J_2 = \int_{L/2}^{L} \sin \frac{n\pi x}{L} \, dx = -\frac{L}{n\pi} \left[\cos \frac{n\pi x}{L} \right]_{L/2}^{L}$$

$$= -\frac{L}{n\pi} \left[(-1)^n - \cos \frac{n\pi}{2} \right],$$

$$J_3 = \left[-\frac{xL}{n\pi} \cos \frac{n\pi x}{L} \right]_{L/2}^{L} + \int_{L/2}^{L} \frac{L}{n\pi} \cos \frac{n\pi x}{L} \, dx$$

$$= \frac{L^2}{2\pi n} \cos \frac{n\pi}{2} - \frac{L^2}{n\pi} (-1)^n - \frac{L^2}{n^2\pi^2} \sin \frac{n\pi}{2}.$$

Thus

$$J_1 - J_3 = -\frac{2L^2}{2\pi n} \cos \frac{n\pi}{2} + \frac{L^2}{n\pi} (-1)^n + \frac{2L^2}{n^2\pi^2} \sin \frac{n\pi}{2}$$

$$= -L J_2 + \frac{2L^2}{n^2\pi^2} \sin \frac{n\pi}{2},$$

and so it follows that

$$\frac{C_n L}{2} = \frac{2A}{L}(J_1 - J_3 + LJ_2) = 2A\frac{2L}{n^2\pi^2}\sin\frac{n\pi}{2}.$$

This is zero if n is even and $C_n = 8A(-1)^{(n-1)/2}/(n^2\pi^2)$ if n is odd. Write $n = 2m+1$, $m = 0, 1, 2, \ldots$, with $C_{2m+1} = \dfrac{8A(-1)^m}{(2m+1)^2\pi^2}$.

The final solution (in which m is replaced by n, to match the question) is thus

$$u(x, t) = \sum_{n=0}^{\infty} \frac{8A(-1)^n}{\pi^2(2n+1)^2}\sin\left[\frac{(2n+1)\pi x}{L}\right]\cos\left[\frac{(2n+1)\pi ct}{L}\right].$$

The velocity profile derived from this is given by

$$\dot{u}(x, t) = \sum_{n=0}^{\infty} \frac{8A(-1)^n}{\pi^2(2n+1)^2}\left(\frac{-(2n+1)\pi c}{L}\right)$$
$$\times \sin\left[\frac{(2n+1)\pi x}{L}\right]\sin\left[\frac{(2n+1)\pi ct}{L}\right],$$

giving the energy in the $(2n+1)$th mode (evaluated when the time-dependent sine function is maximal) as

$$E_{2n+1} = \int_0^L \tfrac{1}{2}\rho\dot{u}_n^2\, dx$$
$$= \int_0^L \frac{\rho}{2}\frac{(8A)^2 c^2}{L^2(2n+1)^2\pi^2}\sin^2\frac{(2n+1)\pi x}{L}$$
$$= \frac{32A^2\rho c^2}{L^2(2n+1)^2\pi^2}\frac{L}{2}.$$

Therefore

$$E = \sum_{n=0}^{\infty} E_{2n+1} = \frac{16A^2\rho c^2}{\pi^2 L}\sum_{n=0}^{\infty}\frac{1}{(2n+1)^2} = \frac{2A^2\rho c^2}{L}.$$

When the mid-point of the string has been displaced sideways by y ($\ll L$), the net (resolved) restoring force is $2T[y/(L/2)] = 4Ty/L$. Thus the total work done to produce a displacement of A is

$$W = \int_0^A \frac{4Ty}{L}\, dy = \frac{2TA^2}{L} = \frac{2\rho c^2 A^2}{L},$$

i.e. the same as the total energy of the subsequent motion.

21.15 *Prove that the potential for $\rho < a$ associated with a vertical split cylinder of radius a, the two halves of which ($\cos\phi > 0$ and $\cos\phi < 0$) are maintained at equal and opposite potentials $\pm V$, is given by*

$$u(\rho, \phi) = \frac{4V}{\pi} \sum_{n=0}^{\infty} \frac{(-1)^n}{2n+1} \left(\frac{\rho}{a}\right)^{2n+1} \cos(2n+1)\phi.$$

The most general solution of the Laplace equation in cylindrical polar coordinates that is independent of z is

$$T(\rho, \phi) = C \ln \rho + D + \sum_{n=1}^{\infty} (A_n \cos n\phi + B_n \sin n\phi)(C_n \rho^n + D_n \rho^{-n}).$$

The required potential must be single-valued and finite in the space inside the cylinder (which includes $\rho = 0$), and on the cylinder it must take the boundary values $u = V$ for $\cos\phi > 0$ and $u = -V$ for $\cos\phi < 0$, i.e the boundary-value function is a square-wave function with average value zero. Although the function is antisymmetric in $\cos\phi$, it is symmetric in ϕ and so the solution will contain only cosine terms (and no sine terms).

These considerations already determine that $C = D = B_n = D_n = 0$, and so have reduced the solution to the form

$$u(\rho, \phi) = \sum_{n=1}^{\infty} A_n \rho^n \cos n\phi.$$

On $\rho = a$ this must match the stated boundary conditions, and so we are faced with a Fourier cosine series calculation. Multiplying through by $\cos m\phi$ and integrating yields

$$A_m a^m \frac{1}{2} 2\pi = 2 \int_0^{\pi/2} V \cos m\phi \, d\phi + 2 \int_{\pi/2}^{\pi} (-V) \cos m\phi \, d\phi$$

$$= 2V \left[\frac{\sin m\phi}{m}\right]_0^{\pi/2} - 2V \left[\frac{\sin m\phi}{m}\right]_{\pi/2}^{\pi}$$

$$= \frac{2V}{m} \left(\sin \frac{m\pi}{2} + \sin \frac{m\pi}{2}\right)$$

$$= (-1)^{(m-1)/2} \frac{4V}{m} \text{ for } m \text{ odd}, = 0 \text{ for } m \text{ even}.$$

Writing $m = 2n + 1$ gives the solution as

$$u(\rho, \phi) = \frac{4V}{\pi} \sum_{n=0}^{\infty} \frac{(-1)^n}{2n+1} \left(\frac{\rho}{a}\right)^{2n+1} \cos(2n+1)\phi.$$

21.17 *Two identical copper bars are each of length a. Initially, one is at $0\,°C$ and the other at $100\,°C$; they are then joined together end to end and thermally isolated. Obtain in the form of a Fourier series an expression $u(x,t)$ for the temperature at any point a distance x from the join at a later time t. Bear in mind the heat flow conditions at the free ends of the bars.*

Taking $a = 0.5\,m$ estimate the time it takes for one of the free ends to attain a temperature of $55\,°C$. The thermal conductivity of copper is $3.8 \times 10^2\,\mathrm{J\,m^{-1}\,K^{-1}\,s^{-1}}$, and its specific heat capacity is $3.4 \times 10^6\,\mathrm{J\,m^{-3}\,K^{-1}}$.

The equation governing the heat flow is

$$k\frac{\partial^2 u}{\partial x^2} = s\frac{\partial u}{\partial t},$$

which is the diffusion equation with diffusion constant $\kappa = k/s = 3.8 \times 10^2/3.4 \times 10^6 = 1.12 \times 10^{-4}\,\mathrm{m^2\,s^{-1}}$.

Making the usual separation of variables substitution shows that the time variation is of the form $T(t) = T(0)e^{-\kappa\lambda^2 t}$ when the spatial solution is a sinusoidal function of λx. The final common temperature is $50\,°C$ and we make this explicit by writing the general solution as

$$u(x,t) = 50 + \sum_\lambda (A_\lambda \sin \lambda x + B_\lambda \cos \lambda x)e^{-\kappa\lambda^2 t}.$$

This term having been taken out, the summation must be antisymmetric about $x = 0$ and therefore contain no cosine terms, i.e. $B_\lambda = 0$.

The boundary condition is that there is no heat flow at $x = \pm a$; this means that $\partial u/\partial x = 0$ at these points and requires

$$\lambda A_\lambda \cos \lambda x|_{x=\pm a} = 0 \quad \Rightarrow \quad \lambda a = (n + \tfrac{1}{2})\pi \quad \Rightarrow \quad \lambda = \frac{(2n+1)\pi}{2a},$$

where n is an integer. This corresponds to a fundamental Fourier period of $4a$. The solution thus takes the form

$$u(x,t) = 50 + \sum_{n=0}^{\infty} A_n \sin \frac{(2n+1)\pi x}{2a} \exp\left(-\frac{(2n+1)^2\pi^2\kappa t}{4a^2}\right).$$

At $t = 0$, the sum must take the values $+50$ for $0 < x < 2a$ and -50 for $-2a < x < 0$. This is (yet) another square-wave function — one that is antisymmetric about $x = 0$ and has amplitude 50. The calculation will not be repeated here but gives $A_n = 200/[(2n+1)\pi]$, making the complete solution

$$u(x,t) = 50 + \frac{200}{\pi}\sum_{n=0}^{\infty}\frac{1}{2n+1}\sin\frac{(2n+1)\pi x}{2a}\exp\left(-\frac{(2n+1)^2\pi^2\kappa t}{4a^2}\right).$$

For a free end, where $x = a$ and $\sin[(2n + 1)\pi x/2a] = (-1)^n$, to attain $55\,°C$ needs

$$\sum_{n=0}^{\infty} \frac{(-1)^n}{2n + 1} \exp\left(-\frac{(2n + 1)^2\,\pi^2\,1.12 \times 10^{-4}}{4 \times 0.25}t\right) = \frac{5\pi}{200} = 0.0785.$$

In principle this is an insoluble equation but, because the RHS $\ll 1$, the $n = 0$ term alone will give a good approximation to t:

$$\exp(-1.105 \times 10^{-3}t) \approx 0.0785 \quad \Rightarrow \quad t \approx 2300 \text{ s.}$$

21.19 *For an infinite metal bar that has an initial ($t = 0$) temperature distribution $f(x)$ along its length, the temperature distribution at a general time t can be shown to be given by*

$$u(x, t) = \frac{1}{\sqrt{4\pi\kappa t}} \int_{-\infty}^{\infty} \exp\left[-\frac{(x - \xi)^2}{4\kappa t}\right] f(\xi)\,d\xi.$$

Find an explicit expression for $u(x, t)$ given that $f(x) = \exp(-x^2/a^2)$.

The given initial distribution is $f(\xi) = \exp(-\xi^2/a^2)$ and so

$$u(x, t) = \frac{1}{\sqrt{4\pi\kappa t}} \int_{-\infty}^{\infty} \exp\left[-\frac{(x - \xi)^2}{4\kappa t}\right] \exp\left(-\frac{\xi^2}{a^2}\right) d\xi.$$

Now consider the exponent in the integrand, writing $1 + \dfrac{4\kappa t}{a^2}$ as τ^2 for compactness:

$$\text{exponent} = -\frac{\xi^2\tau^2 - 2\xi x + x^2}{4\kappa t}$$

$$= -\frac{(\xi\tau - x\tau^{-1})^2 - x^2\tau^{-2} + x^2}{4\kappa t}$$

$$\equiv -\eta^2 + \frac{x^2\tau^{-2} - x^2}{4\kappa t}, \quad \text{defining } \eta,$$

$$\text{with } d\eta = \frac{\tau\,d\xi}{\sqrt{4\kappa t}}.$$

With a change of variable from ξ to η, the integral becomes

$$u(x, t) = \frac{1}{\sqrt{4\pi\kappa t}} \exp\left(\frac{x^2\tau^{-2} - x^2}{4\kappa t}\right) \int_{-\infty}^{\infty} \exp(-\eta^2) \frac{\sqrt{4\kappa t}}{\tau}\,d\eta$$

$$= \frac{1}{\sqrt{\pi}} \frac{1}{\tau} \exp\left(x^2 \frac{1 - \tau^2}{4\kappa t\,\tau^2}\right) \sqrt{\pi}$$

$$= \frac{a}{\sqrt{a^2 + 4\kappa t}} \exp\left(-\frac{x^2}{a^2 + 4\kappa t}\right).$$

In words, although it retains a Gaussian shape, the initial distribution spreads symmetrically about the origin, its variance increasing linearly with time ($a^2 \rightarrow a^2 + 4\kappa t$). As is typical with diffusion processes, for large enough times the width varies as $\sqrt{t}$.

21.21 *In the region $-\infty < x, y < \infty$ and $-t \leq z \leq t$, a charge-density wave $\rho(\mathbf{r}) = A \cos qx$, in the x-direction, is represented by*

$$\rho(\mathbf{r}) = \frac{e^{iqx}}{\sqrt{2\pi}} \int_{-\infty}^{\infty} \tilde{\rho}(\alpha) e^{i\alpha z} \, d\alpha.$$

The resulting potential is represented by

$$V(\mathbf{r}) = \frac{e^{iqx}}{\sqrt{2\pi}} \int_{-\infty}^{\infty} \tilde{V}(\alpha) e^{i\alpha z} \, d\alpha.$$

Determine the relationship between $\tilde{V}(\alpha)$ and $\tilde{\rho}(\alpha)$, and hence show that the potential at the point $(0,0,0)$ is given by

$$\frac{A}{\pi \epsilon_0} \int_{-\infty}^{\infty} \frac{\sin kt}{k(k^2 + q^2)} \, dk.$$

Poisson's equation,

$$\nabla^2 V(\mathbf{r}) = -\frac{\rho(\mathbf{r})}{\epsilon_0},$$

provides the link between a charge density and the potential it produces.

Taking $V(\mathbf{r})$ in the form of its Fourier representation gives $\nabla^2 V$ as

$$\frac{\partial^2 V(\mathbf{r})}{\partial x^2} + \frac{\partial^2 V(\mathbf{r})}{\partial y^2} + \frac{\partial^2 V(\mathbf{r})}{\partial z^2} = \frac{e^{iqx}}{\sqrt{2\pi}} \int_{-\infty}^{\infty} (-q^2 - \alpha^2) \tilde{V}(\alpha) e^{i\alpha z} \, d\alpha,$$

with the $-q^2$ arising from the x-differentiation and the $-\alpha^2$ from the z-differentiation; the $\partial^2 V/\partial y^2$ term contributes nothing.

Comparing this with the integral expression for $-\rho(\mathbf{r})/\epsilon_0$ shows that

$$-\tilde{\rho}(\alpha) = \epsilon_0(-q^2 - \alpha^2)\tilde{V}(\alpha).$$

With the charge-density wave confined in the z-direction to $-t \leq z \leq t$, the expression for $\rho(\mathbf{r})$ in Cartesian coordinates is (in terms of Heaviside functions)

$$\rho(\mathbf{r}) = A e^{iqx}[H(z+t) - H(z-t)].$$

347

The Fourier transform $\tilde{\rho}(\alpha)$ is therefore given by

$$
\begin{aligned}
\tilde{\rho}(\alpha) &= \frac{1}{\sqrt{2\pi}} \int_{-\infty}^{\infty} A[H(z+t) - H(z-t)] e^{-i\alpha z} \, dz \\
&= \frac{A}{\sqrt{2\pi}} \int_{-t}^{t} e^{-i\alpha z} \, dz \\
&= \frac{A}{\sqrt{2\pi}} \frac{e^{-i\alpha t} - e^{i\alpha t}}{-i\alpha} \\
&= \frac{A}{\sqrt{2\pi}} \frac{2\sin \alpha t}{\alpha}.
\end{aligned}
$$

Now,
$$
\begin{aligned}
V(x,0,z) &= \frac{e^{iqx}}{\sqrt{2\pi}} \int_{-\infty}^{\infty} \frac{\tilde{\rho}(\alpha)}{\epsilon_0(q^2+\alpha^2)} e^{i\alpha z} \, d\alpha \\
&= \frac{e^{iqx}}{\sqrt{2\pi}} \int_{-\infty}^{\infty} \frac{e^{i\alpha z}}{\epsilon_0(q^2+\alpha^2)} \frac{A}{\sqrt{2\pi}} \frac{2\sin \alpha t}{\alpha} \, d\alpha, \\
\Rightarrow \quad V(0,0,0) &= \frac{A}{\pi\epsilon_0} \int_{-\infty}^{\infty} \frac{\sin \alpha t}{\alpha(\alpha^2 + q^2)} \, d\alpha,
\end{aligned}
$$

as stated in the question.

21.23 *Find the Green's function $G(\mathbf{r}, \mathbf{r}_0)$ in the half-space $z > 0$ for the solution of $\nabla^2 \Phi = 0$ with Φ specified in cylindrical polar coordinates (ρ, ϕ, z) on the plane $z = 0$ by*

$$
\Phi(\rho, \phi, z) = \begin{cases} 1 & \text{for } \rho \leq 1, \\ 1/\rho & \text{for } \rho > 1. \end{cases}
$$

Determine the variation of $\Phi(0,0,z)$ along the z-axis.

For the half-space $z > 0$ the bounding surface consists of the plane $z = 0$ and the (hemi-spherical) surface at infinity; the Green's function must take zero value on these surfaces. In order to ensure this when a unit point source is introduced at $\mathbf{r} = \mathbf{y}$, we must place a compensating negative unit source at $\mathbf{y}$'s reflection point in the plane. If, in cylindrical polar coordinates, $\mathbf{y} = (\rho, \phi, z_0)$, then the image charge has to be at $\mathbf{y}' = (\rho, \phi, -z_0)$. The resulting Green's function $G(\mathbf{x}, \mathbf{y})$ is given by

$$
G(\mathbf{x}, \mathbf{y}) = -\frac{1}{4\pi|\mathbf{x} - \mathbf{y}|} + \frac{1}{4\pi|\mathbf{x} - \mathbf{y}'|}.
$$

The solution to the problem with a given potential distribution $f(\rho, \phi)$ on the $z = 0$ part of the bounding surface S is given by

$$
\Phi(\mathbf{y}) = \int_S f(\rho, \phi) \left(-\frac{\partial G}{\partial z} \right) \rho \, d\phi \, d\rho,
$$

the minus sign arising because the outward normal to the region is in the negative z-direction. Calculating these functions explicitly gives

$$G(\mathbf{x}, \mathbf{y}) = -\frac{1}{4\pi[\rho^2 + (z - z_0)^2]^{1/2}} + \frac{1}{4\pi[\rho^2 + (z + z_0)^2]^{1/2}},$$

$$\frac{\partial G}{\partial z} = \frac{z - z_0}{4\pi[\rho^2 + (z - z_0)^2]^{3/2}} - \frac{(z + z_0)}{4\pi[\rho^2 + (z + z_0)^2]^{3/2}},$$

$$-\frac{\partial G}{\partial z}\bigg|_{z=0} = -\frac{-2z_0}{4\pi[\rho^2 + z_0^2]^{3/2}}.$$

Substituting the various factors into the general integral gives

$$\Phi(0, 0, z_0) = \int_0^\infty f(\rho) \frac{2z_0}{4\pi[\rho^2 + z_0^2]^{3/2}} 2\pi \rho \, d\rho$$

$$= \int_0^1 \frac{z_0 \rho}{(\rho^2 + z_0^2)^{3/2}} \, d\rho + \int_1^\infty \frac{z_0}{(\rho^2 + z_0^2)^{3/2}} \, d\rho$$

$$= -z_0 \left[(\rho^2 + z_0^2)^{-1/2} \right]_0^1 + \int_\theta^{\pi/2} \frac{z_0^2 \sec^2 u}{z_0^3 \sec^3 u} \, du,$$

where, in the second integral, we have set $\rho = z_0 \tan u$ with $d\rho = z_0 \sec^2 u \, du$ and $\theta = \tan^{-1}(1/z_0)$. The integral can now be obtained in closed form as

$$\Phi(0, 0, z_0) = -\frac{z_0}{(1 + z_0^2)^{1/2}} + 1 + \frac{1}{z_0} \left[\sin u \right]_\theta^{\pi/2}$$

$$= 1 - \frac{z_0}{(1 + z_0^2)^{1/2}} + \frac{1}{z_0} - \frac{1}{z_0(1 + z_0^2)^{1/2}}.$$

Thus the variation of Φ along the z-axis is given by

$$\Phi(0, 0, z) = \frac{z(1 + z^2)^{1/2} - z^2 + (1 + z^2)^{1/2} - 1}{z(1 + z^2)^{1/2}}.$$

21.25 *Find, in the form of an infinite series, the Green's function of the ∇^2 operator for the Dirichlet problem in the region $-\infty < x < \infty$, $-\infty < y < \infty$, $-c \leq z \leq c$.*

The fundamental solution in three dimensions of $\nabla^2 \psi = \delta(\mathbf{r})$ is $\psi(\mathbf{r}) = -1/(4\pi r)$.

For the given problem, $G(\mathbf{r}, \mathbf{r}_0)$ has to take the value zero on $z = \pm c$ and $\to 0$ for $|x| \to \infty$ and $|y| \to \infty$. Image charges have to be added in the regions $z > c$ and $z < -c$ to bring this about after a charge q has been placed at $\mathbf{r}_0 = (x_0, y_0, z_0)$ with $-c < z_0 < c$. Clearly all images will be on the line $x = x_0$, $y = y_0$.

Each image placed at $z = \xi$ in the region $z > c$ will require a further image of the same strength but opposite sign at $z = -c - \xi$ (in the region $z < -c$) so as to maintain the plane $z = -c$ as an equipotential. Likewise, each image placed at $z = -\chi$ in the region $z < -c$ will require a further image of the same strength but opposite sign at $z = c + \chi$ (in the region $z > c$) so as to maintain the plane $z = c$ as an equipotential. Thus successive image charges appear as follows:

$$
\begin{array}{ccc}
-q & 2c - z_0 & -2c - z_0 \\
+q & -3c + z_0 & 3c + z_0 \\
-q & 4c - z_0 & -4c - z_0 \\
+q & \text{etc.} & \text{etc.}
\end{array}
$$

The terms in the Green's function that are additional to the fundamental solution,

$$
-\frac{1}{4\pi}[(x - x_0)^2 + (y - y_0)^2 + (z - z_0)^2]^{-1/2},
$$

are therefore

$$
-\frac{(-1)}{4\pi} \sum_{n=2}^{\infty} \left\{ \frac{(-1)^n}{[(x - x_0)^2 + (y - y_0)^2 + (z + (-1)^n z_0 - nc)^2]^{1/2}} \right.
$$

$$
\left. + \frac{(-1)^n}{[(x - x_0)^2 + (y - y_0)^2 + (z + (-1)^n z_0 + nc)^2]^{1/2}} \right\}.
$$

21.27 *Determine the Green's function for the Klein–Gordon equation in a half-space as follows.*

(a) *By applying the divergence theorem to the volume integral*

$$
\int_V \left[\phi(\nabla^2 - m^2)\psi - \psi(\nabla^2 - m^2)\phi \right] \, dV,
$$

obtain a Green's function expression, as the sum of a volume integral and a surface integral, for the function $\phi(\mathbf{r}')$ that satisfies

$$
\nabla^2 \phi - m^2 \phi = \rho
$$

in V and takes the specified form $\phi = f$ on S, the boundary of V. The Green's function, $G(\mathbf{r}, \mathbf{r}')$, to be used satisfies

$$
\nabla^2 G - m^2 G = \delta(\mathbf{r} - \mathbf{r}')
$$

and vanishes when $\mathbf{r}$ is on S.

(b) *When V is all space, $G(\mathbf{r}, \mathbf{r}')$ can be written as $G(t) = g(t)/t$, where $t = |\mathbf{r} - \mathbf{r}'|$ and $g(t)$ is bounded as $t \to \infty$. Find the form of $G(t)$.*

(c) *Find $\phi(\mathbf{r})$ in the half-space $x > 0$ if $\rho(\mathbf{r}) = \delta(\mathbf{r} - \mathbf{r}_1)$ and $\phi = 0$ both on $x = 0$ and as $r \to \infty$.*

(a) For general ϕ and ψ we have

$$\int_V \left[\phi(\nabla^2 - m^2)\psi - \psi(\nabla^2 - m^2)\phi\right] dV = \int_V \left[\phi\nabla^2\psi - \psi\nabla^2\phi\right] dV$$

$$= \int_V \nabla \cdot (\phi\nabla\psi - \psi\nabla\phi) dV$$

$$= \int_S (\phi\nabla\psi - \psi\nabla\phi) \cdot \mathbf{n}\, dS.$$

Now take ϕ as ϕ, with $\nabla^2\phi - m^2\phi = \rho$ and $\phi = f$ on the surface S, and ψ as $G(\mathbf{r}, \mathbf{r}')$ with $\nabla^2 G - m^2 G = \delta(\mathbf{r} - \mathbf{r}')$ and $G(\mathbf{r}, \mathbf{r}') = 0$ on S:

$$\int_V \left[\phi(\mathbf{r})\delta(\mathbf{r} - \mathbf{r}') - G(\mathbf{r}, \mathbf{r}')\rho(\mathbf{r})\right] dV = \int_S \left[f(\mathbf{r})\nabla G(\mathbf{r}, \mathbf{r}') - \mathbf{0}\right] \cdot \mathbf{n}\, dS,$$

which, on rearrangement, gives

$$\phi(\mathbf{r}') = \int_V G(\mathbf{r}, \mathbf{r}')\rho(\mathbf{r})\, dV + \int_S f(\mathbf{r})\nabla G(\mathbf{r}, \mathbf{r}') \cdot \mathbf{n}\, dS.$$

(b) In the following calculation we start by formally integrating the defining Green's equation,

$$\nabla^2 G - m^2 G = \delta(\mathbf{r} - \mathbf{r}'),$$

over a sphere of radius t centred on $\mathbf{r}'$. Having replaced the volume integral of $\nabla^2 G$ with the corresponding surface integral given by the divergence theorem, we move the origin to $\mathbf{r}'$, denote $|\mathbf{r} - \mathbf{r}'|$ by t' and integrate both sides of the equation from $t' = 0$ to $t' = t$:

$$\int_V \nabla^2 G\, dV - \int_V m^2 G\, dV = \int_V \delta(\mathbf{r} - \mathbf{r}')\, dV,$$

$$\int_S \nabla G \cdot \mathbf{n}\, dS - m^2 \int_V G\, dV = 1,$$

$$4\pi t^2 \frac{dG}{dt} - m^2 \int_0^t G(t')4\pi t'^2\, dt' = 1, \qquad (*)$$

$$4\pi t^2 G'' + 8\pi t G' - 4\pi m^2 t^2 G = 0, \text{ from differentiating w.r.t. } t,$$

$$tG'' + 2G' - m^2 tG = 0.$$

With $G(t) = g(t)/t$,

$$G' = -\frac{g}{t^2} + \frac{g'}{t} \quad \text{and} \quad G'' = \frac{2g}{t^3} - \frac{2g'}{t^2} + \frac{g''}{t},$$

and the equation becomes

$$0 = \frac{2g}{t^2} - \frac{2g'}{t} + g'' - \frac{2g}{t^2} + \frac{2g'}{t} - m^2 g,$$

$$0 = g'' - m^2 g,$$

$$\Rightarrow \quad g(t) = Ae^{-mt}, \text{ since } g \text{ is bounded as } t \to \infty.$$

The value of A is determined by resubstituting into $(*)$, which then reads

$$4\pi t^2 \left(-\frac{Ae^{-mt}}{t^2} - \frac{mAe^{-mt}}{t} \right) - m^2 \int_0^t \frac{Ae^{-mt'}}{t'} 4\pi t'^2 \, dt' = 1,$$

$$-4\pi Ae^{-mt}(1+mt) - 4\pi Am^2 \left(-\frac{te^{-mt}}{m} + \frac{1-e^{-mt}}{m^2} \right) = 1,$$

$$-4\pi A = 1,$$

making the solution

$$G(\mathbf{r}, \mathbf{r}') = -\frac{e^{-mt}}{4\pi t}, \text{ where } t = |\mathbf{r} - \mathbf{r}'|.$$

(c) For the situation in which $\rho(\mathbf{r}) = \delta(\mathbf{r} - \mathbf{r}_1)$, i.e. a unit positive charge at $\mathbf{r}_1 = (x_1, y_1, z_1)$, and $\phi = 0$ on the plane $x = 0$, we must have a unit negative image charge at $\mathbf{r}_2 = (-x_1, y_1, z_1)$. The solution in the region $x > 0$ is then

$$\phi(\mathbf{r}) = -\frac{1}{4\pi} \left(\frac{e^{-m|\mathbf{r}-\mathbf{r}_1|}}{|\mathbf{r} - \mathbf{r}_1|} - \frac{e^{-m|\mathbf{r}-\mathbf{r}_2|}}{|\mathbf{r} - \mathbf{r}_2|} \right).$$

22

Calculus of variations

22.1 *A surface of revolution, whose equation in cylindrical polar coordinates is* $\rho = \rho(z)$, *is bounded by the circles* $\rho = a$, $z = \pm c$ ($a > c$). *Show that the function that makes the surface integral* $I = \int \rho^{-1/2}\, dS$ *stationary with respect to small variations is given by* $\rho(z) = k + z^2/(4k)$, *where* $k = [a \pm (a^2 - c^2)^{1/2}]/2$.

The surface element lying between z and $z + dz$ is given by

$$dS = 2\pi\rho\, [(d\rho)^2 + (dz)^2\,]^{1/2} = 2\pi\rho\,(1 + \rho'^2)^{1/2}\, dz$$

and the integral to be made stationary is

$$I = \int \rho^{-1/2}\, dS = 2\pi \int_{-c}^{c} \rho^{-1/2}\rho\,(1 + \rho'^2)^{1/2}\, dz.$$

The integrand $F(\rho', \rho, z)$ does not in fact contain z explicitly, and so a first integral of the E–L equation, symbolically given by $F - \rho'\partial F/\partial \rho' = k$, is

$$\rho^{1/2}(1 + \rho'^2)^{1/2} - \rho' \left[\frac{\rho^{1/2}\rho'}{(1 + \rho'^2)^{1/2}} \right] = A,$$

$$\frac{\rho^{1/2}}{(1 + \rho'^2)^{1/2}} = A.$$

On rearrangement and subsequent integration this gives

$$\frac{d\rho}{dz} = \left(\frac{\rho - A^2}{A^2} \right)^{1/2},$$

$$\int \frac{d\rho}{\sqrt{\rho - A^2}} = \int \frac{dz}{A},$$

$$2\sqrt{\rho - A^2} = \frac{z}{A} + C.$$

353

Now, $\rho(\pm c) = a$ implies both that $C = 0$ and that $a - A^2 = \dfrac{c^2}{4A^2}$. Thus, writing A^2 as k,

$$4k^2 - 4ka + c^2 = 0 \quad \Rightarrow \quad k = \tfrac{1}{2}[a \pm (a^2 - c^2)^{1/2}].$$

The two stationary functions are therefore

$$\rho = \frac{z^2}{4k} + k,$$

with k as given above. A simple sketch shows that the positive sign in k corresponds to a smaller value of the integral.

22.3 *The refractive index n of a medium is a function only of the distance r from a fixed point O. Prove that the equation of a light ray, assumed to lie in a plane through O, travelling in the medium satisfies (in plane polar coordinates)*

$$\frac{1}{r^2}\left(\frac{dr}{d\phi}\right)^2 = \frac{r^2\,n^2(r)}{a^2\,n^2(a)} - 1,$$

where a is the distance of the ray from O at the point at which $dr/d\phi = 0$.

If $n = [1 + (\alpha^2/r^2)]^{1/2}$ and the ray starts and ends far from O, find its deviation (the angle through which the ray is turned), if its minimum distance from O is a.

An element of path length is $ds = [(dr)^2 + (r\,d\phi)^2]^{1/2}$ and the time taken for the light to traverse it is $n(r)\,ds/c$, where c is the speed of light *in vacuo*. Fermat's principle then implies that the light follows the curve that minimises

$$T = \int \frac{n(r)\,ds}{c} = \int \frac{n(r'^2 + r^2)^{1/2}}{c}\,d\phi,$$

where $r' = dr/d\phi$. Since the integrand does not contain ϕ explicitly, the E–L equation integrates to (see exercise 22.1)

$$n(r'^2 + r^2)^{1/2} - r'\frac{nr'}{(r'^2 + r^2)^{1/2}} = A,$$

$$\frac{nr^2}{(r'^2 + r^2)^{1/2}} = A.$$

Since $r' = 0$ when $r = a$, $A = n(a)a^2/a$, and the equation is as follows:

$$a^2 n^2(a)(r'^2 + r^2) = n^2(r)r^4,$$

$$r'^2 = \frac{n^2(r)r^4}{n^2(a)a^2} - r^2,$$

$$\Rightarrow \quad \frac{1}{r^2}\left(\frac{dr}{d\phi}\right)^2 = \frac{n^2(r)r^2}{n^2(a)a^2} - 1.$$

If $n(r) = [1 + (\alpha/r)^2]^{1/2}$, the minimising curve satisfies

$$\left(\frac{dr}{d\phi}\right)^2 = \frac{r^2(r^2 + \alpha^2)}{a^2 + \alpha^2} - r^2$$

$$= \frac{r^2(r^2 - a^2)}{a^2 + \alpha^2},$$

$$\Rightarrow \quad \frac{d\phi}{(a^2 + \alpha^2)^{1/2}} = \pm \frac{dr}{r\sqrt{r^2 - a^2}}.$$

By symmetry,

$$\frac{\Delta\phi}{(a^2 + \alpha^2)^{1/2}} \equiv \frac{\phi_{\text{final}} - \phi_{\text{initial}}}{(a^2 + \alpha^2)^{1/2}}$$

$$= 2 \int_a^\infty \frac{dr}{r\sqrt{r^2 - a^2}}, \quad \text{set } r = a\cosh\psi,$$

$$= 2 \int_0^\infty \frac{a\sinh\psi}{a^2 \cosh\psi \sinh\psi} \, d\psi$$

$$= \frac{2}{a} \int_0^\infty \text{sech } \psi \, d\psi, \quad \text{set } e^\psi = z,$$

$$= \frac{2}{a} \int_1^\infty \frac{z^{-1} \, dz}{\frac{1}{2}(z + z^{-1})}$$

$$= \frac{2}{a} \int_1^\infty \frac{2 \, dz}{z^2 + 1}$$

$$= \frac{4}{a} \left[\tan^{-1} z \right]_1^\infty$$

$$= \frac{4}{a} \left(\frac{\pi}{2} - \frac{\pi}{4}\right) = \frac{\pi}{a}.$$

If the refractive index were everywhere unity ($\alpha = 0$), $\Delta\phi$ would be π (no deviation). Thus the deviation is given by

$$\frac{\pi}{a}(a^2 + \alpha^2)^{1/2} - \pi.$$

22.5 *Prove the following results about general systems.*

(a) *For a system described in terms of coordinates q_i and t, show that if t does not appear explicitly in the expressions for x, y and z ($x = x(q_i, t)$, etc.) then the kinetic energy T is a homogeneous quadratic function of the $\dot{q}_i$ (it may also involve the q_i). Deduce that $\sum_i \dot{q}_i(\partial T/\partial \dot{q}_i) = 2T$.*

(b) *Assuming that the forces acting on the system are derivable from a potential V, show, by expressing dT/dt in terms of q_i and $\dot{q}_i$, that $d(T + V)/dt = 0$.*

To save space we will use the summation convention for summing over the index of the q_i.

(a) The space variables x, y and z are not explicit functions of t and the kinetic energy, T, is given by

$$T = \tfrac{1}{2}(\alpha_x \dot{x}^2 + \alpha_y \dot{y}^2 + \alpha_z \dot{z}^2)$$

$$= \frac{1}{2}\left[\alpha_x \left(\frac{\partial x}{\partial q_i} \dot{q}_i \right)^2 + \alpha_y \left(\frac{\partial y}{\partial q_j} \dot{q}_j \right)^2 + \alpha_z \left(\frac{\partial z}{\partial q_k} \dot{q}_k \right)^2 \right]$$

$$= A_{mn} \dot{q}_m \dot{q}_n,$$

with

$$A_{mn} = \frac{1}{2}\left(\alpha_x \frac{\partial x}{\partial q_m} \frac{\partial x}{\partial q_n} + \alpha_y \frac{\partial y}{\partial q_m} \frac{\partial y}{\partial q_n} + \alpha_z \frac{\partial z}{\partial q_m} \frac{\partial z}{\partial q_n} \right) = A_{nm}.$$

Hence T is a homogeneous quadratic function of the $\dot{q}_i$ (though the A_{mn} may involve the q_i). Further,

$$\frac{\partial T}{\partial \dot{q}_i} = A_{in} \dot{q}_n + A_{mi} \dot{q}_m = 2 A_{mi} \dot{q}_m$$

$$\text{and} \quad \dot{q}_i \frac{\partial T}{\partial \dot{q}_i} = 2 \dot{q}_i A_{mi} \dot{q}_m = 2T.$$

(b) The Lagrangian is $L = T - V$, with $T = T(q_i, \dot{q}_i)$ and $V = V(q_i)$. Thus

$$\frac{dT}{dt} = \frac{\partial T}{\partial q_i} \dot{q}_i + \frac{dT}{d\dot{q}_i} \ddot{q}_i \quad \text{and} \quad \frac{dV}{dt} = \frac{\partial V}{\partial q_i} \dot{q}_i. \qquad (*)$$

Hamilton's principle requires that

$$\frac{d}{dt}\left(\frac{\partial L}{\partial \dot{q}_i} \right) = \frac{\partial L}{\partial q_i},$$

$$\Rightarrow \quad \frac{d}{dt}\left(\frac{\partial T}{\partial \dot{q}_i} \right) = \frac{\partial T}{\partial q_i} - \frac{\partial V}{\partial q_i}. \qquad (**)$$

But, from part (a),

$$2T = \dot{q}_i \frac{\partial T}{\partial \dot{q}_i},$$

$$\frac{d}{dt}(2T) = \ddot{q}_i \frac{\partial T}{\partial \dot{q}_i} + \dot{q}_i \frac{d}{dt}\left(\frac{\partial T}{\partial \dot{q}_i} \right)$$

$$= \ddot{q}_i \frac{\partial T}{\partial \dot{q}_i} + \dot{q}_i \frac{\partial T}{\partial q_i} - \dot{q}_i \frac{\partial V}{\partial q_i}, \quad \text{using } (**),$$

$$= \frac{dT}{dt} - \frac{dV}{dt}, \qquad \text{using } (*).$$

This can be rearranged as

$$\frac{d}{dt}(T + V) = 0.$$

22.7 *In cylindrical polar coordinates, the curve $(\rho(\theta), \theta, \alpha\rho(\theta))$ lies on the surface of the cone $z = \alpha\rho$. Show that geodesics (curves of minimum length joining two points) on the cone satisfy*

$$\rho^4 = c^2[\beta^2\rho'^2 + \rho^2],$$

where c is an arbitrary constant, but β has to have a particular value. Determine the form of $\rho(\theta)$ and hence find the equation of the shortest path on the cone between the points $(R, -\theta_0, \alpha R)$ and $(R, \theta_0, \alpha R)$.

[You will find it useful to determine the form of the derivative of $\cos^{-1}(u^{-1})$.]

In cylindrical polar coordinates the element of length is given by

$$(ds)^2 = (d\rho)^2 + (\rho\, d\theta)^2 + (dz)^2,$$

and the total length of a curve between two points parameterised by θ_0 and θ_1 is

$$s = \int_{\theta_0}^{\theta_1} \sqrt{\left(\frac{d\rho}{d\theta}\right)^2 + \rho^2 + \left(\frac{dz}{d\theta}\right)^2}\, d\theta$$

$$= \int_{\theta_0}^{\theta_1} \sqrt{\rho^2 + (1 + \alpha^2)\left(\frac{d\rho}{d\theta}\right)^2}\, d\theta, \text{ since } z = \alpha\rho.$$

Since the independent variable θ does not occur explicitly in the integrand, a first integral of the E–L equation is

$$\sqrt{\rho^2 + (1 + \alpha^2)\rho'^2} - \rho' \frac{(1 + \alpha^2)\rho'}{\sqrt{\rho^2 + (1 + \alpha^2)\rho'^2}} = c.$$

After being multiplied through by the square root, this can be arranged as follows:

$$\rho^2 + (1 + \alpha^2)\rho'^2 - (1 + \alpha^2)\rho'^2 = c\sqrt{\rho^2 + (1 + \alpha^2)\rho'^2},$$

$$\rho^4 = c^2[\rho^2 + (1 + \alpha^2)\rho'^2].$$

This is the given equation of the geodesic, in which c is arbitrary but β^2 must have the value $1 + \alpha^2$.

Guided by the hint, we first determine the derivative of $y(u) = \cos^{-1}(u^{-1})$:

$$\frac{dy}{du} = \frac{-1}{\sqrt{1 - u^{-2}}} \frac{-1}{u^2} = \frac{1}{u\sqrt{u^2 - 1}}.$$

Now, returning to the geodesic, rewrite it as

$$\rho^4 - c^2\rho^2 = c^2\beta^2\rho'^2,$$

$$\rho(\rho^2 - c^2)^{1/2} = c\beta\frac{d\rho}{d\theta}.$$

Setting $\rho = cu$,

$$uc^2(u^2 - 1)^{1/2} = c^2\beta\frac{du}{d\theta},$$

$$d\theta = \frac{\beta\,du}{u(u^2 - 1)^{1/2}},$$

which integrates to

$$\theta = \beta\cos^{-1}\left(\frac{1}{u}\right) + k,$$

using the result from the hint.

Since the geodesic must pass through both $(R, -\theta_0, \alpha R)$ and $(R, \theta_0, \alpha R)$, we must have $k = 0$ and

$$\cos\frac{\theta_0}{\beta} = \frac{c}{R}.$$

Further, at a general point on the geodesic,

$$\cos\frac{\theta}{\beta} = \frac{c}{\rho}.$$

Eliminating c then shows that the geodesic on the cone that joins the two given points is

$$\rho(\theta) = \frac{R\cos(\theta_0/\beta)}{\cos(\theta/\beta)}.$$

22.9 *You are provided with a line of length $\pi a/2$ and negligible mass and some lead shot of total mass M. Use a variational method to determine how the lead shot must be distributed along the line if the loaded line is to hang in a circular arc of radius a when its ends are attached to two points at the same height. Measure the distance s along the line from its centre.*

We first note that the total mass of shot available is merely a scaling factor and not a constraint on the minimisation process.

The length of string is sufficient to form one-quarter of a complete circle of radius a, and so the ends of the string must be fixed to points that are $2a\sin(\pi/4) = \sqrt{2}a$ apart.

We take the distribution of shot as $\rho = \rho(s)$ and have to minimise the integral $\int gy(s)\rho(s)\,ds$, but subject to the requirement $\int dx = a/\sqrt{2}$. Expressed as an integral over s, this requirement can be written

$$\frac{a}{\sqrt{2}} = \int_{s=0}^{s=\pi a/4} dx = \int_0^{\pi a/4}(1 - y'^2)^{1/2}\,ds,$$

358

where the derivative y' of y is with respect to s (not x).

We therefore consider the minimisation of $\int F(y, y', s)\, ds$, where

$$F(y, y', s) = gy\rho + \lambda\sqrt{1 - y'^2}.$$

The E–L equation takes the form

$$\frac{d}{ds}\left(\frac{\partial F}{\partial y'}\right) = \frac{\partial F}{\partial y},$$

$$\lambda\frac{d}{ds}\left(\frac{-y'}{\sqrt{1 - y'^2}}\right) = g\rho(s),$$

$$\frac{-\lambda y'}{\sqrt{1 - y'^2}} = \int_0^s g\rho(s')\, ds' \equiv gP(s),$$

since $y'(0) = 0$ by symmetry.

Now we require $P(s)$ to be such that the solution to this equation takes the form of an arc of a circle, $y(s) = y_0 - a\cos(s/a)$. If this is so, then $y'(s) = \sin(s/a)$ and

$$\frac{-\lambda\sin(s/a)}{\cos(s/a)} = gP(s).$$

When $s = \pi a/4$, $P(s)$ must have the value $M/2$, implying that $\lambda = -Mg/2$ and that, consequently,

$$P(s) = \frac{M}{2}\tan\left(\frac{s}{a}\right).$$

The required distribution $\rho(s)$ is recovered by differentiating this to obtain

$$\rho(s) = \frac{dP}{ds} = \frac{M}{2a}\sec^2\left(\frac{s}{a}\right).$$

22.11 *A general result is that light travels through a variable medium by a path that minimises the travel time (this is an alternative formulation of Fermat's principle). With respect to a particular cylindrical polar coordinate system (ρ, ϕ, z), the speed of light $v(\rho, \phi)$ is independent of z. If the path of the light is parameterised as $\rho = \rho(z), \phi = \phi(z)$, show that*

$$v^2(\rho'^2 + \rho^2\phi'^2 + 1)$$

is constant along the path.

For the particular case when $v = v(\rho) = b(a^2 + \rho^2)^{1/2}$, show that the two Euler–Lagrange equations have a common solution in which the light travels along a helical path given by $\phi = Az + B$, $\rho = C$, provided that A has a particular value.

In cylindrical polar coordinates with $\rho = \rho(z)$ and $\phi = \phi(z)$,

$$ds = \left[1 + \left(\frac{d\rho}{dz} \right)^2 + \rho^2 \left(\frac{d\phi}{dz} \right)^2 \right]^{1/2} dz.$$

The total travel time of the light is therefore given by

$$\tau = \int \frac{(1 + \rho'^2 + \rho^2 \phi'^2)^{1/2}}{v(\rho, \phi)} \, dz.$$

Since z does not appear explicitly in the integrand, we have from the general first integral of the E–L equations for more than one dependent variable that

$$\frac{(1 + \rho'^2 + \rho^2 \phi'^2)^{1/2}}{v(\rho, \phi)} - \frac{1}{v} \frac{\rho'^2}{(1 + \rho'^2 + \rho^2 \phi'^2)^{1/2}} - \frac{1}{v} \frac{\rho^2 \phi'^2}{(1 + \rho'^2 + \rho^2 \phi'^2)^{1/2}} = k.$$

Rearranging this gives

$$1 + \rho'^2 + \rho^2 \phi'^2 - \rho'^2 - \rho^2 \phi'^2 = kv(1 + \rho'^2 + \rho^2 \phi'^2)^{1/2},$$
$$1 = kv(1 + \rho'^2 + \rho^2 \phi'^2)^{1/2},$$
$$\Rightarrow \quad v^2(1 + \rho'^2 + \rho^2 \phi'^2) = c, \text{ along the path.}$$

Denoting $(1 + \rho'^2 + \rho^2 \phi'^2)$ by $(**)$ for brevity, the E–L equations for ρ and ϕ are, respectively,

$$\frac{\rho \phi'^2}{v(**)^{1/2}} - \frac{(**)^{1/2}}{v^2} \frac{\partial v}{\partial \rho} = \frac{d}{dz} \left[\frac{\rho'}{v(**)^{1/2}} \right], \quad (1)$$

and

$$-\frac{(**)^{1/2}}{v^2} \frac{\partial v}{\partial \phi} = \frac{d}{dz} \left[\frac{\rho^2 \phi'}{v(**)^{1/2}} \right]. \quad (2)$$

Now, if $v = b(a^2 + \rho^2)^{1/2}$, the only dependence on z in a possible solution $\phi = Az + B$ with $\rho = C$ is through the first of these equations. To see this we note that the square brackets on the RHS's of the two E–L equations do not contain any undifferentiated ϕ-terms and so the derivatives (with respect to z) of both are zero. Since $\partial v / \partial \phi$ is also zero, equation (2) is identically satisfied as $0 = 0$. This leaves only (1), which reads

$$\frac{CA^2}{b(a^2 + C^2)^{1/2}(1 + 0 + C^2 A^2)^{1/2}} - \frac{(1 + 0 + C^2 A^2)^{1/2} bC}{b^2(a^2 + C^2)(a^2 + C^2)^{1/2}} = 0.$$

This is satisfied provided

$$A^2(a^2 + C^2) = 1 + C^2 A^2,$$
$$\text{i.e.} \quad A = a^{-1}.$$

Thus, a solution in the form of a helix is possible provided that the helix has a particular pitch, $2\pi a$.

22.13 *A dam of capacity V (less than $\pi b^2 h/2$) is to be constructed on level ground next to a long straight wall which runs from $(-b, 0)$ to $(b, 0)$. This is to be achieved by joining the ends of a new wall, of height h, to those of the existing wall. Show that, in order to minimise the length L of new wall to be built, it should form part of a circle, and that L is then given by*

$$\int_{-b}^{b} \frac{dx}{(1 - \lambda^2 x^2)^{1/2}},$$

where λ is found from

$$\frac{V}{hb^2} = \frac{\sin^{-1} \mu}{\mu^2} - \frac{(1 - \mu^2)^{1/2}}{\mu}$$

and $\mu = \lambda b$.

The objective is to chose the wall profile, $y = y(x)$, so as to minimise

$$L = \int_{-b}^{b} \sqrt{(dx)^2 + (dy)^2} = \int_{-b}^{b} (1 + y'^2)^{1/2} \, dx$$

subject to the constraint that the capacity of the dam formed is

$$V = h \int_{-b}^{b} y \, dx.$$

For this constrained variation problem we consider the minimisation of

$$K = \int_{-b}^{b} [(1 + y'^2)^{1/2} - \lambda y] \, dx,$$

where λ is a Lagrange multiplier.

Since x does not appear in the integrand, a first integral of the E–L equation is

$$(1 + y'^2)^{1/2} - \lambda y - y' \frac{y'}{(1 + y'^2)^{1/2}} = k,$$

$$\frac{1}{(1 + y'^2)^{1/2}} = k + \lambda y.$$

Rearranging this and integrating gives

$$\frac{1}{(k + \lambda y)^2} - 1 = y'^2,$$

$$\frac{(k + \lambda y) \, dy}{\sqrt{1 - (k + \lambda y)^2}} = dx,$$

$$\Rightarrow \quad -\frac{\sqrt{1 - (k + \lambda y)^2}}{\lambda} = x + c.$$

This result can be arranged in a more familiar form as

$$\lambda^2(x+c)^2 + (k+\lambda y)^2 = 1.$$

This is the equation of a circle that is centred on $(-c, -k/\lambda)$; from symmetry $c = 0$. Further, since $(\pm b, 0)$ lies on the curve, we must have

$$\lambda^2 b^2 + k^2 = 1, \tag{$*$}$$

giving a connection between the Lagrange multiplier and one of the constants of integration. The length of the wall is given by

$$L = \int_{-b}^{b} (1 + y'^2)^{1/2}\, dx = \int_{-b}^{b} \frac{1}{k + \lambda y}\, dx = \int_{-b}^{b} \frac{1}{(1 - \lambda^2 x^2)^{1/2}}\, dx.$$

The remaining constraint determines the value of λ and is that

$$\frac{V}{h} = \int_{-b}^{b} y\, dx = \frac{1}{\lambda} \int_{-b}^{b} \left(\sqrt{1 - \lambda^2 x^2} - k \right) dx$$

$$= \frac{1}{\lambda} \int_{-b}^{b} \left(\sqrt{1 - \lambda^2 x^2} - \sqrt{1 - \lambda^2 b^2} \right) dx, \text{ using}(*),$$

$$\frac{\lambda V}{h} = \left[x\sqrt{1 - \lambda^2 x^2} \right]_{-b}^{b} - \int_{-b}^{b} \frac{-\lambda^2 x\, x}{\sqrt{1 - \lambda^2 x^2}}\, dx - \left[x\sqrt{1 - \lambda^2 b^2} \right]_{-b}^{b}$$

$$= \lambda^2 \int_{-b}^{b} \frac{x^2}{\sqrt{1 - \lambda^2 x^2}}\, dx.$$

To evaluate this integral we set $\lambda x = \sin\theta$ and $\mu = \lambda b = \sin\phi$, to give

$$\frac{\lambda V}{h} = \int_{-\phi}^{\phi} \frac{\sin^2\theta\,\cos\theta}{\lambda \cos\theta}\, d\theta,$$

$$\frac{\lambda^2 V}{h} = \int_{-\phi}^{\phi} \frac{1}{2}(1 - \cos 2\theta)\, d\theta$$

$$= \phi - \frac{2}{4} \sin 2\phi,$$

$$\frac{\mu^2 V}{hb^2} = \sin^{-1}\mu - \frac{1}{2} 2\mu(1 - \mu^2)^{1/2},$$

$$\frac{V}{hb^2} = \frac{\sin^{-1}\mu}{\mu^2} - \frac{(1 - \mu^2)^{1/2}}{\mu}.$$

This equation determines μ and hence λ.

> **22.15** *The Schwarzchild metric for the static field of a non-rotating spherically symmetric black hole of mass M is given by*
>
> $$(ds)^2 = c^2 \left(1 - \frac{2GM}{c^2 r}\right)(dt)^2 - \frac{(dr)^2}{1 - 2GM/(c^2 r)} - r^2 (d\theta)^2 - r^2 \sin^2 \theta \, (d\phi)^2.$$
>
> *Considering only motion confined to the plane $\theta = \pi/2$, and assuming that the path of a small test particle is such as to make $\int ds$ stationary, find two first integrals of the equations of motion. From their Newtonian limits, in which GM/r, $\dot{r}^2$ and $r^2 \dot{\phi}^2$ are all $\ll c^2$, identify the constants of integration.*

For motion confined to the plane $\theta = \pi/2$, $d\theta = 0$ and the corresponding term in the metric can be ignored. With this simplification, we can write

$$ds = \left\{ c^2 \left(1 - \frac{2GM}{c^2 r}\right) - \frac{\dot{r}^2}{1 - (2GM)/(c^2 r)} - r^2 \dot{\phi}^2 \right\}^{1/2} dt.$$

Writing the terms in braces as $\{**\}$, the E–L equation for ϕ reads

$$\frac{d}{dt}\left(\frac{-r^2 \dot{\phi}}{\{**\}^{1/2}}\right) - 0 = 0,$$

$$\Rightarrow \quad \frac{r^2 \dot{\phi}}{\{**\}^{1/2}} = A.$$

In the Newtonian limit $\{**\} \to c^2$ and the equation becomes $r^2 \dot{\phi} = Ac$. Thus, Ac is a measure of the angular momentum of the particle about the origin.

The E–L equation for r is more complicated but, because ds does not contain t explicitly, we can use the general result for the first integral of the E–L equations when there is more than one dependent variable: $F - \sum_i \dot{q}_i \frac{\partial F}{\partial \dot{q}_i} = k$. This gives us a second equation as follows:

$$F - \dot{r}\frac{\partial F}{\partial \dot{r}} - \dot{\phi}\frac{\partial F}{\partial \dot{\phi}} = B,$$

$$\{**\}^{1/2} + \frac{\dot{r}}{\{**\}^{1/2}}\frac{\dot{r}}{[1 - (2GM)/(c^2 r)]} + \frac{\dot{\phi}}{\{**\}^{1/2}} r^2 \dot{\phi} = B.$$

Multiplying through by $\{**\}^{1/2}$ and cancelling the terms in $\dot{r}^2$ and $\dot{\phi}^2$ now gives

$$c^2 - \frac{2GM}{r} = B\left\{c^2 - \frac{2GM}{r} - \frac{\dot{r}^2}{[1 - (2GM)/(c^2 r)]} - r^2 \dot{\phi}^2\right\}^{1/2}.$$

In the Newtonian limits, in which GM/r, $\dot{r}^2$ and $r^2 \dot{\phi}^2$ are all $\ll c^2$, the equation can be rearranged and the braces expanded to first order in small quantities to give

$$B = \left(c^2 - \frac{2GM}{r} \right) \left\{ c^2 - \frac{2GM}{r} - \frac{\dot{r}^2}{[1 - (2GM)/(c^2 r)]} - r^2 \dot{\phi}^2 \right\}^{-1/2},$$

$$cB = c^2 - \frac{2GM}{r} + \frac{c^2 GM}{c^2 r} + \frac{c^2 \dot{r}^2}{2c^2} + \frac{c^2 r^2 \dot{\phi}^2}{2c^2} + \cdots,$$

$$= c^2 - \frac{GM}{r} + \frac{1}{2}(\dot{r}^2 + r^2 \dot{\phi}^2) + \cdots,$$

which can be read as 'total energy = rest mass energy + gravitational energy + radial and azimuthal kinetic energy'. Thus Bc is a measure of the total energy of the test particle.

22.17 *Determine the minimum value that the integral*

$$J = \int_0^1 [x^4 (y'')^2 + 4x^2 (y')^2] \, dx$$

can have, given that y is not singular at $x = 0$ and that $y(1) = y'(1) = 1$. Assume that the Euler–Lagrange equation gives the lower limit and verify retrospectively that your solution satisfies the end-point condition

$$\left[\eta \frac{\partial F}{\partial y'} \right]_a^b = 0,$$

where $F = F(y', y, x)$ and $\eta(x)$ is the variation from the minimising curve.

We first set $y'(x) = u(x)$ with $u(1) = y'(1) = 1$. The integral then becomes

$$J = \int_0^1 [x^4 (u')^2 + 4x^2 u^2] \, dx. \qquad (*)$$

This will be stationary if (using the E–L equation)

$$\frac{d}{dx}(2x^4 u') - 8x^2 u = 0,$$

$$8x^3 u' + 2x^4 u'' - 8x^2 u = 0,$$

$$x^2 u'' + 4xu' - 4u = 0.$$

As this is a homogeneous equation, we try $u(x) = Ax^n$, obtaining

$$n(n - 1) + 4n - 4 = 0 \quad \Rightarrow \quad n = -4, \text{ or } n = 1.$$

The form of $y'(x)$ is thus

$$y'(x) = u(x) = \frac{A}{x^4} + Bx \quad \text{with} \quad A + B = 1.$$

Further,

$$y(x) = -\frac{A}{3x^3} + \frac{Bx^2}{2} + C.$$

Since y is not singular at $x = 0$ and $y(1) = 1$, we have that $A = 0$, $B = 1$ and $C = \frac{1}{2}$, yielding $y(x) = \frac{1}{2}(1 + x^2)$. The minimal value of J is thus

$$J_{\min} = \int_0^1 [x^4(1)^2 + 4x^2(x)^2] \, dx = \int_0^1 5x^4 \, dx = [x^5]_0^1 = 1.$$

In (*) the integrand is $G(u', u, x)$ and so the end-point condition reads

$$\left[\eta \frac{\partial G}{\partial u'}\right]_0^1 = 0.$$

At the upper limit $\eta(1) = 0$, since $u(1) = y'(1) = 1$ is fixed. At the lower limit,

$$\left.\frac{\partial G}{\partial u'}\right|_{x=0} = 2x^4 u'\big|_{x=0} = 0.$$

Thus the contributions at the two limits are individually zero and the boundary condition is satisfied in the simplest way.

22.19 *Find an appropriate but simple trial function and use it to estimate the lowest eigenvalue λ_0 of Stokes' equation*

$$\frac{d^2 y}{dx^2} + \lambda x y = 0, \qquad y(0) = y(\pi) = 0.$$

Explain why your estimate must be strictly greater than λ_0.

Stokes' equation is an S–L equation with $p = 1$, $q = 0$ and $\rho = x$. For the given boundary conditions the obvious trial function is $y(x) = \sin x$. The lowest eigenvalue $\lambda_0 \leq I/J$, where

$$I = \int_0^\pi p y'^2 \, dx = \int_0^\pi \cos^2 x \, dx = \frac{\pi}{2}$$

and

$$J = \int_0^\pi \rho y^2 \, dx = \int_0^\pi x \sin^2 x \, dx$$

$$= \int_0^\pi \tfrac{1}{2} x (1 - \cos 2x) \, dx$$

$$= \left[\frac{x^2}{4}\right]_0^\pi - \left[\frac{x}{2}\frac{\sin 2x}{2}\right]_0^\pi + \frac{1}{2}\int_0^\pi \frac{\sin 2x}{2} \, dx$$

$$= \frac{\pi^2}{4}.$$

Thus $\lambda_0 \leq (\frac{1}{2}\pi)/(\frac{1}{4}\pi^2) = 2/\pi$.

However, if we substitute the trial function directly into the equation we obtain

$$-\sin x + \frac{2}{\pi} x \sin x = 0,$$

which is clearly not satisfied. Thus the trial function is not an eigenfunction, and the actual lowest eigenvalue must be strictly less than the estimate of $2/\pi$.

22.21 *A drumskin is stretched across a fixed circular rim of radius a. Small transverse vibrations of the skin have an amplitude $z(\rho, \phi, t)$ that satisfies*

$$\nabla^2 z = \frac{1}{c^2} \frac{\partial^2 z}{\partial t^2}$$

in plane polar coordinates. For a normal mode independent of azimuth, in which case $z = Z(\rho)\cos\omega t$, find the differential equation satisfied by $Z(\rho)$. By using a trial function of the form $a^v - \rho^v$, with adjustable parameter v, obtain an estimate for the lowest normal mode frequency.

[The exact answer is $(5.78)^{1/2}c/a$.]

In cylindrical polar coordinates, (ρ, ϕ), the wave equation,

$$\nabla^2 z = \frac{1}{c^2} \frac{\partial^2 z}{\partial t^2},$$

has azimuth-independent solutions (i.e. independent of ϕ) of the form $z(\rho, t) = Z(\rho)\cos\omega t$, and reduces to

$$\frac{1}{\rho}\frac{d}{d\rho}\left(\rho\frac{dZ}{d\rho}\right)\cos\omega t = -\frac{Z\omega^2}{c^2}\cos\omega t,$$

$$\frac{d}{d\rho}\left(\rho\frac{dZ}{d\rho}\right) + \frac{\omega^2}{c^2}\rho Z = 0.$$

The boundary conditions require that $Z(a) = 0$ and, so that there is no physical discontinuity in the slope of the drumskin at the origin, $Z'(0) = 0$.

This is an S–L equation with $p = \rho$, $q = 0$ and weight function $w = \rho$. A suitable trial function is $Z(\rho) = a^v - \rho^v$, which automatically satisfies $Z(a) = 0$ and, provided $v > 1$, has $Z'(0) = -v\rho^{v-1}|_{\rho=0} = 0$.

We recall that the lowest eigenfrequency satisfies the general formula

$$\frac{\omega^2}{c^2} \leq \frac{\int_0^a [(pZ')^2 - qZ^2]\,d\rho}{\int_0^a wZ^2\,d\rho}.$$

In this case

$$\frac{\omega^2}{c^2} \le \frac{\int_0^a \rho\, v^2 \rho^{2v-2}\, d\rho}{\int_0^a \rho(a^v - \rho^v)^2\, d\rho}$$

$$= \frac{\int_0^a v^2 \rho^{2v-1}\, d\rho}{\int_0^a (\rho a^{2v} - 2\rho^{v+1}a^v + \rho^{2v+1})\, d\rho}$$

$$= \frac{(v^2 a^{2v})/2v}{\dfrac{a^{2v+2}}{2} - \dfrac{2a^{2v+2}}{v+2} + \dfrac{a^{2v+2}}{2v+2}}$$

$$= \frac{1}{a^2} \frac{v(v+2)(2v+2)}{(v+2)(2v+2) - 4(2v+2) + 2(v+2)}$$

$$= \frac{(v+2)(v+1)}{v a^2}.$$

Since v is an adjustable parameter and we know that, however we choose it, the resulting estimate can *never* be less than the lowest true eigenvalue, we choose the value that minimises the above estimate. Differentiating the estimate with respect to v gives

$$v(2v+3) - (v^2 + 3v + 2) = 0 \quad \Rightarrow \quad v^2 - 2 = 0 \quad \Rightarrow \quad v = \sqrt{2}.$$

Thus the least upper bound to be found with this parameterisation is

$$\omega^2 \le \frac{c^2}{a^2} \frac{(\sqrt{2}+2)(\sqrt{2}+1)}{\sqrt{2}} = \frac{c^2}{2a^2}(\sqrt{2}+2)^2 \quad \Rightarrow \quad \omega = (5.83)^{1/2}\frac{c}{a}.$$

As noted, the actual lowest eigenfrequency is very little below this.

22.23 *For the boundary conditions given below, obtain a functional $\Lambda(y)$ whose stationary values give the eigenvalues of the equation*

$$(1+x)\frac{d^2 y}{dx^2} + (2+x)\frac{dy}{dx} + \lambda y = 0, \qquad y(0) = 0,\ y'(2) = 0.$$

Derive an approximation to the lowest eigenvalue λ_0 using the trial function $y(x) = xe^{-x/2}$. For what value(s) of γ would

$$y(x) = xe^{-x/2} + \beta \sin \gamma x$$

be a suitable trial function for attempting to obtain an improved estimate of λ_0?

Since the derivative of $1 + x$ is not equal to $2 + x$, the given equation is not in self-adjoint form and an integrating factor for the standard form equation,

$$\frac{d^2y}{dx^2} + \frac{2+x}{1+x}\frac{dy}{dx} + \frac{\lambda y}{1+x} = 0,$$

is needed. This will be

$$\exp\left\{\int^x \frac{2+u}{1+u}\,du\right\} = \exp\left\{\int^x \left(1 + \frac{1}{1+u}\right)\,du\right\} = e^x(1+x).$$

Thus, after multiplying through by this IF, the equation takes the S–L form

$$[(1+x)e^x y']' + \lambda e^x y = 0,$$

with $p(x) = (1+x)e^x$, $q(x) = 0$ and $\rho(x) = e^x$.

The required functional is therefore

$$\Lambda(y) = \frac{\int_0^2 [(1+x)e^x y'^2 + 0]\,dx}{\int_0^2 y^2 e^x\,dx},$$

provided that, for the eigenfunctions y_i of the equation, $\left[y_i p(x)y_j'(x)\right]_0^2 = 0$; this condition is automatically satisfied with the given boundary conditions.

For the trial function $y(x) = xe^{-x/2}$, clearly $y(0) = 0$ and, less obviously, $y'(x) = (1 - \frac{1}{2}x)e^{-x/2}$, making $y'(2) = 0$. The functional takes the following form:

$$\Lambda = \frac{\int_0^2 (1+x)e^x\,(1 - \frac{1}{2}x)^2 e^{-x}\,dx}{\int_0^2 x^2 e^{-x} e^x\,dx}$$

$$= \frac{\int_0^2 (1+x)(1 - \frac{1}{2}x)^2\,dx}{\int_0^2 x^2\,dx}$$

$$= \frac{\int_0^2 (1 - x^2 + \frac{1}{4}x^2 + \frac{1}{4}x^3)\,dx}{8/3}$$

$$= \frac{3}{8}\left(2 - \frac{3}{4}\frac{8}{3} + \frac{16}{16}\right) = \frac{3}{8}.$$

Thus the lowest eigenvalue is $\leq \frac{3}{8}$.

We already know that $xe^{-x/2}$ is a suitable trial function and thus $y_2(x) = \sin\gamma x$ can be considered on its own. It satisfies $y_2(0) = 0$, but must also satisfy $y_2'(2) = \gamma\cos(2\gamma) = 0$. This requires that $\gamma = \frac{1}{2}(n + \frac{1}{2})\pi$ for some integer n; trial functions with γ of this form can be used to try to obtain a better bound on λ_0 by choosing the best value for n and adjusting the parameter β.

22.25 *The unnormalised ground-state (i.e. the lowest-energy) wavefunction of the simple harmonic oscillator of classical frequency ω is $\exp(-\alpha x^2)$, where $\alpha = m\omega/2\hbar$. Take as a trial function the orthogonal wavefunction $x^{2n+1}\exp(-\alpha x^2)$, using the integer n as a variable parameter, and apply either Sturm–Liouville theory or the Rayleigh–Ritz principle to show that the energy of the second lowest state of a quantum harmonic oscillator is $\leq 3\hbar\omega/2$.*

We first note that, for n a non-negative integer,

$$\int_{-\infty}^{\infty} x^{2n+1} e^{-\alpha x^2} e^{-\alpha x^2}\, dx = 0$$

on symmetry grounds and so confirm that the ground-state wavefunction, $\exp(-\alpha x^2)$, and the trial function, $\psi_{2n+1} = x^{2n+1}\exp(-\alpha x^2)$, are orthogonal with respect to a unit weight function.

The Hamiltonian for the quantum harmonic oscillator in one-dimension is given by

$$H = -\frac{\hbar^2}{2m}\frac{d^2}{dx^2} + \frac{k}{2}x^2.$$

This means that to prepare the elements required for a Rayleigh–Ritz analysis we will need to find the second derivative of the trial function and evaluate integrals with integrands of the form $x^n \exp(-2\alpha x^2)$. To this end, define

$$I_n = \int_{-\infty}^{\infty} x^n e^{-2\alpha x^2}\, dx, \quad \text{with recurrence relation } I_n = \frac{n-1}{4\alpha}I_{n-2}.$$

Using Leibnitz' formula shows that

$$\frac{d^2\psi_{2n+1}}{dx^2} = \left[2n(2n+1)x^{2n-1} + 2(2n+1)(-2\alpha)x^{2n+1} \right.$$
$$\left. + (4\alpha^2 x^2 - 2\alpha)x^{2n+1} \right] e^{-\alpha x^2}$$
$$= \left[2n(2n+1)x^{2n-1} - 2(4n+3)\alpha x^{2n+1} + 4\alpha^2 x^{2n+3} \right] e^{-\alpha x^2}.$$

Hence, we find that $\langle H \rangle$ is given by

$$-\frac{\hbar^2}{2m}\int_{-\infty}^{\infty} x^{2n+1}e^{-\alpha x^2}\frac{d^2\psi_{2n+1}}{dx^2}\, dx + \frac{k}{2}\int_{-\infty}^{\infty} x^2 x^{4n+2}e^{-2\alpha x^2}\, dx$$

$$= -\frac{\hbar^2}{2m}\left[2n(2n+1)I_{4n} - 2(4n+3)\alpha I_{4n+2} + 4\alpha^2 I_{4n+4} \right] + \frac{k}{2}I_{4n+4}$$

$$= I_{4n+2}\left\{ -\frac{\hbar^2}{2m}\left[\frac{2n(2n+1)4\alpha}{4n+1} - 2(4n+3)\alpha + \frac{4\alpha^2(4n+3)}{4\alpha} \right] + \frac{k(4n+3)}{8\alpha} \right\},$$

where we have used the recurrence relation to express all integrals in terms of

I_{4n+2}. This has been done because the denominator of the Rayleigh–Ritz quotient is this (same) normalisation integral, namely

$$\int_{-\infty}^{\infty} \psi_{2n+1}^* \psi_{2n+1} \, dx = I_{4n+2}.$$

Thus, the estimate $E_{2n+1} = \langle H \rangle / I_{4n+2}$ is given by

$$E_{2n+1} = -\frac{\hbar^2 \alpha}{2m} \left(\frac{16n^2 + 8n - 16n^2 - 16n - 3}{4n + 1} \right) + \frac{k(4n + 3)}{8\alpha}$$

$$= \frac{\hbar^2 \alpha}{2m} \frac{8n + 3}{4n + 1} + \frac{k(4n + 3)}{8\alpha}.$$

Using $\omega^2 = \dfrac{k}{m}$ and $\alpha = \dfrac{m\omega}{2\hbar}$ then yields

$$E_{2n+1} = \frac{\hbar\omega}{4} \left(\frac{8n + 3}{4n + 1} + 4n + 3 \right) = \frac{\hbar\omega}{2} \frac{8n^2 + 12n + 3}{4n + 1}.$$

For non-negative integers n (the orthogonality requirement is not satisfied for non-integer values), this has a minimum value of $\frac{3}{2}\hbar\omega$ when $n = 0$. Thus the second lowest energy level is less than or equal to this value. In fact, it is equal to this value, as can be shown by substituting ψ_1 into $H\psi = E\psi$.

22.27 *The upper and lower surfaces of a film of liquid, which has surface energy per unit area (surface tension) γ and density ρ, have equations $z = p(x)$ and $z = q(x)$, respectively. The film has a given volume V (per unit depth in the y-direction) and lies in the region $-L < x < L$, with $p(0) = q(0) = p(L) = q(L) = 0$. The total energy (per unit depth) of the film consists of its surface energy and its gravitational energy, and is expressed by*

$$E = \tfrac{1}{2}\rho g \int_{-L}^{L} (p^2 - q^2) \, dx + \gamma \int_{-L}^{L} \left[(1 + p'^2)^{1/2} + (1 + q'^2)^{1/2} \right] dx.$$

(a) *Express V in terms of p and q.*

(b) *Show that, if the total energy is minimised, p and q must satisfy*

$$\frac{p'^2}{(1 + p'^2)^{1/2}} - \frac{q'^2}{(1 + q'^2)^{1/2}} = \text{constant}.$$

(c) *As an approximate solution, consider the equations*

$$p = a(L - |x|), \qquad q = b(L - |x|),$$

where a and b are sufficiently small that a^3 and b^3 can be neglected compared with unity. Find the values of a and b that minimise E.

(a) The total volume constraint is given simply by

$$V = \int_{-L}^{L} [p(x) - q(x)] \, dx.$$

(b) To take account of the constraint, consider the minimisation of $E - \lambda V$, where λ is an undetermined Lagrange multiplier. The integrand does not contain x explicitly and so we have two first integrals of the E–L equations, one for $p(x)$ and the other for $q(x)$. They are

$$\frac{1}{2}\rho g(p^2 - q^2) + \gamma(1 + p'^2)^{1/2} + \gamma(1 + q'^2)^{1/2} - \lambda(p - q) - p'\frac{\gamma p'}{(1 + p'^2)^{1/2}} = k_1$$

and

$$\frac{1}{2}\rho g(p^2 - q^2) + \gamma(1 + p'^2)^{1/2} + \gamma(1 + q'^2)^{1/2} - \lambda(p - q) - q'\frac{\gamma q'}{(1 + q'^2)^{1/2}} = k_2.$$

Subtracting these two equations gives

$$\frac{p'^2}{(1 + p'^2)^{1/2}} - \frac{q'^2}{(1 + q'^2)^{1/2}} = \text{constant}.$$

(c) If

$$p = a(L - |x|), \qquad q = b(L - |x|),$$

the derivatives of p and q only take the values $\pm a$ and $\pm b$, respectively, and the volume constraint becomes

$$V = \int_{-L}^{L} (a - b)(L - |x|) \, dx = (a - b)L^2 \quad \Rightarrow \quad b = a - \frac{V}{L^2}.$$

The total energy can now be expressed entirely in terms of a and the given parameters, as follows:

$$\begin{aligned}
E &= \frac{1}{2}\rho g \int_{-L}^{L} (a^2 - b^2)(L - |x|)^2 \, dx + 2\gamma L(1 + a^2)^{1/2} + 2\gamma L(1 + b^2)^{1/2} \\
&= \frac{1}{2}\rho g(a^2 - b^2)\frac{2L^3}{3} + 2\gamma L(1 + \tfrac{1}{2}a^2 + 1 + \tfrac{1}{2}b^2) + O(a^4) + O(b^4) \\
&\approx \frac{\rho g L^3}{3}\left[a^2 - \left(a - \tfrac{V}{L^2}\right)^2 \right] + 2\gamma L\left[2 + \tfrac{1}{2}a^2 + \tfrac{1}{2}\left(a - \tfrac{V}{L^2}\right)^2 \right] \\
&= \frac{\rho g L^3}{3}\left(\frac{2aV}{L^2} - \frac{V^2}{L^4} \right) + 2\gamma L\left(2 + a^2 - \frac{aV}{L^2} + \frac{V^2}{2L^4} \right).
\end{aligned}$$

371

This is minimised with respect to a when

$$\frac{2\rho g L^3 V}{3L^2} + 4\gamma La - \frac{2\gamma LV}{L^2} = 0,$$

$$\Rightarrow \quad a = \frac{V}{2L^2} - \frac{\rho g V}{6\gamma},$$

$$\Rightarrow \quad b = -\frac{V}{2L^2} - \frac{\rho g V}{6\gamma}.$$

As might be expected, $|b| > |a|$ and there is more of the liquid below the $z = 0$ plane than there is above it.

22.29 *The 'stationary value of an integral' approach to finding the eigenvalues of a Sturm–Liouville equation can be extended to two independent variables, x and z, with little modification. In the integral to be minimised, y'^2 is replaced by $(\nabla y)^2$ and the integrals of the various functions of $y(x, z)$ become two-dimensional, i.e. the infinitesimal is $dx\, dz$.*

The vibrations of a trampoline 4 units long and 1 unit wide satisfy the equation

$$\nabla^2 y + k^2 y = 0.$$

By taking the simplest possible permissible polynomial as a trial function, show that the lowest mode of vibration has $k^2 \leq 10.63$ and, by direct solution, that the actual value is 10.49.

Written explicitly, the equation is

$$\frac{\partial^2 y}{\partial x^2} + \frac{\partial^2 y}{\partial z^2} + k^2 y = 0.$$

This is an extended S–L equation with $p(x, z) = 1$, $q(x, z) = 0$, $\rho(x, z) = 1$ and eigenvalue λ. We therefore consider the stationary values of $\Lambda = I/J$, where

$$I = \int \int \left[\left(\frac{\partial y}{\partial x} \right)^2 + \left(\frac{\partial y}{\partial z} \right)^2 \right] dx\, dz$$

and J is the normalisation integral $\int y^2(x, z)\, dx\, dz$.

For a trampoline 4 units long and 1 unit wide, the simplest trial function that satisfies $y(0, z) = y(4, z) = y(x, 0) = y(x, 1) = 0$ is

$$y(x, z) = x(4 - x)z(1 - z).$$

For this function,

$$\frac{\partial y}{\partial x} = (4 - 2x)z(1 - z) \quad \text{and} \quad \frac{\partial y}{\partial z} = (1 - 2z)x(4 - x).$$

Thus, I is given by

$$\int_0^4 (4-2x)^2\, dx \int_0^1 z^2(1-z)^2\, dz + \int_0^4 x^2(4-x)^2\, dx \int_0^1 (1-2z)^2\, dz$$

$$= \left[16(4) - 16\left(\frac{16}{2}\right) + 4\left(\frac{64}{3}\right) \right] \left[\left(\frac{1}{3}\right) - 2\left(\frac{1}{4}\right) + \left(\frac{1}{5}\right) \right]$$

$$+ \left[16\left(\frac{64}{3}\right) - 8\left(\frac{256}{4}\right) + \left(\frac{1024}{5}\right) \right] \left[1(1) - 4\left(\frac{1}{2}\right) + 4\left(\frac{1}{3}\right) \right]$$

$$= 64\left(1 - 2 + \frac{4}{3}\right)\frac{1}{30} + \frac{1024}{30}\left(1 - 2 + \frac{4}{3}\right) = \frac{1088}{90}.$$

Similarly, J is given by

$$\int_0^4 x^2(4-x)^2\, dx \int_0^1 z^2(1-z)^2\, dz$$

$$= \left[16\left(\frac{64}{3}\right) - 8\left(\frac{256}{4}\right) + \left(\frac{1024}{5}\right) \right] \left[\left(\frac{1}{3}\right) - 2\left(\frac{1}{4}\right) + \left(\frac{1}{5}\right) \right]$$

$$= 1024\left(\frac{1}{3} - \frac{2}{4} + \frac{1}{5}\right)^2 = \frac{1024}{900}.$$

Thus the lowest eigenvalue $k^2 \le (1088/90) \div (1024/900) = 10.63$.

The obvious direct solution satisfying the boundary conditions is

$$y(x, z) = A \sin\frac{\pi x}{4} \sin \pi z.$$

Substituting this into the original equation gives

$$-\frac{\pi^2}{16} y(x, z) - \pi^2 y(x, z) + k^2 y(x, z) = 0,$$

which is clearly satisfied if

$$k^2 = \frac{17\pi^2}{16} = 10.49.$$

Integral equations

Since $\cos uv$ is an even function of v, we will make $y(-v) = y(v)$ so that the complete integrand is also an even function of v. The integral I on the LHS can then be written as

$$I = \frac{1}{2}\int_{-\infty}^\infty \cos(xv)y(v)\,dv = \frac{1}{2}\mathrm{Re}\int_{-\infty}^\infty e^{ixv}y(v)\,dv = \frac{1}{2}\int_{-\infty}^\infty e^{ixv}y(v)\,dv,$$

the last step following because $y(v)$ is symmetric in v. The integral is now $\sqrt{2\pi}\,\times$ a Fourier transform, and it follows from the inversion theorem for Fourier transforms applied to

$$\frac{1}{2}\int_{-\infty}^\infty e^{ixv}y(v)\,dv = \exp(-x^2/2)$$

that

$$
\begin{aligned}
y(x) &= \frac{2}{2\pi}\int_{-\infty}^\infty e^{-u^2/2}e^{-iux}\,du \\
&= \frac{1}{\pi}\int_{-\infty}^\infty e^{-(u+ix)^2/2}e^{-x^2/2}\,dx \\
&= \frac{1}{\pi}\sqrt{2\pi}\,e^{-x^2/2} \\
&= \sqrt{\frac{2}{\pi}}\,e^{-x^2/2}.
\end{aligned}
$$

Although, as noted in the question, $y(x)$ is arbitrary for $x < 0$, because its form in this range does not affect the value of the integral, for $x > 0$ it *must* have the form given. This is tricky to prove formally, but any second solution $w(x)$ has to satisfy

$$\int_0^\infty \cos(xv)[y(v) - w(v)] \, dv = 0$$

for all $x > 0$. Intuitively, this implies that $y(x)$ and $w(x)$ are identical functions.

23.3 *Convert*

$$f(x) = \exp x + \int_0^x (x - y)f(y) \, dy$$

into a differential equation, and hence show that its solution is

$$(\alpha + \beta x) \exp x + \gamma \exp(-x),$$

where α, β, γ are constants that should be determined.

We differentiate the integral equation twice and obtain

$$f'(x) = e^x + (x - x)f(x) + \int_0^x f(y) \, dy,$$
$$f''(x) = e^x + f(x).$$

Expressed in the usual differential equation form, this last equation is

$$f''(x) - f(x) = e^x, \text{ for which the CF is } f(x) = Ae^x + Be^{-x}.$$

Since the complementary function contains the RHS of the equation, we try as a PI $f(x) = Cxe^x$:

$$Cxe^x + 2Ce^x - Cxe^x = e^x \quad \Rightarrow \quad \beta = C = \tfrac{1}{2}.$$

The general solution is therefore $f(x) = Ae^x + Be^{-x} + \tfrac{1}{2}xe^x$.

The boundary conditions needed to evaluate A and B are constructed by considering the integral equation and its derivative(s) at $x = 0$, because with $x = 0$ the integral on the RHS contributes nothing. We have

$$f(0) = e^0 + 0 = 1 \quad \Rightarrow \quad A + B = 1$$

and $\qquad f'(0) = e^0 + 0 = 1 \quad \Rightarrow \quad A - B + \tfrac{1}{2} = 1.$

Solving these yields $\alpha = A = \tfrac{3}{4}$ and $\gamma = B = \tfrac{1}{4}$ and makes the complete solution

$$f(x) = \tfrac{3}{4}e^x + \tfrac{1}{4}e^{-x} + \tfrac{1}{2}xe^x.$$

23.5 *Solve for $\phi(x)$ the integral equation*

$$\phi(x) = f(x) + \lambda \int_0^1 \left[\left(\frac{x}{y}\right)^n + \left(\frac{y}{x}\right)^n \right] \phi(y)\, dy,$$

where $f(x)$ is bounded for $0 < x < 1$ and $-\frac{1}{2} < n < \frac{1}{2}$, expressing your answer in terms of the quantities $F_m = \int_0^1 f(y)y^m\, dy$.

    (a) *Give the explicit solution when $\lambda = 1$.*
    (b) *For what values of λ are there no solutions unless $F_{\pm n}$ are in a particular ratio? What is this ratio?*

This equation has a symmetric degenerate kernel, and so we set

$$\phi(x) = f(x) + a_1 x^n + a_2 x^{-n},$$

giving

$$\frac{\phi(x) - f(x)}{\lambda} = \int_0^1 \left(\frac{x^n}{y^n} + \frac{y^n}{x^n} \right) [f(y) + a_1 y^n + a_2 y^{-n}]\, dy$$

$$= x^n \int_0^1 \frac{f(y)}{y^n}\, dy + x^{-n} \int_0^1 y^n f(y)\, dy + a_1 x^n$$

$$+ a_2 x^{-n} + a_1 x^{-n} \int_0^1 y^{2n}\, dy + a_2 x^n \int_0^1 y^{-2n}\, dy$$

$$= x^n \left(F_{-n} + a_1 + \frac{a_2}{1 - 2n} \right) + x^{-n} \left(F_n + a_2 + \frac{a_1}{2n + 1} \right).$$

This is consistent with the assumed form of $\phi(x)$, provided

$$a_1 = \lambda \left(F_{-n} + a_1 + \frac{a_2}{1 - 2n} \right) \quad \text{and} \quad a_2 = \lambda \left(F_n + a_2 + \frac{a_1}{2n + 1} \right).$$

These two simultaneous linear equations can now be solved for a_1 and a_2.

(a) For $\lambda = 1$, the equations simplify and decouple to yield

$$a_2 = -(1 - 2n)F_{-n} \quad \text{and} \quad a_1 = -(1 + 2n)F_n,$$

respectively, giving as the explicit solution

$$\phi(x) = f(x) - (1 + 2n)F_n x^n - (1 - 2n)F_{-n} x^{-n}.$$

(b) For a general value of λ,

$$(1 - \lambda)a_1 - \frac{\lambda}{1 - 2n} a_2 = \lambda F_{-n},$$

$$-\frac{\lambda}{1 + 2n} a_1 + (1 - \lambda)a_2 = \lambda F_n.$$

The case $\lambda = 0$ is trivial, with $\phi(x) = f(x)$, and so suppose that $\lambda \neq 0$. Then, after being divided through by λ, the equations can be written in the matrix and vector form $\mathsf{Aa} = \mathsf{F}$:

$$\begin{pmatrix} \dfrac{1}{\lambda} - 1 & -\dfrac{1}{1 - 2n} \\ -\dfrac{1}{1 + 2n} & \dfrac{1}{\lambda} - 1 \end{pmatrix} \begin{pmatrix} a_1 \\ a_2 \end{pmatrix} = \begin{pmatrix} F_{-n} \\ F_n \end{pmatrix}.$$

In general, this matrix equation will have no solution if $|\mathsf{A}| = 0$. This will be the case if

$$\left(\frac{1}{\lambda} - 1 \right)^2 - \frac{1}{1 - 4n^2} = 0,$$

which, on rearrangement, shows that λ would have to be given by

$$\frac{1}{\lambda} = 1 \pm \frac{1}{\sqrt{1 - 4n^2}}.$$

We note that this value for λ is real because n lies in the range $-\frac{1}{2} < n < \frac{1}{2}$. In fact $-\infty < \lambda < \frac{1}{2}$. Even for these two values of λ, however, if *either* $F_n = F_{-n} = 0$ *or* the matrix equation

$$\begin{pmatrix} \pm \dfrac{1}{\sqrt{1 - 4n^2}} & -\dfrac{1}{1 - 2n} \\ -\dfrac{1}{1 + 2n} & \pm \dfrac{1}{\sqrt{1 - 4n^2}} \end{pmatrix} \begin{pmatrix} a_1 \\ a_2 \end{pmatrix} = \begin{pmatrix} F_{-n} \\ F_n \end{pmatrix}$$

is equivalent to two linear equations that are multiples of each other, there will still be a solution. In this latter case, we must have

$$\frac{F_n}{F_{-n}} = \mp \sqrt{\frac{1 - 2n}{1 + 2n}}.$$

Again we note that, because of the range in which n lies, this ratio is real; this condition can, however, require any value in the range $-\infty$ to ∞ for F_n/F_{-n}.

23.7 *The kernel of the integral equation*

$$\psi(x) = \lambda \int_a^b K(x,y)\psi(y)\,dy$$

has the form

$$K(x,y) = \sum_{n=0}^{\infty} h_n(x)g_n(y),$$

where the $h_n(x)$ form a complete orthonormal set of functions over the interval $[a,b]$.

(a) *Show that the eigenvalues λ_i are given by*

$$|\mathsf{M} - \lambda^{-1}\mathsf{I}| = 0,$$

where M is the matrix with elements

$$M_{kj} = \int_a^b g_k(u)h_j(u)\,du.$$

If the corresponding solutions are $\psi^{(i)}(x) = \sum_{n=0}^{\infty} a_n^{(i)} h_n(x)$, find an expression for $a_n^{(i)}$.

(b) *Obtain the eigenvalues and eigenfunctions over the interval $[0, 2\pi]$ if*

$$K(x,y) = \sum_{n=1}^{\infty} \frac{1}{n} \cos nx \cos ny.$$

(a) We write the ith eigenfunction as

$$\psi^{(i)}(x) = \sum_{n=0}^{\infty} a_n^{(i)} h_n(x).$$

From the orthonormality of the $h_n(x)$, it follows immediately that

$$a_m^{(i)} = \int_a^b h_m(x)\psi^{(i)}(x)\,dx.$$

However, the coefficients $a_m^{(i)}$ have to be found as the components of the eigen-vectors $\mathbf{a}^{(i)}$ defined below, since the $\psi^{(i)}$ are not initially known.

Substituting this assumed form of solution, we obtain

$$\sum_{m=0}^{\infty} a_m^{(i)} h_m(x) = \lambda_i \int_a^b \sum_{n=0}^{\infty} h_n(x)g_n(y) \sum_{l=0}^{\infty} a_l^{(i)} h_l(y)\,dy$$

$$= \lambda_i \sum_{n,l} a_l^{(i)} M_{nl} h_n(x).$$

Since the $\{h_n\}$ are an orthonormal set, it follows that

$$a_m^{(i)} = \lambda_i \sum_{n,l} a_l^{(i)} M_{nl} \delta_{mn} = \lambda_i \sum_{l=0}^{\infty} M_{ml} a_l^{(i)},$$

i.e. $\quad (\mathsf{M} - \lambda_i^{-1} \mathsf{I}) \mathbf{a}^{(i)} = 0.$

Thus, the allowed values of λ_i are given by $|\mathsf{M} - \lambda^{-1}\mathsf{I}| = 0$, and the expansion coefficients $a_m^{(i)}$ by the components of the corresponding eigenvectors.

(b) To make the set $\{h_n(x) = \cos nx\}$ into a complete orthonormal set we need to add the set of functions $\{\eta_v(x) = \sin vx\}$ and then normalise all the functions by multiplying them by $1/\sqrt{\pi}$. For this particular kernel the general functions $g_n(x)$ are given by $g_n(x) = n^{-1}\sqrt{\pi}\cos nx$.

The matrix elements are then

$$M_{kj} = \int_0^{2\pi} \frac{1}{\sqrt{\pi}} \cos ju \, \frac{\sqrt{\pi}}{k} \cos ku \, du = \frac{\pi}{k}\delta_{kj},$$

$$M_{kv} = \int_0^{2\pi} \frac{1}{\sqrt{\pi}} \sin vu \, \frac{\sqrt{\pi}}{k} \cos ku \, du = 0.$$

Thus the matrix M is diagonal and particularly simple. The eigenvalue equation reads

$$\sum_{j=0}^{\infty} \left(\frac{\pi}{k}\delta_{kj} - \lambda_i^{-1}\delta_{kj} \right) a_j^{(i)} = 0,$$

giving the immediate result that $\lambda_k = k/\pi$ with $a_k^{(k)} = 1$ and all other $a_j^{(k)} = a_v^{(k)} = 0$. The eigenfunction corresponding to eigenvalue k/π is therefore

$$\psi^{(k)}(x) = h_k(x) = \frac{1}{\sqrt{\pi}} \cos kx.$$

23.9 *For $f(t) = \exp(-t^2/2)$, use the relationships of the Fourier transforms of $f'(t)$ and $tf(t)$ to that of $f(t)$ itself to find a simple differential equation satisfied by $\tilde{f}(\omega)$, the Fourier transform of $f(t)$, and hence determine $\tilde{f}(\omega)$ to within a constant. Use this result to solve for $h(t)$ the integral equation*

$$\int_{-\infty}^{\infty} e^{-t(t-2x)/2} h(t) \, dt = e^{3x^2/8}.$$

As a standard result,

$$\mathscr{F}\left[f'(t) \right] = i\omega \tilde{f}(\omega),$$

though we will not need this relationship in the following solution.

From its definition,

$$\mathscr{F}\left[tf(t)\right] = \frac{1}{\sqrt{2\pi}}\int_{-\infty}^{\infty} tf(t)\,e^{-i\omega t}\,dt$$

$$= \frac{1}{-i}\frac{d}{d\omega}\left(\frac{1}{\sqrt{2\pi}}\int_{-\infty}^{\infty} f(t)\,e^{-i\omega t}\,dt\right) = i\frac{d\tilde{f}}{d\omega}.$$

Now, for the particular given function,

$$\tilde{f}(\omega) = \frac{1}{\sqrt{2\pi}}\int_{-\infty}^{\infty} e^{-t^2/2}\,e^{-i\omega t}\,dt$$

$$= \frac{1}{\sqrt{2\pi}}\left[\frac{e^{-t^2/2}\,e^{-i\omega t}}{-i\omega}\right]_{-\infty}^{\infty} + \frac{1}{\sqrt{2\pi}}\int_{-\infty}^{\infty}\frac{te^{-t^2/2}\,e^{-i\omega t}}{-i\omega}\,dt$$

$$= 0 - \frac{1}{i\omega}\,i\frac{d\tilde{f}}{d\omega}.$$

Hence,

$$\frac{d\tilde{f}}{d\omega} = -\omega\tilde{f} \quad\Rightarrow\quad \ln\tilde{f} = -\tfrac{1}{2}\omega^2 + k \quad\Rightarrow\quad \tilde{f} = Ae^{-\omega^2/2}, \qquad (*)$$

giving $\tilde{f}(\omega)$ to within a multiplicative constant.

Now, we are given

$$\int_{-\infty}^{\infty} e^{-t(t-2x)/2}\,h(t)\,dt = e^{3x^2/8},$$

$$\Rightarrow\quad \int_{-\infty}^{\infty} e^{-(t-x)^2/2}\,e^{x^2/2}\,h(t)\,dt = e^{3x^2/8},$$

$$\Rightarrow\quad \int_{-\infty}^{\infty} e^{-(x-t)^2/2}\,h(t)\,dt = e^{-x^2/8}. \qquad (**)$$

The LHS of (**) is a convolution integral, and so applying the convolution theorem for Fourier transforms and result (*), used twice, yields

$$\sqrt{2\pi}Ae^{-\omega^2/2}\tilde{h}(\omega) = \mathscr{F}\left[e^{-(x/2)^2/2}\right] = Ae^{-(2\omega)^2/2},$$

$$\Rightarrow\quad \sqrt{2\pi}\,\tilde{h}(\omega) = e^{-3\omega^2/2} = e^{-(\sqrt{3}\omega)^2/2},$$

$$\Rightarrow\quad h(t) = \frac{1}{\sqrt{2\pi A}}e^{-(t/\sqrt{3})^2/2} = \frac{1}{\sqrt{2\pi A}}e^{-t^2/6}.$$

We now substitute in (**) to determine A:

$$\int_{-\infty}^{\infty} e^{-(x-t)^2/2} \frac{1}{\sqrt{2\pi A}} e^{-t^2/6} \, dt = e^{-x^2/8},$$

$$\frac{1}{\sqrt{2\pi A}} \int_{-\infty}^{\infty} e^{-2t^2/3} e^{xt} e^{-x^2/2} e^{x^2/8} \, dt = 1,$$

$$\frac{1}{\sqrt{2\pi A}} \int_{-\infty}^{\infty} \exp\left[-\frac{2}{3}\left(t - \frac{3x}{4} \right)^2 \right] dt = 1.$$

From the normalisation of the Gaussian integral, this implies that

$$\frac{1}{\sqrt{2\pi A}} = \frac{2}{\sqrt{2\pi}\sqrt{3}},$$

which in turn means $A = \sqrt{3}/2$, giving finally that

$$h(t) = \sqrt{\frac{2}{3\pi}} e^{-t^2/6}.$$

This solution can be checked by resubstitution.

23.11 *At an international 'peace' conference a large number of delegates are seated around a circular table with each delegation sitting near its allies and diametrically opposite the delegation most bitterly opposed to it. The position of a delegate is denoted by θ, with $0 \leq \theta \leq 2\pi$. The fury $f(\theta)$ felt by the delegate at θ is the sum of his own natural hostility $h(\theta)$ and the influences on him of each of the other delegates; a delegate at position ϕ contributes an amount $K(\theta - \phi)f(\phi)$. Thus*

$$f(\theta) = h(\theta) + \int_0^{2\pi} K(\theta - \phi)f(\phi) \, d\phi.$$

Show that if $K(\psi)$ takes the form $K(\psi) = k_0 + k_1 \cos \psi$ then

$$f(\theta) = h(\theta) + p + q \cos \theta + r \sin \theta$$

and evaluate p, q and r. A positive value for k_1 implies that delegates tend to placate their opponents but upset their allies, whilst negative values imply that they calm their allies but infuriate their opponents. A walkout will occur if $f(\theta)$ exceeds a certain threshold value for some θ. Is this more likely to happen for positive or for negative values of k_1?

Given that $K(\psi) = k_0 + k_1 \cos \psi$, we try a solution $f(\theta) = h(\theta) + p + q \cos \theta + r \sin \theta$, reducing the equation to

$$p + q \cos \theta + r \sin \theta$$
$$= \int_0^{2\pi} [k_0 + k_1(\cos \theta \cos \phi + \sin \theta \sin \phi)]$$
$$\times [h(\phi) + p + q \cos \phi + r \sin \phi] \, d\phi$$
$$= k_0(H + 2\pi p) + k_1(H_c \cos \theta + H_s \sin \theta + \pi q \cos \theta + \pi r \sin \theta),$$

where $H = \int_0^{2\pi} h(z) \, dz$, $H_c = \int_0^{2\pi} h(z) \cos z \, dz$ and $H_s = \int_0^{2\pi} h(z) \sin z \, dz$.

Thus, on equating the constant terms and the coefficients of $\cos \theta$ and $\sin \theta$, we have

$$p = k_0 H + 2\pi k_0 p \quad \Rightarrow \quad p = \frac{k_0 H}{1 - 2\pi k_0},$$
$$q = k_1 H_c + k_1 \pi q \quad \Rightarrow \quad q = \frac{k_1 H_c}{1 - k_1 \pi},$$
$$r = k_1 H_s + k_1 \pi r \quad \Rightarrow \quad r = \frac{k_1 H_s}{1 - k_1 \pi}.$$

And so the full solution for $f(\theta)$ is given by

$$f(\theta) = h(\theta) + \frac{k_0 H}{1 - 2\pi k_0} + \frac{k_1 H_c}{1 - k_1 \pi} \cos \theta + \frac{k_1 H_s}{1 - k_1 \pi} \sin \theta$$
$$= h(\theta) + \frac{k_0 H}{1 - 2\pi k_0} + \frac{k_1}{1 - k_1 \pi} (H_c^2 + H_s^2)^{1/2} \cos(\theta - \alpha),$$

where $\tan \alpha = H_s / H_c$.

Clearly, the maximum value of $f(\theta)$ will depend upon $h(\theta)$ and its various integrals, but it is most likely to exceed any particular value if k_1 is positive and $\approx \pi^{-1}$. Stick with your friends!

23.13 The operator $\mathcal{M}$ is defined by

$$\mathcal{M} f(x) \equiv \int_{-\infty}^{\infty} K(x, y) f(y) \, dy,$$

where $K(x, y) = 1$ inside the square $|x| < a$, $|y| < a$ and $K(x, y) = 0$ elsewhere. Consider the possible eigenvalues of $\mathcal{M}$ and the eigenfunctions that correspond to them; show that the only possible eigenvalues are 0 and 2a and determine the corresponding eigenfunctions. Hence find the general solution of

$$f(x) = g(x) + \lambda \int_{-\infty}^{\infty} K(x, y) f(y) \, dy.$$

From the given properties of $K(x, y)$ we can assert the following.

(i) No matter what the form of $f(x)$, $\mathcal{M}f(x) = 0$ if $|x| > a$.

(ii) All functions for which both $\int_{-a}^{a} f(y)\,dy = 0$ and $f(x) = 0$ for $|x| > a$ are eigenfunctions corresponding to eigenvalue 0.

(iii) For any function $f(x)$, the integral $\int_{-a}^{a} f(y)\,dy$ is equal to a constant whose value is independent of x; thus $f(x)$ can only be an eigenfunction if it is equal to a constant, μ, for $-a \le x \le a$ and is zero otherwise. For this case $\int_{-a}^{a} f(y)\,dy = 2a\mu$ and the eigenvalue is $2a$.

Point (iii) gives the only possible non-zero eigenvalue, whilst point (ii) shows that eigenfunctions corresponding to zero eigenvalues do exist.

Denote by $S(x, a)$ the function that has unit value for $|x| \le a$ and zero value otherwise; $K(x, y)$ could be expressed as $K(x, y) = S(x, a)S(y, a)$. Substitute the trial solution $f(x) = g(x) + kS(x, a)$ into

$$f(x) = g(x) + \lambda \int_{-\infty}^{\infty} K(x, y)f(y)\,dy.$$

This gives

$$g(x) + kS(x, a) = g(x) + \lambda \int_{-\infty}^{\infty} K(x, y)[g(y) + kS(y, a)]\,dy,$$

$$kS(x, a) = \lambda S(x, a) \int_{-a}^{a} g(y)\,dy + \lambda k\, 2aS(x, a).$$

Here, having replaced $K(x, y)$ by $S(x, a)S(y, a)$, we use the factor $S(y, a)$ to reduce the limits of the y-integration from $\pm\infty$ to $\pm a$. As this result is to hold for all x we must have

$$k = \frac{\lambda G}{1 - 2a\lambda}, \quad \text{where } G = \int_{-a}^{a} g(y)\,dy.$$

The general solution is thus

$$f(x) = \begin{cases} g(x) + \dfrac{\lambda G}{1 - 2a\lambda} & \text{for } |x| \le a, \\ g(x) & \text{for } |x| > a. \end{cases}$$

23.15 *Use Fredholm theory to show that, for the kernel*

$$K(x, z) = (x + z) \exp(x - z)$$

over the interval $[0, 1]$, *the resolvent kernel is*

$$R(x, z; \lambda) = \frac{\exp(x - z)[(x + z) - \lambda(\frac{1}{2}x + \frac{1}{2}z - xz - \frac{1}{3})]}{1 - \lambda - \frac{1}{12}\lambda^2},$$

and hence solve

$$y(x) = x^2 + 2 \int_0^1 (x + z) \exp(x - z) y(z) \, dz,$$

expressing your answer in terms of I_n, *where* $I_n = \int_0^1 u^n \exp(-u) \, du$.

We calculate successive values of d_n and $D_n(x, z)$ using the Fredholm recurrence relations:

$$d_n = \int_a^b D_{n-1}(x, x) \, dx,$$

$$D_n(x, z) = K(x, z)d_n - n \int_a^b K(x, z_1)D_{n-1}(z_1, z) \, dz_1,$$

starting from $d_0 = 1$ and $D_0(x, z) = (x + z)e^{x-z}$. In the first iteration we obtain

$$d_1 = \int_0^1 (u + u)e^{u-u} \, du = 1,$$

$$D_1(x, z) = (x + z)e^{x-z}(1) - 1 \int_0^1 (x + u)e^{x-u}(u + z)e^{u-z} \, du$$

$$= (x + z)e^{x-z} - e^{x-z} \int_0^1 [xz + (x + z)u + u^2] \, du$$

$$= e^{x-z}[\tfrac{1}{2}(x + z) - xz - \tfrac{1}{3}].$$

Performing the second iteration gives

$$d_2 = \int_0^1 e^{u-u}(u - u^2 - \tfrac{1}{3}) \, du = \tfrac{1}{2} - \tfrac{1}{3} - \tfrac{1}{3} = -\tfrac{1}{6},$$

$$D_2(x, z) = (x + z)e^{x-z}(-\tfrac{1}{6})$$

$$- 2 \int_0^1 (x + u)e^{x-u}e^{u-z} \left[\tfrac{1}{2}(u + z) - uz - \tfrac{1}{3} \right] du$$

$$= e^{x-z} \left\{ -\tfrac{1}{6}(x + z) - 2 \left[x \left(\frac{1}{4} + \frac{z}{2} - \frac{z}{2} - \frac{1}{3} \right) + \left(\frac{1}{6} + \frac{z}{4} - \frac{z}{3} - \frac{1}{6} \right) \right] \right\}$$

$$= e^{x-z} \left\{ -\tfrac{1}{6}(x + z) - 2 \left[-\frac{x}{12} - \frac{z}{12} \right] \right\} = 0.$$

Since $D_2(x, z) = 0$, $d_3 = 0$, $D_3(x, z) = 0$, etc. Consequently both $D(x, z; \lambda)$ and $d(\lambda)$ are finite, rather than infinite, series:

$$D(x, z; \lambda) = (x + z)e^{x-z} - \lambda \left[\tfrac{1}{2}(x + z) - xz - \tfrac{1}{3} \right] e^{x-z},$$

$$d(\lambda) = 1 - \lambda + \left(-\tfrac{1}{6} \right) \frac{\lambda^2}{2!} = 1 - \lambda - \tfrac{1}{12} \lambda^2.$$

The resolvent kernel $R(x, z; \lambda)$, given by the ratio $D(x, z; \lambda)/d(\lambda)$, is therefore as stated in the question.

For the particular integral equation, $\lambda = 2$ and $f(x) = x^2$. It follows that

$$d(\lambda) = 1 - 2 - \tfrac{4}{12} = -\tfrac{4}{3} \quad \text{and} \quad D(x, z : \lambda) = (2xz + \tfrac{2}{3})e^{x-z}.$$

The solution is therefore given by

$$y(x) = f(x) + \lambda \int_0^1 R(x, z; \lambda)f(z)\, dz$$

$$= x^2 + 2 \int_0^1 \frac{(2xz + \tfrac{2}{3})z^2 e^{x-z}}{-\tfrac{4}{3}}\, dz$$

$$= x^2 - \int_0^1 (3xz^3 + z^2)e^{x-z}\, dz$$

$$= x^2 - (3xI_3 + I_2)e^x.$$

24

Complex variables

24.1 *Find an analytic function of $z = x + iy$ whose imaginary part is*

$$(y \cos y + x \sin y) \exp x.$$

If the required function is $f(z) = u + iv$, with $v = (y \cos y + x \sin y) \exp x$, then, from the Cauchy–Riemann equations,

$$\frac{\partial v}{\partial x} = e^x(y \cos y + x \sin y + \sin y) = -\frac{\partial u}{\partial y}.$$

Integrating with respect to y gives

$$u = -e^x \int (y \cos y + x \sin y + \sin y)\, dy + f(x)$$

$$= -e^x \left(y \sin y - \int \sin y\, dy - x \cos y - \cos y \right) + f(x)$$

$$= -e^x(y \sin y + \cos y - x \cos y - \cos y) + f(x)$$

$$= e^x(x \cos y - y \sin y) + f(x).$$

We determine $f(x)$ by applying the second Cauchy–Riemann equation, which equates $\partial u/\partial x$ with $\partial v/\partial y$:

$$\frac{\partial u}{\partial x} = e^x(x \cos y - y \sin y + \cos y) + f'(x),$$

$$\frac{\partial v}{\partial y} = e^x(\cos y - y \sin y + x \cos y).$$

By comparison, $f'(x) = 0 \quad \Rightarrow \quad f(x) = k,$

where k is a real constant that can be taken as zero. Hence, the analytic function is given by

$$f(z) = u + iv = e^x(x \cos y - y \sin y + iy \cos y + ix \sin y)$$
$$= e^x[(\cos y + i \sin y)(x + iy)]$$
$$= e^x e^{iy}(x + iy)$$
$$= ze^z.$$

The final line confirms explicitly that this is a function of z alone (as opposed to a function of both z and z^*).

24.3 *Find the radii of convergence of the following Taylor series:*

$$\text{(a)} \sum_{n=2}^{\infty} \frac{z^n}{\ln n}, \quad \text{(b)} \sum_{n=1}^{\infty} \frac{n! z^n}{n^n},$$

$$\text{(c)} \sum_{n=1}^{\infty} z^n n^{\ln n}, \quad \text{(d)} \sum_{n=1}^{\infty} \left(\frac{n+p}{n}\right)^{n^2} z^n, \text{ with } p \text{ real.}$$

In each case we consider the series as $\sum_n a_n z^n$ and apply the formula

$$\frac{1}{R} = \lim_{n \to \infty} |a_n|^{1/n}$$

derived from considering the Cauchy root test for absolute convergence.

(a) $\quad \dfrac{1}{R} = \lim_{n \to \infty} \left(\dfrac{1}{\ln n}\right)^{1/n} = 1$, since $-n^{-1} \ln \ln n \to 0$ as $n \to \infty$.

Thus $R = 1$. For interest, we also note that at the point $z = 1$ the series is

$$\sum_{n=2}^{\infty} \frac{1}{\ln n} > \sum_{n=2}^{\infty} \frac{1}{n},$$

which diverges. This shows that the given series diverges at this point on its circle of convergence.

(b) $\quad \dfrac{1}{R} = \lim_{n \to \infty} \left(\dfrac{n!}{n^n}\right)^{1/n}.$

Since the nth root of $n!$ tends to n as $n \to \infty$, the limit of this ratio is that of n/n, namely unity. Thus $R = 1$ and the series converges inside the unit circle.

(c) $\quad \dfrac{1}{R} = \lim_{n \to \infty} \left(n^{\ln n}\right)^{1/n} = \lim_{n \to \infty} n^{(\ln n)/n}$

$$= \lim_{n \to \infty} \exp\left[\frac{\ln n}{n} \ln n\right] = \exp(0) = 1.$$

Thus $R = 1$ and the series converges inside the unit circle. It is obvious that the series diverges at the point $z = 1$.

(d)
$$\frac{1}{R} = \lim_{n \to \infty} \left[\left(\frac{n+p}{n} \right)^{n^2} \right]^{1/n} = \lim_{n \to \infty} \left(\frac{n+p}{n} \right)^n$$
$$= \lim_{n \to \infty} \left(1 + \frac{p}{n} \right)^n = e^p.$$

Thus $R = e^{-p}$ and the series converges inside a circle of this radius centred on the origin $z = 0$.

24.5 *Determine the types of singularities (if any) possessed by the following functions at $z = 0$ and $z = \infty$:*

(a) $(z-2)^{-1}$,   (b) $(1+z^3)/z^2$,   (c) $\sinh(1/z)$,

(d) e^z/z^3,   (e) $z^{1/2}/(1+z^2)^{1/2}$.

(a) Although $(z-2)^{-1}$ has a simple pole at $z = 2$, at both $z = 0$ and $z = \infty$ it is well behaved and analytic.

(b) Near $z = 0$, $f(z) = (1+z^3)/z^2$ behaves like $1/z^2$ and so has a double pole there. It is clear that as $z \to \infty$ $f(z)$ behaves as z and so has a simple pole there; this can be made more formal by setting $z = 1/\xi$ to obtain $g(\xi) = \xi^2 + \xi^{-1}$ and considering $\xi \to 0$. This leads to the same conclusion.

(c) As $z \to \infty$, $f(z) = \sinh(1/z)$ behaves like $\sinh \xi$ as $\xi \to 0$, i.e. analytically. However, the definition of the sinh function involves an infinite series — in this case an infinite series of inverse powers of z. Thus, no finite n for which

$$\lim_{z \to 0} [z^n f(z)] \text{ is finite}$$

can be found, and $f(z)$ has an essential singularity at $z = 0$.

(d) Near $z = 0$, $f(z) = e^z/z^3$ behaves as $1/z^3$ and has a pole of order 3 at the origin. At $z = \infty$ it has an obvious essential singularity; formally, the series expansion of $e^{1/\xi}$ about $\xi = 0$ contains arbitrarily high inverse powers of ξ.

(e) Near $z = 0$, $f(z) = z^{1/2}/(1+z^2)^{1/2}$ behaves as $z^{1/2}$ and therefore has a branch point there. To investigate its behaviour as $z \to \infty$, we set $z = 1/\xi$ and obtain

$$f(z) = g(\xi) = \left(\frac{\xi^{-1}}{1 + \xi^{-2}} \right)^{1/2} = \left(\frac{\xi}{\xi^2 + 1} \right)^{1/2} \sim \xi^{1/2} \text{ as } \xi \to 0.$$

Hence $f(z)$ also has a branch point at $z = \infty$.

24.7 *Find the real and imaginary parts of the functions* (i) z^2, (ii) e^z, *and* (iii) $\cosh \pi z$. *By considering the values taken by these parts on the boundaries of the region $x \geq 0$, $y \leq 1$, determine the solution of Laplace's equation in that region that satisfies the boundary conditions*

$$\phi(x,0) = 0, \qquad\qquad \phi(0,y) = 0,$$
$$\phi(x,1) = x, \qquad\qquad \phi(1,y) = y + \sin \pi y.$$

Writing $f_k(z) = u_k(x,y) + iv_k(x,y)$, we have

(i) $f_1(z)$ $=$ $z^2 = (x+iy)^2$

$\Rightarrow$ $u_1 = x^2 - y^2$ and $v_1 = 2xy$,

(ii) $f_2(z)$ $=$ $e^z = e^{x+iy} = e^x(\cos y + i \sin y)$

$\Rightarrow$ $u_2 = e^x \cos y$ and $v_2 = e^x \sin y$,

(iii) $f_3(z)$ $=$ $\cosh \pi z = \cosh \pi x \cos \pi y + i \sinh \pi x \sin \pi y$

$\Rightarrow$ $u_3 = \cosh \pi x \cos \pi y$ and $v_3 = \sinh \pi x \sin \pi y$.

All of these u and v are necessarily solutions of Laplace's equation (this follows from the Cauchy–Riemann equations), and, since Laplace's equation is linear, we can form any linear combination of them and it will also be a solution. We need to choose the combination that matches the given boundary conditions.

Since the third and fourth conditions involve x and $\sin \pi y$, and these appear only in v_1 and v_3, respectively, let us try a linear combination of them:

$$\phi(x,y) = A(2xy) + B(\sinh \pi x \sin \pi y).$$

The requirement $\phi(x,0) = 0$ is clearly satisfied, as is $\phi(0,y) = 0$. The condition $\phi(x,1) = x$ becomes $2Ax + 0 = x$, requiring $A = \frac{1}{2}$, and the ,remaining condition, $\phi(1,y) = y + \sin \pi y$, takes the form $y + B \sinh \pi \sin \pi y = y + \sin \pi y$, thus determining B as $1/\sinh \pi$.

With ϕ a solution of Laplace's equation and all of the boundary conditions satisfied, the uniqueness theorem guarantees that

$$\phi(x,y) = xy + \frac{\sinh \pi x \sin \pi y}{\sinh \pi}$$

is the correct solution.

24.9 *The* fundamental theorem of algebra *states that, for a complex polynomial* $p_n(z)$ *of degree n, the equation* $p_n(z) = 0$ *has precisely n complex roots. By applying* Liouville's theorem, *which reads*

If $f(z)$ is analytic and bounded for all z then f is a constant,

to $f(z) = 1/p_n(z)$, *prove that* $p_n(z) = 0$ *has at least one complex root. Factor out that root to obtain* $p_{n-1}(z)$ *and, by repeating the process, prove the fundamental theorem.*

We prove this result by the method of contradiction. Suppose $p_n(z) = 0$ has no roots in the complex plane, then $f_n(z) = 1/p_n(z)$ is bounded for all z and, by Liouville's theorem, is therefore a constant. It follows that $p_n(z)$ is also a constant and that $n = 0$. However, if $n > 0$ we have a contradiction and it was wrong to suppose that $p_n(z) = 0$ has no roots; it must have at least one. Let one of them be $z = z_1$; i.e. $p_n(z)$, being a polynomial, can be written $p_n(z) = (z - z_1)p_{n-1}(z)$.

Now, by considering $f_{n-1}(z) = 1/p_{n-1}(z)$ in just the same way, we can conclude that either $n - 1 = 0$ or a further reduction is possible. It is clear that n such reductions are needed to make f_0 a constant, thus establishing that $p_n(z) = 0$ has precisely n (complex) roots.

Many of the remaining exercises in this chapter involve contour integration and the choice of a suitable contour. In order to save the space taken by drawing several broadly similar contours that differ only in notation, the positions of poles, the values of lengths or angles, or other minor details, we show in figure 24.1 a number of typical contour types to which reference can be made.

24.11 *The function*

$$f(z) = (1 - z^2)^{1/2}$$

of the complex variable z is defined to be real and positive on the real axis for $-1 < x < 1$. *Using cuts running along the real axis for* $1 < x < +\infty$ *and* $-\infty < x < -1$, *show how* $f(z)$ *is made single-valued and evaluate it on the upper and lower sides of both cuts.*

Use these results and a suitable contour in the complex z-plane to evaluate the integral

$$I = \int_1^\infty \frac{dx}{x(x^2 - 1)^{1/2}}.$$

Confirm your answer by making the substitution $x = \sec\theta$.

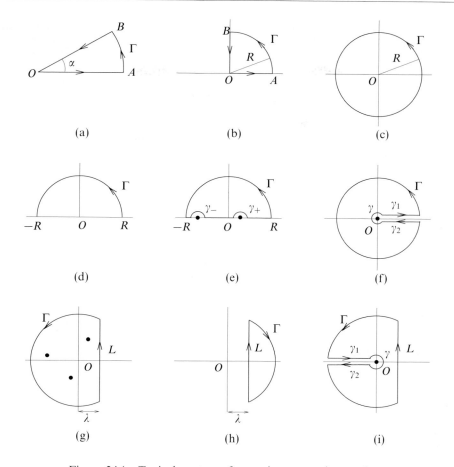

Figure 24.1   Typical contours for use in contour integration.

As usual when dealing with branch cuts aimed at making a multi-valued function into a single-valued one, we introduce polar coordinates centred on the branch points. For $f(z)$ the branch points are at $z = \pm 1$, and so we define r_1 as the distance of z from the point 1 and θ_1 as the angle the line joining 1 to z makes with the part of the x-axis for which $1 < x < +\infty$, with $0 \le \theta_1 \le 2\pi$. Similarly, r_2 and θ_2 are centred on the point -1, but θ_2 lies in the range $-\pi \le \theta_2 \le \pi$.

With these definitions,

$$f(z) = (1 - z^2)^{1/2} = (1 - z)^{1/2}(1 + z)^{1/2}$$
$$= \left[(-r_1 e^{i\theta_1})(r_2 e^{i\theta_2})\right]^{1/2}$$
$$= (r_1 r_2)^{1/2} e^{i(\theta_1 + \theta_2 - \pi)/2}.$$

In the final line the choice between $\exp(+i\pi)$ and $\exp(-i\pi)$ for dealing with the

391

minus sign appearing before r_1 in the second line was resolved by the requirement that $f(z)$ is real and positive when $-1 < x < 1$ with $y = 0$. For these values of z, $r_1 = 1 - x$, $r_2 = 1 + x$, $\theta_1 = \pi$ and $\theta_2 = 0$. Thus,

$$f(z) = [(1 - x)(1 + x)]^{1/2} e^{(\pi+0-\pi)/2} = (1 - x^2)^{1/2} e^{i0} = +(1 - x^2)^{1/2},$$

as required.

Now applying the same prescription to points lying just above and just below each of the cuts, we have

$$x > 1, y = 0_+ \qquad r_1 = x - 1 \quad r_2 = x + 1 \quad \theta_1 = 0 \quad \theta_2 = 0$$

$$\Rightarrow \quad f(z) = (x^2 - 1)^{1/2} e^{i(0+0-\pi)/2} = -i(x^2 - 1)^{1/2},$$

$$x > 1, y = 0_- \qquad r_1 = x - 1 \quad r_2 = x + 1 \quad \theta_1 = 2\pi \quad \theta_2 = 0$$

$$\Rightarrow \quad f(z) = (x^2 - 1)^{1/2} e^{i(2\pi+0-\pi)/2} = i(x^2 - 1)^{1/2},$$

$$x < -1, y = 0_+ \qquad r_1 = 1 - x \quad r_2 = -x - 1 \quad \theta_1 = \pi \quad \theta_2 = \pi$$

$$\Rightarrow \quad f(z) = (x^2 - 1)^{1/2} e^{i(\pi+\pi-\pi)/2} = i(x^2 - 1)^{1/2},$$

$$x < -1, y = 0_- \qquad r_1 = 1 - x \quad r_2 = -x - 1 \quad \theta_1 = \pi \quad \theta_2 = -\pi$$

$$\Rightarrow \quad f(z) = (x^2 - 1)^{1/2} e^{i(\pi-\pi-\pi)/2} = -i(x^2 - 1)^{1/2}.$$

To use these results to evaluate the given integral I, consider the contour integral

$$J = \int_C \frac{dz}{z(1 - z^2)^{1/2}} = \int_c \frac{dz}{z f(z)}.$$

Here C is a large circle (consisting of arcs Γ_1 and Γ_2 in the upper and lower half-planes, respectively) of radius R centred on the origin but indented along the positive and negative x-axes by the cuts considered earlier. At the ends of the cuts are two small circles γ_1 and γ_2 that enclose the branch points $z = 1$ and $z = -1$, respectively. Thus the complete closed contour, starting from γ_1 and moving along the positive real axis, consists of, in order, circle γ_1, cut C_1, arc Γ_1, cut C_2, circle γ_2, cut C_3, arc Γ_2 and cut C_4, leading back to γ_1.

On the arcs Γ_1 and Γ_2 the integrand is $O(R^{-2})$ and the contributions to the contour integral $\to 0$ as $R \to \infty$. For the small circle γ_1, where we can set $z = 1 + \rho e^{i\phi}$ with $dz = i\rho e^{i\phi} d\phi$, we have

$$\int_{\gamma_1} \frac{dz}{z(1 + z)^{1/2}(1 - z)^{1/2}} = \int_0^{2\pi} \frac{i\rho e^{i\phi}}{(1 + \rho e^{i\phi})(2 + \rho e^{i\phi})^{1/2}(-\rho e^{i\phi})^{1/2}} d\phi,$$

and this $\to 0$ as $\rho \to 0$. Similarly, the small circle γ_2 contributes nothing to the contour integral. This leaves only the contributions from the four arms of the branch cuts. To relate these to I we use our previous results about the value of $f(z)$ on the various arms:

$$\text{on } C_1, z = x \text{ and } \int_{C_1} = \int_1^\infty \frac{dx}{x[-i(x^2-1)^{1/2}]} = iI;$$

$$\text{on } C_2, z = -x \text{ and } \int_{C_2} = \int_\infty^1 \frac{-dx}{-x[i(x^2-1)^{1/2}]} = iI;$$

$$\text{on } C_3, z = -x \text{ and } \int_{C_3} = \int_1^\infty \frac{-dx}{-x[-i(x^2-1)^{1/2}]} = iI;$$

$$\text{on } C_4, z = x \text{ and } \int_{C_1} = \int_\infty^1 \frac{dx}{x[i(x^2-1)^{1/2}]} = iI.$$

So the full contour integral around C has the value $4iI$. But, this must be the same as $2\pi i$ times the residue of $z^{-1}(1-z^2)^{-1/2}$ at $z = 0$, which is the only pole of the integrand inside the contour. The residue is clearly unity, and so we deduce that $I = \pi/2$.

This particular integral can be evaluated much more simply using elementary methods. Setting $x = \sec\theta$ with $dx = \sec\theta\tan\theta\, d\theta$ gives

$$
\begin{aligned}
I &= \int_1^\infty \frac{dx}{x(x^2-1)^{1/2}} \\
&= \int_0^{\pi/2} \frac{\sec\theta\tan\theta\, d\theta}{\sec\theta\,(\sec^2\theta-1)^{1/2}} = \int_0^{\pi/2} d\theta = \frac{\pi}{2},
\end{aligned}
$$

and so verifies the result obtained by contour integration.

24.13 *Prove that if $f(z)$ has a simple zero at z_0 then $1/f(z)$ has residue $1/f'(z_0)$ there. Hence evaluate*

$$\int_{-\pi}^\pi \frac{\sin\theta}{a-\sin\theta}\, d\theta,$$

where a is real and > 1.

If $f(z)$ is analytic and has a simple zero at $z = z_0$ then it can be written as

$$f(z) = \sum_{n=1}^\infty a_n(z-z_0)^n, \qquad \text{with } a_1 \neq 0.$$

Using a binomial expansion,

$$\frac{1}{f(z)} = \frac{1}{a_1(z-z_0)\left(1 + \sum_{n=2}^{\infty} \frac{a_n}{a_1}(z-z_0)^{n-1}\right)}$$

$$= \frac{1}{a_1(z-z_0)}(1 + b_1(z-z_0) + b_2(z-z_0)^2 + \cdots),$$

for some coefficients, b_i. The residue at $z = z_0$ is clearly a_1^{-1}.

But, from differentiating the Taylor expansion,

$$f'(z) = \sum_{n=1}^{\infty} n a_n (z-z_0)^{n-1},$$

$$\Rightarrow \quad f'(z_0) = a_1 + 0 + 0 + \cdots = a_1,$$

i.e. the residue $= \dfrac{1}{a_1}$ can also be expressed as $\dfrac{1}{f'(z_0)}$.

Denote the required integral by I and consider the contour integral

$$J = \int_C \frac{dz}{a - \frac{1}{2i}(z - z^{-1})} = \int_C \frac{2iz\,dz}{2aiz - z^2 + 1},$$

where C is the unit circle, i.e. contour (c) of figure 24.1 with $R = 1$. The denominator has simple zeros at $z = ai \pm \sqrt{-a^2 + 1} = i(a \pm \sqrt{a^2-1})$. Since a is strictly greater than 1, $\alpha = i(a - \sqrt{a^2-1})$ lies strictly inside the unit circle, whilst $\beta = i(a + \sqrt{a^2-1})$ lies strictly outside it (and need not be considered further).

Extending the previous result to the case of $h(z) = g(z)/f(z)$, where $g(z)$ is analytic at z_0, the residue of $h(z)$ at $z = z_0$ can be seen to be $g(z_0)/f'(z_0)$. Applying this, we find that the residue of the integrand at $z = \alpha$ is given by

$$\left.\frac{2iz}{2ai - 2z}\right|_{z=\alpha} = \frac{i\alpha}{ai - ai + i\sqrt{a^2-1}} = \frac{\alpha}{\sqrt{a^2-1}}.$$

Now on the unit circle, $z = e^{i\theta}$ with $dz = i e^{i\theta} d\theta$, and J can be written as

$$J = \int_{-\pi}^{\pi} \frac{i e^{i\theta}\,d\theta}{a - \frac{1}{2i}(e^{i\theta} - e^{-i\theta})} = \int_{-\pi}^{\pi} \frac{i(\cos\theta + i\sin\theta)\,d\theta}{a - \sin\theta}.$$

Hence,

$$I = -\operatorname{Re} J = -\operatorname{Re} 2\pi i \frac{i(a - \sqrt{a^2-1})}{\sqrt{a^2-1}}$$

$$= 2\pi \left(\frac{a}{\sqrt{a^2-1}} - 1\right).$$

394

Although it is not asked for, we can also deduce from the fact that the residue at $z = \alpha$ is purely imaginary that

$$\int_{-\pi}^{\pi} \frac{\cos \theta}{a - \sin \theta} \, d\theta = 0,$$

a result that can also be obtained by more elementary means, when it is noted that the numerator of the integrand is the derivative of the denominator.

24.15 *Prove that*

$$\int_0^\infty \frac{\cos mx}{4x^4 + 5x^2 + 1} \, dx = \frac{\pi}{6} \left(4e^{-m/2} - e^{-m} \right) \quad \textit{for } m > 0.$$

Since, when z is on the real axis, the integrand is equal to

$$\text{Re} \, \frac{e^{imz}}{(z^2 + 1)(4z^2 + 1)} = \text{Re} \, \frac{e^{imz}}{(z + i)(z - i)(2z + i)(2z - i)},$$

we consider the integral of $f(z) = \dfrac{e^{imz}}{(z + i)(z - i)(2z + i)(2z - i)}$ around contour (d) in figure 24.1.

As $|f(z)| \sim |z|^{-4}$ as $z \to \infty$ and $m > 0$, all the conditions for Jordan's lemma to hold are satisfied and the integral around the large semicircle contributes nothing. For this integrand there are two poles inside the contour, at $z = i$ and at $z = \frac{1}{2}i$. The respective residues are

$$\frac{e^{-m}}{2i \, 3i \, i} = \frac{ie^{-m}}{6} \quad \text{and} \quad \frac{e^{-m/2}}{\frac{3i}{2} \left(-\frac{i}{2} \right) 2i} = \frac{-2ie^{-m/2}}{3}.$$

The residue theorem therefore reads

$$\int_{-\infty}^{\infty} \frac{e^{imx}}{4x^4 + 5x^2 + 1} \, dx + 0 = 2\pi i \left(\frac{ie^{-m}}{6} - \frac{2ie^{-m/2}}{3} \right),$$

and the stated result follows from equating real parts and changing the lower integration limit, recognising that the integrand is symmetric about $x = 0$ and so the integral from 0 to ∞ is equal to half of that from $-\infty$ to ∞.

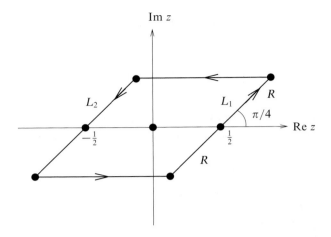

Figure 24.2   The contour used in exercise 24.17.

24.17   *The following is an alternative (and roundabout!) way of evaluating the Gaussian integral.*

(a) *Prove that the integral of* $[\exp(i\pi z^2)]\operatorname{cosec}\pi z$ *around the parallelogram with corners* $\pm 1/2 \pm R\exp(i\pi/4)$ *has the value* $2i$.

(b) *Show that the parts of the contour parallel to the real axis give no contribution when* $R \to \infty$.

(c) *Evaluate the integrals along the other two sides by putting* $z' = r\exp(i\pi/4)$ *and working in terms of* $z' + \frac{1}{2}$ *and* $z' - \frac{1}{2}$. *Hence by letting* $R \to \infty$ *show that*

$$\int_{-\infty}^{\infty} e^{-\pi r^2}\, dr = 1.$$

The integral is

$$\int_C e^{i\pi z^2} \operatorname{cosec}\pi z\, dz = \int_C \frac{e^{i\pi z^2}}{\sin \pi z}\, dz$$

and the suggested contour C is shown in figure 24.2.

(a) The integrand has (simple) poles only on the real axis at $z = n$, where n is an integer. The only such pole enclosed by C is at $z = 0$. The residue there is

$$a_{-1} = \lim_{z \to 0} \frac{z e^{i\pi z^2}}{\sin \pi z} = \frac{1}{\pi}.$$

The value of the integral around C is therefore $2\pi i \times (\pi^{-1}) = 2i$.

(b) On the parts of C parallel to the real axis, $z = \pm Re^{i\pi/4} + x'$, where $-\frac{1}{2} \leq x' \leq \frac{1}{2}$. The integrand is thus given by

$$
\begin{aligned}
f(z) &= \frac{1}{\sin \pi z} \exp \left[i\pi(\pm Re^{i\pi/4} + x')^2 \right] \\
&= \frac{1}{\sin \pi z} \exp \left[i\pi \left(R^2 e^{i\pi/2} \pm 2Rx'e^{i\pi/4} + x'^2 \right) \right] \\
&= \frac{1}{\sin \pi z} \exp \left[-\pi R^2 \pm \frac{2\pi i Rx'}{\sqrt{2}}(1 + i) + i\pi x'^2 \right] \\
&= O \left(\exp[-\pi R^2 \mp \sqrt{2}\pi Rx'] \right) \\
&\to 0 \text{ as } R \to \infty.
\end{aligned}
$$

Since the integration range is finite $(-\frac{1}{2} \leq x' \leq \frac{1}{2})$, the integrals $\to 0$ as $R \to \infty$.

(c) On the first of the other two sides, let us set $z = \frac{1}{2} + re^{i\pi/4}$ with $-R \leq r \leq R$. The corresponding integral I_1 is

$$
\begin{aligned}
I_1 &= \int_{L_1} e^{i\pi z^2} \operatorname{cosec} \pi z \, dz \\
&= \int_{-R}^{R} \frac{\exp \left[i\pi(\frac{1}{2} + re^{i\pi/4})^2 \right]}{\sin \left[\pi(\frac{1}{2} + re^{i\pi/4}) \right]} e^{i\pi/4} \, dr \\
&= \int_{-R}^{R} \frac{e^{i\pi/4} \exp(i\pi re^{i\pi/4}) \exp(i\pi r^2 i)e^{i\pi/4}}{\cos(\pi re^{i\pi/4})} \, dr \\
&= \int_{-R}^{R} \frac{i \exp(i\pi re^{i\pi/4}) e^{-\pi r^2}}{\cos(\pi re^{i\pi/4})} \, dr.
\end{aligned}
$$

Similarly (remembering the sense of integration), the remaining side contributes

$$
I_2 = -\int_{-R}^{R} \frac{i \exp(-i\pi re^{i\pi/4}) e^{-\pi r^2}}{-\cos(\pi re^{i\pi/4})} \, dr.
$$

Adding together all four contributions gives

$$
0 + 0 + \int_{-R}^{R} \frac{i[\exp(i\pi re^{i\pi/4}) + \exp(-i\pi re^{i\pi/4})]e^{-\pi r^2}}{\cos(\pi re^{i\pi/4})} \, dr,
$$

which simplifies to

$$
\int_{-R}^{R} 2ie^{-\pi r^2} \, dr.
$$

From part (a), this must be equal to $2i$ as $R \to \infty$, and so $\displaystyle\int_{-\infty}^{\infty} e^{-\pi r^2} \, dr = 1$.

24.19 *Using a suitable cut plane, prove that if α is real and $0 < \alpha < 1$ then*

$$\int_0^\infty \frac{x^{-\alpha}}{1+x} \, dx$$

has the value $\pi \operatorname{cosec} \pi\alpha$.

As α is not an integer, the complex form of the integrand $f(z) = \dfrac{z^{-\alpha}}{1+z}$ is not single-valued. We therefore need to perform the contour integration in a cut plane; contour (f) of figure 24.1 is a suitable contour. We will be making use of the fact that, because the integrand takes different values on γ_1 and γ_2, the contributions coming from these two parts of the complete contour, although related, do *not* cancel.

The contributions from γ and Γ are both zero because:

(i) around γ, $|zf(z)| \sim \dfrac{z\,z^{-\alpha}}{1} = z^{1-\alpha} \to 0$ as $|z| \to 0$, since $\alpha < 1$;

(ii) around Γ, $|zf(z)| \sim \dfrac{z\,z^{-\alpha}}{z} = z^{-\alpha} \to 0$ as $|z| \to \infty$, since $\alpha > 0$.

Therefore, the only contributions come from the cut; on γ_1, $z = xe^{0i}$, whilst on γ_2, $z = xe^{2\pi i}$. The only pole inside the contour is a simple one at $z = -1 = e^{i\pi}$, where the residue is $e^{-i\pi\alpha}$. The residue theorem now reads

$$0 + \int_0^\infty \frac{x^{-\alpha}}{1+x} \, dx + 0 - \int_0^\infty \frac{x^{-\alpha}e^{-2\pi i\alpha}}{1+xe^{2\pi i}} \, dx = 2\pi i\, e^{-i\pi\alpha},$$

$$\Rightarrow \quad (1 - e^{-2\pi i\alpha}) \int_0^\infty \frac{x^{-\alpha}}{1+x} \, dx = 2\pi i\, e^{-i\pi\alpha}.$$

This can be rearranged to read

$$\int_0^\infty \frac{x^{-\alpha}}{1+x} \, dx = \frac{2\pi i\, e^{-i\pi\alpha}}{(1 - e^{-2\pi i\alpha})} = \frac{2\pi i}{e^{i\pi\alpha} - e^{-i\pi\alpha}} = \frac{\pi}{\sin \pi\alpha},$$

thus establishing the stated result.

24.21 *By integrating a suitable function around a large semicircle in the upper half plane and a small semicircle centred on the origin, determine the value of*

$$I = \int_0^\infty \frac{(\ln x)^2}{1+x^2} \, dx$$

and deduce, as a by-product of your calculation, that

$$\int_0^\infty \frac{\ln x}{1+x^2} \, dx = 0.$$

The suggested contour is that shown in figure 24.1(e), but with only one inden-tation γ on the real axis (at $z = 0$) and with $R = \infty$. The appropriate complex function is

$$f(z) = \frac{(\ln z)^2}{1 + z^2}.$$

The only pole inside the contour is at $z = i$, and the residue there is given by

$$\frac{(\ln i)^2}{i + i} = \frac{(\ln 1 + i(\pi/2))^2}{2i} = -\frac{\pi^2}{8i}.$$

To evaluate the integral around γ, we set $z = \rho e^{i\theta}$ with $\ln z = \ln \rho + i\theta$ and $dz = i\rho e^{i\theta} d\theta$; the integral becomes

$$\int_\pi^0 \frac{\ln^2 \rho + 2i\theta \ln \rho - \theta^2}{1 + \rho^2 e^{2i\theta}} i\rho e^{i\theta} d\theta, \text{ which} \to 0 \text{ as } \rho \to 0.$$

Thus γ contributes nothing. Even more obviously, on Γ, $|zf(z)| \sim z^{-1}$ and tends to zero as $|z| \to \infty$, showing that Γ also contributes nothing.

On γ_+, $z = xe^{i0}$ and the contribution is equal to I.

On γ_-, $z = xe^{i\pi}$ and the contribution is (remembering that the contour actually runs from $x = \infty$ to $x = 0$) given by

$$I_- = -\int_0^\infty \frac{(\ln x + i\pi)^2}{1 + x^2} e^{i\pi} dx$$

$$= I + 2i\pi \int_0^\infty \frac{\ln x}{1 + x^2} dx - \pi^2 \int_0^\infty \frac{1}{1 + x^2} dx.$$

The residue theorem for the complete closed contour thus reads

$$0 + I + 0 + I + 2i\pi \int_0^\infty \frac{\ln x}{1 + x^2} dx - \pi^2 \left[\tan^{-1} x \right]_0^\infty = 2\pi i \left(\frac{-\pi^2}{8i} \right).$$

Equating the real parts $\quad \Rightarrow \quad 2I - \frac{1}{2}\pi^3 = -\frac{1}{4}\pi^3 \quad \Rightarrow \quad I = \frac{1}{8}\pi^3.$

Equating the imaginary parts gives the stated by-product.

25

Applications of complex variables

Many of the exercises in this chapter involve contour integration and the choice of a suitable contour. In order to save the space taken by drawing several broadly similar contours that differ only in notation, the positions of poles, the values of lengths or angles, or other minor details, we make reference to figure 24.1 which shows a number of typical contour types.

25.1 *In the method of complex impedances for a.c. circuits, an inductance L is represented by a complex impedance $Z_L = i\omega L$ and a capacitance C by $Z_C = 1/(i\omega C)$. Kirchhoff's circuit laws,*

$$\sum_i I_i = 0 \text{ at a node and } \sum_i Z_i I_i = \sum_j V_j \text{ around any closed loop,}$$

are then applied as if the circuit were a d.c. one.

Apply this method to the a.c. bridge connected as in figure 25.1 to show that if the resistance R is chosen as $R = (L/C)^{1/2}$ then the amplitude of the current I_R through it is independent of the angular frequency ω of the applied a.c. voltage $V_0 e^{i\omega t}$.

Determine how the phase of I_R, relative to that of the voltage source, varies with the angular frequency ω.

Omitting the common factor $e^{i\omega t}$ from all currents and voltages, let the current drawn from the voltage source be (the complex quantity) I and the current flowing from A to D be I_1. Then the currents in the remaining branches are $AE : I - I_1$, $DB : I_1 - I_R$ and $EB : I - I_1 + I_R$.

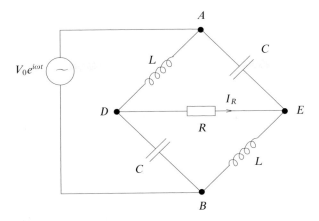

Figure 25.1   The inductor–capacitor–resistor network for exercise 25.1.

Applying $\sum_i Z_i I_i = \sum_j V_j$ to three separate loops yields

loop $ADBA$
$$i\omega L I_1 + \frac{1}{i\omega C}(I_1 - I_R) = V_0,$$

loop $ADEA$
$$i\omega L I_1 + R I_R - \frac{1}{i\omega C}(I - I_1) = 0,$$

loop $DBED$
$$\frac{1}{i\omega C}(I_1 - I_R) - i\omega L(I - I_1 + I_R) - R I_R = 0.$$

Now, denoting $(LC)^{-1}$ by ω_0^2 and choosing R as $(L/C)^{1/2} = (\omega_0 C)^{-1}$, we can write these equations as follows:

$$\left(1 - \frac{\omega^2}{\omega_0^2}\right) I_1 - I_R = i\omega C V_0,$$

$$-I + \left(1 - \frac{\omega^2}{\omega_0^2}\right) I_1 + i\frac{\omega}{\omega_0} I_R = 0,$$

$$\frac{\omega^2}{\omega_0^2} I + \left(1 - \frac{\omega^2}{\omega_0^2}\right) I_1 + \left(-1 + \frac{\omega^2}{\omega_0^2} - i\frac{\omega}{\omega_0}\right) I_R = 0.$$

Eliminating I from the last two of these yields

$$\left(1 + \frac{\omega^2}{\omega_0^2}\right)\left(1 - \frac{\omega^2}{\omega_0^2}\right) I_1 - \left(\frac{i\omega}{\omega_0} + 1\right)\left(1 - \frac{\omega^2}{\omega_0^2}\right) I_R = 0.$$

Thus,

$$I_R = \frac{1 + \dfrac{\omega^2}{\omega_0^2}}{1 + i\dfrac{\omega}{\omega_0}} I_1 = \frac{\omega_0^2 + \omega^2}{\omega_0(\omega_0 + i\omega)}\frac{\omega_0^2(i\omega C V_0 + I_R)}{\omega_0^2 - \omega^2}.$$

After some cancellation and rearrangement,

$$(\omega_0^2 - \omega^2) I_R = \omega_0(\omega_0 - i\omega)(i\omega C V_0 + I_R),$$
$$(i\omega\omega_0 - \omega^2) I_R = \omega_0\omega(i\omega_0 + \omega) C V_0,$$

and so

$$I_R = \omega_0 C V_0 \frac{i\omega_0 + \omega}{i\omega_0 - \omega} = \omega_0 C V_0 \frac{(i\omega_0 + \omega)(-i\omega_0 - \omega)}{(i\omega_0 - \omega)(-i\omega_0 - \omega)}$$

$$= \omega_0 C V_0 \frac{\omega_0^2 - \omega^2 - 2i\omega\omega_0}{\omega_0^2 + \omega^2}.$$

From this we can read off

$$|I_R| = \omega_0 C V_0 \frac{\left[(\omega^2 - \omega_0^2)^2 + 4\omega^2\omega_0^2\right]^{1/2}}{\omega_0^2 + \omega^2} = \omega_0 C V_0, \text{ i.e. independent of } \omega,$$

and

$$\phi = \text{phase of } I_R = \tan^{-1} \frac{-2\omega\omega_0}{\omega_0^2 - \omega^2}.$$

Thus I_R (which was arbitrarily and notionally defined as flowing from D to E in the equivalent d.c. circuit) has an imaginary part that is always negative but a real part that changes sign as ω passes through ω_0. Its phase ϕ, relative to that of the voltage source, therefore varies from 0 when ω is small to $-\pi$ when ω is large.

25.3 *For the function*

$$f(z) = \ln\left(\frac{z+c}{z-c}\right),$$

where c is real, show that the real part u of f is constant on a circle of radius $c\,\text{cosech}\,u$ *centred on the point* $z = c\coth u$. *Use this result to show that the electrical capacitance per unit length of two parallel cylinders of radii a, placed with their axes 2d apart, is proportional to* $[\cosh^{-1}(d/a)]^{-1}$.

From

$$f(z) = \ln\left(\frac{z+c}{z-c}\right) = \ln\left|\frac{z+c}{z-c}\right| + i\arg\left(\frac{z+c}{z-c}\right),$$

we have that

$$u = \ln\left|\frac{z+c}{z-c}\right| = \frac{1}{2}\ln\frac{(x+c)^2 + y^2}{(x-c)^2 + y^2} \quad\Rightarrow\quad e^{2u} = \frac{(x+c)^2 + y^2}{(x-c)^2 + y^2}.$$

The curve upon which $u(x,y)$ is constant is therefore given by

$$(x^2 - 2cx + c^2 + y^2)e^{2u} = x^2 + 2xc + c^2 + y^2.$$

This can be rewritten as

$$x^2(e^{2u} - 1) - 2xc(e^{2u} + 1) + y^2(e^{2u} - 1) + c^2(e^{2u} - 1) = 0,$$

$$x^2 - 2xc\frac{e^{2u} + 1}{e^{2u} - 1} + y^2 + c^2 = 0,$$

$$x^2 - 2xc \coth u + y^2 + c^2 = 0,$$

which, in conic-section form, becomes

$$(x - c \coth u)^2 + y^2 = c^2 \coth^2 u - c^2 = c^2 \operatorname{cosech}^2 u.$$

This is a circle with centre $(c \coth u, 0)$ and radius $|c \operatorname{cosech} u|$.

Now consider two such circles with the same value of $|c \operatorname{cosech} u|$, equal to a, but different values of u satisfying $c \coth u_1 = -d$ and $c \coth u_2 = +d$. These two equations imply that $u_1 = -u_2$, corresponding physically to equal but opposite charges $-Q$ and $+Q$ placed on identical cylindrical conductors that coincide with the circles; the conductors are raised to potentials u_1 and u_2.

We have already established that we need $c \coth u_2 = d$ and $c \operatorname{cosech} u_2 = a$. Dividing these two equations gives $\cosh u_2 = d/a$.

The capacitance (per unit length) of the arrangement is given by the magnitude of the charge on one conductor divided by the potential difference between the conductors that results from the presence of that charge, i.e.

$$C = \frac{Q}{u_2 - u_1} \propto \frac{1}{2u_2} = \frac{1}{2 \cosh^{-1}(d/a)},$$

as stated in the question.

25.5 *By considering in turn the transformations*

$$z = \tfrac{1}{2}c(w + w^{-1}) \quad and \quad w = \exp \zeta,$$

where $z = x + iy$, $w = r \exp i\theta$, $\zeta = \xi + i\eta$ and c is a real positive constant, show that $z = c \cosh \zeta$ maps the strip $\xi \geq 0$, $0 \leq \eta \leq 2\pi$, onto the whole z-plane. Which curves in the z-plane correspond to the lines $\xi = $ constant and $\eta = $ constant? Identify those corresponding to $\xi = 0$, $\eta = 0$ and $\eta = 2\pi$.

The electric potential ϕ of a charged conducting strip $-c \leq x \leq c$, $y = 0$, satisfies

$$\phi \sim -k \ln(x^2 + y^2)^{1/2} \text{ for large values of } (x^2 + y^2)^{1/2},$$

with ϕ constant on the strip. Show that $\phi = \operatorname{Re}[-k \cosh^{-1}(z/c)]$ and that the magnitude of the electric field near the strip is $k(c^2 - x^2)^{-1/2}$.

We first note that the combined transformation is given by

$$z = \frac{c}{2}(e^{\zeta} + e^{-\zeta}) = c \cosh \zeta \quad \Rightarrow \quad \zeta = \cosh^{-1} \frac{z}{c}.$$

The successive connections linking the strip in the ζ-plane and its image in the z-plane are

$$z = c \cosh \zeta = c \cosh(\xi + i\eta)$$
$$= c \cosh \xi \cos \eta + ic \sinh \xi \sin \eta, \text{ with } \xi > 0, 0 \le \eta \le 2\pi,$$

$$re^{i\theta} = w = e^{\zeta} = e^{\xi}e^{i\eta}, \text{ with the strip as } 1 < r < \infty, 0 \le \theta \le 2\pi,$$

$$x + iy = z = \frac{c}{2}(w + w^{-1})$$
$$= \frac{c}{2}[r(\cos\theta + i\sin\theta) + r^{-1}(\cos\theta - i\sin\theta)]$$
$$= \frac{c}{2}\left(r + \frac{1}{r}\right)\cos\theta + i\frac{c}{2}\left(r - \frac{1}{r}\right)\sin\theta.$$

This last expression for z and the previous specification of the strip in terms of r and θ show that both x and y can take all values, i.e. that the original strip in the ζ-plane is mapped onto the whole of the z-plane. From the two expressions for z we also see that $x = c \cosh \xi \cos \eta$ and $y = c \sinh \xi \sin \eta$.

For ξ constant, the contour in the xy-plane, obtained by eliminating η, is

$$\frac{x^2}{c^2 \cosh^2 \xi} + \frac{y^2}{c^2 \sinh^2 \xi} = 1, \quad \text{i.e. an ellipse.}$$

The eccentricity of the ellipse is given by

$$e = \left(\frac{c^2 \cosh^2 \xi - c^2 \sinh^2 \xi}{c^2 \cosh^2 \xi}\right)^{1/2} = \frac{1}{\cosh \xi}.$$

The foci of the ellipse are at $\pm e \times$ the major semi-axis, i.e. $\pm 1/\cosh \xi \times c \cosh \xi = \pm c$. This is independent of ξ and so all the ellipses are confocal.

Similarly, for η constant, the contour is

$$\frac{x^2}{c^2 \cos^2 \eta} - \frac{y^2}{c^2 \sin^2 \eta} = 1.$$

This is one of a set of confocal hyperbolae.

(i) $\xi = 0 \quad \Rightarrow \quad y = 0, x = c \cos \eta$.
This is the finite line (degenerate ellipse) on the x-axis, $-c \le x \le c$.

(ii) $\eta = 0 \quad \Rightarrow \quad y = 0, x = c \cosh \xi$.
This is a part of the x-axis not covered in (i), $c < x < \infty$. The other part, $-\infty < x < -c$, corresponds to $\eta = \pi$.

(iii) This is the same as (the first case) in (ii).

Now, in the ζ-plane, consider the real part of the function $F(\zeta) = -k\zeta$, with k real. On $\xi = 0$ [case (i) above] it reduces to Re $\{-ik\eta\}$, which is zero for all η, i.e. a constant. This implies that the real part of the transformed function will be a constant (actually zero) on $-c \leq x \leq c$ in the z-plane.

Further,

$$(x^2 + y^2)^{1/2} = (c^2 \cosh^2 \xi \cos^2 \eta + c^2 \sinh^2 \xi \sin^2 \eta)^{1/2}$$
$$\approx \tfrac{1}{2}ce^\xi \text{ for large } \xi,$$
$$\Rightarrow \quad \xi \approx \ln(x^2 + y^2)^{1/2} + \text{fixed constant.}$$

Hence,

$$\text{Re } \{-k\zeta\} = -k\xi \approx -k \ln(x^2 + y^2)^{1/2} \text{ for large } (x^2 + y^2)^{1/2}.$$

Thus, the transformation

$$F(\zeta) = -k\zeta \quad \rightarrow \quad f(z) = -k \cosh^{-1} \frac{z}{c}$$

produces a function in the z-plane that satisfies the stated boundary conditions (as well as satisfying Laplace's equation). It is therefore the required solution.

The electric field near the conducting strip, where $y = 0$ and $z^2 = x^2$, can have no component in the x-direction (except at the points $x = \pm c$), but its magnitude is still given by

$$E = |f'(z)| = \left| -\frac{k}{\sqrt{z^2 - c^2}} \right| = \frac{k}{(c^2 - x^2)^{1/2}}.$$

25.7 *Use contour integration to answer the following questions about the complex zeros of a polynomial equation.*

(a) *Prove that $z^8 + 3z^3 + 7z + 5$ has two zeros in the first quadrant.*
(b) *Find in which quadrants the zeros of $2z^3 + 7z^2 + 10z + 6$ lie. Try to locate them.*

(a) Consider the principle of the argument applied to the integral of $f(z) = z^8 + 3z^3 + 7z + 5$ around contour (b) in figure 24.1.

On OA $f(z)$ is always real and $\Delta_{AB} \arg(f) = 0$.

On AB the argument of f increases by $8 \times \tfrac{1}{2}\pi = 4\pi$.

On BO $z = iy$ and $f(z) = h(y) = y^8 - 3iy^3 + 7iy + 5$. The argument of $h(y)$ is therefore

$$\tan^{-1} \frac{-3y^3 + 7y}{y^8 + 5}.$$

The appropriate choice at $y = \infty$ for this multi-valued function is 4π, as we have just shown. As y decreases from ∞ the argument initially decreases, but passes through 4π again when $y = \sqrt{7/3}$. After that it remains greater than 4π until returning to that value at $y = 0$. Further, since $y^8 + 5$ has no zeros for real y, $\arg(h)$ can reach neither $\frac{7}{2}\pi$ nor $\frac{9}{2}\pi$. Consequently, we deduce that $\Delta_{BO} \arg(f) = 0$.

In summary, $\Delta \arg(f)$ around the closed contour is 4π, and it follows from the principle of the argument that the first quadrant must contain 2 zeros of $f(z)$.

(b) For $f(z) = 2z^3 + 7z^2 + 10z + 6$ we initially follow the same procedure as in part (a), although, as it is a cubic with all of its coefficients positive, we know that it must have at least one negative real zero.

It is straightforward to conclude that $\Delta_{OA} \arg(f) = 0$ and that around the curve AB the change of argument is $\Delta_{AB} \arg(f) = \frac{3}{2}\pi$. On BO,

$$\arg(f) = \tan^{-1} \frac{10y - 2y^3}{6 - 7y^2}.$$

At $y = \infty$ this is $\frac{3}{2}\pi$ (as we have just established) and, as y decreases towards 0, it passes through π at $y = \sqrt{5}$ and $\frac{1}{2}\pi$ at $y = \sqrt{6/7}$, and finally becomes zero at $y = 0$. Thus the net change around the whole closed contour is zero, and we conclude that there are no zeros in the first quadrant. Since the zeros of polynomials with real coefficients occur in complex conjugate pairs, it follows that the fourth quadrant also contains no zeros. This shows that the complex conjugate zeros of $f(z)$ are located in the second and third quadrants.

We start our search for the negative real zero by tabulating some easy-to-calculate values of $f(x)$, the choice of successive values of x being guided by previous results:

z	0	-1	-2	-1.5
$f(z)$	6	1	-2	0

By chance, we have hit upon an exact zero, $z = -\frac{3}{2}$. It follows that $(2z + 3)$ is a factor of $f(z)$, which can be written

$$f(z) = (2z + 3)(z^2 + 2z + 2).$$

The other two zeros are therefore

$$z = -1 \pm \sqrt{1 - 2} = -1 \pm i.$$

As expected, these are in the second and third quadrants.

25.9 *Prove that*

$$\sum_{-\infty}^{\infty} \frac{1}{n^2 + \frac{3}{4}n + \frac{1}{8}} = 4\pi.$$

Carry out the summation numerically, say between -4 and 4, and note how much of the sum comes from values near the poles of the contour integration.

In order to evaluate this sum, we must first find a function of z that takes the value of the corresponding term in the sum whenever z is an integer. Clearly this is

$$\frac{1}{z^2 + \frac{3}{4}z + \frac{1}{8}}.$$

Further, too make use of the properties of contour integrals, we need to multiply this function by one that has simple poles at the same points, each with unit residue. An appropriate choice of integrand is therefore

$$f(z) = \frac{\pi \cot \pi z}{z^2 + \frac{3}{4}z + \frac{1}{8}} = \frac{\pi \cot \pi z}{(z + \frac{1}{2})(z + \frac{1}{4})}.$$

The contour to be used must enclose all integer values of z, both positive and negative and, in practical terms, must give zero contribution for $|z| \to \infty$, except possibly on the real axis. A large circle C, centred on the origin (see contour (c) in figure 24.1) suggests itself.

As $|zf(z)| \to 0$ on C, the contour integral has value zero. This implies that the residues at the enclosed poles add up to zero. The residues are

$$\frac{\pi \cot(-\frac{1}{2}\pi)}{-\frac{1}{2} + \frac{1}{4}} = 0 \qquad \text{at } z = -\frac{1}{2},$$

$$\frac{\pi \cot(-\frac{1}{4}\pi)}{-\frac{1}{4} + \frac{1}{2}} = -4\pi \qquad \text{at } z = -\frac{1}{4},$$

$$\sum_{n=-\infty}^{\infty} \frac{1}{(n + \frac{1}{2})(n + \frac{1}{4})} \qquad \text{at } z = n, \ -\infty < n < \infty.$$

The quoted result follows immediately.

For the rough numerical summation we tabulate n, $D(n) = n^2 + \frac{3}{4}n + \frac{1}{8}$ and the

407

nth term of the series, $1/D(n)$:

n	$D(n)$	$1/D(n)$
-4	13.125	0.076
-3	6.875	0.146
-2	2.625	0.381
-1	0.375	2.667
0	0.125	8.000
1	1.875	0.533
2	5.625	0.178
3	11.375	0.088
4	19.125	0.052

The total of these nine terms is 12.121; this is to be compared with the total for the entire infinite series (of positive terms), which is $4\pi = 12.566$. It will be seen that the sum is dominated by the terms for $n = 0$ and $n = -1$. These two values bracket the positions on the real axis of the poles at $z = -\frac{1}{2}$ and $z = -\frac{1}{4}$.

25.11 *By considering the integral of*

$$\left(\frac{\sin \alpha z}{\alpha z}\right)^2 \frac{\pi}{\sin \pi z}, \qquad \alpha < \frac{\pi}{2},$$

around a circle of large radius, prove that

$$\sum_{m=1}^{\infty} (-1)^{m-1} \frac{\sin^2 m\alpha}{(m\alpha)^2} = \frac{1}{2}.$$

Denote the given function by $f(z)$ and consider its integral around contour (c) in figure 24.1.

As $|z| \to \infty$, $\sin \alpha z \sim e^{\alpha|z|}$, and so $f(z) \sim |z|^{-2} e^{2\alpha|z|} e^{-\pi|z|} = z^{-2} e^{(2\alpha-\pi)|z|}$, and, since $\alpha < \frac{1}{2}\pi$, $|zf(z)\,dz| \to 0$ as $|z| \to \infty$ and the integral around the contour has value zero for $R = \infty$.

The function $f(z)$ has simple poles at $z = n$, where n is an integer, $-\infty < n < \infty$. The pole at $z = 0$ is only a first-order pole as the term in parentheses $\to 1$ as $z \to 0$ and has no singularity there. It follows that the sum of the residues of $f(z)$ at all of its poles is zero. For $n \neq 0$, that residue is

$$\pi \left(\frac{\sin n\alpha}{n\alpha}\right)^2 \left(\frac{d(\sin \pi z)}{dz}\bigg|_{z=n}\right)^{-1} = \left(\frac{\sin n\alpha}{n\alpha}\right)^2 \frac{1}{\cos \pi n}$$

$$= (-1)^n \left(\frac{\sin n\alpha}{n\alpha}\right)^2.$$

For $n = 0$ the residue is 1.

Since the general residue is an even function of n, the sum for $-\infty < n \leq -1$ is equal to that for $1 \leq n < \infty$, and the zero sum of the residues can be written

$$1 + 2 \sum_{n=1}^{\infty} (-1)^n \left(\frac{\sin n\alpha}{n\alpha} \right)^2 = 0,$$

leading immediately to the stated result.

25.13 *Find the function $f(t)$ whose Laplace transform is*

$$\bar{f}(s) = \frac{e^{-s} - 1 + s}{s^2}.$$

Consider first the Taylor series expansion of $\bar{f}(s)$ about $s = 0$:

$$\bar{f}(s) = \frac{e^{-s} - 1 + s}{s^2} = \frac{(1 - s + \frac{1}{2}s^2 + \cdots) - 1 + s}{s^2} \sim \frac{1}{2} + O(s).$$

Thus $\bar{f}$ has no pole at $s = 0$, and λ in the Bromwich integral can be as small as we wish (but > 0). When the integration line is made part of a closed contour C, the inversion integral becomes

$$f(t) = \int_C \frac{e^{-s}e^{st} - e^{st} + se^{st}}{s^2} \, ds.$$

For $t < 0$, all the terms $\to 0$ as $\mathrm{Re}\, s \to \infty$, and so we close the contour in the right half-plane, as in contour (h) of figure 24.1. On Γ, s times the integrand $\to 0$, and, as the contour encloses no poles, it follows that the integral along L is zero. Thus $f(t) = 0$ for $t < 0$.

For $t > 1$, all terms $\to 0$ as $\mathrm{Re}\, s \to -\infty$, and so we close the contour in the left half-plane, as in contour (g) of figure 24.1. On Γ, s times the integrand again $\to 0$, and, as this contour also encloses no poles, it again follows that the integral along L is zero. Thus $f(t) = 0$ for $t > 1$, as well as for $t < 0$.

For $0 < t < 1$, we need to separate the Bromwich integral into two parts (guided by the different ways in which the parts behave as $|s| \to \infty$):

$$f(t) = \int_L \frac{e^{-s}e^{st}}{s^2} \, ds + \int_L \frac{(s-1)e^{st}}{s^2} \, ds \equiv I_1 + I_2.$$

For I_1 the exponent is $s(t - 1)$; $t - 1$ is negative and so, as in the case $t < 0$, we close the contour in the right half-plane [contour (h)]. No poles are included in this contour, and we conclude that $I_1 = 0$.

For I_2 the exponent is st, indicating that (g) is the appropriate contour. However,

$(s-1)/s^2$ does have a pole at $s = 0$ and that is inside the contour. The integral around Γ contributes nothing (that is why it was chosen), and the integral along L must be equal to the residue of $(s-1)e^{st}/s^2$ at $s = 0$. Now,

$$\frac{(s-1)e^{st}}{s^2} = \left(\frac{1}{s} - \frac{1}{s^2}\right)\left(1 + st + \frac{s^2 t^2}{2!} + \cdots\right) = -\frac{1}{s^2} + \frac{1}{s}(1-t) + \cdots.$$

The residue, and hence the value of I_2, is therefore $1 - t$. Since I_1 has been shown to have value 0, $1 - t$ is also the expression for $f(t)$ for $0 < t < 1$.

25.15 *Use contour* (i) *in figure 24.1 to show that the function with Laplace transform* $s^{-1/2}$ *is* $(\pi x)^{-1/2}$.

[*For an integrand of the form* $r^{-1/2}\exp(-rx)$, *change variable to* $t = r^{1/2}$.]

With the suggested contour no poles of $s^{-1/2}e^{sx}$ are enclosed and so the integral of $(2\pi i)^{-1}s^{-1/2}e^{sx}$ around the closed curve must have the value zero. It is also clear that the integral along Γ will be zero since Re $s < 0$ on Γ.

For the small circle γ enclosing the origin, set $s = \rho e^{i\theta}$, with $ds = i\rho e^{i\theta}\,d\theta$, and consider

$$\lim_{\rho \to 0} \int_0^{2\pi} \rho^{-1/2}e^{-i\theta/2}\exp(x\rho e^{i\theta})i\rho e^{i\theta}\,d\theta.$$

This $\to 0$ as $\rho \to 0$ (as $\rho^{1/2}$).

On the upper cut, γ_1, $s = re^{i\pi}$ and the contribution to the integral is

$$\frac{1}{2\pi i}\int_\infty^0 \frac{e^{-i\pi/2}}{r^{1/2}}\exp(rxe^{i\pi})e^{i\pi}\,dr,$$

whilst, on the lower cut, γ_2, $s = re^{-i\pi}$, and its contribution to the integral is

$$\frac{1}{2\pi i}\int_0^\infty \frac{e^{i\pi/2}}{r^{1/2}}\exp(rxe^{-i\pi})e^{-i\pi}\,dr.$$

Combining the two (and making both integrals run over the same range) gives

$$-\frac{1}{2\pi i}\int_0^\infty \frac{2i}{r^{1/2}}e^{-rx}\,dr = -\frac{1}{\pi}\int_0^\infty \frac{1}{t}e^{-t^2 x}2t\,dt, \text{ after setting } r = t^2,$$

$$= -\frac{2}{\pi}\frac{\sqrt{\pi}}{2\sqrt{x}}.$$

Since this must add to the Bromwich integral along L to make zero, it follows that the function with Laplace transform $s^{-1/2}$ is $(\pi x)^{-1/2}$.

25.17 *The equation*

$$\frac{d^2y}{dz^2} + \left(v + \frac{1}{2} - \frac{1}{4}z^2\right)y = 0,$$

sometimes called the Weber–Hermite equation, has solutions known as parabolic cylinder functions. Find, to within (possibly complex) multiplicative constants, the two W.K.B. solutions of this equation that are valid for large $|z|$. In each case, determine the leading term and show that the multiplicative correction factor is of the form $1 + O(v^2/z^2)$.

Identify the Stokes and anti-Stokes lines for the equation. On which of the Stokes lines is the W.K.B. solution that tends to zero for z large, real and negative, the dominant solution?

If we consider the equation to be of the generic form

$$\frac{d^2y}{dz^2} + f(z)y = 0,$$

then the W.K.B. solutions are, to within a constant multiplier,

$$y_\pm(z) = \frac{1}{[f(z)]^{1/4}} \exp\left\{\pm i \int^z \sqrt{f(u)}\,du\right\}.$$

In this particular case, writing $v + \frac{1}{2}$ as μ for the time being, these solutions are

$$y_\pm(z) = \frac{1}{(\mu - \frac{1}{4}z^2)^{1/4}} \exp\left\{\pm i \int^z \sqrt{\mu - \frac{u^2}{4}}\,du\right\}.$$

Now we seek solutions for large z, and, in this spirit, make binomial expansions of both roots in inverse powers of the relevant variable, z or u. This enables us to write, for a succession of multiplicative complex constants and working to $O(z^{-2})$,

$$y_\pm(z) = \frac{A}{(\frac{1}{4}z^2 - \mu)^{1/4}} \exp\left\{\pm i^2 \int^z \sqrt{\frac{u^2}{4} - \mu}\,du\right\}$$

$$= \frac{B}{\sqrt{z}}\left(1 + \frac{\mu}{z^2} + \cdots\right) \exp\left\{\pm i^2 \int^z \frac{u}{2}\left(1 - \frac{4\mu}{u^2}\right)^{1/2} du\right\}$$

$$= \frac{B}{\sqrt{z}}\left(1 + \frac{\mu}{z^2}\right) \exp\left\{\mp \int^z \left(\frac{u}{2} - \frac{\mu}{u} - \frac{\mu^2}{u^3} + \cdots\right) du\right\}.$$

411

Performing the indefinite integral in the exponent yields

$$y_\pm(z) = \frac{B}{\sqrt{z}} \left(1 + \frac{\mu}{z^2}\right) \exp\left\{\mp\left(\frac{z^2}{4} - \mu \ln z + \frac{\mu^2}{2z^2} + \cdots\right)\right\}$$

$$= \frac{B}{\sqrt{z}} \left(1 + \frac{\mu}{z^2}\right) e^{\mp z^2/4} z^{\pm\mu} \left(1 \mp \frac{\mu^2}{2z^2} + \cdots\right)$$

$$= \frac{B}{\sqrt{z}} e^{\mp z^2/4} z^{\pm\mu} \left(1 + \frac{2\mu \mp \mu^2}{2z^2} + \cdots\right).$$

Replacing μ by $\nu + \frac{1}{2}$ and writing the two solutions separately, we have

$$y_1(z) = C\, e^{-z^2/4}\, z^\nu \left[1 + O\left(\frac{\nu^2}{z^2}\right)\right], \qquad y_2(z) = D\, e^{z^2/4}\, z^{-(\nu+1)} \left[1 + O\left(\frac{\nu^2}{z^2}\right)\right].$$

The Stokes lines are determined by the argument(s) of z that make the exponent in the solutions purely real, resulting in one solution being very large (dominant) and one very small (subdominant). As the exponent is proportional to z^2, the Stokes lines are given by $\arg z$ equals 0, $\pi/2$, π or $3\pi/2$. For z large, real and negative, the solution that tends to zero is $y_1(z) \propto e^{-z^2/4}$. This is dominant when z^2 is real and negative, i.e. when z lies on either $\arg z = \pi/2$ or $\arg z = 3\pi/2$.

The anti-Stokes lines, on which the exponent is purely imaginary and consequently the two solutions are comparable in magnitude, are clearly given by the four lines $\arg z = (2n+1)\pi/4$ for $n = 0, 1, 2, 3$.

25.19 *The function $h(z)$ of the complex variable z is defined by the integral*

$$h(z) = \int_{-i\infty}^{i\infty} \exp(t^2 - 2zt)\, dt.$$

(a) *Make a change of integration variable, $t = iu$, and evaluate $h(z)$ using a standard integral. Is your answer valid for all finite z?*

(b) *Evaluate the integral using the method of steepest descents, considering in particular the cases (i) z is real and positive, (ii) z is real and negative and (iii) z is purely imaginary and equal to $i\beta$, where β is real. In each case sketch the corresponding contour in the complex t-plane.*

(c) *Evaluate the integral for the same three cases as specified in part (b) using the method of stationary phases. To determine an appropriate contour that passes through a saddle point $t = t_0$, write $t = t_0 + (u + iv)$ and apply the criterion for determining a level line. Sketch the relevant contour in each case, indicating what freedom there is to distort it.*

Comment on the accuracy of the results obtained using the approximate methods adopted in (b) and (c).

Before we consider the three different methods of evaluating the integral, we note that its limits lie one in each of the $\pi/2$ sectors of the complex t-plane that are centred on the negative and positive parts of the imaginary axis. All contours that we employ must do the same, though it will not matter exactly where in these sectors they formally end, as, within them, the integrand, which behaves like $\exp(-|t|^2)$, goes (rapidly) to zero as $|t| \to \infty$.

(a) Making the change of integration variable $t = iu$ with $dt = i\,du$ gives $h(z)$ as

$$h(z) = \int_{-\infty}^{\infty} \exp(-u^2 - 2izu)\,i\,du$$

$$= \int_{-\infty}^{\infty} \exp[-(u + iz)^2]\,\exp(-z^2)\,i\,du$$

$$= i\sqrt{\pi}\,e^{-z^2}.$$

It is the behaviour of the dominant term in the exponent that determines the convergence or otherwise of the integral. In this case, the t^2 term dominates the term containing z, and, since, as discussed above, it produces convergence, the result is valid for all (finite) values of z.

(b) We first identify the saddle point(s) t_0 of the integrand by setting the derivative of the exponent equal to zero:

$$0 = \frac{d}{dt}(t^2 - 2zt) = 2t - 2z \quad \Rightarrow \quad t_0 = z; \text{ only one saddle point.}$$

The second derivative of the exponent is 2 (independent of the value of z in this case), and so, in the standard notation $f''(t_0) = Ae^{i\alpha}$, we have $A = 2$ and $\alpha = 0$. The value of $f_0 \equiv f(t_0)$ is $t_0^2 - 2zt_0 = -z^2$.

The remaining task is to determine the orientation and direction of traversal of the saddle point. With $t - t_0 = se^{i\theta}$, the possible lines of steepest descents (l.s.d.) are given by $2\theta + \alpha = 0$, $\pm\pi$ or 2π. Of these, the need for $\frac{1}{2}As^2 \cos(2\theta + \alpha)$ to be negative picks out $\theta = \pm\frac{1}{2}\pi$. Thus the l.s.d. through the saddle point is parallel to the imaginary axis and the direction of traversal is $+\frac{1}{2}\pi$. Since this lies (just) in the range $-\frac{1}{2}\pi < \theta \leq \frac{1}{2}\pi$, we take the positive sign from the general formula

$$\pm \left(\frac{2\pi}{A}\right)^{1/2} \exp(f_0)\exp[\tfrac{1}{2}i(\pi - \alpha)]$$

and obtain

$$h(z) = +\left(\frac{2\pi}{2}\right)^{1/2} \exp(-z^2)\exp[\tfrac{1}{2}i(\pi - 0)]$$

$$= i\sqrt{\pi}\,e^{-z^2}.$$

The conclusion about the orientation and sense of traversal of the saddle point did not depend upon the value of z (because $f''(t_0)$ did not). Consequently the

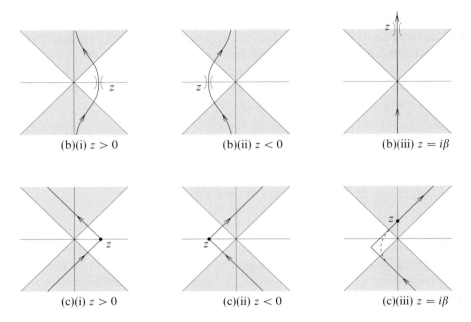

Figure 25.2   The contours following (b) the lines of steepest descents and (c) the lines of stationary phase for the integral in exercise 25.19.

value of the integral is the same for all three cases, though the path in the complex t-plane is determined by z, as is shown in the upper row of sketches in figure 25.2.

(c) We know from general theory that the directions of the level lines at a saddle point make an angle of $\pi/4$ with the l.s.d. through the point. From this and the results of part (b) we can say that the level lines at $t_0 = z$ have directions $\theta = \pm\pi/4$ and $\pm 3\pi/4$.

The same conclusion can be reached, and an indication of suitable contours obtained, by writing $t = t_0 + u + iv$ and requiring that the resulting integrand has a constant magnitude for all u and v. That magnitude must be the same as it is at the saddle point, i.e. when $u = v = 0$.

We consider first cases (i) and (ii) in which $t_0 = z$ is real and $t = (z + u) + iv$. The integrand is then

$$g(u, v) = \exp(t^2 - 2zt) = \exp[(z + u)^2 - v^2 + 2iv(z + u) - 2z(z + u + iv)],$$

with $g(0, 0) = \exp(-z^2)$. For the integrand to have a constant magnitude, the real part of the exponent must not depend upon u and v. The u- and v-dependent part of the real part of the exponent is $2zu + u^2 - v^2 - 2zu$, and this must therefore have the value 0 for all u and v, i.e. $v = \pm u$. These are the same lines as $\theta = \pm \pi/4$ and $\theta = \pm 3\pi/4$.

Now, although the saddle point at $t_0 = z$ lies outside both of the regions in which the contour must begin and end, the contour must go through it. It is therefore necessary for the contour to turn through a right angle at the saddle point; it transfers from one of the level lines that pass through the saddle to the other one. As will be seen from sketches (c)(i) and (c)(ii), the contour in case (i), $z > 0$, turns to the left by $\pi/2$ as it passes through the saddle; that for $z < 0$ turns to the right through $\pi/2$.

The formula for the total contribution to the integral from integrating through the saddle point along a level line is the same as that for an l.s.d. evaluation, though the former is a Fresnel integral and the latter is an error integral. The stationary phase calculation therefore also yields the value $i\sqrt{\pi} \exp(-z^2)$ for $h(z)$.

In both of the present cases, the sharp turn through a right angle at the saddle point means that the vector diagram for the integral consists of one-half from each of two Cornu spirals that are mirror images of each other. Each is broken at its centre point where the phase of the integrand is stationary. The two half spirals join at right angles at the point that is midway between their 'winding points'.

We now turn to case (iii), in which $z = i\beta$ is imaginary. In this case the saddle point lies within one of the two regions that each contain one end of the contour. However, a parallel analysis to that for cases (i) and (ii), setting $t = u + i(\beta + v)$, yields the same conclusion, namely that $v = \pm u$ are appropriate level lines through the saddle.

It is a matter of choice whether the solid line shown in sketch (c)(iii), or its mirror image in the imaginary axis, is chosen; the calculated value for the integral will be the same. The result for $h(z)$ will also be the same as for cases (i) and (ii), i.e. $i\sqrt{\pi} \exp(-z^2)$, or, more explicitly in this case, $i\sqrt{\pi} \exp(\beta^2)$. Since the contour does not have to go through any particular point other than $t = i\beta$ (and does not need to take a right-angled turn there) and the integrand is analytic, the contour in the end-region not containing z can follow almost any path. One variation from two intersecting straight lines is shown dashed in figure 25.2(c)(iii).

Finally, we note that the fact that all methods give the same answer for $h(z)$, even though the l.s.d. and stationary phase calculations are, in general, approximations, can be put down to the particular form of the integrand. The exponent, $t^2 - 2zt$, is a quadratic function, and so its Taylor series terminates after three terms (of which the second vanishes at the saddle point). Consequently, the l.s.d. and stationary phase approaches which ignore the cubic and higher terms in the Taylor series are *not* approximations. This, together with the fact that there is only one saddle point in the whole t-plane, means that the methods produce exact results for this form of integrand.

25.21 *The stationary phase approximation to an integral of the form*

$$F(v) = \int_a^b g(t)e^{ivf(t)}\,dt, \qquad |v| \gg 1,$$

where $f(t)$ is a real function of t and $g(t)$ is a slowly varying function (when compared with the argument of the exponential), can be written as

$$F(v) \sim \left(\frac{2\pi}{|v|}\right)^{1/2} \sum_{n=1}^{N} \frac{g(t_n)}{\sqrt{A_n}} \exp\left\{i\left[vf(t_n) + \frac{\pi}{4}\,\mathrm{sgn}\!\left(vf''(t_n)\right)\right]\right\},$$

where the t_n are the N stationary points of $f(t)$ that lie in $a < t_1 < t_2 < \cdots < t_N < b$, $A_n = |f''(t_n)|$, and $\mathrm{sgn}(x)$ is the sign of x.

Use this result to find an approximation, valid for large positive values of v, to the integral

$$F(v,z) = \int_{-\infty}^{\infty} \frac{1}{1+t^2}\,\cos[\,(2t^3 - 3zt^2 - 12z^2t)v\,]\,dt,$$

where z is a real positive parameter.

Since the argument of the cosine function is everywhere real, we can consider the required integral as the real part of

$$\int_{-\infty}^{\infty} \frac{1}{1+t^2}\,\exp\{\,i\,[\,(2t^3 - 3zt^2 - 12z^2t)v\,]\}\,dt,$$

to which we can apply the stated approximation directly. To do so, we need to calculate values for all of the terms appearing in the quoted 'omnibus' formula.

We start by determining the stationary points involved, given by

$$0 = f'(t) = 6t^2 - 6zt - 12z^2 \quad \Rightarrow \quad (t+z)(t-2z) = 0$$
$$\Rightarrow \quad t_1 = -z \text{ and } t_2 = 2z.$$

Thus $N = 2$ and the required second derivatives, $f''(t) = 12t - 6z$, and values, $f_n = f(t_n)$, are given by

$$f_1 = -2z^3 - 3z^3 + 12z^3 = 7z^3, \quad f_2 = 16z^3 - 12z^3 - 24z^3 = -20z^3,$$
$$f''(t_1) = -12z - 6z = -18z, \quad f''(t_2) = 24z - 6z = 18z.$$

The two corresponding values of the multiplicative function $g(t) = (1+t^2)^{-1}$ are

$$g(t_1) = (1 + z^2)^{-1} \quad \text{and} \quad g(t_2) = (1 + 4z^2)^{-1}.$$

Substituting all of these gives

$$F(v,z) \sim \text{Re}\ \left(\frac{2\pi}{v}\right)^{1/2} \left\{ \frac{\exp[\,i(7vz^3 - \frac{1}{4}\pi)\,]}{\sqrt{18z}\,(1 + z^2)} + \frac{\exp[\,i(-20vz^3 + \frac{1}{4}\pi)\,]}{\sqrt{18z}\,(1 + 4z^2)} \right\}$$

$$= \left(\frac{\pi}{9zv}\right)^{1/2} \left[\frac{\cos(7vz^3 - \frac{1}{4}\pi)}{1 + z^2} + \frac{\cos(20vz^3 - \frac{1}{4}\pi)}{1 + 4z^2} \right]$$

as the stationary phase approximation to the integral.

25.23 *Use the method of steepest descents to find an asymptotic approximation, valid for z large, real and positive, to the function defined by*

$$F_v(z) = \int_C \exp(-iz \sin t + ivt)\, dt,$$

where v is real and non-negative and C is a contour that starts at $t = -\pi + i\infty$ and ends at $t = -i\infty$.

Let us denote the integrand by $g(t)$ and the exponent by $f(t)$; thus $g(t) = \exp[\,f(t)\,]$.

We first check that the integrand $\to 0$ at the two end-points; if it did not, the method could not be even approximately correct. As the end-points involve $\pm\infty$, we should formally consider a limiting process, but in practice we need only identify the dominant term in each expression and determine its behaviour as $t \to \infty$.

At $t = -\pi + i\infty$,

$$\sin t = \frac{1}{2i}\, [e^{i(-\pi + i\infty)} - e^{-i(-\pi + i\infty)}] = \frac{1}{2i}\, (-0 + e^{\infty}) = \frac{1}{2i}\, e^{\infty}.$$

Thus

$$g(-\pi + i\infty) = \exp[\,-iz(e^{\infty}/2i) + iv(-\pi + i\infty)\,] = 0,$$

for z real and > 0 and for all v. Similarly, at $t = -i\infty$,

$$\sin(-i\infty) = \frac{1}{2i}\, (e^{-i^2\infty} - e^{i^2\infty}) = \frac{1}{2i}\, e^{\infty}$$

and

$$g(-i\infty) = \exp[\,-iz(e^{\infty}/2i) + iv(-i\infty)\,] = 0.$$

In each case, the behaviour of $f(t)$ is dominated by the exponentiation appearing in the sine term; as this produces a negative exponent for the exponential function determining $g(t)$, the latter $\to 0$ at both end-points.

We next determine the position(s) and properties of the saddle points. These are given by

$$0 = \frac{df}{dt} = -iz \cos t + iv$$

$$\Rightarrow \quad t_0 = \cos^{-1} \frac{v}{z} \text{ (real) with } -\pi < t_0 < 0 \text{ and } z > v,$$

$$\frac{d^2 f}{dt^2} = iz \sin t,$$

$$f''(t_0) = iz \sin \left(\cos^{-1} \frac{v}{z} \right) = \frac{-iz\sqrt{z^2 - v^2}}{z}$$

$$= -i\sqrt{z^2 - v^2}$$

$$\equiv A e^{i\alpha} \text{ with } A = \sqrt{z^2 - v^2} \text{ and } \alpha = \frac{3\pi}{2},$$

$$f_0 \equiv f(t_0) = -iz \sin \left(\cos^{-1} \frac{v}{z} \right) + iv \left(\cos^{-1} \frac{v}{z} \right)$$

$$= i\sqrt{z^2 - v^2} + iv \left(\cos^{-1} \frac{v}{z} \right).$$

Thus the only saddle point is at $t_0 = \cos^{-1}(v/z)$, and the values of $f(t_0)$ and $f''(t_0)$ are given above.

The final step before evaluating the approximate expression for the integral is to determine the direction of the contour through the saddle point. A line of steepest descents (l.s.d.), on which the phase of $f(t)$ is constant, is given by $\sin(2\theta + \alpha) = 0$, where $t - t_0 = se^{i\theta}$ and α is as determined above by $f''(t_0)$. Thus $2\theta + 3\pi/2 = 0, \pm\pi, 2\pi$ are possible lines, and, of the resulting possible values of $\pm\pi/4$ and $\pm 3\pi/4$ for θ, it is clear that approaching from the direction $\theta = 3\pi/4$ and leaving in the direction $\theta = -\pi/4$ is appropriate. This can be verified by considering the first non-constant, non-vanishing term in the Taylor expansion of $f(t)$, namely

$$\frac{1}{2!} (t - t_0)^2 f''(t_0) = \frac{1}{2!} s^2 e^{-2i\pi/4}(-i\sqrt{z^2 - v^2}) = -\frac{s^2 \sqrt{z^2 - v^2}}{2}.$$

This is real and negative (in both cases, since $e^{-2i\pi/4} = e^{6i\pi/4}$), thus confirming that the standard result for integrating over the saddle point can be used. This is

$$I \approx \pm \left(\frac{2\pi}{A} \right)^{1/2} \exp(f_0) \exp[\tfrac{1}{2} i(\pi - \alpha)],$$

with the $\pm$ choice being resolved by the direction in which the l.s.d. passes through the saddle point; it is positive if $|\theta| < \pi/2$ and negative otherwise. In this particular case, the l.s.d. is traversed in the direction $-\pi/4$ through the saddle point and the plus sign is appropriate.

Finally, inserting all of the specific data for this case into the general formula, we find that

$$F_v(z) \approx \left[\frac{2\pi}{\sqrt{z^2 - v^2}}\right]^{1/2} \exp\left[i\sqrt{z^2 - v^2} + iv\,\cos^{-1}\frac{v}{z}\right] \exp[\tfrac{1}{2}i(\pi - \tfrac{3}{2}\pi)]$$

$$= \left[\frac{2\pi}{\sqrt{z^2 - v^2}}\right]^{1/2} \exp\left[i\left(\sqrt{z^2 - v^2} + v\,\cos^{-1}\frac{v}{z} - \frac{\pi}{4}\right)\right]$$

$$\approx \left[\frac{2\pi}{z}\right]^{1/2} \exp\left[i\left(z - \frac{v\pi}{2} - \frac{\pi}{4}\right)\right], \text{ for } z \gg v.$$

This last approximation enables us to identify the function $F_v(z)$ as probably being a multiple of the Hankel function (Bessel function of the third kind) $H_v^{(1)}(z)$, though, as different functions can have the same asymptotic form, this cannot be certain.

26

Tensors

26.1 *Use the basic definition of a Cartesian tensor to show the following.*

(a) *That for any general, but fixed, ϕ,*

$$(u_1, u_2) = (x_1 \cos \phi - x_2 \sin \phi, \; x_1 \sin \phi + x_2 \cos \phi)$$

are the components of a first-order tensor in two dimensions.

(b) *That*

$$\begin{pmatrix} x_2^2 & x_1 x_2 \\ x_1 x_2 & x_1^2 \end{pmatrix}$$

is not a tensor of order 2. To establish that a single element does not transform correctly is sufficient.

Consider a rotation of the (unprimed) coordinate axes through an angle θ to give the new (primed) axes. Under this rotation,

$$x_1 \to x_1' = x_1 \cos \theta + x_2 \sin \theta,$$
$$x_2 \to x_2' = -x_1 \sin \theta + x_2 \cos \theta,$$
$$x_3 \to x_3' = x_3,$$

and the transformation matrix L_{ij} is given by

$$\mathsf{L} = \begin{pmatrix} \cos \theta & \sin \theta & 0 \\ -\sin \theta & \cos \theta & 0 \\ 0 & 0 & 1 \end{pmatrix}.$$

(a) Denoting $\cos \theta$ and $\sin \theta$ by c and s, respectively, we compare

$$u_1' = x_1' \cos \phi - x_2' \sin \phi = cx_1 \cos \phi + sx_2 \cos \phi + sx_1 \sin \phi - cx_2 \sin \phi$$

with

$$u_1' = cu_1 + su_2 = cx_1 \cos \phi - cx_2 \sin \phi + sx_1 \sin \phi + sx_2 \cos \phi.$$

These two are equal, showing that the first component transforms correctly. However, this alone is not sufficient; for (u_1, u_2) to be the components of a first-order tensor, *all* components must transform correctly. We therefore also compare the remaining transformed component:

$$u_2' = x_1' \sin \phi + x_2' \cos \phi = cx_1 \sin \phi + sx_2 \sin \phi - sx_1 \cos \phi + cx_2 \cos \phi$$

is to be compared with

$$u_2' = -su_1 + cu_2 = -sx_1 \cos \phi + sx_2 \sin \phi + cx_1 \sin \phi + cx_2 \cos \phi.$$

These two are also equal, showing that both components do transform correctly and that (u_1, u_2) are indeed the components of a first-order tensor.

We note, in passing, that $u_1 + iu_2$ is the complex vector obtained by rotating the 'base vector' $x_1 + ix_2$ through an angle ϕ in the complex plane:

$$\begin{aligned} u_1 + iu_2 &= e^{i\phi}(x_1 + ix_2) \\ &= (\cos \phi + i \sin \phi)(x_1 + ix_2) \\ &= (x_1 \cos \phi - x_2 \sin \phi) + i(x_1 \sin \phi + x_2 \cos \phi). \end{aligned}$$

In view of this observation, and of the definition of a first-order tensor as a set of objects 'that transform in the same way as a position vector', it is perhaps not surprising to find that the given expressions form the components of a tensor.

(b) Consider the transform of the first element $u_{11} = x_2^2$. This becomes

$$u_{11}' = (x_2')^2 = (-sx_1 + cx_2)^2 = s^2 x_1^2 - 2scx_1 x_2 + c^2 x_2^2.$$

If it transforms as a component of a tensor, then it must also be the case that

$$u_{11}' = L_{1k} L_{1l} u_{kl} = c^2 x_2^2 + csx_1 x_2 + scx_1 x_2 + s^2 x_1^2.$$

But, these two RHSs are not equal, and it follows that the given set of expressions cannot form the components of a tensor of order 2. It is not necessary to consider any more u_{ij}; failure of any one element to transform correctly rules out the possibility of the set being a tensor.

26.3 *In the usual approach to the study of Cartesian tensors the system is considered fixed and the coordinate axes are rotated. The transformation matrix used is therefore that for components relative to rotated coordinate axes. An alternative view is that of taking the coordinate axes as fixed and rotating the components of the system; this is equivalent to reversing the signs of all rotation angles.*

Using this alternative view, determine the matrices representing (a) a positive rotation of $\pi/4$ about the x-axis and (b) a rotation of $-\pi/4$ about the y-axis. Determine the initial vector $\mathbf{r}$ which, when subjected to (a) followed by (b), finishes at $(3, 2, 1)$.

The normal notation for the two rotation matrices would be

$$
A = \begin{pmatrix} 1 & 0 & 0 \\ 0 & \cos\theta & \sin\theta \\ 0 & -\sin\theta & \cos\theta \end{pmatrix} \quad \text{and } B = \begin{pmatrix} \cos\phi & 0 & \sin\phi \\ 0 & 1 & 0 \\ -\sin\phi & 0 & \cos\phi \end{pmatrix},
$$

with $\theta = \phi = \pi/4$.

In the alternative view (denoted by ") they would have the same forms but with $\theta = \phi = -\pi/4$, namely

$$
A" = \frac{1}{\sqrt{2}} \begin{pmatrix} \sqrt{2} & 0 & 0 \\ 0 & 1 & -1 \\ 0 & 1 & 1 \end{pmatrix} \quad \text{and } B" = \frac{1}{\sqrt{2}} \begin{pmatrix} 1 & 0 & -1 \\ 0 & \sqrt{2} & 0 \\ 1 & 0 & 1 \end{pmatrix}.
$$

The matrix representing (a) followed by (b) in this alternative view is thus

$$
B"A" = \frac{1}{2} \begin{pmatrix} 1 & 0 & -1 \\ 0 & \sqrt{2} & 0 \\ 1 & 0 & 1 \end{pmatrix} \begin{pmatrix} \sqrt{2} & 0 & 0 \\ 0 & 1 & -1 \\ 0 & 1 & 1 \end{pmatrix}
$$

$$
= \frac{1}{2} \begin{pmatrix} \sqrt{2} & -1 & -1 \\ 0 & \sqrt{2} & -\sqrt{2} \\ \sqrt{2} & 1 & 1 \end{pmatrix}.
$$

The required point is the solution of

$$
\frac{1}{2} \begin{pmatrix} \sqrt{2} & -1 & -1 \\ 0 & \sqrt{2} & -\sqrt{2} \\ \sqrt{2} & 1 & 1 \end{pmatrix} \begin{pmatrix} x \\ y \\ z \end{pmatrix} = \begin{pmatrix} 3 \\ 2 \\ 1 \end{pmatrix}.
$$

Using the fact that B"A" is orthogonal, and therefore its inverse is simply its transpose, this can be solved directly as

$$
\begin{pmatrix} x \\ y \\ z \end{pmatrix} = \frac{1}{2} \begin{pmatrix} \sqrt{2} & 0 & \sqrt{2} \\ -1 & \sqrt{2} & 1 \\ -1 & -\sqrt{2} & 1 \end{pmatrix} \begin{pmatrix} 3 \\ 2 \\ 1 \end{pmatrix} = \begin{pmatrix} 2\sqrt{2} \\ -1+\sqrt{2} \\ -1-\sqrt{2} \end{pmatrix}.
$$

As a partial check, we compute $|\mathbf{r}_{\text{initial}}|^2 = 8 + (3 - 2\sqrt{2}) + (3 + 2\sqrt{2}) = 14 = 3^2 + 2^2 + 1^2 = |\mathbf{r}_{\text{final}}|^2$, i.e. the length of the vector is unchanged by the rotations, as it should be.

26.5 *Use the quotient law for tensors to show that the array*

$$\begin{pmatrix} y^2 + z^2 - x^2 & -2xy & -2xz \\ -2yx & x^2 + z^2 - y^2 & -2yz \\ -2zx & -2zy & x^2 + y^2 - z^2 \end{pmatrix}$$

forms a second-order tensor.

To test whether the array is a second-order tensor we need to contract it with an arbitrary known second-order tensor. By 'arbitrary tensor' we mean a tensor in which any one component can be made to be the only non-zero component.

Since any second-order tensor can always be written as the sum of a symmetric and an anti-symmetric tensor, and all operations are linear, it will be sufficient to prove the result for one known symmetric tensor and one known antisymmetric tensor. The simplest symmetric second-order tensor S_{ij} is the (symmetric) outer product of the (by definition) first-order tensor (x, y, z) with itself, i.e $S_{ij} = x_i x_j$. Denoting the given array by B_{ij}, we consider

$$\begin{aligned} B_{ij} S_{ij} &= B_{ij} x_i x_j \\ &= x^2(y^2 + z^2 - x^2) + y^2(x^2 + z^2 - y^2) + z^2(x^2 + y^2 - z^2) \\ &\quad + 2[xy(-2xy) + xz(-2xz) + yz(-2yz)] \\ &= -2x^2 y^2 - 2x^2 z^2 - 2y^2 z^2 - x^4 - y^4 - z^4 \\ &= -(x^2 + y^2 + z^2)^2 = -|\mathbf{x}|^4. \end{aligned}$$

The term in parentheses in the last line is formally $x_i x_i$, i.e the contracted product of a first-order tensor with itself, and therefore an invariant zero-order tensor. Squaring an invariant or multiplying it by a constant (-1) leaves it as an invariant, leading to the conclusion that $B_{ij} S_{ij}$ is a zero-order tensor.

We now turn to an antisymmetric tensor, where a suitable second-order tensor A_{ij} is the contraction of the third-order tensor ϵ_{ijk} with the first-order tensor x_i. Thus A_{ij} has the form

$$A = \begin{pmatrix} 0 & z & -y \\ -z & 0 & x \\ y & -x & 0 \end{pmatrix}$$

and the contracted tensor is

$$B_{ij} A_{ij} = 0 - 2xyz + 2yxz + 2zyx + 0 - 2xyz - 2yzx + 2xzy + 0 = 0.$$

Now, 0 is an even more obvious invariant than $|\mathbf{x}|^2$ and so $B_{ij}A_{ij}$ is also a zero-order tensor.

Taking the results of the last two paragraphs together, it follows from the quotient law that B_{ij} is a second-order tensor.

26.7 *Use tensor methods to establish that*

$$grad\ \tfrac{1}{2}(\mathbf{u} \cdot \mathbf{u}) = \mathbf{u} \times curl\ \mathbf{u} + (\mathbf{u} \cdot grad)\mathbf{u}.$$

Now use this result and the general divergence theorem for tensors to show that, for a vector field $\mathbf{A}$,

$$\int_S [\mathbf{A}(\mathbf{A} \cdot d\mathbf{S}) - \tfrac{1}{2}A^2 d\mathbf{S}] = \int_V [\mathbf{A}\ div\mathbf{A} - \mathbf{A} \times curl\ \mathbf{A}]\ dV,$$

where S *is the surface enclosing the volume* V.

We start with the most complicated of the terms in the identity:

$$[\mathbf{u} \times (\nabla \times \mathbf{u})]_i = \epsilon_{ijk}u_j(\nabla \times \mathbf{u})_k = \epsilon_{ijk}u_j\,\epsilon_{klm}\frac{\partial u_m}{\partial x_l}$$

$$= (\delta_{il}\delta_{jm} - \delta_{im}\delta_{jl})u_j\frac{\partial u_m}{\partial x_l} = u_j\frac{\partial u_j}{\partial x_i} - u_j\frac{\partial u_i}{\partial x_j}$$

$$= \frac{1}{2}\frac{\partial}{\partial x_i}(u_j u_j) - (\mathbf{u} \cdot \nabla)u_i = [\tfrac{1}{2}\nabla(\mathbf{u} \cdot \mathbf{u}) - (\mathbf{u} \cdot \nabla)\mathbf{u}]_i,$$

which establishes the first result. To establish the second result we first note that

$$\frac{\partial}{\partial x_j}(A_i A_j) = A_i\frac{\partial A_j}{\partial x_j} + A_j\frac{\partial A_i}{\partial x_j} = [\mathbf{A}\nabla \cdot \mathbf{A} + (\mathbf{A} \cdot \nabla)\mathbf{A}]_i. \qquad (*)$$

Next we consider the ith component of the integrand on the RHS of the putative equation and use the first result to replace $\mathbf{A} \times (\nabla \times \mathbf{A})$.

$$[\mathrm{RHS}]_i = A_i(\nabla \cdot \mathbf{A}) - [\mathbf{A} \times (\nabla \times \mathbf{A})]_i$$

$$= A_i(\nabla \cdot \mathbf{A}) - \tfrac{1}{2}\nabla_i A^2 + (\mathbf{A} \cdot \nabla)A_i = \frac{\partial}{\partial x_j}(A_i A_j) - \frac{1}{2}\frac{\partial(A^2)}{\partial x_i}, \qquad \text{using } (*).$$

We can now integrate this equation over the volume V and apply the divergence theorem for tensors to both terms individually:

$$\int_V [\mathrm{RHS}]_i\, dV = \int_V \frac{\partial}{\partial x_j}(A_i A_j)\, dV - \frac{1}{2}\int_V \frac{\partial(A^2)}{\partial x_i}\, dV$$

$$= \int_S A_i A_j\, dS_j - \frac{1}{2}\int_S A^2\, dS_i = \int_S [\mathbf{A}(\mathbf{A} \cdot d\mathbf{S}) - \tfrac{1}{2}A^2 d\mathbf{S}]_i.$$

This concludes the proof.

26.9 *The equation*

$$|A|\epsilon_{lmn} = A_{li}A_{mj}A_{nk}\epsilon_{ijk} \qquad (*)$$

is a more general form of the expression for the determinant of a 3×3 matrix A. *This would normally be written as*

$$|A| = \epsilon_{ijk}A_{i1}A_{j2}A_{k3},$$

but the form (*) *removes the explicit mention of* $1, 2, 3$ *at the expense of an additional Levi–Civita symbol. The* (*) *form of expression for a determinant can be readily extended to cover a general $N \times N$ matrix.*

The following is a list of some of the common properties of determinants.

- (a) *Determinant of the transpose.* The transpose matrix A^T has the same determinant as A itself, i.e.

  $$|A^T| = |A|.$$

- (b) *Interchanging two rows or two columns.* If two rows (or columns) of A are interchanged, its determinant changes sign but is unaltered in magnitude.
- (c) *Identical rows or columns.* If any two rows (or columns) of A are identical or are multiples of one another, then it can be shown that $|A| = 0$.
- (d) *Adding a constant multiple of one row (or column) to another.* The determinant of a matrix is unchanged in value by adding to the elements of one row (or column) any fixed multiple of the elements of another row (or column).
- (e) *Determinant of a product.* If A and B are square matrices of the same order then

  $$|AB| = |A||B| = |BA|.$$

  A simple extension of this property gives, for example,

  $$|AB \cdots G| = |A||B| \cdots |G| = |A||G| \cdots |B| = |A \cdots GB|,$$

  which shows that the determinant is invariant to permutations of the matrices in a multiple product.

Use the form given in (*) *to prove the above properties. For definiteness take* $N = 3$, *but convince yourself that your methods of proof would be valid for any positive integer $N > 1$.*

(a) We write the expression for $|A^T|$ using the given formalism, recalling that $(A^T)_{ij} = (A)_{ji}$. We then contract both sides with ϵ_{lmn}:

$$|A^T|\epsilon_{lmn} = A_{il}A_{jm}A_{kn}\epsilon_{ijk},$$
$$|A^T|\epsilon_{lmn}\epsilon_{lmn} = A_{il}A_{jm}A_{kn}\epsilon_{lmn}\epsilon_{ijk},$$
$$= |A|\epsilon_{ijk}\epsilon_{ijk},$$
$$|A^T| = |A|.$$

In the third line we have used the definition of $|A|$ (with the roles of the sets of dummy variables $\{i, j, k\}$ and $\{l, m, n\}$ interchanged), and in the fourth line, we have cancelled the scalar quantity $\epsilon_{lmn}\epsilon_{lmn} = \epsilon_{ijk}\epsilon_{ijk}$; the value of this scalar is $N(N-1)$, but that is irrelevant here.

(b) Every non-zero term on the RHS of (∗) contains any particular row index once and only once. The same can be said for the Levi–Civita symbol on the LHS. Thus interchanging two rows is equivalent to interchanging two of the subscripts of ϵ_{lmn} and thereby reversing its sign. Consequently, the whole RHS changes sign and the magnitude of $|A|$ remains the same, though its sign is changed.

(c) If, say, $A_{pi} = \lambda A_{pj}$, for some particular pair of values i and j and all p, then in the (multiple) summation on the RHS of (∗) each A_{nk} appears multiplied by (with no summation over i and j)

$$\epsilon_{ijk}A_{li}A_{mj} + \epsilon_{jik}A_{lj}A_{mi} = \epsilon_{ijk}\lambda A_{lj}A_{mj} + \epsilon_{jik}A_{lj}\lambda A_{mj} = 0,$$

since $\epsilon_{ijk} = -\epsilon_{jik}$. Consequently, grouped in this way, all pairs of terms contribute nothing to the sum and $|A| = 0$.

(d) Consider the matrix B whose m, jth element is defined by $B_{mj} = A_{mj} + \lambda A_{pj}$, where $p \neq m$. The only case that needs detailed analysis is when l, m and n are all different. Since $p \neq m$ it must be the same as either l or n; suppose that $p = l$. The determinant of B is given by

$$|B|\epsilon_{lmn} = A_{li}(A_{mj} + \lambda A_{lj})A_{nk}\epsilon_{ijk}$$
$$= A_{li}A_{mj}A_{nk}\epsilon_{ijk} + \lambda A_{li}A_{lj}A_{nk}\epsilon_{ijk}$$
$$= |A|\epsilon_{lmn} + \lambda 0,$$

where we have used the row equivalent of the intermediate result obtained for columns in (c). Thus we conclude that $|B| = |A|$.

(e) If $X = AB$, then

$$|X|\epsilon_{lmn} = A_{lx}B_{xi}A_{my}B_{yj}A_{nz}B_{zk}\epsilon_{ijk}.$$

Contract both sides with ϵ_{lmn}:

$$|X|\epsilon_{lmn}\epsilon_{lmn} = \epsilon_{lmn}A_{lx}A_{my}A_{nz}\,\epsilon_{ijk}B_{xi}B_{yj}B_{zk}$$
$$= \epsilon_{xyz}|A^{\mathrm{T}}|\epsilon_{xyz}|B|,$$
$$\Rightarrow \quad |X| = |A^{\mathrm{T}}||B| = |A||B|, \quad \text{using result (a).}$$

To obtain the last line we have cancelled the non-zero scalar $\epsilon_{lmn}\epsilon_{lmn} = \epsilon_{xyz}\epsilon_{xyz}$ from both sides, as we did in the proof of result (a).

The extension to the product of any number of matrices is obvious. Replacing B by CD or by DC and applying the result just proved extends it to a product of three matrices. Extension to any higher number is done in the same way.

26.11 *Given a non-zero vector* **v**, *find the value that should be assigned to* α *to make*

$$P_{ij} = \alpha v_i v_j \quad and \quad Q_{ij} = \delta_{ij} - \alpha v_i v_j$$

into parallel and orthogonal projection tensors, respectively, i.e. tensors that satisfy, respectively, $P_{ij}v_j = v_i$, $P_{ij}u_j = 0$ *and* $Q_{ij}v_j = 0$, $Q_{ij}u_j = u_i$, *for any vector* **u** *that is orthogonal to* **v**.

Show, in particular, that Q_{ij} *is unique, i.e. that if another tensor* T_{ij} *has the same properties as* Q_{ij} *then* $(Q_{ij} - T_{ij})w_j = 0$ *for any vector* **w**.

Consider

$$P_{ij}v_j = \alpha v_i v_j v_j = \alpha |\mathbf{v}|^2 v_i, \text{ and}$$
$$P_{ij}u_j = \alpha v_i v_j u_j = \alpha v_i (v_j u_j) = 0, \text{ as } \mathbf{u} \text{ is orthogonal to } \mathbf{v}.$$

For $P_{ij}v_j = v_i$ it is clearly necessary that $\alpha = |\mathbf{v}|^{-2}$.

With this choice,

$$Q_{ij}v_j = (\delta_{ij} - \alpha v_i v_j)v_j = v_i - \alpha(v_j v_j)v_i = v_i - |\mathbf{v}|^{-2}(v_j v_j)v_i = 0, \text{ and}$$
$$Q_{ij}u_j = (\delta_{ij} - \alpha v_i v_j)u_j = u_i - \alpha(v_j u_j)v_i = u_i - 0v_i = u_i.$$

Thus the one assigned value for α gives both P_{ij} and Q_{ij} the required properties.

Let $\mathbf{u}^{(1)}$ and $\mathbf{u}^{(2)}$ be any two linearly independent non-zero vectors orthogonal to **v**. Then *any* vector **w** can be expressed as $\lambda\mathbf{v} + \mu\mathbf{u}^{(1)} + \nu\mathbf{u}^{(2)}$.

Now suppose that T_{ij} has the same properties as Q_{ij} and consider

$$(Q_{ij} - T_{ij})w_j = (Q_{ij} - T_{ij})(\lambda v_j + \mu u_j^{(1)} + \nu u_j^{(2)})$$
$$= \lambda 0 + \mu u_j^{(1)} + \nu u_j^{(2)} - \lambda T_{ij}v_j - \mu T_{ij}u_j^{(1)} - \nu T_{ij}u_j^{(2)}$$
$$= 0 + \mu u_j^{(1)} + \nu u_j^{(2)} - 0 - \mu u_j^{(1)} - \nu u_j^{(2)} = 0.$$

In this sense, Q_{ij} is unique.

26.13 *In a certain crystal the unit cell can be taken as six identical atoms lying at the corners of a regular octahedron. Convince yourself that these atoms can also be considered as lying at the centres of the faces of a cube and hence that the crystal has cubic symmetry. Use this result to prove that the conductivity tensor for the crystal,* σ_{ij}, *must be isotropic.*

It is easiest to start with a cube and then join the centre points of any pair of faces that have a common edge. The network of 12 lines so formed are the edges of a regular octahedron.

The crystal has cubic symmetry and must therefore be invariant under rotations that leave a cube unchanged (apart from the labelling of its corners). One such symmetry operation is rotation (by $2\pi/3$ about a body diagonal; this relabels the axes $O123$ as axes $O3'1'2'$ in the rotated system. The (orthogonal) base-vector transformation matrix S has as its i, jth component the ith component of e'_j with respect to the basis $\{e_k\}$. The coordinate transformation matrix L is the transpose of this. For the rotation under consideration,

$$S = \begin{pmatrix} 0 & 0 & 1 \\ 1 & 0 & 0 \\ 0 & 1 & 0 \end{pmatrix} \quad \text{and} \quad L = S^T = \begin{pmatrix} 0 & 1 & 0 \\ 0 & 0 & 1 \\ 1 & 0 & 0 \end{pmatrix}.$$

The conductivity tensor is a second-order tensor and so $\sigma'_{ij} = L_{ik} L_{jm} \sigma_{km}$ or, in matrix form,

$$\sigma' = L\sigma L^T$$
$$= \begin{pmatrix} 0 & 1 & 0 \\ 0 & 0 & 1 \\ 1 & 0 & 0 \end{pmatrix} \begin{pmatrix} \sigma_{11} & \sigma_{12} & \sigma_{13} \\ \sigma_{21} & \sigma_{22} & \sigma_{23} \\ \sigma_{31} & \sigma_{32} & \sigma_{33} \end{pmatrix} \begin{pmatrix} 0 & 0 & 1 \\ 1 & 0 & 0 \\ 0 & 1 & 0 \end{pmatrix}$$
$$= \begin{pmatrix} \sigma_{22} & \sigma_{23} & \sigma_{21} \\ \sigma_{32} & \sigma_{33} & \sigma_{31} \\ \sigma_{12} & \sigma_{13} & \sigma_{11} \end{pmatrix}.$$

This must be the same tensor as σ and so requires that

$$\sigma_{11} = \sigma_{22} = \sigma_{33}; \quad \sigma_{12} = \sigma_{23} = \sigma_{31}; \quad \sigma_{21} = \sigma_{32} = \sigma_{13}.$$

We also note that the transformed tensor is the original one, but with $1 \to 2$, $2 \to 3$ and $3 \to 1$.

Now, restarting from the original situation, consider a rotation of $\pi/2$ about the 3-axis. This clearly carries the $O1$-axis onto the original $O2$-axis and the $O2$-axis onto the original negative $O1$-axis. Therefore, by the substitutions $1 \to 2$, $2 \to -1$ and $3 \to 3$ (where a component changes sign for each minus sign on its subscripts) or by a matrix calculation similar to the previous one, the new transformed conductivity tensor is

$$\sigma' = \begin{pmatrix} \sigma_{22} & -\sigma_{21} & \sigma_{23} \\ -\sigma_{12} & \sigma_{11} & -\sigma_{13} \\ \sigma_{32} & -\sigma_{31} & \sigma_{33} \end{pmatrix}.$$

Again the invariance of σ imposes requirements. In this case,

$$\sigma_{11} = \sigma_{22}; \quad \sigma_{13} = \sigma_{23} = -\sigma_{13}.$$

428

The last set of equalities requires that $\sigma_{13} = \sigma_{23} = 0$ and hence, by the previous result, that $\sigma_{ij} = 0$ whenever $i \neq j$. Since $\sigma_{11} = \sigma_{22} = \sigma_{33}$, σ is a multiple of the unit matrix, and it follows that σ_{ij} is an isotropic tensor.

Either by direct calculation or by noting that any rotational symmetry of a cube can be represented as an ordered sequence of the two rotations already used, it can be shown that other symmetries do not impose any further constraint on the remaining non-zero elements of the conductivity tensor. Intuitively this must be so, since σ now contains only one free parameter, the common value of σ_{11}, σ_{22} and σ_{33}, and this is required to describe the level of conductivity, which must vary from one crystal to another, and certainly between crystals of different elements.

26.15 *In a certain system of units the electromagnetic stress tensor M_{ij} is given by*

$$M_{ij} = E_i E_j + B_i B_j - \tfrac{1}{2}\delta_{ij}(E_k E_k + B_k B_k),$$

where the electric and magnetic fields, $\mathbf{E}$ and $\mathbf{B}$, are first-order tensors. Show that M_{ij} is a second-order tensor.

Consider a situation in which $|\mathbf{E}| = |\mathbf{B}|$ but the directions of $\mathbf{E}$ and $\mathbf{B}$ are not parallel. Show that $\mathbf{E} \pm \mathbf{B}$ are principal axes of the stress tensor and find the corresponding principal values. Determine the third principal axis and its corresponding principal value.

In the calculation of the transformed RHS, $\mathbf{E}$ and $\mathbf{B}$ transform with a single 'L-matrix', but δ_{ij}, being a second-order tensor, requires two. It may simply be noticed that $E_k E_k$ and $B_k B_k$ are scalars and therefore unaltered in the transformation; but, if not, then the orthogonal properties of L, $L_{ik}L_{jk} = \delta_{ij}$ and $L_{ki}L_{kj} = \delta_{ij}$, are needed:

$$M_{ij} = E_i E_j + B_i B_j - \tfrac{1}{2}\delta_{ij}(E_k E_k + B_k B_k),$$
$$M'_{ij} = L_{im}E_m L_{jn}E_n + L_{im}B_m L_{jn}B_n$$
$$\qquad - \tfrac{1}{2}L_{ip}L_{jq}\delta_{pq}(L_{kr}E_r L_{ks}E_s + L_{kr}B_r L_{ks}B_s)$$
$$\qquad = L_{im}L_{jn}(E_m E_n + B_m B_n) - \tfrac{1}{2}L_{ip}L_{jq}\delta_{pq}(\delta_{rs}E_r E_s + \delta_{rs}B_r B_s)$$
$$\qquad = L_{im}L_{jn}[E_m E_n + B_m B_n - \tfrac{1}{2}\delta_{mn}(E_r E_r + B_r B_r)]$$
$$\qquad = L_{im}L_{jn}M_{mn}.$$

To obtain the penultimate line we relabelled the dummy suffices p and q as m and n. Thus M_{ij} transforms as a second-order tensor; it is real and symmetric and will therefore have orthogonal eigenvectors.

429

For the case $|\mathbf{E}| = |\mathbf{B}|$, i.e. $E^2 = B^2$, denote $E_i \pm B_i$ by v_i and consider

$$
\begin{aligned}
M_{ij}v_j &= M_{ij}(E_j \pm B_j) \\
&= E_iE_j(E_j \pm B_j) + B_iB_j(E_j \pm B_j) - \tfrac{1}{2}\delta_{ij}(E^2 + B^2)(E_j \pm B_j) \\
&= E_iE^2 \pm E_i(\mathbf{E} \cdot \mathbf{B}) + B_i(\mathbf{B} \cdot \mathbf{E}) \pm B_iB^2 \\
&\quad - \tfrac{1}{2}(E^2 + B^2)(E_i \pm B_i) \\
&= (E_i \pm B_i)[\,E^2 \pm (\mathbf{E} \cdot \mathbf{B}) - \tfrac{1}{2}2E^2\,], \text{ using } E^2 = B^2, \\
&= \pm(\mathbf{E} \cdot \mathbf{B})(E_i \pm B_i) \\
&= \pm(\mathbf{E} \cdot \mathbf{B})v_i.
\end{aligned}
$$

This shows that $\mathbf{E} \pm \mathbf{B}$ are eigenvectors of M_{ij} (i.e. its principal axes) with principal values $\pm(\mathbf{E} \cdot \mathbf{B})$.

The third principal axis is orthogonal to both of these and is therefore in the direction

$$(\mathbf{E} + \mathbf{B}) \times (\mathbf{E} - \mathbf{B}) = 0 + (\mathbf{B} \times \mathbf{E}) - (\mathbf{E} \times \mathbf{B}) - 0 = 2(\mathbf{B} \times \mathbf{E}).$$

To determine its principal value, consider

$$
\begin{aligned}
M_{ij}(\mathbf{B} \times \mathbf{E})_j &= M_{ij}\epsilon_{jlm}B_lE_m \\
&= E_iE_j\epsilon_{jlm}B_lE_m + B_iB_j\epsilon_{jlm}B_lE_m - \tfrac{1}{2}\delta_{ij}2E^2\epsilon_{jlm}B_lE_m \\
&= 0 + 0 - E^2(\mathbf{B} \times \mathbf{E})_i, \text{ since } \epsilon_{jlm}X_lX_j = 0.
\end{aligned}
$$

Thus, the third principal value is $-E^2$ (or $-B^2$). This value could have been deduced from the trace of $M_{ij} = E^2 + B^2 - \tfrac{3}{2}(E^2 + B^2) = -E^2$, since the two eigenvalues found previously are $\pm\mathbf{E} \cdot \mathbf{B}$, which sum to zero. The three eigenvalues together must add up to the trace; hence, the third one is $-E^2$.

26.17 *A rigid body consists of eight particles, each of mass m, held together by light rods. In a certain coordinate frame the particles are at positions*

$$\pm a(3, 1, -1), \quad \pm a(1, -1, 3), \quad \pm a(1, 3, -1), \quad \pm a(-1, 1, 3).$$

Show that, when the body rotates about an axis through the origin, if the angular velocity and angular momentum vectors are parallel then their ratio must be $40ma^2$, $64ma^2$ or $72ma^2$.

Because the particles are symmetrically placed in pairs with respect to the origin, the inertia tensor, given by

$$I_{ij} = \sum_{\text{particles}} m(r^2\delta_{ij} - x_ix_j),$$

will be twice that calulated for the + signs alone. It components are therefore

$$I_{11} = 2ma^2(2 + 10 + 10 + 10) = 64ma^2,$$
$$I_{12} = I_{21} = -2ma^2(3 - 1 + 3 - 1) = -8ma^2,$$
$$I_{13} = I_{31} = -2ma^2(-3 + 3 - 1 - 3) = 8ma^2,$$
$$I_{22} = 2ma^2(10 + 10 + 2 + 10) = 64ma^2,$$
$$I_{23} = I_{32} = -2ma^2(-1 - 3 - 3 + 3) = 8ma^2,$$
$$I_{33} = 2ma^2(10 + 2 + 10 + 2) = 48ma^2.$$

The resulting tensor is

$$8ma^2 \begin{pmatrix} 8 & -1 & 1 \\ -1 & 8 & 1 \\ 1 & 1 & 6 \end{pmatrix}$$

and its principal moments are $8ma^2\lambda$, where

$$
\begin{aligned}
0 &= \begin{vmatrix} 8 - \lambda & -1 & 1 \\ -1 & 8 - \lambda & 1 \\ 1 & 1 & 6 - \lambda \end{vmatrix} \\
&= (8 - \lambda)(\lambda^2 - 14\lambda + 47) + (-7 + \lambda) + (-9 + \lambda) \\
&= (8 - \lambda)(\lambda^2 - 14\lambda + 47 - 2) \\
&= (8 - \lambda)(\lambda - 9)(\lambda - 5).
\end{aligned}
$$

Thus the principal moments are $40ma^2$, $64ma^2$ and $72ma^2$. As a partial check: $40 + 64 + 72 = 8(8 + 8 + 6)$.

If the angular velocity ω and the angular momentum $\mathbf{J} = I\omega$ are parallel, then the body is rotating about one of its principal axes (the eigenvectors of I); their ratio is the principal moment about that axis and is thus one of the three values calculated above.

26.19 *A block of wood contains a number of thin soft-iron nails (of constant permeability). A unit magnetic field directed eastwards induces a magnetic moment in the block having components $(3, 1, -2)$, and similar fields directed northwards and vertically upwards induce moments $(1, 3, -2)$ and $(-2, -2, 2)$ respectively. Show that all the nails lie in parallel planes.*

The magnetic moment M, the permeability μ and the magnetic field H for iron of constant pemeability are connected by $M_i = \mu_{ij}H_j$. Taking the 1-, 2- and

3-directions as East, North and vertical, μ has the form

$$\begin{pmatrix} 3 & 1 & -2 \\ 1 & 3 & -2 \\ -2 & -2 & 2 \end{pmatrix}.$$

By adding the first two columns to twice the third one, it can be seen that this matrix has zero determinant. The matrix therefore has at least one zero eigenvalue. (The same conclusion can be reached using the routine method for finding eigenvalues; they are 0, 2 and 6.)

Thus, a field parallel to the eigenvector corresponding to this zero eigenvalue will induce no moment in the block. Physically this means that all the nails lie in planes to which this direction is a normal. To find the direction we solve

$$\begin{pmatrix} 3 & 1 & -2 \\ 1 & 3 & -2 \\ -2 & -2 & 2 \end{pmatrix}\begin{pmatrix} x \\ y \\ z \end{pmatrix} = \begin{pmatrix} 0 \\ 0 \\ 0 \end{pmatrix} \Rightarrow \begin{pmatrix} x \\ y \\ z \end{pmatrix} = \begin{pmatrix} 1 \\ 1 \\ 2 \end{pmatrix}.$$

We conclude that all the nails lie at right angles to this direction.

26.21 *For a general isotropic medium, the stress tensor p_{ij} and strain tensors e_{ij} are related by*

$$p_{ij} = \frac{\sigma E}{(1+\sigma)(1-2\sigma)}\, e_{kk}\delta_{ij} + \frac{E}{1+\sigma}\, e_{ij},$$

where E is Young's modulus and σ is Poisson's ratio.

By considering an isotropic body subjected to a uniform hydrostatic pressure (no shearing stress), show that the bulk modulus k, defined by the ratio of the pressure to the fractional decrease in volume, is given by $k = E/[3(1-2\sigma)]$.

Consider a small rectangular parallelepiped, with one corner at the origin and the opposite one at (a_1, a_2, a_3), subjected to a uniform hydrostatic pressure. The isotropy of the pressure means that all forces are normal to the surfaces on which they act and that the stress and strain tensor components $p_{ij} = e_{ij} = 0$ for $i \neq j$. Furthermore, because of the symmetry of the situation, when $i \neq j$, not only is e_{ij} zero, but so are the individual $\partial u_i/\partial x_j$ that are its constituents.

In the current situation, $p_{11} = p_{22} = p_{33} = -p$ and so, writing $\sum_k e_{kk}$ as θ, we have, for each i ($i = 1, 2, 3$) with no summation over i implied, that

$$-p = p_{ii} = \frac{E}{(1+\sigma)(1-2\sigma)}\, [\sigma\theta + (1-2\sigma)e_{ii}].$$

Adding the three equations together gives

$$-3p = \frac{E}{(1+\sigma)(1-2\sigma)} [3\sigma\theta + (1-2\sigma)\theta] = \frac{E\theta}{1-2\sigma}.$$

Now the fractional increase f in the volume of the parallelepiped is given by

$$\frac{1}{a_1 a_2 a_3} \left(a_1 + \frac{\partial u_1}{\partial x_i} a_i + \cdots \right) \left(a_2 + \frac{\partial u_2}{\partial x_i} a_i + \cdots \right) \left(a_3 + \frac{\partial u_3}{\partial x_i} a_i + \cdots \right) - 1.$$

Since $\partial u_i / \partial x_j = 0$ for $i \neq j$, the only three non-zero first-order terms are

$$f = \frac{\partial u_1}{\partial x_1} + \frac{\partial u_2}{\partial x_2} + \frac{\partial u_3}{\partial x_3} = e_{11} + e_{22} + e_{33} = 0.$$

We conclude that the bulk modulus, k, is given by

$$k = \frac{p}{-f} = \frac{-E\theta}{3(1-2\sigma)} \frac{1}{(-\theta)} = \frac{E}{3(1-2\sigma)}.$$

26.23 *A fourth-order tensor T_{ijkl} has the properties*

$$T_{jikl} = -T_{ijkl}, \qquad T_{ijlk} = -T_{ijkl}.$$

Prove that for any such tensor there exists a second-order tensor K_{mn} such that

$$T_{ijkl} = \epsilon_{ijm}\epsilon_{kln}K_{mn}$$

and give an explicit expression for K_{mn}. Consider two (separate) special cases, as follows.

(a) *Given that T_{ijkl} is isotropic and $T_{ijji} = 1$, show that T_{ijkl} is uniquely determined and express it in terms of Kronecker deltas.*

(b) *If now T_{ijkl} has the additional property*

$$T_{klij} = -T_{ijkl},$$

show that T_{ijkl} has only three linearly independent components and find an expression for T_{ijkl} in terms of the vector

$$V_i = -\tfrac{1}{4}\epsilon_{jkl}T_{ijkl}.$$

As K_{mn} is to be a second-order tensor, we need to construct such a tensor from T_{ijkl}. Since the latter is of fourth order, it needs to be contracted n times with a tensor of order $2n - 2$ for some positive integer n. In view of the final stated expression for T_{ijkl}, involving $\epsilon_{ijm}\epsilon_{kln}$, i.e. a sixth-order tensor, we try $n = 4$ and, starting from $T_{ijkl} = \epsilon_{ijm}\epsilon_{kln}K_{mn}$, consider

$$\begin{aligned}
\epsilon_{pij}\epsilon_{qkl}T_{ijkl} &= \epsilon_{pij}\epsilon_{qkl}\epsilon_{ijm}\epsilon_{kln}K_{mn} \\
&= (\delta_{jj}\delta_{pm} - \delta_{jm}\delta_{pj})(\delta_{ll}\delta_{qn} - \delta_{ln}\delta_{ql})K_{mn} \\
&= (3\delta_{pm} - \delta_{pm})(3\delta_{qn} - \delta_{qn})K_{mn} \\
&= 4K_{pq}.
\end{aligned}$$

Clearly, $K_{mn} = \frac{1}{4}\epsilon_{mij}\epsilon_{nkl}T_{ijkl}$ has the required property.

(a) Given that T_{ijkl} is isotropic, and noting that ϵ_{mij} and ϵ_{nkl} are also isotropic, we conclude that K_{mn} must itself be isotropic. It must therefore be some multiple of δ_{mn} (as this is the most general isotropic second-order tensor), i.e. $K_{mn} = \lambda\delta_{mn}$ for one or more values of λ. Thus,

$$\begin{aligned}
T_{ijkl} &= \epsilon_{ijm}\epsilon_{kln}\lambda\delta_{mn} \\
&= \lambda\epsilon_{ijm}\epsilon_{klm} \\
&= \lambda(\delta_{ik}\delta_{jl} - \delta_{il}\delta_{jk}).
\end{aligned}$$

Now, since $T_{ijji} = 1$,

$$\begin{aligned}
1 &= \lambda(\delta_{ij}\delta_{ji} - \delta_{ii}\delta_{jj}) \\
&= \lambda[\delta_{ii} - (\delta_{ii})^2] = \lambda(3 - 9) \quad \Rightarrow \quad \lambda = -\tfrac{1}{6}.
\end{aligned}$$

We conclude that λ, and therefore also T_{ijkl}, is unique with $T_{ijkl} = \frac{1}{6}(\delta_{il}\delta_{jk} - \delta_{ik}\delta_{jl})$.

(b) To examine the implications of the antisymmetry indicated by $T_{klij} = -T_{ijkl}$, we interchange the pair of dummy suffices $\{i, j\}$ with the pair $\{k, l\}$ to obtain the third line below — and then switch them back again in the fourth line using the antisymmetry:

$$\begin{aligned}
K_{mn} &= \tfrac{1}{4}\epsilon_{mij}\epsilon_{nkl}T_{ijkl}, \\
K_{nm} &= \tfrac{1}{4}\epsilon_{nij}\epsilon_{mkl}T_{ijkl} \\
&= \tfrac{1}{4}\epsilon_{nkl}\epsilon_{mij}T_{klij} \\
&= -\tfrac{1}{4}\epsilon_{nkl}\epsilon_{mij}T_{ijkl} \\
&= -K_{mn}.
\end{aligned}$$

Thus K_{mn} is antisymmetric. It therefore has zeros on its leading diagonal and only three linearly independent components as non-diagonal elements. Since T_{ijkl} is uniquely defined in terms of K_{mn}, it too has only three linearly independent components.

Now consider

$$\epsilon_{jkl} T_{ijkl} = \epsilon_{jkl} \epsilon_{ijm} \epsilon_{kln} K_{mn},$$
$$-4V_i = (\delta_{km}\delta_{li} - \delta_{ki}\delta_{lm})\epsilon_{kln} K_{mn}$$
$$= (\epsilon_{min} - \epsilon_{imn}) K_{mn}$$
$$= 2\epsilon_{min} K_{mn}.$$

To 'invert' this relationship, consider

$$\epsilon_{irs} V_i = -\tfrac{1}{2}\epsilon_{irs} \epsilon_{min} K_{mn}$$
$$= -\tfrac{1}{2}(\delta_{rn}\delta_{sm} - \delta_{rm}\delta_{sn}) K_{mn}$$
$$= -\tfrac{1}{2}(K_{sr} - K_{rs})$$
$$= K_{rs} \quad \Rightarrow \quad K_{mn} = \epsilon_{pmn} V_p.$$

Finally, expressing T_{ijkl}, as given in the question, explicitly in terms of the vector V_i, using the result obtained above, we have

$$T_{ijkl} = \epsilon_{ijm} \epsilon_{kln} \epsilon_{pmn} V_p$$
$$= (\delta_{in}\delta_{jp} - \delta_{ip}\delta_{jn})\epsilon_{kln} V_p$$
$$= \epsilon_{kli} V_j - \epsilon_{klj} V_i.$$

26.25 *In a general coordinate system u^i, $i = 1, 2, 3$, in three-dimensional Euclidean space, a volume element is given by*

$$dV = |\mathbf{e}_1 \, du^1 \cdot (\mathbf{e}_2 \, du^2 \times \mathbf{e}_3 \, du^3)|.$$

Show that an alternative form for this expression, written in terms of the determinant g of the metric tensor, is given by

$$dV = \sqrt{g} \, du^1 \, du^2 \, du^3.$$

Show that under a general coordinate transformation to a new coordinate system u'^i, the volume element dV remains unchanged, i.e. show that it is a scalar quantity.

Working in terms of the Cartesian bases vectors $\mathbf{i}$, $\mathbf{j}$ and $\mathbf{k}$, let

$$\mathbf{e}_m = \lambda_{mx} \mathbf{i} + \lambda_{my} \mathbf{j} + \lambda_{mz} \mathbf{k}, \text{ for } m = 1, 2, 3.$$

Then,

$$\mathbf{e}_2 \, du^2 \times \mathbf{e}_3 \, du^3 = du^2 \, du^3 (\lambda_{2x}\lambda_{3y} \mathbf{k} - \lambda_{2x}\lambda_{3z} \mathbf{j} - \lambda_{2y}\lambda_{3x} \mathbf{k}$$
$$+ \lambda_{2y}\lambda_{3z} \mathbf{i} + \lambda_{2z}\lambda_{3x} \mathbf{j} - \lambda_{2z}\lambda_{3y} \mathbf{i}),$$

and it follows that

$$dV = \mathbf{e}_1 \, du^1 \cdot (\mathbf{e}_2 \, du^2 \times \mathbf{e}_3 \, du^3)$$
$$= du^1 \, du^2 \, du^3 \left[\lambda_{1x}(\lambda_{2y}\lambda_{3z} - \lambda_{2z}\lambda_{3y}) + \lambda_{1y}(\lambda_{2z}\lambda_{3x} - \lambda_{2x}\lambda_{3z}) \right. $$
$$\left. + \lambda_{1z}(\lambda_{2x}\lambda_{3y} - \lambda_{2y}\lambda_{3x}) \right]$$
$$= du^1 \, du^2 \, du^3 \begin{vmatrix} \lambda_{1x} & \lambda_{1y} & \lambda_{1z} \\ \lambda_{2x} & \lambda_{2y} & \lambda_{2z} \\ \lambda_{3x} & \lambda_{3y} & \lambda_{3z} \end{vmatrix}$$
$$\equiv du^1 \, du^2 \, du^3 |\mathsf{A}|, \text{ thus defining } \mathsf{A}.$$

Now consider an element of the matrix $\mathsf{A}\mathsf{A}^{\mathrm{T}}$:

$$(\mathsf{A}\mathsf{A}^{\mathrm{T}})_{mn} = \sum_r \mathsf{A}_{mr}\mathsf{A}_{nr} = \lambda_{mx}\lambda_{nx} + \lambda_{my}\lambda_{ny} + \lambda_{mz}\lambda_{nz}.$$

But the elements of the metric tensor are given by

$$\mathsf{g}_{mn} = \mathbf{e}_m \cdot \mathbf{e}_n = \lambda_{mx}\lambda_{nx} + \lambda_{my}\lambda_{ny} + \lambda_{mz}\lambda_{nz}.$$

Hence $\mathsf{A}\mathsf{A}^{\mathrm{T}} = \mathsf{g}$ and, in particular, $|\mathsf{A}| \, |\mathsf{A}^{\mathrm{T}}| = |\mathsf{g}|$. Since $|\mathsf{A}| = |\mathsf{A}^{\mathrm{T}}|$, it follows that $|\mathsf{A}| = |\mathsf{g}|^{1/2} = \sqrt{g}$ and

$$dV = du^1 \, du^2 \, du^3 \, |\mathsf{A}| = \sqrt{g} \, du^1 \, du^2 \, du^3.$$

For a transformation $u'^i = u'^i(u^1, u^2, u^3)$,

$$du'^1 \, du'^2 \, du'^3 = \left| \frac{\partial u'}{\partial u} \right| du^1 \, du^2 \, du^3,$$

and the covariant components of the second-order tensor g_{ij} transform as

$$g'_{ij} = \frac{\partial u^k}{\partial u'^i} \frac{\partial u^l}{\partial u'^j} g_{kl},$$

$$\Rightarrow \quad g' = \left| \frac{\partial u}{\partial u'} \right| \left| \frac{\partial u}{\partial u'} \right| g \text{ (on taking determinants)},$$

$$\Rightarrow \quad \sqrt{g'} = \left| \frac{\partial u}{\partial u'} \right| \sqrt{g}.$$

Thus, the new volume element is

$$dV' = \sqrt{g'} \, du'^1 \, du'^2 \, du'^3$$
$$= \left| \frac{\partial u}{\partial u'} \right| \sqrt{g} \left| \frac{\partial u'}{\partial u} \right| du^1 \, du^2 \, du^3$$
$$= \sqrt{g} \, du^1 \, du^2 \, du^3 = dV.$$

This shows that dV is a scalar quantity.

26.27 *Find an expression for the second covariant derivative, written in semi-colon notation as $v_{i;jk} \equiv (v_{i;j})_{;k}$, of a vector v_i. By interchanging the order of differentiation and then subtracting the two expressions, we define the components R^l_{ijk} of the Riemann tensor as*

$$v_{i;jk} - v_{i;kj} \equiv R^l_{ijk} v_l.$$

Show that in a general coordinate system u^i these components are given by

$$R^l_{ijk} = \frac{\partial \Gamma^l_{ik}}{\partial u^j} - \frac{\partial \Gamma^l_{ij}}{\partial u^k} + \Gamma^m_{ik} \Gamma^l_{mj} - \Gamma^m_{ij} \Gamma^l_{mk}.$$

By first considering Cartesian coordinates, show that all the components $R^l_{ijk} \equiv 0$ for any coordinate system in three-dimensional Euclidean space.
In such a space, therefore, we may change the order of the covariant derivatives without changing the resulting expression.

For the covariant derivative of the covariant components of a vector, we have

$$v_{i;j} = \frac{\partial v_i}{\partial u^j} - \Gamma^k_{ij} v_k,$$

where Γ^k_{ij} is a Christoffel symbol of the second kind. Hence,

$$v_{i;jk} \equiv (v_{i;j})_{;k}$$

$$= \left(\frac{\partial v_i}{\partial u^j} - \Gamma^l_{ij} v_l \right)_{;k}$$

$$= \frac{\partial}{\partial u^k} \left(\frac{\partial v_i}{\partial u^j} - \Gamma^l_{ij} v_l \right) - \Gamma^m_{ik} \left(\frac{\partial v_m}{\partial u^j} - \Gamma^l_{mj} v_l \right)$$

$$= \frac{\partial^2 v_i}{\partial u^k \partial u^j} - \Gamma^l_{ij} \frac{\partial v_l}{\partial u^k} - v_l \frac{\partial \Gamma^l_{ij}}{\partial u^k} - \Gamma^m_{ik} \frac{\partial v_m}{\partial u^j} + \Gamma^m_{ik} \Gamma^l_{mj} v_l.$$

Interchanging subscripts j and k,

$$v_{i;kj} = \frac{\partial^2 v_i}{\partial u^j \partial u^k} - \Gamma^l_{ik} \frac{\partial v_l}{\partial u^j} - v_l \frac{\partial \Gamma^l_{ik}}{\partial u^j} - \Gamma^m_{ij} \frac{\partial v_m}{\partial u^k} + \Gamma^m_{ij} \Gamma^l_{mk} v_l.$$

When these two expressions are subtracted to define the Riemann tensor, the first, second and fourth terms (the second of one with the fourth of the other and vice versa) on the two RHSs cancel in pairs to yield

$$R^l_{ijk} v_l \equiv v_{i;jk} - v_{i;kj} = \left(\frac{\partial \Gamma^l_{ik}}{\partial u^j} - \frac{\partial \Gamma^l_{ij}}{\partial u^k} + \Gamma^m_{ik} \Gamma^l_{mj} - \Gamma^m_{ij} \Gamma^l_{mk} \right) v_l.$$

Now, in three-dimensional Euclidean space, one possible coordinate system is the Cartesian one. In this system $g = 1$ and all of its derivatives are zero. Thus all Christoffel symbols and their derivatives are zero, as are all components of the Riemann tensor. As *all* the components vanish in this Cartesian coordinate system, they must do so in *any* coordinate system in this space.

26.29 *We may define Christoffel symbols of the first kind by*

$$\Gamma_{ijk} = g_{il}\Gamma^l_{jk}.$$

Show that these are given by

$$\Gamma_{ijk} = \frac{1}{2}\left(\frac{\partial g_{ki}}{\partial u^j} + \frac{\partial g_{ij}}{\partial u^k} - \frac{\partial g_{jk}}{\partial u^i}\right).$$

By permuting indices, verify that

$$\frac{\partial g_{ij}}{\partial u^k} = \Gamma_{ijk} + \Gamma_{jik}.$$

Using the fact that $\Gamma^l_{jk} = \Gamma^l_{kj}$, *show that*

$$g_{ij;k} \equiv 0,$$

i.e. that the covariant derivative of the metric tensor is identically zero in all coordinate systems.

Starting from Christoffel symbols of the second kind, we have

$$\Gamma_{ijk} = g_{il}\Gamma^l_{jk}$$

$$= \frac{1}{2}g_{il}g^{ln}\left(\frac{\partial g_{kn}}{\partial u^j} + \frac{\partial g_{nj}}{\partial u^k} - \frac{\partial g_{jk}}{\partial u^n}\right)$$

$$= \frac{1}{2}\delta^n_i\left(\frac{\partial g_{kn}}{\partial u^j} + \frac{\partial g_{nj}}{\partial u^k} - \frac{\partial g_{jk}}{\partial u^n}\right)$$

$$= \frac{1}{2}\left(\frac{\partial g_{ki}}{\partial u^j} + \frac{\partial g_{ij}}{\partial u^k} - \frac{\partial g_{jk}}{\partial u^i}\right).$$

Next, forming the symmetric sum of two Christoffel symbols:

$$\Gamma_{ijk} + \Gamma_{jik} = \frac{1}{2}\left(\frac{\partial g_{ki}}{\partial u^j} + \frac{\partial g_{ij}}{\partial u^k} - \frac{\partial g_{jk}}{\partial u^i}\right) + \frac{1}{2}\left(\frac{\partial g_{kj}}{\partial u^i} + \frac{\partial g_{ji}}{\partial u^k} - \frac{\partial g_{ik}}{\partial u^j}\right)$$

$$= \frac{1}{2}\left(\frac{\partial g_{ij}}{\partial u^k} + \frac{\partial g_{ji}}{\partial u^k}\right) + \frac{1}{2}\left(\frac{\partial g_{ki}}{\partial u^j} - \frac{\partial g_{ik}}{\partial u^j}\right)$$

$$+ \frac{1}{2}\left(-\frac{\partial g_{jk}}{\partial u^i} + \frac{\partial g_{kj}}{\partial u^i}\right)$$

$$= \frac{\partial g_{ij}}{\partial u^k} + 0 + 0.$$

To obtain the last line we have used the fact that the metric tensor is symmetric, $g_{ij} = g_{ji}$.

Further, since $\Gamma^l_{jk} = \Gamma^l_{kj}$, and therefore $g_{il}\Gamma^l_{jk} = g_{il}\Gamma^l_{kj}$, we have that $\Gamma_{ijk} = \Gamma_{ikj}$, i.e Christoffel symbols of the first kind are symmetric under the interchange of the last two indices.

Finally, forming the covariant derivative of g_{ij}:

$$
\begin{aligned}
g_{ij;k} &= \frac{\partial}{\partial u^k}(g_{ij}\, \mathbf{e}^i \otimes \mathbf{e}^j) \\
&= \frac{\partial g_{ij}}{\partial u^k}\, \mathbf{e}^i \otimes \mathbf{e}^j + g_{ij}\, \frac{\partial \mathbf{e}^i}{\partial u^k} \otimes \mathbf{e}^j + g_{ij}\, \mathbf{e}^i \otimes \frac{\partial \mathbf{e}^j}{\partial u^k} \\
&= \frac{\partial g_{ij}}{\partial u^k}\, \mathbf{e}^i \otimes \mathbf{e}^j + g_{ij}\, (-\Gamma^i_{lk}\mathbf{e}^l) \otimes \mathbf{e}^j + g_{ij}\, \mathbf{e}^i \otimes (-\Gamma^j_{mk}\mathbf{e}^m) \\
&= \frac{\partial g_{ij}}{\partial u^k}\, \mathbf{e}^i \otimes \mathbf{e}^j - \Gamma_{jlk}\mathbf{e}^l \otimes \mathbf{e}^j - \Gamma_{imk}\mathbf{e}^i \otimes \mathbf{e}^m, \text{ since } g_{ij} = g_{ji}, \\
&= \left(\frac{\partial g_{ij}}{\partial u^k} - \Gamma_{jik} - \Gamma_{ijk} \right) \mathbf{e}^i \otimes \mathbf{e}^j, \text{ renaming dummy suffices,} \\
&= 0\, (\mathbf{e}^i \otimes \mathbf{e}^j), \text{ from the previous result.}
\end{aligned}
$$

Thus the covariant derivative of the metric tensor is identically zero in all coordinate systems.

27

Numerical methods

To provide a satisfactory iteration scheme, the equation must be rearranged in the form $x = f(x)$, where $f(x)$ is a slowly varying function of x; we then use $x_{n+1} = f(x_n)$ as the iteration scheme.

In the present case the rearrangement is straightforward, as, by taking logarithms, we can write the equation as $x = \ln 40x$. Since $\ln z$ is a slowly varying function of z, we can take $x_{n+1} = \ln 40x_n$ as the iteration scheme.

We start with the (poor) guess that $x = 1$. The successive values generated by the scheme are (to 5 s.f.)

$$1, \ 3.6889, \ 4.9942, \ 5.2972, \ 5.3560, \ 5.3671, \ 5.3691, \ 5.3696, \ 5.3696, \ \dots .$$

Thus to 4 s.f. we give the answer as $x = 5.370$. In fact, after 15 iterations the calculated value is stable to 10 s.f. at 5.369640395.

> **27.3** Show the following results about rearrangement schemes for polynomial equations.
>
> (a) That if a polynomial equation $g(x) \equiv x^m - f(x) = 0$, where $f(x)$ is a polynomial of degree less than m and for which $f(0) \neq 0$, is solved using a rearrangement iteration scheme $x_{n+1} = [f(x_n)]^{1/m}$, then, in general, the scheme will have only first-order convergence.
> (b) By considering the cubic equation
>
> $$x^3 - ax^2 + 2abx - (b^3 + ab^2) = 0$$
>
> for arbitrary non-zero values of a and b, demonstrate that, in special cases, the same rearrangement scheme can give second- (or higher-) order convergence.

(a) If we represent the iteration scheme as $x_{n+1} = F(x_n)$ then the scheme will have only first-order convergence unless $F'(\xi) = 0$, where ξ is the solution to the original equation satisfying $\xi^m = f(\xi)$ or, equivalently, $\xi = F(\xi)$.

In this case $F(x) = [f(x)]^{1/m}$ and

$$F'(\xi) = \frac{1}{m}[f(\xi)]^{(1-m)/m} f'(\xi).$$

Since $f(0) \neq 0$, $x = 0$ cannot be one of the solutions ξ of the original equation. Now, $f(\xi) = \xi^m$ and so the first two factors in the expression for $F'(\xi)$ have the value $m^{-1}(\xi^m)^{(1-m)/m} = m^{-1}\xi^{1-m}$. This is neither zero nor infinite and so $F'(\xi)$ can only be zero if $f'(\xi) = 0$; in general this will not be the case and the convergence will be only of first order.

(b) For the given equation $m = 3$ and $f(x) = ax^2 - 2abx + (b^3 + ab^2)$. It follows that $f'(x) = 2ax - 2ab$ and that $f'(x) = 0$ when $x = b$. However, $x = b$, also satisfies the original equation

$$b^3 - ab^2 + 2ab^2 - b^3 - ab^2 = 0,$$

and therefore, in the terminology used in part (a), $\xi = b$ and $F'(\xi) = F'(b) = 0$. This shows that the convergence will be of second (or higher) order.

In fact, further differentiation shows that $F''(b) = 2a/3b^2$ and, as this is non-zero, the convergence is only of second order.

27.5 *Solve the following set of simultaneous equations using Gaussian elimination (including interchange where it is formally desirable):*

$$x_1 + 3x_2 + 4x_3 + 2x_4 = 0,$$
$$2x_1 + 10x_2 - 5x_3 + x_4 = 6,$$
$$4x_2 + 3x_3 + 3x_4 = 20,$$
$$-3x_1 + 6x_2 + 12x_3 - 4x_4 = 16.$$

Since the largest (in magnitude) coefficient of x_1 appears in the final equation, we reorder them to make it first (labelled I) and divide through by -1 to make the coefficient of x_1 positive:

$$3x_1 - 6x_2 - 12x_3 + 4x_4 = -16. \qquad \text{(I)}$$

The first and second equations now have $\frac{1}{3}$ and $\frac{2}{3}$ (respectively) of (I) subtracted from them to eliminate x_1. The third equation does not contain x_1 and so is left unchanged:

$$5x_2 + 8x_3 + \tfrac{2}{3}x_4 = \tfrac{16}{3}, \qquad \text{(a)}$$
$$14x_2 + 3x_3 - \tfrac{5}{3}x_4 = \tfrac{50}{3}, \qquad \text{(b)}$$
$$4x_2 + 3x_3 + 3x_4 = 20. \qquad \text{(c)}$$

Equation (b) is now the one with the largest coefficient of x_2, and so we take as the second finalised equation

$$14x_2 + 3x_3 - \tfrac{5}{3}x_4 = \tfrac{50}{3}, \qquad \text{(II)}$$

and subtract the needed fractions of this from (a) and (c) to eliminate x_2 from them:

$$\tfrac{97}{14}x_3 + (\tfrac{2}{3} + \tfrac{25}{42})x_4 = \tfrac{16}{3} - \tfrac{250}{42}, \qquad \text{(d)}$$
$$(3 - \tfrac{12}{14})x_3 + (3 + \tfrac{20}{42})x_4 = 20 - \tfrac{200}{42}. \qquad \text{(e)}$$

Rationalising these two equations we have

$$291x_3 + 53x_4 = -26, \qquad (\text{d}') \equiv \text{(III)}$$
$$90x_3 + 146x_4 = 640. \qquad (\text{e}')$$

Finally, eliminating x_3 from (e′) gives

$$(146 - \tfrac{90}{291}\,53)x_4 = 640 - \tfrac{90}{291}(-26),$$
$$37716x_4 = 188580,$$
$$x_4 = 5. \qquad \text{(IV)}$$

Resubstitution then gives

from (III), $x_3 = \frac{-26-(53\times5)}{291} = -1$,

from (II), $x_2 = \frac{1}{14}(\frac{50}{3} + \frac{5\times5}{3} - 3(-1)) = 2$,

from (I), $x_1 = \frac{1}{3}(-16 - (4 \times 5) + 12(-1) + 6(2)) = -12$,

making the solution $x_1 = -12$, $x_2 = 2$, $x_3 = -1$ and $x_4 = 5$.

27.7 *Simultaneous linear equations that result in tridiagonal matrices can sometimes be solved in the same way as three-term recurrence relations. Consider the tridiagonal simultaneous equations*

$$x_{i-1} + 4x_i + x_{i+1} = 3(\delta_{i+1,0} - \delta_{i-1,0}), \quad i = 0, \pm1, \pm2, \dots .$$

Prove that for $i > 0$ the equations have a general solution of the form $x_i = \alpha p^i + \beta q^i$, where p and q are the roots of a certain quadratic equation. Show that a similar result holds for $i < 0$. In each case express x_0 in terms of the arbitrary constants $\alpha, \beta, \dots$.

Now impose the condition that x_i is bounded as $i \to \pm\infty$ and obtain a unique solution.

We substitute the trial solution $x_i = \alpha p^i + \beta q^i$ into the given equation for $i \geq 2$ and obtain

$$\alpha(p^{i-1} + 4p^i + p^{i+1}) + \beta(q^{i-1} + 4q^i + q^{i+1}) = 3(0 - 0) = 0;$$

this is satisfied for arbitrary α and β if p and q are the two roots of the quadratic equation $1 + 4r + r^2 = 0$.

Using the same form for $i = 1$, but with these specific values for p and q, we have

$$x_0 + 4x_1 + x_2 = 3(0 - 1),$$
$$x_0 + 4(\alpha p + \beta q) + \alpha p^2 + \beta q^2 = -3,$$
$$x_0 + \alpha(4p + p^2) + \beta(4q + q^2) = -3,$$
$$x_0 + \alpha(-1) + \beta(-1) = -3.$$

To obtain the final line we used the fact that both p and q satisfy $4r + r^2 = -1$. Similarly, for $i \leq -1$ the solution is $x_i = \alpha' p^i + \beta' q^i$, with $x_0 - \alpha' - \beta' = +3$.

In addition, for $i = 0$, we have from the original equation that

$$\frac{\alpha'}{p} + \frac{\beta'}{q} + 4x_0 + \alpha p + \beta q = 0.$$

443

The values of p and q are $-2 \pm \sqrt{4-1}$, with, say, $p = -2 + \sqrt{3}$ and $|p| < 1$, and $q = -2 - \sqrt{3}$ and $|q| > 1$.

Now, the solution is to be bounded as $i \to \pm\infty$. The fact that $|q| > 1$ and the condition at $+\infty$ together require that $\beta = 0$, whilst $|p|^{-1} > 1$ and the condition at $-\infty$ imply that $\alpha' = 0$. We are left with three equations for three unknowns:

$$x_0 - \alpha + 3 = 0,$$
$$x_0 - \beta' - 3 = 0,$$
$$\frac{\beta'}{-2 - \sqrt{3}} + 4x_0 + \alpha(-2 + \sqrt{3}) = 0.$$

We now rearrange the last of these and substitute from the first two:

$$\beta' + 4(-2 - \sqrt{3})x_0 + \alpha = 0,$$
$$\Rightarrow \quad (x_0 - 3) - (8 + 4\sqrt{3})x_0 + (x_0 + 3) = 0,$$

and $x_0 = 0$, $\alpha = 3$, $\beta' = -3$. The solution is thus

$$x_i = \begin{cases} 3(-2 + \sqrt{3})^i & i \geq 1, \\ 0 & i = 0, \\ -3(-2 - \sqrt{3})^i & i \leq -1. \end{cases}$$

The final entry could be written as $-3(-2 + \sqrt{3})^{-i}$.

27.9 *Although it can easily be shown, by direct calculation, that*

$$\int_0^\infty e^{-x} \cos(kx)\, dx = \frac{1}{1 + k^2},$$

the form of the integrand is also appropriate for a Gauss–Laguerre numerical integration. Using a 5-point formula, investigate the range of values of k for which the formula gives accurate results. At about what value of k do the results become inaccurate at the 1% level?

The integrand is an even function of k and so only positive k need be considered. The points and weights for the 5-point Gauss–Laguerre integration are

x_i	w_i
0.26356 03197	0.52175 56106
1.41340 30591	0.39866 68111
3.59642 57710	0.07594 24497
7.08581 00059	0.00361 17587
12.6408 00844	0.00002 33700

The table below gives the exact and calculated results to four places of decimals, as well as the percentage error in the calculated result. It shows that the error is not more than 1% for $|k|$ less than about 1.1.

k	Exact	Calculated	% error
0.0	1.0000	1.0000	0.0
0.5	0.8000	0.8000	0.0
0.8	0.6098	0.6097	0.0
1.0	0.5000	0.5005	0.1
1.1	0.4525	0.4545	0.4
1.2	0.4098	0.4145	1.1
1.3	0.3717	0.3800	2.2
1.5	0.3077	0.3200	4.0
1.7	0.2571	0.2535	−1.4
2.0	0.2000	0.1184	−40.8
3.0	0.1000	0.1674	67.4

27.11 Consider the integrals I_p defined by

$$I_p = \int_{-1}^{1} \frac{x^{2p}}{\sqrt{1-x^2}}\, dx.$$

(a) By setting $x = \sin\theta$ and using the recurrence relation quoted below, show that I_p has the value

$$I_p = 2\,\frac{2p-1}{2p}\,\frac{2p-3}{2p-2}\cdots\frac{1}{2}\,\frac{\pi}{2}.$$

Recurrence relation: If $J(n)$ is defined for a non-negative integer n by

$$J(n) = \int_0^{\pi/2} \sin^n\theta\, d\theta,$$

then, for $n > 2$,

$$J(n) = \frac{n-1}{n}J(n-2).$$

(b) Evaluate I_p for $p = 1, 2, \ldots, 6$ using 5- and 6-point Gauss–Chebyshev integration (conveniently run on a spreadsheet such as Excel) and compare the results with those in (a). In particular, show that, as expected, the 5-point scheme first fails to be accurate when the order of the polynomial numerator $(2p)$ exceeds $(2 \times 5) - 1 = 9$. Likewise, verify that the 6-point scheme evaluates I_5 accurately but is in error for I_6.

(a) Setting $x = \sin\theta$ with $dx = \cos\theta$ converts I_p to

$$I_p = \int_{-\pi/2}^{\pi/2} \frac{\sin^{2p}\theta}{\cos\theta} \cos\theta \, d\theta = 2\int_0^{\pi/2} \sin^{2p}\theta \, d\theta = 2J(2p),$$

using the given definition of $J(n)$. Applying the reduction formula then gives

$$I_p = 2\frac{2p-1}{2p}\frac{2p-3}{2p-2}\cdots\frac{1}{2}\frac{\pi}{2},$$

where we have used the obvious result $J(0) = \pi/2$.

(b) The points and weights needed for a Gauss–Chebyshev integration are given analytically by

$$x_i = \cos\frac{(i - \frac{1}{2})\pi}{n}, \qquad w_i = \frac{\pi}{n}, \qquad \text{for } i = 1, \ldots, n.$$

Here we have to take the cases $n = 5$ and $n = 6$. The following table gives the exact result calculated in (a) and the values obtained using the n-point Gauss–Chebyshev formula.

p	Exact	$n = 5$	$n = 6$
1	1.570796	1.570796	1.570796
2	1.178097	1.178097	1.178097
3	0.981748	0.981748	0.981748
4	0.859029	0.859029	0.859029
5	0.773126	0.766990	0.773126
6	0.708699	0.690291	0.707165

It will be seen that, as stated in the question, the $p = 5$, $n = 5$ and both the $p = 6$ values diverge from the exact result. The discrepancy is of the order of 1% when $p = n$, i.e. when the order of the polynomial in the numerator of I_p first exceeds $2n - 1$.

27.13 *Given a random number η uniformly distributed on $(0, 1)$, determine the function $\xi = \xi(\eta)$ that would generate a random number ξ distributed as*

(a) 2ξ on $0 \le \xi < 1$,

(b) $\frac{3}{2}\sqrt{\xi}$ on $0 \le \xi < 1$,

(c) $\dfrac{\pi}{4a}\cos\dfrac{\pi\xi}{2a}$ on $-a \le \xi < a$,

(d) $\frac{1}{2}\exp(-|\xi|)$ on $-\infty < \xi < \infty$.

For each required distribution $f(t)$ in the range (a, b) we need to determine the cumulative distribution function $F(y) = \int_a^y dt$ and then take $F(y)$ as uniformly distributed on $(0, 1)$. A correctly normalised distribution has $F(b) = 1$. For any given random number η, the corresponding variable, distributed as $f(\xi)$, is $\xi = F^{-1}(\eta)$.

(a) For $f(t) = 2t$,

$$F(y) = \int_0^y 2t \, dt = y^2 \quad \Rightarrow \quad \eta = \xi^2 \quad \Rightarrow \quad \xi = \sqrt{\eta}.$$

(b) For $f(t) = \frac{3}{2}\sqrt{t}$,

$$F(y) = \int_0^y \frac{3}{2}\sqrt{t} \, dt = y^{3/2} \quad \Rightarrow \quad \eta = \xi^{3/2} \quad \Rightarrow \quad \xi = \eta^{2/3}.$$

(c) For $f(t) = \dfrac{\pi}{4a} \cos \dfrac{\pi t}{2a}$,

$$F(y) = \frac{\pi}{4a} \int_{-a}^y \cos\left(\frac{\pi t}{2a}\right) dt = \frac{1}{2}\left[\sin\left(\frac{\pi y}{2a}\right) + 1\right],$$

$$\Rightarrow \quad \eta = \frac{1}{2}\left[\sin\left(\frac{\pi \xi}{2a}\right) + 1\right] \quad \Rightarrow \quad \xi = \frac{2a}{\pi}\sin^{-1}(2\eta - 1).$$

(d) For $f(t) = \frac{1}{2}\exp(-|t|)$,

for $y < 0$, $\quad F(y) = \displaystyle\int_{-\infty}^y \frac{e^t}{2} \, dt = \frac{e^y}{2}$,

for $y > 0$, $\quad F(y) = \displaystyle\int_{-\infty}^0 \frac{e^t}{2} \, dt + \int_0^y \frac{e^{-t}}{2} \, dt$

$$= \frac{1}{2} + \frac{1 - e^{-y}}{2} = \frac{1}{2}(2 - e^{-y}).$$

It follows that

$$\eta = \begin{cases} \frac{1}{2}e^{\xi} & \xi \leq 0, \\ 1 - \frac{1}{2}e^{-\xi} & \xi > 0, \end{cases} \quad \text{and } \xi = \begin{cases} \ln 2\eta & \eta \leq 0.5, \\ -\ln(2 - 2\eta) & 0.5 < \eta < 1. \end{cases}$$

27.15 *Use a Taylor series to solve the equation*

$$\frac{dy}{dx} + xy = 0, \qquad y(0) = 1,$$

evaluating $y(x)$ for $x = 0.0$ to 0.5 in steps of 0.1.

In order to construct the Taylor series we need to find the derivatives $y^{(n)} \equiv$

$d^{(n)}y/dx^n$ up to, say, $n = 6$ and evaluate them at $x = 0$. We will also need $y(0) = 1$. The derivatives are

$$y' = -xy \quad \Rightarrow \quad y^{(1)}(0) = 0,$$
$$y^{(2)} = = -y - xy'(x) = -y + x^2 y \quad \Rightarrow \quad y^{(2)}(0) = -1,$$
$$y^{(3)} = 2xy + (-1 + x^2)y'(x) = 3xy - x^3 y \quad \Rightarrow \quad y^{(3)}(0) = 0,$$
$$y^{(4)} = 3y - 3x^2 y + (3x - x^3)y'(x)$$
$$= 3y - 6x^2 y + x^4 y \quad \Rightarrow \quad y^{(4)}(0) = 3,$$
$$y^{(5)} = -12xy + 4x^3 y + (3 - 6x^2 + x^4)y'(x)$$
$$= -15xy + 10x^3 y - x^5 y \quad \Rightarrow \quad y^{(5)}(0) = 0,$$
$$y^{(6)} = -15y + 30x^2 y - 5x^4 y + (-15x + 10x^3 - x^5)y'(x)$$
$$= -15y + 45x^2 y - 15x^4 y + x^6 y \quad \Rightarrow \quad y^{(6)}(0) = -15.$$

Thus, the Taylor series for an expansion about $x = 0$ is given by

$$y(x) = 1 - \frac{x^2}{2!} + \frac{3x^4}{4!} - \frac{15x^6}{6!} + O(x^8)$$
$$= 1 - \frac{x^2}{2} + \frac{x^4}{8} - \frac{x^6}{48} + O(x^8).$$

To four significant figures the values of $y(x)$ calculated using this Taylor series are $y(0.1) = 0.9950$, $y(0.2) = 0.9802$, $y(0.3) = 0.9560$, $y(0.4) = 0.9231$ and $y(0.5) = 0.8825$.

For interest, we note that the exact solution of the differential equation, which is separable, is given by

$$\frac{dy}{y} = -x \, dx \quad \Rightarrow \quad \ln y = -\frac{x^2}{2} + c$$
$$y(0) = 1 \quad \Rightarrow \quad c = 0 \quad \Rightarrow \quad y(x) = e^{-x^2/2},$$

which has the Taylor series

$$y(x) = 1 - \frac{x^2}{2^1 1!} + \frac{x^4}{2^2 2!} - \frac{x^6}{2^3 3!} + \cdots .$$

As expected, this is the same as that found directly from the differential equation, up to the last term calculated; clearly the next term is $O(x^8)$.

To four significant figures the exact solution and the Taylor expansion give the same values over the given range of x; for $x = 0.6$ they differ by 1 in the fourth decimal place.

27.17 *A more refined form of the Adams predictor–corrector method for solving the first-order differential equation*

$$\frac{dy}{dx} = f(x, y)$$

is known as the Adams–Moulton–Bashforth scheme. At any stage (say the nth) in an Nth-order scheme, the values of x and y at the previous N solution points are first used to predict the value of y_{n+1}. This approximate value of y at the next solution point, x_{n+1}, denoted by $\bar{y}_{n+1}$, is then used together with those at the previous $N-1$ solution points to make a more refined (corrected) estimation of $y(x_{n+1})$. The calculational procedure for a third-order scheme is summarised by the following two equations:

$$\bar{y}_{n+1} = y_n + h(a_1 f_n + a_2 f_{n-1} + a_3 f_{n-2}) \qquad (predictor),$$
$$y_{n+1} = y_n + h(b_1 f(x_{n+1}, \bar{y}_{n+1}) + b_2 f_n + b_3 f_{n-1}) \qquad (corrector).$$

(a) *Find Taylor series expansions for f_{n-1} and f_{n-2} in terms of the function $f_n = f(x_n, y_n)$ and its derivatives at x_n.*

(b) *Substitute them into the predictor equation and, by making that expression for $\bar{y}_{n+1}$ coincide with the true Taylor series for y_{n+1} up to order h^3, establish simultaneous equations that determine the values of a_1, a_2 and a_3.*

(c) *Find the Taylor series for f_{n+1} and substitute it and that for f_{n-1} into the corrector equation. Make the corrected prediction for y_{n+1} coincide with the true Taylor series by choosing the weights b_1, b_2 and b_3 appropriately.*

(d) *The values of the numerical solution of the differential equation*

$$\frac{dy}{dx} = \frac{2(1+x)y + x^{3/2}}{2x(1+x)}$$

at three values of x are given in the following table.

x	0.1	0.2	0.3
$y(x)$	0.030628	0.084107	0.150328

Use the above predictor–corrector scheme to find the value of $y(0.4)$ and compare your answer with the accurate value, 0.225577.

(a) 'Taylor series' expansions, using increments in x of $-h$ and $-2h$, give

$$f_{n-1} = f_n - h f'_n + \tfrac{1}{2} h^2 f''_n - \tfrac{1}{6} h^3 f_n^{(3)} + \cdots,$$
$$f_{n-2} = f_n - 2h f'_n + \tfrac{4}{2} h^2 f''_n - \tfrac{8}{6} h^3 f_n^{(3)} + \cdots.$$

These expansions are not true Taylor series as the only derivatives used are those with respect to x; however, the same is true of all subsequent expansions.

(b) Substitution in the predictor equation gives

$$\overline{y_{n+1}} = y_n + h(a_1 f_n + a_2 f_{n-1} + a_3 f_{n-2})$$
$$= y_n + h[(a_1 + a_2 + a_3)f_n + h(-a_2 - 2a_3)f_n'$$
$$+ h^2(\tfrac{1}{2}a_2 + 2a_3)f_n'' + \cdots].$$

Now, the accurate Taylor series for y_{n+1} is

$$y_{n+1} = y_n + hf_n + \tfrac{1}{2}h^2 f_n' + \tfrac{1}{6}h^3 f_n'' + \cdots.$$

To make these two expressions coincide up to order h^3, we need

$$\left.\begin{array}{rl} a_1 + a_2 + a_3 &= 1 \\ -a_2 - 2a_3 &= \tfrac{1}{2} \\ \tfrac{1}{2}a_2 + 2a_3 &= \tfrac{1}{6} \end{array}\right\} a_2 = -\tfrac{4}{3},\ a_3 = \tfrac{5}{12},\ a_1 = \tfrac{23}{12}.$$

(c) In the same way as in part (a),

$$f_{n+1} = f_n + hf_n' + \tfrac{1}{2}h^2 f_n'' + \tfrac{1}{6}h^3 f_n^{(3)} + \cdots,$$

and substitution in the corrector equation gives

$$y_{n+1} = y_n + h(b_1 f(x_{n+1}, \overline{y_{n+1}}) + b_2 f_n + b_3 f_{n-1})$$
$$= y_n + h[b_1 f(x_{n+1}, y_{n+1}) + b_2 f_n + b_3 f_{n-1}], \text{ to order } h^3$$
$$\equiv y_n + h(b_1 f_{n+1} + b_2 f_n + b_3 f_{n-1}), \text{ to order } h^3$$
$$= y_n + h[(b_1 + b_2 + b_3)f_n + h(b_1 - b_3)f_n'$$
$$+ h^2(\tfrac{1}{2}b_1 + \tfrac{1}{2}b_3)f_n'' + \cdots].$$

To make this coincide with the accurate Taylor series up to order h^3, we need

$$\left.\begin{array}{rl} b_1 + b_2 + b_3 &= 1 \\ b_1 - b_3 &= \tfrac{1}{2} \\ \tfrac{1}{2}b_1 + \tfrac{1}{2}b_3 &= \tfrac{1}{6} \end{array}\right\} b_1 = \tfrac{5}{12},\ b_3 = -\tfrac{1}{12},\ b_2 = \tfrac{2}{3}.$$

(d) We repeat the given table, indexing it and adding a line giving the values of $f(x, y)$.

n	1	2	3
x_n	0.1	0.2	0.3
$y_n(x_n)$	0.030628	0.084107	0.150328
$f_n(x_n, y_n)$	0.450020	0.606874	0.711756

Now, taking $n = 3$, we apply the predictor formula with the calculated values for the a_i and find $\overline{y_4} = 0.224582$. This allows us to calculate $f(x_4, \overline{y_4})$ as 0.787332. Finally, applying the corrector formula, using the calculated values for the b_i, we find the corrected value $y_4 = 0.225527$. This is to be compared with the accurate value of 0.225577 (and the predicted, but uncorrected, value of 0.224582).

27.19 *To solve the ordinary differential equation*

$$\frac{du}{dt} = f(u, t)$$

for $f = f(t)$, the explicit two-step finite difference scheme

$$u_{n+1} = \alpha u_n + \beta u_{n-1} + h(\mu f_n + v f_{n-1})$$

may be used. Here, in the usual notation, h is the time step, $t_n = nh$, $u_n = u(t_n)$ and $f_n = f(u_n, t_n)$; α, β, μ, and v are constants.

> (a) *A particular scheme has $\alpha = 1$, $\beta = 0$, $\mu = 3/2$ and $v = -1/2$. By considering Taylor expansions about $t = t_n$ for both u_{n+j} and f_{n+j}, show that this scheme gives errors of order h^3.*
> (b) *Find the values of α, β, μ and v that will give the greatest accuracy.*

We will need the Taylor expansions of $u_{n\pm 1}$ and f_{n-1}. They are given by

$$u_{n\pm 1} = u_n \pm h u_n' + \frac{1}{2!} h^2 u_n'' \pm \frac{1}{3!} h^3 u_n^{(3)} + \cdots,$$

$$f_{n-1} = u_{n-1}' = u_n' - h u_n'' + \frac{1}{2!} h^2 u_n^{(3)} - \frac{1}{3!} h^3 u_n^{(4)} + \cdots.$$

(a) This scheme calculates u_{n+1} as

$$u_{n+1} = u_n + h \left(\frac{3}{2} f_n - \frac{1}{2} f_{n-1} \right)$$

$$= u_n + h \left[\frac{3}{2} u_n' - \frac{1}{2} \left(u_n' - h u_n'' + \frac{1}{2!} h^2 u_n^{(3)} - \frac{1}{3!} h^3 u_n^{(4)} + \cdots \right) \right].$$

This is to be compared with

$$u_{n+1} = u_n + h u_n' + \frac{1}{2!} h^2 u_n'' + \frac{1}{3!} h^3 u_n^{(3)} + \cdots.$$

Omitting terms that appear in both expressions, we have

$$\frac{1}{3!} h^3 u_n^{(3)} + \cdots \approx -\frac{1}{4} h^3 u_n^{(3)} + \cdots,$$

showing that the error is

$$\left(\frac{1}{3!} + \frac{1}{4} \right) h^3 u_n^{(3)} = \frac{5}{12} h^3 u_n^{(3)} + O(h^4).$$

451

(b) For the best accuracy we require that

$$u_{n+1} = u_n + hu'_n + \frac{1}{2!} h^2 u''_n + \frac{1}{3!} h^3 u_n^{(3)} + \cdots$$

and

$$\alpha u_n + \beta \left(u_n - hu'_n + \frac{1}{2!} h^2 u''_n - \frac{1}{3!} h^3 u_n^{(3)} + \cdots \right) + h\mu u'_n$$

$$+ h v \left(u'_n - hu''_n + \frac{1}{2!} h^2 u_n^{(3)} - \frac{1}{3!} h^3 u_n^{(4)} + \cdots \right)$$

should match up to as high a positive power of h as possible.

With four parameters available, we can expect to match terms in h^n up to $n = 3$:

$$h^0 : 1 = \alpha + \beta,$$
$$h^1 : 1 = -\beta + \mu + v,$$
$$h^2 : \tfrac{1}{2} = \tfrac{1}{2}\beta - v,$$
$$h^3 : \tfrac{1}{6} = -\tfrac{1}{6}\beta + \tfrac{1}{2}v.$$

The final two equations are equivalent to $\beta = 1 + 2v$ and $1 + \beta = 3v$, yielding $v = 2$ and $\beta = 5$; it then follows that $\mu = 4$ and $\alpha = -4$. With this set of values, the finite difference scheme,

$$u_{n+1} = -4u_n + 5u_{n-1} + h(4f_n + 2f_{n-1}),$$

has errors of order h^4.

27.21 *Write a computer program that would solve, for a range of values of λ, the differential equation*

$$\frac{dy}{dx} = \frac{1}{\sqrt{x^2 + \lambda y^2}}, \qquad y(0) = 1,$$

using a third-order Runge–Kutta scheme. Consider the difficulties that might arise when $\lambda < 0$.

The relevant equations for a third-order Runge–Kutta scheme are

$$y_{i+1} = y_i + \tfrac{1}{6}(b_1 + 4b_2 + b_3),$$

where

$$b_1 = hf(x_i, y_i),$$
$$b_2 = hf(x_i + \tfrac{1}{2}h, \; y_i + \tfrac{1}{2}b_1),$$
$$b_3 = hf(x_i + h, \; y_i + 2b_2 - b_1).$$

The function $f(x, y)$, in this case, is $(x^2 + \lambda y^2)^{-1/2}$.

This calculation can be set up easily on a spreadsheet such as *Excel*, and it is immediately apparent that, with the given boundary value $y(0) = 1$, no significant finesse is needed. For positive values of λ the solution y is a monotonically (and boringly!) increasing function of x with values lying between 1 and ∞, the latter being approached rapidly only when λ is very small. Even with λ as small as 0.01, a step size Δx of 0.1 is adequate unless great precision is needed.

The difficulties that might arise for $\lambda < 0$ do not need much consideration; there is no real solution for *any* negative value of λ. The reason for this is easy to see. At the initial point, $x = 0$, $y = 1$ and λy^2 is negative and so the square root does not yield a real value for the derivative dy/dx.

More interesting results arise if the initial value is given elsewhere than at $x = 0$. For example, if $f(1) = 1$ then a solution can be calculated for negative values of λ greater than about -0.582 and if $f(1) = 2$ then a solution exists for $\lambda > -0.2057$.

27.23 *For some problems, numerical or algebraic experimentation may suggest the form of the complete solution. Consider the problem of numerically integrating the first-order wave equation*

$$\frac{\partial u}{\partial t} + A\frac{\partial u}{\partial x} = 0,$$

in which A is a positive constant. A finite difference scheme for this partial differential equation is

$$\frac{u(p, n+1) - u(p, n)}{\Delta t} + A\frac{u(p, n) - u(p-1, n)}{\Delta x} = 0,$$

where $x = p\Delta x$ and $t = n\Delta t$, with p any integer and n a non-negative integer. The initial values are $u(0, 0) = 1$ and $u(p, 0) = 0$ for $p \neq 0$.

(a) *Carry the difference equation forward in time for two or three steps and attempt to identify the pattern of solution. Establish the criterion for the method to be numerically stable.*

(b) *Suggest a general form for $u(p, n)$, expressing it in generator function form, i.e. as '$u(p, n)$ is the coefficient of s^p in the expansion of $G(n, s)$'.*

(c) *Using your form of solution (or that given in the answers!), obtain an explicit general expression for $u(p, n)$ and verify it by direct substitution into the difference equation.*

(d) *An analytic solution of the original PDE indicates that an initial disturbance propagates undistorted. Under what circumstances would the difference scheme reproduce that behaviour?*

If we write $A\Delta t/\Delta x$ as c, the equation becomes

$$u(p, n+1) - u(p, n) + c[u(p, n) - u(p-1, n)] = 0,$$

with $u(0,0) = 1$ and $u(p,0) = 0$ for $p \neq 0$.

(a) For calculational purposes we rearrange the equation and then substitute trial values:

$$u(p, n+1) = (1-c)u(p, n) + cu(p-1, n), \qquad (*)$$
$$u(0, 1) = (1-c)u(0, 0) + cu(-1, 0) = 1 - c,$$
$$u(1, 1) = (1-c)u(1, 0) + cu(0, 0) = c,$$
$$u(m, 1) = (1-c)u(m, 0) + cu(m-1, 0) = 0 \text{ for } m > 1,$$
$$u(0, 2) = (1-c)u(0, 1) + cu(-1, 1) = (1-c)^2,$$
$$u(1, 2) = (1-c)u(1, 1) + cu(0, 1) = 2c(1-c),$$
$$u(2, 2) = (1-c)u(2, 1) + cu(1, 1) = c^2,$$
$$u(m, 2) = (1-c)u(m, 1) + cu(m-1, 1) = 0 \text{ for } m > 2.$$

By now the pattern is clear, as is the condition for numerical stability, namely $c < 1$.

(b) For the nth time-step, the $n + 1$ values of $u(p, n)$, $p = 0, 1, \dots, n$ appear to be given by the terms in the binomial expansion of $[(1-c) + cs]^n$. Using the language of generating functions, we would say that '$u(p, n)$ is the coefficient of s^p in the expansion of $[(1-c) + cs]^n$.

(c) If this conjecture is correct, then

$$u(p, n) = \frac{n!\,(1-c)^{n-p}\,c^p}{p!\,(n-p)!}.$$

Substituting this form into the difference equation $(*)$ yields

$$\frac{(n+1)!\,(1-c)^{n+1-p}\,c^p}{p!\,(n+1-p)!} = \frac{(1-c)\,n!\,(1-c)^{n-p}\,c^p}{p!\,(n-p)!} + \frac{c\,n!\,(1-c)^{n+1-p}\,c^{p-1}}{(p-1)!\,(n+1-p)!}.$$

Multiplying through by $p!\,(n+1-p)!$ and dividing by $n!\,(1-c)^{n+1-p}\,c^p$ gives

$$(n+1) = (n-p+1) + p.$$

This is satisfied for all n and p, showing that the proposed solution satisfies the equation. It also gives $u(0, 0) = 1$, confirming that it is the required solution.

(d) For the special case $c = 1$, the recurrence relation reduces to

$$u(p, n+1) = u(p-1, n),$$

i.e. the disturbance u at the point $p\Delta x$ at time $(n+1)\Delta t$ is exactly the same as that at position $(p-1)\Delta x$ one time-step earlier. In other words, the disturbance propagates undistorted at speed A.

454

From the point of view of the numerical integration, this situation ($c = 1$ exactly) is both on the edge of instability and unlikely to be realised in practice.

27.25 *Laplace's equation,*

$$\frac{\partial^2 V}{\partial x^2} + \frac{\partial^2 V}{\partial y^2} = 0,$$

is to be solved for the region and boundary conditions shown in figure 27.1.

Starting from the given initial guess for the potential values V and using the simplest possible form of relaxation, obtain a better approximation to the actual solution. Do not aim to be more accurate than ± 0.5 units and so terminate the process when subsequent changes would be no greater than this.

We start by imposing a coordinate grid symmetrically on the region, so that the initial guess is

$$V(0, 1) = V(\pm 1, 1) = 20,$$
$$V(i, 2) = 40 \text{ for all } i,$$

and the fixed boundary conditions are

$$V(i, 0) = 0 \text{ for } |i| < 2,$$
$$V(i, 1) = 0 \text{ for all } |i| \geq 2,$$
$$V(i, 3) = 80 \text{ for all } i.$$

On symmetry grounds, we need consider only non-negative values of i.

We now apply the simplest relaxation scheme,

$$V_{i,j} \rightarrow \tfrac{1}{4}(V_{i+1,j} + V_{i-1,j} + V_{i,j+1} + V_{i,j-1}),$$

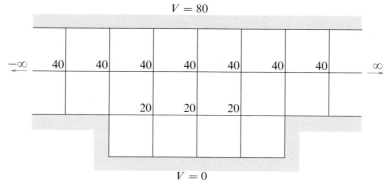

Figure 27.1   Region, boundary values and initial guessed solution values.

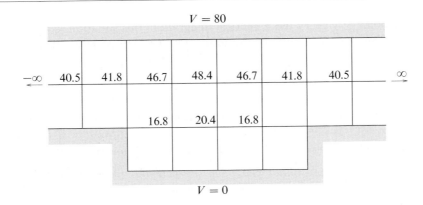

Figure 27.2   The solution to exercise 27.25.

for each point (i, j) that does not lie on the boundaries, where V_{ij} is prescribed and cannot be changed. The very simplest scheme would use only values from the previous iteration, but there is no additional labour involved in using previously calculated values from the current iteration when evaluating the RHS of the relationship. For this scheme the first few iterations produce the following results (to 3 s.f.):

$V_{0,1}$	$V_{1,1}$	$V_{0,2}$	$V_{1,2}$	$V_{2,2}$	$V_{3,2}$
20.0	20.0	40.0	40.0	40.0	40.0
20.0	15.0	45.0	45.0	41.3	40.3
18.8	15.9	47.2	46.1	41.6	40.4
19.8	16.5	48.0	46.5	41.7	40.4
20.2	16.7	48.3	46.7	41.8	40.4
20.4	16.8	48.4	46.8	41.8	40.4

The value at $(0, 1)$ is the one most likely to show the largest change at each iteration, as it is the one 'furthest from the fixed boundaries'. As the most recent changes have been 0.4 and 0.2, the process can be halted at this point, although the monotonic behaviour of values after the second iteration makes it harder to be sure that the differences between the final values and the current ones are within any given range.

The correct self-consistent solution (again to 3 s.f.) has corresponding values 20.6, 16.8, 48.5, 46.8, 41.8 and 40.5. This set of values is reached after nine iterations and is shown in figure 27.2. If the values from the previous iteration (rather than the most recently calculated ones) are used, the same ultimate result is reached (as expected), but about 17 iterations are needed to achieve the same self-consistency.

27.27 *The Schrödinger equation for a quantum mechanical particle of mass m moving in a one-dimensional harmonic oscillator potential* $V(x) = kx^2/2$ *is*

$$-\frac{\hbar^2}{2m}\frac{d^2\psi}{dx^2} + \frac{kx^2\psi}{2} = E\psi.$$

For physically acceptable solutions the wavefunction $\psi(x)$ must be finite at $x = 0$, tend to zero as $x \to \pm\infty$ and be normalised, so that $\int |\psi|^2\,dx = 1$. In practice, these constraints mean that only certain (quantised) values of E, the energy of the particle, are allowed. The allowed values fall into two groups, those for which $\psi(0) = 0$ and those for which $\psi(0) \neq 0$.

Show that if the unit of length is taken as $[\hbar^2/(mk)]^{1/4}$ and the unit of energy as $\hbar(k/m)^{1/2}$ then the Schrödinger equation takes the form

$$\frac{d^2\psi}{dy^2} + (2E' - y^2)\psi = 0.$$

Devise an outline computerised scheme, using Runge–Kutta integration, that will enable you to:

- *determine the three lowest allowed values of E;*
- *tabulate the normalised wavefunction corresponding to the lowest allowed energy.*

You should consider explicitly:

- *the variables to use in the numerical integration;*
- *how starting values near $y = 0$ are to be chosen;*
- *how the condition on ψ as $y \to \pm\infty$ is to be implemented;*
- *how the required values of E are to be extracted from the results of the integration;*
- *how the normalisation is to be carried out.*

We start by setting $x = \alpha y$, where α is the new unit of length; then $d/dx = \alpha^{-1}d/dy$ and

$$-\frac{\hbar^2}{2m}\alpha^{-2}\frac{d^2\psi}{dy^2} + \frac{k\alpha^2 y^2}{2} = E\psi,$$

$$\frac{d^2\psi}{dy^2} - \alpha^4\frac{mk}{\hbar^2}y^2\psi + \alpha^2\frac{2mE}{\hbar^2}\psi = 0.$$

Although, strictly, it should be given a new symbol, we continue to denote the required solution by ψ, now taken as a function of y rather than of x.

Now if α is chosen as $(\hbar^2/mk)^{1/4}$ and E is written as $E = \beta E'$, where $\beta = \hbar(k/m)^{1/2}$, this equation becomes

$$\frac{d^2\psi}{dy^2} + (2E' - y^2)\psi = 0.$$

We note that this is a Sturm–Liouville equation with $p = 1$, $q = -y^2$, unit weight function and eigenvalue $2E'$; we therefore expect its solutions for different values of E' to be orthogonal.

To keep the notation the same as that normally used when describing numerical integration, we rewrite the equation as

$$\frac{d^2y}{dx^2} + (\lambda - x^2)y = 0. \qquad (*)$$

So that this second-order equation can be handled using an R–K routine, it has to be written as two first-order equations using an auxiliary variable. We make the simplest choice of $z \equiv dy/dx$, thus making a (two-component) 'vector' of dependent variables $(y, z)^T$ with governing equations

$$\frac{dy}{dx} = z \qquad \text{and} \qquad \frac{dz}{dx} = (x^2 - \lambda)y.$$

The computer program will need to contain a subroutine that, given an input vector $(x, y, z)^T$, returns an output vector $(dy/dx, dz/dx)^T$ calculated as $(z, (x^2 - \lambda)y)^T$. This is used to calculate the function $f(x_i, u_i)$ that appears on the (four) RHSs of, say, a fourth-order RK routine:

$$u_{i+1} = u_i + \tfrac{1}{6}(c_1 + 2c_2 + 2c_3 + c_4),$$

where

$$c_1 = hf(x_i, u_i),$$
$$c_2 = hf(x_i + \tfrac{1}{2}h, \ u_i + \tfrac{1}{2}c_1),$$
$$c_3 = hf(x_i + \tfrac{1}{2}h, \ u_i + \tfrac{1}{2}c_2),$$
$$c_4 = hf(x_i + h, \ u_i + c_3).$$

Here, at each stage in the calculation of a one-step advance, u_i stands for y_i and z_i in turn.

Since equation $(*)$ is unchanged under the substitution $x \to -x$, and the boundary conditions, $y \to 0$ at $\pm\infty$, can be considered as both symmetric and antisymmetric, we can expect to find solutions that are either purely symmetric or purely antisymmetric. Consequently, we need only consider positive values of x, starting with $y(0) = 1$ for symmetric solutions and $y(0) = 0$ for antisymmetric ones. What will distinguish one potential solution from another is the value assigned to the initial slope $z(0)$. Clearly, one combination to be avoided is $y(0) = z(0) = 0$; such a computation will 'never get off the ground'. Intuition suggests that the initial slope should be zero if $y(0) = 1$ and non-zero if $y(0) = 0$.

As the formal range of x is infinite, we need to investigate the likely behaviour of a computed solution for large x; we want it to tend to zero for acceptable solutions. For large x equation $(*)$ approximates to

$$\frac{d^2y}{dx^2} = x^2y,$$

and if we substitute a trial function $y = e^{\gamma x^n}$ with $n > 0$ we find that

$$\frac{d^2y}{dx^2} = n(n-1)\gamma x^{n-2}e^{\gamma x^n} + n^2\gamma^2 x^{2n-2}e^{\gamma x^n}.$$

The dominant term in this expression is the second one; if this is to match x^2y then $n = 2$ and $\gamma = \pm\frac{1}{2}$. The case $\gamma = -\frac{1}{2}$ is clearly the one that is required, but, inevitably, even if the appropriate eigenvalue could be hit upon exactly, rounding errors are bound to introduce some of the $\gamma = +\frac{1}{2}$ solution. Thus we could never expect a computed solution actually to tend to zero and remain close to it however many steps are taken.

A more practical way to implement the boundary condition is to require y (and hence necessarily z) to remain within some specified narrow (but empirical) band about zero over, say, the interval $5 < x < 6$ — chosen because $5^n \exp(-5^2/2)$ is less than $\sim 10^{-3}$ for any moderate value of n and we cannot hope to achieve better accuracy than one part in a thousand without using more sophisticated techniques. Thus, in practice, the integration has to be over a finite range.

A crude technique is therefore to run the integration routine from $x = 0$ up to $x = 5$ for a mesh of values for λ (≥ 0) and $z(0)$ (in the ranges discussed above) and so evaluate the solution $v(\lambda, z(0)) = y(5)$. If all v have the same sign and vary smoothly with λ and $z(0)$, then a larger range of λ is indicated. However, if the v have mixed signs, interpolated values of λ and $z(0)$ should be tried, aiming to produce $v(\lambda, z(0)) \approx 0$. When this has been achieved, the test in the previous paragraph should be implemented to give further refinement. A graphical screen display of the calculated solution would be a considerable advantage in following what is happening.

Once a value of λ that results in a solution that approaches and stays near zero over the test range has been found, the corresponding values of y need to be divided by the square root of the value of the integral $\int_{-\infty}^{\infty} y^2\,dx$, so as to normalise the solution; they can then be tabulated. The integral can be evaluated well enough using the trapezium or Simpson's rule formulae over the finite range $0 < x < 5$ and doubling the result.

In order to be reasonably certain of finding the three lowest allowed values of E, the search should start from λ (i.e. $2E$) equal to zero and incremented in amounts $\Delta\lambda$ less than, but not negligible compared with, the average values of x^2 to be covered. The latter are of order unity, and so $\Delta\lambda = 0.1$ is reasonable. The step length h in the x-variable might be chosen in the range 0.01 to 0.1 with the smaller values used when near a potential solution in the $(\lambda, z(0))$ grid.

[As has been indicated in several exercises in previous chapters, the actual eigenvalues λ are 1, 3, 5, ..., $2n + 1$, ... and the corresponding solutions are $\exp(-x^2/2)$ multiplied by a Hermite polynomial $H_n(x)$.]

28

Group theory

28.1 *For each of the following sets, determine whether they form a group under the operation indicated (where it is relevant you may assume that matrix multiplication is associative):*

(a) *the integers (mod 10) under addition;*
(b) *the integers (mod 10) under multiplication;*
(c) *the integers* 1, 2, 3, 4, 5, 6 *under multiplication (mod 7);*
(d) *the integers* 1, 2, 3, 4, 5 *under multiplication (mod 6);*
(e) *all matrices of the form*

$$\begin{pmatrix} a & a-b \\ 0 & b \end{pmatrix},$$

where a and b are integers (mod 5) and $a \neq 0 \neq b$, under matrix multiplication;
(f) *those elements of the set in* (e) *that are of order 1 or 2 (taken together);*
(g) *all matrices of the form*

$$\begin{pmatrix} 1 & 0 & 0 \\ a & 1 & 0 \\ b & c & 1 \end{pmatrix},$$

where a, b, c are integers, under matrix multiplication.

In all cases we need to establish whether the prescribed combination law is associative and whether, under it, the set possesses the properties of (i) closure, (ii) having an identity element and (iii) containing an inverse for every element present. If any one of these conditions fails, the set cannot form a group under the given law.

(a) Addition is associative and the set $\{0, 1, 2, \ldots, 9\}$ is closed under addition (mod 10), e.g. $7 + 6 = 3$. The identity is 0 and every element has an inverse, e.g. $(7)^{-1} = 3$. The set does form a group.

(b) For the set $\{0, 1, 2, \ldots, 9\}$ under multiplication the identity can only be 1. However, for any element $X \neq 1$ the set does *not* contain an inverse Y such that $XY = 1$. As a specific example, if $X = 2$ then 0.5 would need to be in the set – but it is not. The set does *not* form a group under multiplication.

(c) Multiplication is associative and the group table would be as below. The entries are calculated by expressing each product modulo 7. For example, $4 \times 5 = 20 = (2 \times 7) + 6 = 6 \pmod{7}$

	1	2	3	4	5	6
1	1	2	3	4	5	6
2	2	4	6	1	3	5
3	3	6	2	5	1	4
4	4	1	5	2	6	3
5	5	3	1	6	4	2
6	6	5	4	3	2	1

This demonstrates (i) closure, (ii) the existence of an identity element (1) and (iii) an inverse for each element (1 appears in every row). The set does form a group.

(d) The set is not closed under multiplication (mod 6) and *cannot* form a group. For example, $2 \times 3 = 0 \pmod{6}$ and 0 is not in the given set.

(e) With the associativity of matrix multiplication assumed and $a = b = 1$ yielding a unit element, consider

$$\begin{pmatrix} a & a-b \\ 0 & b \end{pmatrix} \begin{pmatrix} c & c-d \\ 0 & d \end{pmatrix} = \begin{pmatrix} ac & ac - ad + ad - bd \\ 0 & bd \end{pmatrix}$$

$$= \begin{pmatrix} ac & ac - bd \\ 0 & bd \end{pmatrix},$$

implying closure. We also note that interchanging a with c and b with d shows that any two matrices in the set commute.

Since neither a nor b is 0, the determinant of a general matrix in the set is non-zero and its inverse can be constructed as

$$\frac{1}{ab} \begin{pmatrix} b & b-a \\ 0 & a \end{pmatrix} = \begin{pmatrix} a^{-1} & a^{-1} - b^{-1} \\ 0 & b^{-1} \end{pmatrix}.$$

The question then arises as to whether a^{-1} is an integer; in multiplication mod 5 it is. For example, if $a = 3$ then $a^{-1} = 2$ since $3 \times 2 = 6 = 1 \pmod{5}$. The full set of values is: $1^{-1} = 1$, $2^{-1} = 3$, $3^{-1} = 2$ and $4^{-1} = 4$.

Thus each inverse is of the required form and a general one can be verified:

$$\frac{1}{ab}\begin{pmatrix} b & b-a \\ 0 & a \end{pmatrix}\begin{pmatrix} a & a-b \\ 0 & b \end{pmatrix} = \frac{1}{ab}\begin{pmatrix} ab & ab-b^2+b^2-ab \\ 0 & ab \end{pmatrix}$$

$$= \begin{pmatrix} 1 & 0 \\ 0 & 1 \end{pmatrix}.$$

All four requirements are satisfied and the set of matrices is, in fact, a group.

(f) As always, the only element of order 1 is the unit element. Elements of order 2 must satisfy

$$\begin{pmatrix} a & a-b \\ 0 & b \end{pmatrix}\begin{pmatrix} a & a-b \\ 0 & b \end{pmatrix} = \begin{pmatrix} a^2 & a^2-b^2 \\ 0 & b^2 \end{pmatrix} = \begin{pmatrix} 1 & 0 \\ 0 & 1 \end{pmatrix}.$$

Thus a and b must both be elements whose squares are unity (mod 5); each must be either 1 or 4 [since $4^2 = 16 = 1$ (mod 5)]. The four matrices to consider are thus

$$\begin{pmatrix} 1 & 0 \\ 0 & 1 \end{pmatrix}, \quad \begin{pmatrix} 4 & 0 \\ 0 & 4 \end{pmatrix}, \quad \begin{pmatrix} 1 & 2 \\ 0 & 4 \end{pmatrix}, \quad \begin{pmatrix} 4 & 3 \\ 0 & 1 \end{pmatrix}.$$

The identity element is present and, from the way they were defined, each is its own inverse. Only closure remains to be tested. As all matrices in the set commute [see (e) above], we need test only

$$\begin{pmatrix} 4 & 0 \\ 0 & 4 \end{pmatrix}\begin{pmatrix} 1 & 2 \\ 0 & 4 \end{pmatrix} = \begin{pmatrix} 4 & 3 \\ 0 & 1 \end{pmatrix},$$

$$\begin{pmatrix} 4 & 0 \\ 0 & 4 \end{pmatrix}\begin{pmatrix} 4 & 3 \\ 0 & 1 \end{pmatrix} = \begin{pmatrix} 1 & 2 \\ 0 & 4 \end{pmatrix},$$

$$\begin{pmatrix} 1 & 2 \\ 0 & 4 \end{pmatrix}\begin{pmatrix} 4 & 3 \\ 0 & 1 \end{pmatrix} = \begin{pmatrix} 4 & 0 \\ 0 & 4 \end{pmatrix}.$$

Each product is one of the set of four. So closure is established and the set does form a group – a subgroup, of order 4, of the group in (e).

(g) The product of two such matrices is

$$\begin{pmatrix} 1 & 0 & 0 \\ a & 1 & 0 \\ b & c & 1 \end{pmatrix}\begin{pmatrix} 1 & 0 & 0 \\ x & 1 & 0 \\ y & z & 1 \end{pmatrix} = \begin{pmatrix} 1 & 0 & 0 \\ a+x & 1 & 0 \\ b+cx+y & c+z & 1 \end{pmatrix}.$$

Since all elements of the original two matrices are integers, so are all elements of the product and closure is established.

Clearly, $a = b = c = 0$ provides the identity element and, since the determinant of each matrix is 1, inverses can be constructed in the usual way, typically

$$\begin{pmatrix} 1 & 0 & 0 \\ -a & 1 & 0 \\ ac-b & -c & 1 \end{pmatrix}.$$

This is of the correct form as can be verified as follows:

$$\begin{pmatrix} 1 & 0 & 0 \\ -a & 1 & 0 \\ ac - b & -c & 1 \end{pmatrix} \begin{pmatrix} 1 & 0 & 0 \\ a & 1 & 0 \\ b & c & 1 \end{pmatrix} = \begin{pmatrix} 1 & 0 & 0 \\ 0 & 1 & 0 \\ 0 & 0 & 1 \end{pmatrix}.$$

Thus, assuming associativity, the group property of the set is established.

28.3 *Define a binary operation* $\bullet$ *on the set of real numbers by*

$$x \bullet y = x + y + rxy,$$

where r is a non-zero real number. Show that the operation $\bullet$ is associative.

Prove that $x \bullet y = -r^{-1}$ if, and only if, $x = -r^{-1}$ or $y = -r^{-1}$. Hence prove that the set of all real numbers excluding $-r^{-1}$ forms a group under the operation $\bullet$.

To demonstrate the associativity we need to show that $x \bullet (y \bullet z)$ is the same thing as $(x \bullet y) \bullet z$. So consider

$$\begin{aligned} x \bullet (y \bullet z) &= x + (y \bullet z) + rx(y \bullet z) \\ &= x + y + z + ryz + rx(y + z + ryz) \\ &= x + y + z + r(yz + xy + xz) + r^2 xyz \end{aligned}$$

and

$$\begin{aligned} (x \bullet y) \bullet z &= (x \bullet y) + z + r(x \bullet y)z \\ &= x + y + rxy + z + r(x + y + rxy)z \\ &= x + y + z + r(xy + xz + yz) + r^2 xyz. \end{aligned}$$

The two RHSs are equal, showing that the operation $\bullet$ is associative.

Firstly, suppose that $x = -r^{-1}$. Then

$$x \bullet y = -\frac{1}{r} + y + r\left(-\frac{1}{r}\right)y = -\frac{1}{r} + y - y = -\frac{1}{r}.$$

Similarly $y = -r^{-1} \quad \Rightarrow \quad x \bullet y = -r^{-1}$.

Secondly, suppose that $x \bullet y = -r^{-1}$. Then

$$\begin{aligned} x + y + rxy &= -r^{-1}, \\ rx + ry + r^2 xy + 1 &= 0, \\ (rx + 1)(ry + 1) &= 0, \\ \Rightarrow \quad \text{either } x = -r^{-1} & \text{ or } y = -r^{-1}. \end{aligned}$$

Thus $x \bullet y = -r^{-1} \iff (x = -r^{-1}$ or $y = -r^{-1})$.

If $\mathcal{S} = \{$real numbers $\neq -r^{-1}\}$, then

(i) Associativity under • has already been shown.

(ii) If x and y belong to S, then $x \bullet y$ is a real number and, in view of the second result above, is $\neq -r^{-1}$. Thus $x \bullet y$ belongs to S and the set is closed under the operation •.

(iii) For any x belonging to S, $x \bullet 0 = x + 0 + rx0 = x$. Thus 0 is an identity element.

(iv) An inverse x^{-1} of x must satisfy $x \bullet x^{-1} = 0$, i.e.

$$x + x^{-1} + rxx^{-1} = 0 \quad \Rightarrow \quad x^{-1} = -\frac{x}{1 + rx}.$$

This is a real (finite) number since $x \neq -r^{-1}$ and, further, $x^{-1} \neq -r^{-1}$, since if it were we could deduce that $1 = 0$. Thus the set S contains an inverse for each of its elements.

These four results together show that S is a group under the operation •.

28.5 *The following is a 'proof' that reflexivity is an unnecessary axiom for an equivalence relation.*

Because of symmetry $X \sim Y$ implies $Y \sim X$. Then by transitivity $X \sim Y$ and $Y \sim X$ imply $X \sim X$. Thus symmetry and transitivity imply reflexivity, which therefore need not be separately required.

Demonstrate the flaw in this proof using the set consisting of all real numbers plus the number i. Show by investigating the following specific cases that, whether or not reflexivity actually holds, it cannot be deduced from symmetry and transitivity alone.

(a) $X \sim Y$ *if* $X + Y$ *is real.*
(b) $X \sim Y$ *if* XY *is real.*

Let elements X and Y be drawn from the set S consisting of the real numbers together with i.

(a) For the definition $X \sim Y$ if $X + Y$ is real, we have

(i) that

$$X \sim Y \quad \Rightarrow \quad X + Y \text{ is real} \quad \Rightarrow \quad Y + X \text{ is real} \quad \Rightarrow \quad Y \sim X,$$

i.e symmetry holds;

(ii) that if $X \sim Y$ then neither X nor Y can be i and, equally, if $Y \sim Z$ then neither Y nor Z can be i. It then follows that $X + Z$ is real and $X \sim Z$, i.e. transitivity holds.

Thus both symmetry and transitivity hold, though it is obvious that $X \not\sim X$ if X is i. Thus symmetry and transitivity do *not* necessarily imply reflexivity, showing that the 'proof' is flawed – in this case the proof fails when X is i because there is no distinct 'Y' available, something assumed in the proof.

(b) For the definition $X \sim Y$ if XY is real, we have

(i) that

$$X \sim Y \quad \Rightarrow \quad XY \text{ is real} \quad \Rightarrow \quad YX \text{ is real} \quad \Rightarrow \quad Y \sim X,$$

i.e symmetry holds;

(ii) that if $X \sim Y$ then neither X nor Y is i. Equally, if $Y \sim Z$ then neither Y nor Z is i. It then follows that XZ is real and $X \sim Z$, i.e. transitivity holds.

Thus both symmetry and transitivity hold and, setting Z equal to X, they do imply the reflexivity property for the real elements of S. However, they cannot establish it for the element i – though it happens to be true in this case as i^2 is real.

28.7 S is the set of all 2×2 matrices of the form

$$A = \begin{pmatrix} w & x \\ y & z \end{pmatrix}, \qquad \text{where } wz - xy = 1.$$

Show that S is a group under matrix multiplication. Which element(s) have order 2? Prove that an element A has order 3 if $w + z + 1 = 0$.

The condition $wz - xy = 1$ is the same as $|A| = 1$; it follows that the set contains an identity element (with $w = z = 1$ and $x = y = 0$). Moreover, each matrix in S has an inverse and, since $|A^{-1}||A| = |I| = 1$ implies that $|A^{-1}| = 1$, the inverses also belong to the set.

If A and B belong to S then, since $|AB| = |A||B| = 1 \times 1 = 1$, their product also belongs to S, i.e. the set is closed.

These observations, together with the associativity of matrix multiplication establish that the set S is, in fact, a group under this operation.

If A is to have order 2 then

$$\begin{pmatrix} w & x \\ y & z \end{pmatrix} \begin{pmatrix} w & x \\ y & z \end{pmatrix} = \begin{pmatrix} 1 & 0 \\ 0 & 1 \end{pmatrix},$$

i.e. $w^2 + xy = 1$, $x(w + z) = 0$, $y(w + z) = 0$, $xy + z^2 = 1$.

These imply that $w^2 = z^2$ and that either $z = -w$ or $x = y = 0$. If $z = -w$, then both

$$w^2 + xy = 1, \qquad \text{from the above condition,}$$

$$\text{and} \quad -w^2 - xy = 1, \qquad \text{from } wz - xy = 1.$$

This is not possible and so we must have $x = y = 0$, implying that w and z are either both $+1$ or both -1. The former gives the identity (of order 1), and so the matrix given by the latter, $A = -I$, is the only element in S of order 2.

If $w + z + 1 = 0$ (as well as $xy = wz - 1$), A^2 can be written as

$$A^2 = \begin{pmatrix} w^2 + xy & x(w+z) \\ y(w+z) & xy + z^2 \end{pmatrix}$$
$$= \begin{pmatrix} w^2 + wz - 1 & -x \\ -y & wz - 1 + z^2 \end{pmatrix}$$
$$= \begin{pmatrix} -w - 1 & -x \\ -y & -z - 1 \end{pmatrix}.$$

Multiplying again by A gives

$$A^3 = \begin{pmatrix} -w - 1 & -x \\ -y & -z - 1 \end{pmatrix} \begin{pmatrix} w & x \\ y & z \end{pmatrix}$$
$$= -\begin{pmatrix} w(w+1) + xy & (w+1)x + xz \\ wy + y(z+1) & xy + z(z+1) \end{pmatrix}$$
$$= -\begin{pmatrix} w(w+1) + wz - 1 & x \times 0 \\ y \times 0 & wz - 1 + z(z+1) \end{pmatrix}$$
$$= -\begin{pmatrix} (w \times 0) - 1 & 0 \\ 0 & (z \times 0) - 1 \end{pmatrix}$$
$$= \begin{pmatrix} 1 & 0 \\ 0 & 1 \end{pmatrix}.$$

Thus A has order 3.

28.9 *If A is a group in which every element other than the identity, I, has order 2, prove that A is Abelian. Hence show that if X and Y are distinct elements of A, neither being equal to the identity, then the set $\{I, X, Y, XY\}$ forms a subgroup of A.*

Deduce that if B is a group of order 2p, with p a prime greater than 2, then B must contain an element of order p.

If every element of A, apart from I, has order 2, then, for any element X, $X^2 = I$. Consider two elements X and Y and let $XY = Z$. Then

$$X^2 Y = XZ \quad \Rightarrow \quad Y = XZ,$$
$$XY^2 = ZY \quad \Rightarrow \quad X = ZY.$$

It follows that $YX = XZZY$. But, since $XY = Z$, Z must belong to A and therefore $Z^2 = I$. Substituting this gives $YX = XY$, proving that the group is Abelian.

Consider the set $S = \{I, X, Y, XY\}$, for which

(i) Associativity holds, since it does for A.
(ii) It is closed, the only products needing non-trivial examinations being $XYX = XXY = X^2Y = Y$ and $YXY = XYY = XY^2 = X$ (here we have twice used the fact that A, and hence S, is Abelian).
(iii) The identity I is contained in the set.
(iv) Since all elements are of order 2 (or 1), each is its own inverse.

Thus the set is a subgroup of A of order 4.

Now consider the group B of order $2p$, where p is prime. Since the order of an element must divide the order of the group, all elements in B must have order 1 (I only) or 2 or p.

Suppose *all* elements, other than I, have order 2. Then, as shown above, B must be Abelian and have a subgroup of order 4. However, the order of any subgroup must divide the order of the group and 4 cannot divide $2p$ since p is prime. It follows that the supposition that all elements can be of order 2 is false, and consequently at least one must have order p.

28.11 *Identify the eight symmetry operations on a square. Show that they form a group D_4 (known to crystallographers as 4mm and to chemists as C_{4v}) having one element of order 1, five of order 2 and two of order 4. Find its proper subgroups and the corresponding cosets.*

The operation of leaving the square alone is a trivial symmetry operation, but an important one, as it is the identity I of the group; it has order 1.

Rotations about an axis perpendicular to the plane of the square by $\pi/4$, $\pi/2$ and $3\pi/2$ each take the square into itself. The first and last of these have to be repeated four times to reproduce the effect of I, and so they have order 4. The rotation by $\pi/2$ clearly has order 2.

Reflections in the two axes parallel to the sides of the square and passing through its centre are also symmetry operations, as are reflections in the two principal diagonals of the square; all of these reflections have order 2.

Using the notation indicated in figure 28.1, R being a rotation of $\pi/2$ about an axis perpendicular to the square, we have: I has order 1; R^2, m_1, m_2, m_3, m_4 have order 2; R, R^3 have order 4.

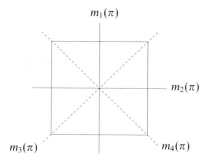

Figure 28.1   The notation for exercise 28.11.

The group multiplication table takes the form

	I	R	R^2	R^3	m_1	m_2	m_3	m_4
I	I	R	R^2	R^3	m_1	m_2	m_3	m_4
R	R	R^2	R^3	I	m_4	m_3	m_1	m_2
R^2	R^2	R^3	I	R	m_2	m_1	m_4	m_3
R^3	R^3	I	R	R^2	m_3	m_4	m_2	m_1
m_1	m_1	m_3	m_2	m_4	I	R^2	R	R^3
m_2	m_2	m_4	m_1	m_3	R^2	I	R^3	R
m_3	m_3	m_2	m_4	m_1	R^3	R	I	R^2
m_4	m_4	m_1	m_3	m_2	R	R^3	R^2	I

Inspection of this table shows the existence of the non-trivial subgroups listed below, and tedious but straightforward evaluation of the products of selected elements of the group with all the elements of any one subgroup provides the cosets of that subgroup. The results are as follows:

subgroup $\{I, R, R^2, R^3\}$ has cosets $\{I, R, R^2, R^3\}$, $\{m_1, m_2, m_3, m_4\}$;

subgroup $\{I, R^2, m_1, m_2\}$ has cosets $\{I, R^2, m_1, m_2\}$, $\{I, R^2, m_3, m_4\}$;

subgroup $\{I, R^2, m_3, m_4\}$ has cosets $\{I, R^2, m_3, m_4\}$, $\{I, R^2, m_1, m_2\}$;

subgroup $\{I, R^2\}$ has cosets $\{I, R^2\}$, $\{R, R^3\}$, $\{m_1, m_2\}$, $\{m_3, m_4\}$;

subgroup $\{I, m_1\}$ has cosets $\{I, m_1\}$, $\{R, m_3\}$, $\{R^2, m_2\}$, $\{R^3, m_4\}$;

subgroup $\{I, m_2\}$ has cosets $\{I, m_2\}$, $\{R, m_4\}$, $\{R^2, m_1\}$, $\{R^3, m_3\}$;

subgroup $\{I, m_3\}$ has cosets $\{I, m_3\}$, $\{R, m_2\}$, $\{R^2, m_4\}$, $\{R^3, m_1\}$;

subgroup $\{I, m_4\}$ has cosets $\{I, m_4\}$, $\{R, m_1\}$, $\{R^2, m_3\}$, $\{R^3, m_2\}$.

28.13 *Find the group G generated under matrix multiplication by the matrices*

$$A = \begin{pmatrix} 0 & 1 \\ 1 & 0 \end{pmatrix}, \qquad B = \begin{pmatrix} 0 & i \\ i & 0 \end{pmatrix}.$$

Determine its proper subgroups, and verify for each of them that its cosets exhaust G.

Before we can draw up a group multiplication table to search for subgroups, we must determine the multiple products of A and B with themselves and with each other:

$$A^2 = \begin{pmatrix} 0 & 1 \\ 1 & 0 \end{pmatrix} \begin{pmatrix} 0 & 1 \\ 1 & 0 \end{pmatrix} = \begin{pmatrix} 1 & 0 \\ 0 & 1 \end{pmatrix} = I.$$

Since $B = iA$, it follows that $B^2 = -I$, that $AB = iI = BA$, and that $B^3 = -B$. In brief, A is of order 2, B is of order 4, and A and B commute. The eight distinct elements of the group are therefore: $I, A, B, B^2, B^3, AB, AB^2$ and AB^3.

The group multiplication table is

	I	A	B	B^2	B^3	AB	AB^2	AB^3
I	I	A	B	B^2	B^3	AB	AB^2	AB^3
A	A	I	AB	AB^2	AB^3	B	B^2	B^3
B	B	AB	B^2	B^3	I	AB^2	AB^3	A
B^2	B^2	AB^2	B^3	I	B	AB^3	A	AB
B^3	B^3	AB^3	I	B	B^2	A	AB	AB^2
AB	AB	B	AB^2	AB^3	A	B^2	B^3	I
AB^2	AB^2	B^2	AB^3	A	B	B^3	I	B
AB^3	AB^3	B^3	A	AB	AB^2	I	B	B^2

By inspection, the subgroups and their cosets are as follows:

$$\{I, A\} : \{I, A\}, \{B, AB\}, \{B^2, AB^2\}, \{B^3, AB^3\};$$
$$\{I, B^2\} : \{I, B^2\}, \{A, AB^2\}, \{B, B^3\}, \{AB, AB^3\};$$
$$\{I, AB^2\} : \{I, AB^2\}, \{A, B^2\}, \{B, AB^3\}, \{B^3, AB\};$$
$$\{I, B, B^2, B^3\} : \{I, B, B^2, B^3\}, \{A, AB, AB^2, AB^3\};$$
$$\{I, AB, B^2, AB^3\} : \{I, AB, B^2, AB^3\}, \{A, B, AB^2, B^3\}.$$

As expected, in each case the cosets exhaust the group, with each element in one and only one coset.

28.15 *Consider the following mappings between a permutation group and a cyclic group.*

(a) *Denote by A_n the subset of the permutation group S_n that contains all the even permutations. Show that A_n is a subgroup of S_n.*

(b) *List the elements of S_3 in cycle notation and identify the subgroup A_3.*

(c) *For each element X of S_3, let $p(X) = 1$ if X belongs to A_3 and $p(X) = -1$ if it does not. Denote by C_2 the multiplicative cyclic group of order 2. Determine the images of each of the elements of S_3 for the following four mappings:*

$$\Phi_1 : S_3 \rightarrow C_2 \qquad X \rightarrow p(X),$$
$$\Phi_2 : S_3 \rightarrow C_2 \qquad X \rightarrow -p(X),$$
$$\Phi_3 : S_3 \rightarrow A_3 \qquad X \rightarrow X^2,$$
$$\Phi_4 : S_3 \rightarrow S_3 \qquad X \rightarrow X^3.$$

(d) *For each mapping, determine whether the kernel K is a subgroup of S_3 and, if so, whether the mapping is a homomorphism.*

(a) With A_n as the subset of S_n that contains all the even permutations, we need to demonstrate that it has the four properties that characterise a group:

(i) If X and Y belong to A_n so does XY, as the product of two even permutations is even. This establishes closure.
(ii) From the definition of an even permutation, the identity I belongs to A_n.
(iii) If X belongs to A_n so does X^{-1}, as the number of pair interchanges needed to change from X to I is the same as the number needed to go in the opposite direction. This establishes the existence, within the subset, of an inverse for each member of the subset.
(iv) Associativity follows from that of the group S_n.

Thus A_n does possess the four properties and is a subgroup of S_n.

(b) (1), (123) and (132) belong to A_3. The permutations (12), (13) and (23) do not belong, as each involves only one pair interchange.

(c) With the given definition of $p(X)$,

$$p(X) = 1 \text{ for } X = (1), (123), (132),$$
$$p(X) = -1 \text{ for } X = (12), (13), (23).$$

C_2 consists of the two elements $+1$ and -1.

For $\Phi_1 : S_3 \rightarrow C_2$, $X \rightarrow p(X)$, elements in A_3 have image $+1$; those not in A_3 have image -1.

For $\Phi_2 : S_3 \rightarrow C_2$, $X \rightarrow -p(X)$, elements in A_3 have image -1; those not in A_3 have image $+1$.

For $\Phi_3 : S_3 \to A_3$, $X \to X^2$

$$(1) \to (1)(1) = (1),$$
$$(123) \to (123)(123) = (132),$$
$$(132) \to (132)(132) = (123),$$
$$(12) \to (12)(12) = (1), \quad \text{similarly, (13) and (23).}$$

For $\Phi_4 : S_3 \to S_3$, $X \to X^3$

$$(1) \to (1)(1) = (1),$$
$$(123) \to (123)(123)(123) = (132)(123) = (1),$$
$$(132) \to (132)(132)(132) = (123)(132) = (1),$$
$$(12) \to (12)(12)(12) = (1)(12) = (12), \quad \text{similarly, (13) and (23).}$$

(d) For Φ_1, the kernel is the set of elements belonging to A_3 and, as already shown, this is a subgroup of S_3.

We note that the product of two even or two odd permutations is an even permutation, whilst the product of an odd and an even permutation is an odd permutation. We also note that $+1 \times +1$ and -1×-1 are both equal to $+1$, whilst $+1 \times -1 = -1$. Since Φ_1 maps even permutations onto $+1$ and odd permutations onto -1, the preceding observations imply that Φ_1 is a homomorphism.

For Φ_2, the kernel is the set of elements *not* belonging to A_3. Since this set does not contain the identity (1), it cannot be a subgroup of S_3.

For Φ_3 the kernel is the set $\{(1), (12), (13), (23)\}$. Since, for example, $(12)(13) = (132)$, the set is not closed and so cannot form a group. It cannot, therefore, be a subgroup of S_3.

For Φ_4 the kernel is the set $\{(1), (123), (132)\}$, i.e. the subgroup A_3. However, for example, $[(12)(13)]' = (132)' = (1)$, whilst $(12)'(13)' = (12)(13) = (132)$; these two results are not equal, showing that the mapping *cannot* be a homomorphism.

28.17 *The group of all non-singular $n \times n$ matrices is known as the general linear group $GL(n)$ and that with only real elements as $GL(n, \mathbf{R})$. If $\mathbf{R}^*$ denotes the multiplicative group of non-zero real numbers, prove that the mapping $\Phi : GL(n, \mathbf{R}) \to \mathbf{R}^*$, defined by $\Phi(M) = \det M$, is a homomorphism.*

Show that the kernel $\mathcal{K}$ of Φ is a subgroup of $GL(n, \mathbf{R})$. Determine its cosets and show that they themselves form a group.

If P and Q are two matrices belonging to $GL(n, \mathbf{R})$ then, under Φ,

$$(PQ)' = |PQ| = |P| \, |Q| = P'Q'.$$

Thus Φ is a homomorphism.

The kernel $\mathcal{K}$ of the mapping consists of all matrices in $GL(n, \mathbf{R})$ that map onto the identity in $\mathbf{R}^*$, i.e all matrices whose determinant is 1.

To determine whether $\mathcal{K}$ is a subgroup of the general linear group, let X and Y belong to $\mathcal{K}$. Then, testing $\mathcal{K}$ for the four group-defining properties, we have

(i) $(XY)' = X'Y' = 1 \times 1 = 1$, i.e. XY also belongs to $\mathcal{K}$, showing the closure of the kernel.
(ii) The associative law holds for the elements of $\mathcal{K}$ since it does so for all elements of $GL(n, \mathbf{R})$.
(iii) $|\mathsf{I}| = 1$ and so I belongs to $\mathcal{K}$.
(iv) Since $\mathsf{X}^{-1}\mathsf{X} = \mathsf{I}$, it follows that $|\mathsf{X}^{-1}||\mathsf{X}| = |\mathsf{I}|$ and $|\mathsf{X}^{-1}| \times 1 = 1$. Hence $|\mathsf{X}^{-1}| = 1$ and so X^{-1} also belongs to $\mathcal{K}$.

This completes the proof that $\mathcal{K}$ is a subgroup of $GL(n, \mathbf{R})$.

Two matrices P and Q in $GL(n, \mathbf{R})$ belong to the same coset of $\mathcal{K}$ if

$$\mathsf{Q} = \mathsf{PX}, \text{ where } \mathsf{X} \text{ is some element in } \mathcal{K}.$$

It then follows that

$$|\mathsf{Q}| = |\mathsf{P}|\,|\mathsf{X}| = |\mathsf{P}| \times 1.$$

Thus the requirement for two matrices to be in the same coset is that they have equal determinants.

Let us denote by C_i the (infinite) set of all matrices whose determinant has the value i; the label i will itself take on an infinite continuum of values, excluding 0. Then,

(i) For *any* $\mathsf{M}_i \in C_i$ and *any* $\mathsf{M}_j \in C_j$ we have

$$|\mathsf{M}_i\mathsf{M}_j| = |\mathsf{M}_i|\,|\mathsf{M}_j| = i \times j,$$

implying that we always have $\mathsf{M}_i\mathsf{M}_j \in C_{(i \times j)}$. Thus the set of cosets is closed, with $C_i \times C_j = C_{(i \times j)}$.
(ii) The associative law holds, since it does so for matrix multiplication in general, and the product of three matrices, and hence its determinant, is independent of the order in which the individual multiplications are carried out.
(iii) Since

$$|\mathsf{M}_i\mathsf{M}_1| = |\mathsf{M}_i|\,|\mathsf{M}_1| = i,$$

$C_i \times C_1 = C_i$, showing that C_1 is an identity element in the set.
(iv) Since

$$|\mathsf{M}_i\mathsf{M}_{1/i}| = |\mathsf{M}_i|\,|\mathsf{M}_{1/i}| = i \times (1/i) = 1,$$

$C_i \times C_{1/i} = C_1$, showing that the set of cosets contains an inverse for each coset.

This completes the proof that the cosets themselves form a group under coset multiplication (and also that $\mathcal{K}$ is a normal subgroup).

28.19 *Given that matrix* M *is a member of the multiplicative group* $GL(3, \mathbf{R})$, *determine, for each of the following additional constraints on* M *(applied separately), whether the subset satisfying the constraint is a subgroup of* $GL(3, \mathbf{R})$:

    (a) $M^T = M$;
    (b) $M^T M = I$;
    (c) $|M| = 1$;
    (d) $M_{ij} = 0$ *for* $j > i$ *and* $M_{ii} \neq 0$.

The matrices belonging to $GL(3, \mathbf{R})$ have the general properties that they are non-singular, possess inverses and have real elements. The operation of matrix multiplication is associative, and this will be assumed in the rest of the exercise, in which A and B are general matrices satisfying the various defining constraints.

(a) $M^T = M$: the set of symmetric matrices.

Now, for two symmetric matrices A and B,

$$(AB)^T = B^T A^T = BA$$

and this is not equal to AB in general, as matrix multiplication is not necessarily commutative. The set is therefore not closed and cannot form a group; equally it cannot be a subgroup of $GL(3, \mathbf{R})$.

(b) $M^T M = I$: the set of orthogonal matrices.

Clearly, the identity I belongs to the set, and furthermore

$$(M^{-1})^T M^{-1} = (M^T)^{-1} M^{-1} = (M^{-1})^{-1} M^{-1} = MM^{-1} = I,$$

i.e. if M belongs to the set then so does M^{-1}. Finally,

$$(AB)^T AB = B^T A^T AB = B^T I B = I,$$

showing that the set is closed. This completes the proof that the orthogonal matrices form a subgroup of $GL(3, \mathbf{R})$.

(c) $|M| = 1$: the set of unimodular matrices.

$$|AB| = |A|\,|B| = 1 \times 1 = 1 \quad \text{closure,}$$
$$|I| = 1 \quad \text{identity,}$$
$$M^{-1}M = I \quad \Rightarrow \quad |M^{-1}|\,|M| = 1 \quad \Rightarrow \quad |M^{-1}| = 1 \quad \text{inverse.}$$

These three results (and associativity) show that the unimodular matrices do form a subgroup of $GL(3, \mathbf{R})$.

(d) $M_{ij} = 0$ for $j > i$ and $M_{ii} \neq 0$: the set of lower diagonal matrices with non-zero diagonal elements.

Taking first the question of closure, consider the matrix product $C = AB$. A typical element of C above the leading diagonal is

$$C_{12} = A_{11}B_{12} + A_{12}B_{22} + A_{13}B_{32}$$
$$= A_{11}0 + 0B_{22} + 0A_{32} = 0,$$

and a typical element on the leading diagonal is

$$C_{11} = A_{11}B_{11} + A_{12}B_{21} + A_{13}B_{31}$$
$$= A_{11}B_{11} + 0B_{21} + 0A_{31} = A_{11}B_{11} \neq 0.$$

That C_{11} is not equal to zero follows from the fact that neither A_{11} nor B_{11} is zero. Similarly, C_{13} and C_{23} are zero, whilst C_{22} and C_{33} are non-zero. Thus C has all the properties defining the set, and so belongs to it. The set is therefore closed.

Clearly the matrix I_3 belongs to the set, which therefore contains an identity element.

Since no diagonal element is zero, the determinant (which is the product of the diagonal elements for lower triangular matrices) of any member of the set cannot be zero. All members must therefore have inverses. For the matrix

$$A = \begin{pmatrix} A_{11} & 0 & 0 \\ A_{21} & A_{22} & 0 \\ A_{31} & A_{32} & A_{33} \end{pmatrix},$$

the inverse is given by

$$A^{-1} = \frac{1}{A_{11}A_{22}A_{33}} \begin{pmatrix} A_{22}A_{33} & 0 & 0 \\ -A_{21}A_{33} & A_{11}A_{33} & 0 \\ A_{21}A_{32} - A_{22}A_{31} & -A_{11}A_{32} & A_{11}A_{22} \end{pmatrix}.$$

We note that each of the diagonal elements of A^{-1} is the product of two non-zero terms, and is therefore itself non-zero. Thus A^{-1} has the correct form for a member of the set – lower diagonal with non-zero diagonal elements – and so belongs to the set, which has now been shown to have all the properties needed to make it a group and hence a subgroup of $GL(3, \mathbf{R})$.

28.21 *Show that* $\mathcal{D}_4$*, the group of symmetries of a square, has two isomorphic subgroups of order 4.*

The quaternion group Q *is the set of elements*

$$\{1, -1, i, -i, j, -j, k, -k\},$$

with $i^2 = j^2 = k^2 = -1$*,* $ij = k$ *and its cyclic permutations, and* $ji = -k$ *and its cyclic permutations. Its multiplication table reads as follows:*

	1	−1	i	$-i$	j	$-j$	k	$-k$
1	1	−1	i	$-i$	j	$-j$	k	$-k$
−1	−1	1	$-i$	i	$-j$	j	$-k$	k
i	i	$-i$	−1	1	k	$-k$	$-j$	j
$-i$	$-i$	i	1	−1	$-k$	k	j	$-j$
j	j	$-j$	$-k$	k	−1	1	i	$-i$
$-j$	$-j$	j	k	$-k$	1	−1	$-i$	i
k	k	$-k$	j	$-j$	$-i$	i	−1	1
$-k$	$-k$	k	$-j$	j	i	$-i$	1	−1

Show that there exists a two-to-one homomorphism from the quaternion group Q *onto one (and hence either) of the two subgroups of* $\mathcal{D}_4$*, and determine its kernel.*

We first reproduce the multiplication table for $\mathcal{D}_4$:

	I	R	R^2	R^3	m_1	m_2	m_3	m_4
I	I	R	R^2	R^3	m_1	m_2	m_3	m_4
R	R	R^2	R^3	I	m_4	m_3	m_1	m_2
R^2	R^2	R^3	I	R	m_2	m_1	m_4	m_3
R^3	R^3	I	R	R^2	m_3	m_4	m_2	m_1
m_1	m_1	m_3	m_2	m_4	I	R^2	R	R^3
m_2	m_2	m_4	m_1	m_3	R^2	I	R^3	R
m_3	m_3	m_2	m_4	m_1	R^3	R	I	R^2
m_4	m_4	m_1	m_3	m_2	R	R^3	R^2	I

Here R is a rotation by $\pi/2$ in the plane of the square, m_1 and m_2 are reflections in the axes parallel to the sides of the square, and m_3 and m_4 are reflections in the square's diagonals.

As shown in exercise 28.11, $\mathcal{D}_4$ has three proper subgroups of order 4. They are $\{I, R, R^2, R^3\}$, $\mathcal{H}_1 = \{I, R^2, m_1, m_2\}$ and $\mathcal{H}_2 = \{I, R^2, m_3, m_4\}$. The first of these is a cyclic subgroup but the other two are not. The group tables for the latter two, extracted from the one above, are as follows

$\mathcal{H}_1$	I	R^2	m_1	m_2
I	I	R^2	m_1	m_2
R^2	R^2	I	m_2	m_1
m_1	m_1	m_2	I	R^2
m_2	m_2	m_1	R^2	I

$\mathcal{H}_2$	I	R^2	m_3	m_4
I	I	R^2	m_3	m_4
R^2	R^2	I	m_4	m_3
m_3	m_3	m_4	I	R^2
m_4	m_4	m_3	R^2	I

Clearly, these two subgroups are isomorphic with $m_1 \leftrightarrow m_3$ and $m_2 \leftrightarrow m_4$ and the other elements unchanged.

Next, we reproduce the group table for the quaternion group, but with the columns and rows reordered (this does not alter the information it carries):

	1	i	j	k	-1	$-i$	$-j$	$-k$
1	1	i	j	k	-1	$-i$	$-j$	$-k$
i	i	-1	k	$-j$	$-i$	1	$-k$	j
j	j	$-k$	-1	i	$-j$	k	1	$-i$
k	k	j	$-i$	-1	$-k$	$-j$	i	1
-1	-1	$-i$	$-j$	$-k$	1	i	j	k
$-i$	$-i$	1	$-k$	j	i	-1	k	$-j$
$-j$	$-j$	k	1	$-i$	j	$-k$	-1	i
$-k$	$-k$	$-j$	i	1	k	j	$-i$	-1

If we now make the two-to-one mapping

$$\Phi \quad : \quad \pm 1 \to I, \quad \pm i \to R^2, \quad \pm j \to m_1, \quad \pm k \to m_2,$$

each quadrant of the table for $\mathcal{Q}$ becomes identical to that for $\mathcal{H}_1$, showing that Φ is a homomorphism of $\mathcal{Q}$ onto $\mathcal{H}_1$. As $\mathcal{H}_1$ and $\mathcal{H}_2$ are isomorphic there exists a similar homomorphism onto $\mathcal{H}_2$.

Finally, the kernel of either mapping contains those elements of $\mathcal{Q}$ that map onto I, namely 1 and -1.

28.23 Find (a) *all the proper subgroups and* (b) *all the conjugacy classes of the symmetry group of a regular pentagon.*

A regular pentagon (see figure 28.2) has rotational symmetries and reflection symmetries about lines that join a vertex to the centre-point of the opposite side. Clearly there are five of the latter, m_i ($i = 1, 2, \ldots, 5$). If R represents a rotation by $2\pi/5$, then the rotational symmetries are R, R^2, R^3 and R^4. To these must be added the 'do nothing' identity I. The symmetry group of the regular pentagon

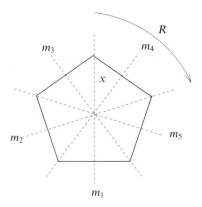

Figure 28.2   The regular pentagon of exercise 28.23.

($\mathcal{C}_{5v}$ in chemical notation) therefore consists of the following ten elements (with their orders):

element	:	I	R	R^2	R^3	R^4	m_1	m_2	m_3	m_4	m_5
order	:	1	5	5	5	5	2	2	2	2	2

(a) As the order of the group is 10, the order of any proper subgroup can only be 2 or 5. As I must be in every subgroup and the order of any element in it must divide the order of the subgroup, it is clear that there is only one subgroup of order 5 and that is $\{I, R, R^2, R^3, R^4\}$. Similarly, there are five subgroups of order 2, namely $\{I, m_i\}$ for m_i $(i = 1, 2, \dots, 5)$.

(b) As always, I is in a class by itself.

We now prove a useful general result about elements in the same conjugacy class: namely, that they have the same order. Let X and Y be in the same class ($Y = g_i X g_i^{-1}$ for some g_i belong to the group) and let X have order m, i.e. $X^m = I$. Then

$$Y^m = g_i X g_i^{-1} \, g_i X g_i^{-1} \cdots g_i X g_i^{-1} = g_i X^m g_i^{-1} = g_i g_i^{-1} = I.$$

This implies that the order of Y divides the order of X. Similarly the order of X divides the order of Y. Therefore X and Y have the same order. Applying this result to the given group, we see that a conjugacy class cannot contain a mixture of rotations and reflections.

We first note the obvious result that $R^p R^q (R^p)^{-1} = R^q$ for any valid p and q. Next, by considering the effects of various combinations of symmetries on a general point x of the pentagon (as marked in the figure), we find that for any i and j, $(i, j = 1, 2 \dots, 5)$,

$$m_i R m_i = R^4 \quad \text{and} \quad m_j R^4 m_j = R.$$

These results, together with that noted above, imply that R and R^4 constitute a class. Similarly R^2 and R^3 make up a (different) class.

Turning to the reflections, we see that the following chain of results, for example, shows that all five reflections must be in the same class (recall that each reflection is its own inverse):

$$m_1 m_2 m_1 = m_5, \quad m_3 m_5 m_3 = m_1, \quad m_2 m_1 m_2 = m_3, \quad m_1 m_3 m_1 = m_4.$$

In summary, there are four conjugacy classes and they are I, (R, R^4), (R^2, R^3) and $(m_1, m_2, m_3, m_4, m_5)$.

29

Representation theory

29.1 *A group G has four elements I, X, Y and Z, which satisfy $X^2 = Y^2 = Z^2 = XYZ = I$. Show that G is Abelian and hence deduce the form of its character table.*

Show that the matrices

$$\mathsf{D}(I) = \begin{pmatrix} 1 & 0 \\ 0 & 1 \end{pmatrix}, \qquad \mathsf{D}(X) = \begin{pmatrix} -1 & 0 \\ 0 & -1 \end{pmatrix},$$

$$\mathsf{D}(Y) = \begin{pmatrix} -1 & -p \\ 0 & 1 \end{pmatrix}, \qquad \mathsf{D}(Z) = \begin{pmatrix} 1 & p \\ 0 & -1 \end{pmatrix},$$

where p is a real number, form a representation D of G. Find its characters and decompose it into irreps.

Since I necessarily commutes with all other elements, we need only consider products such as XY. Now,

$$\begin{aligned} XYZ = I &\quad\Rightarrow\quad X^2 YZ = X &\quad\Rightarrow\quad YZ = X, \\ XYZ = I &\quad\Rightarrow\quad XYZ^2 = Z &\quad\Rightarrow\quad XY = Z \\ &\quad\Rightarrow\quad XY^2 = ZY &\quad\Rightarrow\quad X = ZY. \end{aligned}$$

Thus, $YZ = X = ZY$, showing that Y and Z commute. Similarly, $XY = Z = YX$ and $XZ = Y = ZX$. We conclude that the group is Abelian.

As the group is Abelian, each element is in a class of its own and there are therefore four classes and consequently four irreps $\mathsf{D}^{(\lambda)}$. Since

$$\sum_{\lambda=1}^{4} n_\lambda^2 = g = 4,$$

where n_λ is the dimension of representation λ, the only possibility is that $n_\lambda = 1$ for each λ, i.e the group has four one-dimensional irreps.

As for all sets of irreps, the identity irrep $D^{(1)} = A_1$ must be present and the characters of the others must be orthogonal to the $(1,1,1,1)$ characters of A_1. Further, for each one-dimensional irrep, the identity I must have the character $+1$. The character table must therefore take the form

χ	I	X	Y	Z
A_1	1	1	1	1
$D^{(2)}$	1	1	-1	-1
$D^{(3)}$	1	-1	1	-1
$D^{(4)}$	1	-1	-1	1

For the proposed representation we first need to verify the multiplication properties. Those for $D(I)$ are immediate. For the others:

$$D(X)D(X) = \begin{pmatrix} -1 & 0 \\ 0 & -1 \end{pmatrix} \begin{pmatrix} -1 & 0 \\ 0 & -1 \end{pmatrix} = \begin{pmatrix} 1 & 0 \\ 0 & 1 \end{pmatrix} = D(I),$$

$$D(Y)D(Y) = \begin{pmatrix} -1 & -p \\ 0 & 1 \end{pmatrix} \begin{pmatrix} -1 & -p \\ 0 & 1 \end{pmatrix} = \begin{pmatrix} 1 & 0 \\ 0 & 1 \end{pmatrix} = D(I),$$

$$D(Z)D(Z) = \begin{pmatrix} 1 & p \\ 0 & -1 \end{pmatrix} \begin{pmatrix} 1 & p \\ 0 & -1 \end{pmatrix} = \begin{pmatrix} 1 & 0 \\ 0 & 1 \end{pmatrix} = D(I),$$

$$D(X)D(Y)D(Z) = \begin{pmatrix} -1 & 0 \\ 0 & -1 \end{pmatrix} \begin{pmatrix} -1 & -p \\ 0 & 1 \end{pmatrix} \begin{pmatrix} 1 & p \\ 0 & -1 \end{pmatrix}$$

$$= \begin{pmatrix} -1 & 0 \\ 0 & -1 \end{pmatrix} \begin{pmatrix} -1 & 0 \\ 0 & -1 \end{pmatrix} = D(I),$$

Since the defining relationships for I, X, Y and Z are $X^2 = Y^2 = Z^2 = XYZ = I$, these results show that the matrices form a representation of $\mathcal{G}$, whatever the value of p.

Now $\chi(g_i)$ is equal to the trace of $D(g_i)$ and so the character set for this representation (in the order I, X, Y, Z) is $(2, -2, 0, 0)$. The only rows in the character table that can be added to produce the correct totals for all four elements are the third and fourth. This shows that

$$D = D^{(3)} \oplus D^{(4)}.$$

29.3 *The quaternion group Q (see exercise 28.21) has eight elements $\{\pm 1, \pm i, \pm j, \pm k\}$ obeying the relations*

$$i^2 = j^2 = k^2 = -1, \quad ij = k = -ji.$$

Determine the conjugacy classes of Q and deduce the dimensions of its irreps. Show that Q is homomorphic to the four-element group V, which is generated by two distinct elements a and b with $a^2 = b^2 = (ab)^2 = I$. Find the one-dimensional irreps of V and use these to help determine the full character table for Q.

As always, the identity, $+1$, is in a class by itself and, since it commutes with every other element in the group, so is -1.

Now consider all products of the form $X^{-1}iX$:

$$\begin{array}{ll} 1\,i\,1 = i, & (-1)\,i\,(-1) = i, \\ (-i)\,i\,i = i, & i\,i\,(-i) = i, \\ (-j)\,i\,j = (-j)\,k = -i, & j\,i\,(-j) = (-k)(-j) = -i, \\ (-k)\,i\,k = (-k)(-j) = -i, & k\,i\,(-k) = j\,(-k) = -i. \end{array}$$

These show that i and $-i$ are in the same class. Similarly $\{j, -j\}$ and $\{k, -k\}$ are two other classes. This exhausts the group.

In summary there are five classes, the elements in any one class all having the same order. They are

$$\begin{array}{cccccc} \text{class}: & \{1\} & \{-1\} & \{\pm i\} & \{\pm j\} & \{\pm k\} \\ \text{order}: & 1 & 2 & 4 & 4 & 4 \end{array}$$

It follows that there are five irreps and, since $\sum_{\lambda=1}^{5} n_\lambda^2 = 8$, they can only be one two-dimensional and four one-dimensional irreps.

Turning to the group V,

$$(ab)^2 = I \quad \Rightarrow \quad abab^2 = b \quad \Rightarrow \quad aba = b \quad \Rightarrow \quad aba^2 = ba$$
$$\Rightarrow \quad ab = ba,$$

i.e. a and b commute. Also, it follows that

$$a(ab) = a(ba) = (ab)a \quad \text{and} \quad b(ab) = (ba)b = (ab)b.$$

Thus, all four elements commute, the group V is Abelian, and each of its elements is in a class of its own. As in exercise 29.1, an Abelian group of order 4 must have four irreps and the character table

χ_v	I	a	b	ab
A_1	1	1	1	1
$D^{(2)}$	1	1	-1	-1
$D^{(3)}$	1	-1	1	-1
$D^{(4)}$	1	-1	-1	1

The multiplication table for the quaternion group is, as given in exercise 21 of chapter 28 of the form

Q	1	-1	i	$-i$	j	$-j$	k	$-k$
1	1	-1	i	$-i$	j	$-j$	k	$-k$
-1	-1	1	$-i$	i	$-j$	j	$-k$	k
i	i	$-i$	-1	1	k	$-k$	$-j$	j
$-i$	$-i$	i	1	-1	$-k$	k	j	$-j$
j	j	$-j$	$-k$	k	-1	1	i	$-i$
$-j$	$-j$	j	k	$-k$	1	-1	$-i$	i
k	k	$-k$	j	$-j$	$-i$	i	-1	1
$-k$	$-k$	k	$-j$	j	i	$-i$	1	-1

If each entry ± 1 is replaced by I, each entry $\pm i$ by a, each entry $\pm j$ by b and each entry $\pm k$ by ab, then this table reduces to four copies of the table

$\mathcal{V}$	I	a	b	ab
I	I	a	b	ab
a	a	I	ab	b
b	b	ab	I	a
ab	ab	b	a	I

which is the group multiplication table for group $\mathcal{V}$. The same conclusion can be reached by replacing each 2×2 block containing only ± 1 by I, each 2×2 block containing only $\pm i$ by a, etc.; this results in a single copy. Both approaches lead to the conclusion that there is a two-to-one homomorphism from Q onto $\mathcal{V}$.

Since the homomorphism maps all elements of any one conjugacy class of Q onto the same element of $\mathcal{V}$, the one-dimensional irreps of Q must be the same as those of $\mathcal{V}$. Further, since both of the classes $\{1\}$ and $\{-1\}$ map onto I in $\mathcal{V}$ they will have common characters in each one-dimensional irrep (1 in every case, as it happens). As shown earlier, there will also be a two-dimensional irrep. Its character for I must be 2 (the dimension of the irrep); the other characters can be determined from the requirement of orthogonality to the characters of the other (one-dimensional) irreps.

483

The character table for Q therefore has the form

χ_Q	1	-1	$\pm i$	$\pm j$	$\pm k$
A_1	1	1	1	1	1
$D^{(2)}$	1	1	1	-1	-1
$D^{(3)}$	1	1	-1	1	-1
$D^{(4)}$	1	1	-1	-1	1
$D^{(5)}$	2	w	x	y	z

We require that

$$1(1)(2) + 1(1)(w) + 2(1)(x) + 2(1)y + 2(1)z = 0,$$
$$1(1)(2) + 1(1)(w) + 2(1)(x) + 2(-1)y + 2(-1)z = 0,$$
$$1(1)(2) + 1(1)(w) + 2(-1)(x) + 2(1)y + 2(-1)z = 0,$$
$$1(1)(2) + 1(1)(w) + 2(-1)(x) + 2(-1)y + 2(1)z = 0.$$

These equations have the solution $w = -2$, $x = y = z = 0$, thus completing the full character table for Q.

29.5 *The group of pure rotations (excluding reflections and inversions) that take a cube into itself has 24 elements. The group is isomorphic to the permutation group S_4 and hence has the same character table, once corresponding classes have been established. By counting the number of elements in each class, make the correspondences below (the final two cannot be decided purely by counting, and should be taken as given).*

Permutation class type	*Symbol (physics)*	*Action*
(1)	I	*none*
(123)	3	*rotations about a body diagonal*
(12)(34)	2_z	*rotation of π about the normal to a face*
(1234)	4_z	*rotations of $\pm\pi/2$ about the normal to a face*
(12)	2_d	*rotation of π about an axis through the centres of opposite edges*

Given in table 29.1 is the character table for S_4. Reformulate it in terms of the elements of the rotation symmetry group (432 or O) of a cube and use it when answering exercise 29.7.

The rotational symmetries of the cube are as follows.

(i) 'Do nothing'. There is only one such operation; it is the identity and so corresponds to $(1) \equiv (1)(2)(3)(4)$.

Irrep	Typical element and class size				
	(1)	(12)	(123)	(1234)	(12)(34)
	1	6	8	6	3
A_1	1	1	1	1	1
A_2	1	−1	1	−1	1
E	2	0	−1	0	2
T_1	3	1	0	−1	−1
T_2	3	−1	0	1	−1

Table 29.1   The character table for the permutation group S_4.

(ii) Rotations about a body diagonal. There are four body diagonals and rotations of $2\pi/3$ and $4\pi/3$ are possible about each. Thus there are eight elements and this must correspond to $(123) \equiv (123)(4)$.

(iii) Rotations by π about a face normal. Although there are six faces to the cube, they define only three distinct face normals and hence there are three elements in this set. They therefore correspond to $(12)(34)$.

(iv) Rotations of $\pi/2$ and $3\pi/2$ about a face normal. With three distinct face normals and two possible rotation angles for each, the set contains six elements. These could correspond to $(12) \equiv (12)(3)(4)$ or to (1234).

(v) Rotations of π about axes that join the centres of opposite edges. There are six such axes and hence six elements. As in (iv), these could correspond to $(12) \equiv (12)(3)(4)$ or to (1234).

Taking the identification given in the question for (iv) and (v), the reformulated table (in which only the headings have changed) is given in table 29.2.

Irrep	Typical element and class size				
	I	2_d	3	4_z	2_z
	1	6	8	6	3
A_1	1	1	1	1	1
A_2	1	−1	1	−1	1
E	2	0	−1	0	2
T_1	3	1	0	−1	−1
T_2	3	−1	0	1	−1

Table 29.2   The character table for the symmetry group 432 or O.

We note that the rotational symmetries of a cube can, alternatively, be characterised by the effects they have on the orientations in space of its four body diagonals. For example, a rotation of π about a face normal interchanges them in pairs – represented in cycle notation by the form $(12)(34)$. Using this formulation of the symmetry group, the assignments for (iv) and (v) are unambiguous.

29.7 *In a certain crystalline compound, a thorium atom lies at the centre of a regular octahedron of six sulphur atoms at positions* $(\pm a, 0, 0)$, $(0, \pm a, 0)$, $(0, 0, \pm a)$. *These can be considered as being positioned at the centres of the faces of a cube of side* $2a$. *The sulphur atoms produce at the site of the thorium atom an electric field that has the same symmetry group as a cube* (432 or O).

The five degenerate d-electron orbitals of the thorium atom can be expressed, relative to any arbitrary polar axis, as

$$(3\cos^2\theta - 1)f(r), \qquad e^{\pm i\phi}\sin\theta\cos\theta f(r), \qquad e^{\pm 2i\phi}\sin^2\theta f(r).$$

A rotation about that polar axis through an angle ϕ' *in effect changes* ϕ *to* $\phi - \phi'$. *Use this to show that the character of the rotation in a representation based on the orbital wavefunctions is given by*

$$1 + 2\cos\phi' + 2\cos 2\phi'$$

and hence that the characters of the representation, in the order of the symbols given in exercise 29.5, is 5, -1, 1, -1, 1. *Deduce that the five-fold degenerate level is split into two levels, a doublet and a triplet.*

The electric field at the thorium atom has the symmetries of group 432 and the d-electron orbitals are

$$\psi_1 = (3\cos^2\theta - 1)f(r),$$
$$\psi_{2,3} = e^{\pm i\phi}\sin\theta\cos\theta f(r),$$
$$\psi_{4,5} = e^{\pm 2i\phi}\sin^2\theta f(r).$$

Taking the ψ_i ($i = 1, 2, \ldots, 5$) as a basis, the representation of a rotation by ϕ' is a 5×5 matrix whose diagonal elements are equal to the factor by which each basis function is multiplied when that function is subjected to the rotation.

ψ_1 does not depend upon ϕ and so is unaltered; its entry is 1.

$\psi_{2,3}$ become $\sin\theta\cos\theta f(r) e^{\pm i\phi \mp i\phi'}$; their entries are $e^{-i\phi'}$ and $e^{i\phi'}$.

$\psi_{4,5}$ become $\sin^2\theta f(r) e^{\pm 2i\phi \mp 2i\phi'}$; their entries are $e^{-2i\phi'}$ and $e^{i2\phi'}$.

The trace of the representative matrix, and therefore the character of the rotation, is thus

$$\chi = 1 + e^{-i\phi'} + e^{i\phi'} + e^{-2i\phi'} + e^{2i\phi'} = 1 + 2\cos\phi' + 2\cos 2\phi'.$$

For the symmetry elements in the group 432, the corresponding rotation angles and characters are as follows:

Symmetry	ϕ'	χ
I	0	$1 + 2 + 2 = 5$
3	$\pm 2\pi/3$	$1 + 2(-\frac{1}{2}) + 2(-\frac{1}{2}) = -1$
2_z	π	$1 + 2(-1) + 2(1) = 1$
4_z	$\pm \pi/2$	$1 + 2(0) + 2(-1) = -1$
2_d	π	$1 + 2(-1) + 2(1) = 1$

Rewriting these results in a form similar to that in which the character table of 432 has been previously presented, we have

$$
\begin{array}{cccccc}
\text{Symmetry} & I & 2_d & 3 & 4_z & 2_z \\
\text{Character, } \chi & 5 & 1 & -1 & -1 & 1
\end{array}
$$

We now compare this with table 29.2, compiled in exercise 29.5, and see, or calculate using the equation

$$
m_\mu = \frac{1}{g} \sum_X \left[\chi^{(\mu)}(X) \right]^* \chi(X) = \frac{1}{g} \sum_i c_i \left[\chi^{(\mu)}(X_i) \right]^* \chi(X_i),
$$

that this character set is the direct sum of those for the two dimensional irrep E and the three-dimensional irrep T_1, as given in that table:

	I	2_d	3	4_z	2_z
E	2	0	−1	0	2
T_1	3	1	0	−1	−1
χ	5	1	−1	−1	1

The n (mixed) orbitals transforming according to any particular n-dimensional irrep will all have the same energy, but, barring accidental coincidences, it will be a different energy from that corresponding to a different irrep. Thus the five-fold degenerate level is split into a doublet (E) and a triplet (T_1).

487

29.9 *The hydrogen atoms in a methane molecule CH_4 form a perfect tetrahedron with the carbon atom at its centre. The molecule is most conveniently described mathematically by placing the hydrogen atoms at the points $(1, 1, 1)$, $(1, -1, -1)$, $(-1, 1, -1)$ and $(-1, -1, 1)$. The symmetry group to which it belongs, the tetrahedral group $(\bar{4}3m$ or $T_d)$, has classes typified by I, 3, 2_z, m_d and $\bar{4}_z$, where the first three are as in exercise 29.5, m_d is a reflection in the mirror plane $x - y = 0$ and $\bar{4}_z$ is a rotation of $\pi/2$ about the z-axis followed by an inversion in the origin. A reflection in a mirror plane can be considered as a rotation of π about an axis perpendicular to the plane, followed by an inversion in the origin.*

The character table for the group $\bar{4}3m$ is very similar to that for the group 432, and has the form shown in table 29.3. By following the steps given below, determine how many different internal vibration frequencies the CH_4 molecule has.

(a) *Consider a representation based on the twelve coordinates x_i, y_i, z_i for $i = 1, 2, 3, 4$. For those hydrogen atoms that transform into themselves, a rotation through an angle θ about an axis parallel to one of the coordinate axes gives rise in this natural representation to the diagonal elements 1 for the corresponding coordinate and $2 \cos \theta$ for the two orthogonal coordinates. If the rotation is followed by an inversion then these entries are multiplied by -1. Atoms not transforming into themselves give a zero diagonal contribution. Show that the characters of the natural representation are 12, 0, 0, 0, 2 and hence that its expression in terms of irreps is*

$$A_1 \oplus E \oplus T_1 \oplus 2T_2.$$

(b) *The irreps of the bodily translational and rotational motions are included in this expression and need to be identified and removed. Show that when this is done it can be concluded that there are three different internal vibration frequencies in the CH_4 molecule. State their degeneracies and check that they are consistent with the expected number of normal coordinates needed to describe the internal motions of the molecule.*

(a) We consider each type of rotation in turn and determine how many of the hydrogen atoms are transformed into themselves, i.e. do not change position.

Under I all twelve atoms retain their original positions and so $\chi(I) = 12$.

For the symmetry 3 the rotation angle is $\pm 2\pi/3$ and for each such rotation one atom retains its original place. However, it contributes $1 + 2 \cos(2\pi/3) = 1 + 2(-\frac{1}{2}) = 0$ and so $\chi(3) = 0$.

For the symmetries 2_z and $\bar{4}_z$ no atoms retain their original places and the corresponding characters are both 0.

Irreps	Typical element and class size					Functions transforming according to irrep
	I	3	2_z	$\bar{4}_z$	m_d	
	1	8	3	6	6	
A_1	1	1	1	1	1	$x^2 + y^2 + z^2$
A_2	1	1	1	-1	-1	
E	2	-1	2	0	0	$(x^2 - y^2, 3z^2 - r^2)$
T_1	3	0	-1	1	-1	(R_x, R_y, R_z)
T_2	3	0	-1	-1	1	$(x, y, z); (xy, yz, zx)$

Table 29.3 The character table for the symmetry group $\bar{4}3m$.

Finally, for m_d, a reflection in one of the six mirror planes, there are two atoms that lie in any one of the planes (with the other two atoms placed symmetrically, one on either side of it). Thus two atoms are unchanged. As explained in the question, a reflection in a mirror plane can be considered as a rotation of π about an axis perpendicular to the plane, followed by an inversion in the origin; the latter gives rise to an additional factor of -1. As a result, each of the two atoms contributes $(-1)(1 + 2\cos\pi) = 1$ to the character of m_d. Thus $\chi(m_d) = 2$ and the full character set of the natural representation is $(12, 0, 0, 0, 2)$.

It then follows that

$$m_{A_1} = \frac{1(1)(12) + 8(1)(0) + 3(1)(0) + 6(1)(0) + 6(1)(2)}{24} = 1,$$

$$m_{A_2} = \frac{1(1)(12) + 8(1)(0) + 3(1)(0) + 6(-1)(0) + 6(-1)(2)}{24} = 0,$$

$$m_E = \frac{1(2)(12) + 8(-1)(0) + 3(2)(0) + 6(0)(0) + 6(0)(2)}{24} = 1,$$

$$m_{T_1} = \frac{1(3)(12) + 8(0)(0) + 3(-1)(0) + 6(1)(0) + 6(-1)(2)}{24} = 1,$$

$$m_{T_2} = \frac{1(3)(12) + 8(0)(0) + 3(-1)(0) + 6(-1)(0) + 6(1)(2)}{24} = 2.$$

Thus the irreps present in this representation are

$$A_1 \oplus E \oplus T_1 \oplus 2T_2.$$

(b) Bodily translation of the centre of mass of the molecule is included in this representation since the representation allows all coordinates to vary independently. From table 29.3, the set (x, y, z) transforms as T_2 and so this motion corresponds to one of the two T_2 irreps found above.

Equally a rigid-body rotation of the molecule *about* its centre of mass is included; from the table (R_x, R_y, R_z) transform as T_1 and so this rotation, which contains no internal vibrations, accounts for the T_1 irrep.

After these two irreps are removed we are left with the irreps of the internal vibrations, which are A_1, E, and T_2. They are, respectively, one-, two- and three-dimensional irreps and therefore the corresponding vibration frequencies have degeneracies of 1, 2 and 3. This gives a total of six internal coordinates, in accordance with the twelve original ones, less the three translational coordinates of the centre of mass and the three coordinates needed to specify the direction of the axis of a rigid-body rotation.

29.11 *Use the results of exercise 28.23 to find the character table for the dihedral group* $\mathcal{D}_5$, *the symmetry group of a regular pentagon.*

As shown in exercise 28.23, the group $\mathcal{D}_5$ has ten elements and four classes: $\{I\}$, $\{R, R^4\}$, $\{R^2, R^3\}$ and $\{m_1, m_2, m_3, m_4, m_5\}$. Here R is a rotation through $2\pi/5$.

Since there are ten elements and four classes, and hence four irreps, we must have that the dimensionalities of the irreps satisfy

$$n_1^2 + n_2^2 + n_3^2 + n_4^2 = 10.$$

This has only one (non-zero) integral solution, $n_1 = n_2 = 1$ and $n_3 = n_4 = 2$. The identity irrep, A_1, must be one of the one-dimensional irreps, and the character table must have the form

Irrep	I	R, R^4	R^2, R^3	m_i ($i = 1, 5$)
A_1	1	1	1	1
A_2	1	a	b	c
E_1	2	d	e	f
E_2	2	g	h	j

For A_2 we must have both $1 + 2|a|^2 + 2|b|^2 + 5|c|^2 = 10$ (summation rule) and $1 + 2a + 2b + 5c = 0$ (orthogonality with A_1). Since the m_i have order 2 and A_2 is one-dimensional, c can only be a second root of unity, i.e. either of 1 or -1. The only solution to these simultaneous equations, even allowing a and b to be complex (but restricted to each being a fifth root of unity), is $a = b = 1$ and $c = -1$.

For E_1 (and similarly for E_2),

$$4 + 2|d|^2 + 2|e|^2 + 5|f|^2 = 10.$$

Arguing as previously, we conclude that, because E is two-dimensional, f can only be the sum of two values which are either $+1$ or -1. Hence, only 0 and ± 2 are possible, and the values ± 2 are impossible in this case. This conclusion can be confirmed by noting that the E_1 character set has to be orthogonal to those for both A_1 and A_2. So both

$$1(1)2 + 2(1)d + 2(1)e + 5(1)f = 0$$

and
$$1(1)2 + 2(1)d + 2(1)e + 5(-1)f = 0,$$

implying that $f = 0$.

We are left with

$$|d|^2 + |e|^2 = 3 \text{ and } 1 + d + e = 0.$$

This clearly has no integer solutions, but we attempt to find real solutions before considering complex ones. If d is real then e must also be real. Substituting $e = -1 - d$ into the first equation gives the quadratic equation

$$d^2 + (-1-d)^2 = 3 \quad \Rightarrow \quad d^2 + d - 1 = 0 \quad \Rightarrow \quad d = \frac{-1 \pm \sqrt{5}}{2}.$$

If d is taken as $\frac{1}{2}(-1 + \sqrt{5})$ (the golden mean!) then $e = \frac{1}{2}(-1 - \sqrt{5})$, the other root of the quadratic. This completes the character set for E_1. That for E_2 is obtained by setting $j = 0$ and assigning $\frac{1}{2}(-1 + \sqrt{5})$ to h and $\frac{1}{2}(-1 - \sqrt{5})$ to g; this can be confirmed by checking the orthogonality relation

$$1(2)(2) + 2[\tfrac{1}{2}(-1 + \sqrt{5})\tfrac{1}{2}(-1 - \sqrt{5})]$$
$$+ 2[\tfrac{1}{2}(-1 - \sqrt{5})\tfrac{1}{2}(-1 + \sqrt{5})] + 5(0)(0) = 4 - 2 - 2 + 0 = 0.$$

We also note that, for example,

$$\frac{-1 + \sqrt{5}}{2} = \exp\left(\frac{2\pi i}{5}\right) + \exp\left(\frac{4 \times 2\pi i}{5}\right) = 2\cos\frac{2\pi}{5} = 0.6180,$$

i.e. d and h are each the sum of two fifth roots of unity. The same applies to e and g. The final character table reads

Irrep	I	R, R^4	R^2, R^3	m_i $(i = 1, 5)$
A_1	1	1	1	1
A_2	1	1	1	-1
E_1	2	$\frac{1}{2}(-1 + \sqrt{5})$	$-\frac{1}{2}(1 + \sqrt{5})$	0
E_2	2	$-\frac{1}{2}(1 + \sqrt{5})$	$\frac{1}{2}(-1 + \sqrt{5})$	0

29.13 *Further investigation of the crystalline compound considered in exercise 29.7 shows that the octahedron is not quite perfect but is elongated along the* $(1, 1, 1)$ *direction with the sulphur atoms at positions* $\pm(a+\delta, \delta, \delta)$, $\pm(\delta, a+\delta, \delta)$, $\pm(\delta, \delta, a+\delta)$, *where* $\delta \ll a$. *This structure is invariant under the (crystallographic) symmetry group 32 with three two-fold axes along directions typified by* $(1, -1, 0)$. *The latter axes, which are perpendicular to the* $(1, 1, 1)$ *direction, are axes of two-fold symmetry for the perfect octahedron. The group 32 is really the three-dimensional version of the group 3m and has the same character table. That for 3m is*

3m	I	A, B	C, D, E
A_1	1	1	1
A_2	1	1	-1
E	2	-1	0

Use this to show that, when the distortion of the octahedron is included, the doublet found in exercise 29.7 is unsplit but the triplet breaks up into a singlet and a doublet.

The perfect octahedron is invariant under the operations of group 432, whose character table is as follows:

Irrep	Typical element and class size				
	I	2_d	3	4_z	2_z
	1	6	8	6	3
A_1	1	1	1	1	1
A_2	1	-1	1	-1	1
E	2	0	-1	0	2
T_1	3	1	0	-1	-1
T_2	3	-1	0	1	-1

The distorted octahedron is invariant only under the operations of the smaller group 32, whose character table is

3m	I	R, R^2	m_i
A_1	1	1	1
A_2	1	1	-1
E	2	-1	0

Here R is a rotation through $2\pi/3$ and its class corresponds to the class denoted by '3' in group 432. The reflection symmetries correspond to rotations by π when considered as operations in three dimensions (as opposed to in a plane); thus they correspond to the class 2_d. We are thus concerned with the first three classes in the 432 table, but with the second and third interchanged as compared with the 32 table.

Using the order in the 32 table, E has the characters $(2, -1, 0)$. This two-dimensional irrep also appears in the 432 table, and so the corresponding doublet level in the thorium atom is not split as a result of the distortion of the sulphur octahedron. However, the triplet level, whose components transform as T_1, will be affected by the distortion. The irrep T_1 does not appear in the 32 table but has to be made up from E and A_1; in terms of character sets

$$(3, 0, 1) = (2, -1, 0) + (1, 1, 1).$$

In physical terms, the triplet state in thorium is split by the distorted electric field due to the sulphur atoms into a doublet and a singlet.

Probability

30.1 *By shading or numbering Venn diagrams, determine which of the following are valid relationships between events. For those that are, prove the relationship using de Morgan's laws.*

(a) $\overline{(\bar{X} \cup Y)} = X \cap \bar{Y}$.

(b) $\bar{X} \cup \bar{Y} = \overline{(X \cup Y)}$.

(c) $(X \cup Y) \cap Z = (X \cup Z) \cap Y$.

(d) $X \cup \overline{(Y \cap Z)} = (X \cup \bar{Y}) \cap \bar{Z}$.

(e) $X \cup \overline{(Y \cap Z)} = (X \cup \bar{Y}) \cup \bar{Z}$.

For each part of this question we refer to the corresponding part of figure 30.1.

(a) This relationship is correct as both expressions define the shaded region that is both inside X and outside Y.

(b) This relationship is *not* valid. The LHS specifies the whole sample space *apart from* the region marked with the heavy shading. The RHS defines the region that is lightly shaded. The unmarked regions of X and Y are included in the former but not in the latter.

(c) This relationship is *not* valid. The LHS specifies the sum of the regions marked 2, 3 and 4 in the figure, whilst the RHS defines the sum of the regions marked 1, 3 and 4.

(d) This relationship is *not* valid. On the LHS, $\overline{Y \cap Z}$ is the whole sample space apart from regions 3 and 4. So $X \cup \overline{(Y \cap Z)}$ consists of all regions except for region 3. On the RHS, $X \cup \bar{Y}$ contains all regions except 3 and 7. The events $\bar{Z}$ contain regions 1, 6, 7 and 8 and so $(X \cup \bar{Y}) \cap \bar{Z}$ consists of regions 1, 6 and 8. Thus regions 2, 4, 5 and 7 are in one specification but not in the other.

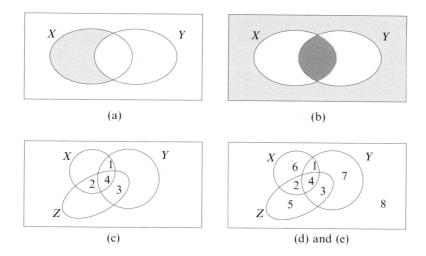

Figure 30.1   The Venn diagrams used in exercise 30.1.

(e) This relationship is valid. The LHS is as found in (d), namely all regions except for region 3. The RHS consists of the union (as opposed to the intersection) of the two subregions found in (d) and thus contains those regions found in either or both of $X \cup \bar{Y}$ (1, 2, 4, 5, 6 and 8) and $\bar{Z}$ (1, 6, 7 and 8). This covers all regions except region 3 – in agreement with those found for the LHS.

For the two valid relationships, their proofs using de Morgan's laws are:

(a) $\quad \overline{(\bar{X} \cup Y)} = \bar{\bar{X}} \cap \bar{Y} = X \cap \bar{Y},$

(e) $\quad X \cup \overline{(Y \cap Z)} = X \cup (\bar{Y} \cup \bar{Z}) = (X \cup \bar{Y}) \cup \bar{Z}.$

30.3 *A and B each have two unbiased four-faced dice, the four faces being numbered 1, 2, 3 and 4. Without looking, B tries to guess the sum x of the numbers on the bottom faces of A's two dice after they have been thrown onto a table. If the guess is correct B receives x^2 euros, but if not he loses x euros.*

Determine B's expected gain per throw of A's dice when he adopts each of the following strategies:

(a) *he selects x at random in the range $2 \le x \le 8$;*

(b) *he throws his own two dice and guesses x to be whatever they indicate;*

(c) *he takes your advice and always chooses the same value for x. Which number would you advise?*

We first calculate the probabilities $p(x)$ and the corresponding gains $g(x) = p(x)x^2 - [1 - p(x)]x$ for each value of the total x. Expressing both in units of $1/16$, they are as follows:

x	2	3	4	5	6	7	8
$p(x)$	1	2	3	4	3	2	1
$g(x)$	-26	-24	-4	40	30	0	-56

(a) If B's guess is random in the range $2 \leq x \leq 8$ then his expected return is

$$\frac{1}{16}\frac{1}{7}(-26 - 24 - 4 + 40 + 30 + 0 - 56) = -\frac{40}{112} = -0.36 \text{ euros.}$$

(b) If he picks by throwing his own dice then his distribution of guesses is the same as that of $p(x)$ and his expected return is

$$\frac{1}{16}\frac{1}{16}[1(-26) + 2(-24) + 3(-4) + 4(40) + 3(30) + 2(0) + 1(-56)]$$

$$= \frac{108}{256} = 0.42 \text{ euros.}$$

(c) As is clear from the tabulation, the best return of $40/16 = 2.5$ euros is expected if B always chooses '5' as his guess. Of course, you should not advise him but offer to take his place!

30.5 *Two duellists, A and B, take alternate shots at each other, and the duel is over when a shot (fatal or otherwise!) hits its target. Each shot fired by A has a probability α of hitting B, and each shot fired by B has a probability β of hitting A. Calculate the probabilities P_1 and P_2, defined as follows, that A will win such a duel: P_1, A fires the first shot; P_2, B fires the first shot.*

If they agree to fire simultaneously, rather than alternately, what is the probability P_3 that A will win, i.e. hit B without being hit himself?

Each shot has only two possible outcomes, a hit or a miss. P_1 is the probability that A will win when it is his turn to fire the next shot, and he is still able to do so (event W). There are three possible outcomes of the first two shots: C_1, A hits with his shot; C_2, A misses but B hits; C_3, both miss. Thus

$$P_1 = \sum_i \Pr(C_i)\Pr(W|C_i)$$

$$= [\alpha \times 1] + [(1 - \alpha)\beta \times 0] + [(1 - \alpha)(1 - \beta) \times P_1]$$

$$\Rightarrow \quad P_1 = \frac{\alpha}{\alpha + \beta - \alpha\beta}.$$

When B fires first but misses, the situation is the one just considered. But if B

hits with his first shot then clearly A's chances of winning are zero. Since these are the only two possible outcomes of B's first shot, we can write

$$P_2 = [\beta \times 0] + [(1 - \beta) \times P_1] \quad \Rightarrow \quad P_2 = \frac{(1 - \beta)\alpha}{\alpha + \beta - \alpha\beta}.$$

When both fire at the same time there are four possible outcomes D_i to the first round: D_1, A hits and B misses; D_2, B hits but A misses; D_3, they both hit; D_4, they both miss. If getting hit, even if you manage to hit your opponent, does not count as a win, then

$$P_3 = \sum_i \Pr(D_i) \Pr(W|D_i)$$

$$= [\alpha(1 - \beta) \times 1] + [(1 - \alpha)\beta \times 0] + [\alpha\beta \times 0] + [(1 - \alpha)(1 - \beta) \times P_3].$$

This can be rearranged as

$$P_3 = \frac{\alpha(1 - \beta)}{\alpha + \beta - \alpha\beta} = P_2.$$

Thus the result is the same as if B had fired first. However, we also note that if all that matters to A is that B is hit, whether or not he is hit himself, then the third bracket takes the value $\alpha\beta \times 1$ and P_3 takes the same value as P_1.

30.7 *A tennis tournament is arranged on a straight knockout basis for 2^n players, and for each round, except the final, opponents for those still in the competition are drawn at random. The quality of the field is so even that in any match it is equally likely that either player will win. Two of the players have surnames that begin with 'Q'. Find the probabilities that they play each other*

(a) *in the final,*
(b) *at some stage in the tournament.*

Let p_r be the probability that *before* the rth round the two players are both still in the tournament (and, by implication, have not met each other). Clearly, $p_1 = 1$.

Before the rth round there are 2^{n+1-r} players left in. For both 'Q' players to still be in before the $(r + 1)$th round, Q_1 must avoid Q_2 in the draw and both must win their matches. Thus

$$p_{r+1} = \frac{2^{n+1-r} - 2}{2^{n+1-r} - 1} \left(\frac{1}{2}\right)^2 p_r.$$

(a) The probability that they meet in the final is p_n, given by

$$p_n = 1 \frac{2^n - 2}{2^n - 1} \frac{1}{4} \frac{2^{n-1} - 2}{2^{n-1} - 1} \frac{1}{4} \cdots \frac{2^2 - 2}{2^2 - 1} \frac{1}{4}$$

$$= \left(\frac{1}{4}\right)^{n-1} 2^{n-1} \left[\frac{(2^{n-1} - 1)(2^{n-2} - 1) \cdots (2^1 - 1)}{(2^n - 1)(2^{n-1} - 1) \cdots (2^2 - 1)}\right]$$

$$= \left(\frac{1}{4}\right)^{n-1} 2^{n-1} \frac{1}{2^n - 1}$$

$$= \frac{1}{2^{n-1}(2^n - 1)}.$$

(b) The more general solution to the recurrence relation derived above is

$$p_r = 1 \frac{2^n - 2}{2^n - 1} \frac{1}{4} \frac{2^{n-1} - 2}{2^{n-1} - 1} \frac{1}{4} \cdots \frac{2^{n+2-r} - 2}{2^{n+2-r} - 1} \frac{1}{4}$$

$$= \left(\frac{1}{4}\right)^{r-1} 2^{r-1} \left[\frac{(2^{n-1} - 1)(2^{n-2} - 1) \cdots (2^{n+1-r} - 1)}{(2^n - 1)(2^{n-1} - 1) \cdots (2^{n+2-r} - 1)}\right]$$

$$= \left(\frac{1}{2}\right)^{r-1} \frac{2^{n+1-r} - 1}{2^n - 1}.$$

Before the rth round, if they are both still in the tournament, the probability that they will be drawn against each other is $(2^{n-r+1} - 1)^{-1}$. Consequently, the chance that they will meet at *some* stage is

$$\sum_{r=1}^{n} p_r \frac{1}{2^{n-r+1} - 1} = \sum_{r=1}^{n} \left(\frac{1}{2}\right)^{r-1} \frac{2^{n+1-r} - 1}{2^n - 1} \frac{1}{2^{n-r+1} - 1}$$

$$= \frac{1}{2^n - 1} \sum_{r=1}^{n} \left(\frac{1}{2}\right)^{r-1}$$

$$= \frac{1}{2^n - 1} \frac{1 - (\frac{1}{2})^n}{1 - \frac{1}{2}} = \frac{1}{2^{n-1}}.$$

This same conclusion can also be reached in the folowing way. The probability that Q_1 is *not* put out of (i.e. wins) the tournament is $(\frac{1}{2})^n$. It follows that the probability that Q_1 is put out is $1 - (\frac{1}{2})^n$ and that the player responsible is Q_2 with probability $[1 - (\frac{1}{2})^n]/(2^n - 1) = 2^{-n}$. Similarly, the probability that Q_2 is put out and that the player responsible is Q_1 is also 2^{-n}. These are exclusive events but cover all cases in which Q_1 and Q_2 meet during the tournament, the probability of which is therefore $2 \times 2^{-n} = 2^{n-1}$.

30.9 *An electronics assembly firm buys its microchips from three different suppliers; half of them are bought from firm X, whilst firms Y and Z supply 30% and 20%, respectively. The suppliers use different quality-control procedures and the percentages of defective chips are 2%, 4% and 4% for X, Y and Z, respectively. The probabilities that a defective chip will fail two or more assembly-line tests are 40%, 60% and 80%, respectively, whilst all defective chips have a 10% chance of escaping detection. An assembler finds a chip that fails only one test. What is the probability that it came from supplier X?*

Since the number of tests failed by a defective chip are mutually exclusive outcomes (0, 1 or ≥ 2), a chip supplied by X has a probability of failing just one test given by $0.02(1 - 0.1 - 0.4) = 0.010$. The corresponding probabilities for chips supplied by Y and Z are $0.04(1 - 0.1 - 0.6) = 0.012$ and $0.04(1 - 0.1 - 0.8) = 0.004$, respectively.

Using '1' to denote failing a single test, Bayes' theorem gives the probability that the chip was supplied by X as

$$\Pr(X|1) = \frac{\Pr(1|X)\,\Pr(X)}{\Pr(1|X)\,\Pr(X) + \Pr(1|Y)\,\Pr(Y) + \Pr(1|Z)\,\Pr(Z)}$$

$$= \frac{0.010 \times 0.5}{0.010 \times 0.5 + 0.012 \times 0.3 + 0.004 \times 0.2} = \frac{50}{94}.$$

30.11 *A boy is selected at random from amongst the children belonging to families with n children. It is known that he has at least two sisters. Show that the probability that he has $k - 1$ brothers is*

$$\frac{(n-1)!}{(2^{n-1} - n)(k-1)!(n-k)!},$$

for $1 \leq k \leq n - 2$ and zero for other values of k. Assume that boys and girls are equally likely.

The boy has $n - 1$ siblings. Let A_j be the event that $j - 1$ of them are brothers, i.e. his family contains j boys and $n - j$ girls. The probability of event A_j is

$$\Pr(A_j) = \frac{{}^{n-1}C_{j-1}\left(\frac{1}{2}\right)^{n-1}}{\sum_{j=1}^{n} {}^{n-1}C_{j-1}\left(\frac{1}{2}\right)^{n-1}} = \frac{(n-1)!}{2^{n-1}(j-1)!(n-j)!}.$$

If B is the event that the boy has at least two sisters, then

$$\Pr(B|A_j) = \begin{cases} 1 & 1 \leq j \leq n - 2, \\ 0 & n - 1 \leq j \leq n. \end{cases}$$

Now we apply Bayes' theorem to give the probability that he has $k - 1$ brothers:

$$\Pr(A_k|B) = \frac{1 \, \Pr(A_k)}{\sum_{j=1}^{n-2} 1 \, \Pr(A_j)},$$

for $1 \le k \le n-2$. The denominator of this expression is the sum $1 = (\frac{1}{2} + \frac{1}{2})^{n-1} = \sum_{j=1}^{n} {}^{n-1}C_{j-1} (\frac{1}{2})^{n-1}$, but omitting the $j = n - 1$ and the $j = n$ terms, and so is equal to

$$1 - \frac{(n-1)!}{2^{n-1}(n-2)!\,1!} - \frac{(n-1)!}{2^{n-1}(n-1)!\,0!} = \frac{1}{2^{n-1}} \left[2^{n-1} - (n-1) - 1 \right].$$

Thus,

$$\Pr(A_k|B) = \frac{(n-1)!}{2^{n-1}(k-1)!(n-k)!} \frac{2^{n-1}}{2^{n-1} - n} = \frac{(n-1)!}{(2^{n-1} - n)(k-1)!(n-k)!},$$

as given in the question.

30.13 *A set of $2N+1$ rods consists of one of each integer length $1, 2, \ldots, 2N, 2N+1$. Three, of lengths a, b and c, are selected, of which a is the longest. By considering the possible values of b and c, determine the number of ways in which a non-degenerate triangle (i.e. one of non-zero area) can be formed (i) if a is even, and (ii) if a is odd. Combine these results appropriately to determine the total number of non-degenerate triangles that can be formed with the $2N + 1$ rods, and hence show that the probability that such a triangle can be formed from a random selection (without replacement) of three rods is*

$$\frac{(N - 1)(4N + 1)}{2(4N^2 - 1)}.$$

Rod a is the longest of the three rods. As no two are the same length, let $a > b > c$. To form a non-degenerate triangle we require that $b + c > a$, and, in consequence, $4 \le a \le 2N + 1$.

(i) With a even. Consider each b $(< a)$ in turn and determine how many values of c allow a triangle to be made:

b	Values of c	Number of c values
$a - 1$	$2, 3, \cdots, a - 2$	$a - 3$
$a - 2$	$3, 4, \cdots, a - 3$	$a - 5$
$\cdots$	$\cdots$	$\cdots$
$\frac{1}{2}a + 1$	$\frac{1}{2}a$	1

Thus, there are $1 + 3 + 5 + \cdots + (a - 3)$ possible triangles when a is even.

(ii) A table for odd a is similar, except that the last line will read $b = \frac{1}{2}(a+3)$, $c = \frac{1}{2}(a-1)$ or $\frac{1}{2}(a+1)$, and the number of c values $= 2$. Thus there are $2 + 4 + 6 + \cdots + (a-3)$ possible triangles when a is odd.

To find the total number $n(N)$ of possible triangles, we group together the cases $a = 2m$ and $a = 2m+1$, where $m = 1, 2, \ldots, N$. Then,

$$n(N) = \sum_{m=2}^{N} [1 + 3 + \cdots + (2m-3)] + [2 + 4 + \cdots + (2m+1-3)]$$

$$= \sum_{m=2}^{N} \sum_{k=1}^{2m-2} k = \sum_{m=2}^{N} \frac{1}{2}(2m-2)(2m-1) = \sum_{m=2}^{N} 2m^2 - 3m + 1$$

$$= 2\left[\frac{1}{6}N(N+1)(2N+1) - 1\right] - 3\left[\frac{1}{2}N(N+1) - 1\right] + N - 1$$

$$= \frac{N}{6}[2(N+1)(2N+1) - 9(N+1) + 6]$$

$$= \frac{N}{6}(4N^2 - 3N - 1) = \frac{N}{6}(4N+1)(N-1).$$

The number of ways that three rods can be drawn at random (without replacement) is $(2N+1)(2N)(2N-1)/3!$ and so the probability that they can form a triangle is

$$\frac{N(4N+1)(N-1)}{6} \frac{3!}{(2N+1)(2N)(2N-1)} = \frac{(N-1)(4N+1)}{2(4N^2-1)},$$

as stated in the question.

30.15 *The duration (in minutes) of a telephone call made from a public call-box is a random variable T. The probability density function of T is*

$$f(t) = \begin{cases} 0 & t < 0, \\ \frac{1}{2} & 0 \le t < 1, \\ ke^{-2t} & t \ge 1, \end{cases}$$

where k is a constant. To pay for the call, 20 pence has to be inserted at the beginning, and a further 20 pence after each subsequent half-minute. Determine by how much the average cost of a call exceeds the cost of a call of average length charged at 40 pence per minute.

From the normalisation of the PDF, we must have

$$1 = \int_0^\infty f(t)\, dt = \frac{1}{2} + \int_1^\infty ke^{-2t}\, dt = \frac{1}{2} + \frac{ke^{-2}}{2} \quad \Rightarrow \quad k = e^2.$$

The average length of a call is given by

$$\bar{t} = \int_0^1 t\,\frac{1}{2}\,dt + \int_1^\infty t\,e^2 e^{-2t}\,dt$$

$$= \frac{1}{2}\frac{1}{2} + \left[\frac{te^2 e^{-2t}}{-2}\right]_1^\infty + \int_1^\infty \frac{e^2 e^{-2t}}{2}\,dt = \frac{1}{4} + \frac{1}{2} + \frac{e^2}{2}\left[\frac{e^{-2t}}{-2}\right]_1^\infty = \frac{3}{4} + \frac{1}{4} = 1.$$

Let $p_n = \Pr\{\frac{1}{2}(n-1) < t < \frac{1}{2}n\}$. The corresponding cost is $c_n = 20n$. Clearly, $p_1 = p_2 = \frac{1}{4}$ and, for $n > 2$,

$$p_n = e^2 \int_{(n-1)/2}^{n/2} e^{-2t}\,dt = e^2 \left[\frac{e^{-2t}}{-2}\right]_{(n-1)/2}^{n/2} = \frac{1}{2}e^2(e-1)e^{-n}.$$

The average cost of a call is therefore

$$\bar{c} = 20\left[\frac{1}{4} + 2\frac{1}{4} + \sum_{n=3}^\infty n\frac{1}{2}e^2(e-1)e^{-n}\right] = 15 + 10e^2(e-1)\sum_{n=3}^\infty ne^{-n}.$$

Now, the final summation might be recognised as part of an arithmetico-geometric series whose sum can be found from the standard formula

$$S = \frac{a}{1-r} + \frac{rd}{(1-r)^2},$$

with $a = 0$, $d = 1$ and $r = e^{-1}$, or could be evaluated directly by noting that as a geometric series,

$$\sum_{n=0}^\infty e^{-nx} = \frac{1}{1 - e^{-x}}.$$

Differentiating this with respect to x and then setting $x = 1$ gives

$$-\sum_{n=0}^\infty ne^{-nx} = -\frac{e^{-x}}{(1 - e^{-x})^2} \quad\Rightarrow\quad \sum_{n=0}^\infty ne^{-n} = \frac{e^{-1}}{(1 - e^{-1})^2}.$$

From either method it follows that

$$\sum_{n=3}^\infty ne^{-n} = \frac{e}{(e-1)^2} - e^{-1} - 2e^{-2}$$

$$= \frac{e - e + 2 - e^{-1} - 2 + 4e^{-1} - 2e^{-2}}{(e-1)^2} = \frac{3e^{-1} - 2e^{-2}}{(e-1)^2}.$$

The total charge therefore exceeds that of a call of average length (1 minute) charged at 40 pence per minute by the amount (in pence)

$$15 + 10e^2(e-1)\frac{3e^{-1} - 2e^{-2}}{(e-1)^2} - 40 = \frac{10(3e-2) - 25e + 25}{e-1} = \frac{5e + 5}{e-1} = 10.82.$$

30.17 *If the scores in a cup football match are equal at the end of the normal period of play, a 'penalty shoot-out' is held in which each side takes up to five shots (from the penalty spot) alternately, the shoot-out being stopped if one side acquires an unassailable lead (i.e. has a lead greater than its opponents have shots remaining). If the scores are still level after the shoot-out a 'sudden death' competition takes place.*

In sudden death each side takes one shot and the competition is over if one side scores and the other does not; if both score, or both fail to score, a further shot is taken by each side, and so on. Team 1, which takes the first penalty, has a probability p_1, which is independent of the player involved, of scoring and a probability q_1 ($= 1 - p_1$) of missing; p_2 and q_2 are defined likewise.

Let $\Pr(i : x, y)$ be the probability that team i has scored x goals after y attempts, and $f(M)$ be the probability that the shoot-out terminates after a total of M shots.

(a) *Prove that the probability that 'sudden death' will be needed is*

$$f(11+) = \sum_{r=0}^{5} (^5C_r)^2 (p_1 p_2)^r (q_1 q_2)^{5-r}.$$

(b) *Give reasoned arguments (preferably without first looking at the expressions involved) which show that, for $N = 3, 4, 5$,*

$$f(M = 2N) = \sum_{r=0}^{2N-6} \left\{ \begin{array}{l} p_2 \Pr(1 : r, N) \Pr(2 : 5 - N + r, N - 1) \\ + q_2 \Pr(1 : 6 - N + r, N) \Pr(2 : r, N - 1) \end{array} \right\}$$

and, for $N = 3, 4$,

$$f(M = 2N + 1) = \sum_{r=0}^{2N-5} \left\{ \begin{array}{l} p_1 \Pr(1 : 5 - N + r, N) \Pr(2 : r, N) \\ + q_1 \Pr(1 : r, N) \Pr(2 : 5 - N + r, N) \end{array} \right\}.$$

(c) *Give an explicit expression for $\Pr(i : x, y)$ and hence show that if the teams are so well matched that $p_1 = p_2 = 1/2$ then*

$$f(2N) = \sum_{r=0}^{2N-6} \left(\frac{1}{2^{2N}} \right) \frac{N!(N-1)!6}{r!(N-r)!(6-N+r)!(2N-6-r)!},$$

$$f(2N+1) = \sum_{r=0}^{2N-5} \left(\frac{1}{2^{2N}} \right) \frac{(N!)^2}{r!(N-r)!(5-N+r)!(2N-5-r)!}.$$

(d) *Evaluate these expressions to show that, expressing $f(M)$ in units of 2^{-8},*

M	6	7	8	9	10	11+
$f(M)$	8	24	42	56	63	63

Give a simple explanation of why $f(10) = f(11+)$.

(a) For 'sudden death' to be needed the scores must be equal after ten shots, five from each side. A score of r goals each has a probability

$$\left({}^5C_r\, p_1^r\, q_1^{5-r} \right) \times \left({}^5C_r\, p_2^r\, q_2^{5-r} \right),$$

and the total probability that the scores are equal after ten shots is obtained by summing this over all possible values of r ($r = 0, 1, \dots, 5$). Thus

$$f(11+) = \sum_{r=0}^{5} ({}^5C_r)^2 (p_1 p_2)^r (q_1 q_2)^{5-r}.$$

(b) For the shoot-out to terminate after $2N$ shots (≤ 10 shots), one team must be $6 - N$ goals ahead and team 2 must just have taken the last shot.

(i) If team 1 won, it was because team 2 failed with their Nth shot and team 1 must have been $6 - N$ goals ahead before the final shot was taken. The probability for this is $q_2\,\mathrm{Pr}(1 : 6 - N + r, N)\,\mathrm{Pr}(2 : r, N - 1)$.
(ii) If team 2 won, it must have been successful with its last shot and, before it, must have been $5 - N$ goals ahead. The probability for this is $p_2\,\mathrm{Pr}(1 : r, N)\,\mathrm{Pr}(2 : 5 - N + r, N - 1)$.

This type of finish can only arise if $N > 5 - N$, i.e. $N = 3$, 4 or 5. Further, since in $\mathrm{Pr}(i : x, y)$ we must have $x \leq y$, the range for r is determined, from (i), by $6 - N + r \leq N$ and, from (ii), by $5 - N + r \leq N - 1$; these both give $0 \leq r \leq 2N - 6$. Thus

$$f(M = 2N) = \sum_{r=0}^{2N-6} \left\{ \begin{array}{l} p_2\,\mathrm{Pr}(1 : r, N)\,\mathrm{Pr}(2 : 5 - N + r, N - 1) \\ {}+ q_2\,\mathrm{Pr}(1 : 6 - N + r, N)\,\mathrm{Pr}(2 : r, N - 1) \end{array} \right\}.$$

For $M = 2N + 1$, the shoot-out terminates after team 1's $(N + 1)$th shot, which must have been successful if it wins, or unsuccessful if team 2 wins.

(i) If team 1 wins, it must now be $6 - N$ goals ahead, i.e. it was $5 - N$ goals ahead before its successful $(N + 1)$th shot. This has probability $p_1\,\mathrm{Pr}(1 : 5 - N + r, N)\,\mathrm{Pr}(2 : r, N)$.
(i) If team 2 wins, it must have been $5 - N$ goals ahead, before team 1's unsuccessful $(N + 1)$th shot. The probability for this is $q_1\,\mathrm{Pr}(1 : r, N)\,\mathrm{Pr}(2 : 5 - N + r, N)$.

This type of ending can only occur if $N > 5 - N$ and $2N + 1 \leq 10$, i.e. $N = 3$ or 4. Arguing as before, we see that both (i) and (ii) require $5 - N + r \leq N$, i.e. $0 \leq r \leq 2N - 5$. Thus

$$f(M = 2N + 1) = \sum_{r=0}^{2N-5} \left\{ \begin{array}{l} p_1\,\mathrm{Pr}(1 : 5 - N + r, N)\,\mathrm{Pr}(2 : r, N) \\ {}+ q_1\,\mathrm{Pr}(1 : r, N)\,\mathrm{Pr}(2 : 5 - N + r, N) \end{array} \right\}.$$

(c) As in part (a), $\mathrm{Pr}(i : x, y)$ is given by the binomial distribution as

$$\mathrm{Pr}(i : x, y) = {}^y C_x\, p_i^x\, q_i^{y-x}.$$

We now set $p_1 = p_2 = q_1 = q_2 = \frac{1}{2}$ and calculate

$$f(2N) = \sum_{r=0}^{2N-6} \left[\frac{1}{2}\,{}^N C_r\,{}^{N-1}C_{5-N+r}\left(\frac{1}{2}\right)^N\left(\frac{1}{2}\right)^{N-1}\right.$$
$$\left. + \frac{1}{2}\,{}^N C_{6-N+r}\,{}^{N-1}C_r\left(\frac{1}{2}\right)^N\left(\frac{1}{2}\right)^{N-1}\right]$$
$$= \sum_{r=0}^{2N-6} \left(\frac{1}{2^{2N}}\right)\left[\frac{N!\,(N-1)!}{r!\,(N-r)!\,(5-N+r)!\,(2N-6-r)!}\right.$$
$$\left. + \frac{N!\,(N-1)!}{(6-N+r)!\,(2N-6-r)!\,r!\,(N-1-r)!}\right]$$
$$= \sum_{r=0}^{2N-6} \left(\frac{1}{2^{2N}}\right)\frac{N!\,(N-1)!\,[6-N+r+N-r]}{r!\,(N-r)!\,(6-N+r)!\,(2N-6-r)!}$$
$$= \sum_{r=0}^{2N-6} \left(\frac{1}{2^{2N}}\right)\frac{N!\,(N-1)!\,6}{r!\,(N-r)!\,(6-N+r)!\,(2N-6-r)!}.$$

The value of $f(2N+1)$ is found in a similar way. But, since $p_1 = p_2 = q_1 = q_2 = \frac{1}{2}$, the two terms contributing to it for any particular value of r are equal and each has the value

$$\frac{1}{2}\,{}^N C_{5-N+r}\left(\frac{1}{2}\right)^N\,{}^N C_r\left(\frac{1}{2}\right)^N.$$

When these terms are added and then summed over r we obtain

$$f(2N+1) = \sum_{r=0}^{2N-5} \left(\frac{1}{2^{2N}}\right)\frac{(N!)^2}{r!\,(N-r)!\,(5-N+r)!\,(2N-5-r)!}.$$

(d) Evaluating these expressions for the allowed values of N, that is 3, 4 and 5 for $f(2N)$, and 3 and 4 for $f(2N+1)$, is straightforward but somewhat tedious. The results, as given in the question, are

M	6	7	8	9	10	11+
$f(M)$	8	24	42	56	63	63

Here $f(M)$ is expressed in units of 2^{-8}. As expected, these probabilities add up to unity, and it can be seen that sudden death is needed in about one-quarter of such shoot-outs.

The equality of $f(10)$ and f(11+) is simply explained by the fact that, if the shoot-out has not been settled by then, team 2 is just as likely ($p_2 = \frac{1}{2}$) to take it into sudden death by scoring with its fifth shot as it is to lose it ($q_2 = \frac{1}{2}$) by missing.

30.19 *A continuous random variable X has a probability density function $f(x)$; the corresponding cumulative probability function is $F(x)$. Show that the random variable $Y = F(X)$ is uniformly distributed between 0 and 1.*

We first note that, as $F(x)$ is a cumulative probability density function, it has values $F(-\infty) = 0$ and $F(\infty) = 1$ and that $y = F(x)$ has a single-valued inverse $x = x(y)$.

With $Y = F(X)$, we have from the standard result for the distribution of single-valued inverse functions that

$$g(Y) = f(X(Y)) \left| \frac{dX}{dY} \right|.$$

However, in this particular case of Y being the cumulative probability function of X, we can evaluate $|dX/dY|$ more explicitly. This is because

$$\frac{dY}{dX} = \frac{d}{dX} F(X) = \frac{d}{dX} \int_{-\infty}^{X} f(u)\, du = f(X),$$

and is non-negative. So,

$$g(Y) = f(X(Y)) \left| \frac{dX}{dY} \right| = \frac{dY}{dX} \left| \frac{dX}{dY} \right| = 1.$$

This shows that Y is uniformly distributed on $(0, 1)$.

30.21 *This exercise is about interrelated binomial trials.*

(a) *In two sets of binomial trials T and t, the probabilities that a trial has a successful outcome are P and p, respectively, with corresponding probabilites of failure of $Q = 1 - P$ and $q = 1 - p$. One 'game' consists of a trial T, followed, if T is successful, by a trial t and then a further trial T. The two trials continue to alternate until one of the T-trials fails, at which point the game ends. The score S for the game is the total number of successes in the t-trials. Find the PGF for S and use it to show that*

$$E[S] = \frac{Pp}{Q}, \qquad V[S] = \frac{Pp(1 - Pq)}{Q^2}.$$

(b) *Two normal unbiased six-faced dice A and B are rolled alternately starting with A; if A shows a 6 the experiment ends. If B shows an odd number no points are scored, one point is scored for a 2 or a 4, and two points are awarded for a 6. Find the average and standard deviation of the score for the experiment and show that the latter is the greater.*

This is a situation in which the score for the game is a variable length sum, the length N being determined by the outcome of the T-trials. The probability that $N = n$ is given by $h_n = P^n Q$, since n T-trials must succeed and then be followed by a failing T-trial. Thus the PGF for the length of each 'game' is given by

$$\chi_N(t) \equiv \sum_{n=0}^{\infty} h_n t^n = \sum_{n=0}^{\infty} P^n Q t^n = \frac{Q}{1 - Pt}.$$

For each permitted Bernoulli t-trial, $X_i = 1$ with probability p and $X_i = 0$ with probability q; its PGF is thus $\Phi_X(t) = q + pt$. The score for the game is $S = \sum_{i=1}^{N} X_i$ and its PGF is given by the compound function

$$\Xi_S(t) = \chi_N(\Phi_X(t))$$
$$= \frac{Q}{1 - P(q + pt)},$$

in which the PGF for a single t-trial forms the argument of the PGF for the length of each 'game'.

It follows that the mean of S is found from

$$\Xi'_S(t) = \frac{QPp}{(1 - Pq - Ppt)^2} \quad \Rightarrow \quad E[S] = \Xi'_S(1) = \frac{QPp}{(1 - P)^2} = \frac{Pp}{Q}.$$

To calculate the variance of S we need to find $\Xi''_S(1)$. This second derivative is

$$\Xi''_S(t) = \frac{2QP^2p^2}{(1 - Pq - Ppt)^3} \quad \Rightarrow \quad \Xi''_S(1) = \frac{2P^2p^2}{Q^2}.$$

The variance is therefore

$$V[S] = \Xi''_S(1) + \Xi'_S(1) - [\Xi'_S(1)]^2$$
$$= \frac{2P^2p^2}{Q^2} + \frac{Pp}{Q} - \frac{P^2p^2}{Q^2}$$
$$= \frac{Pp(Pp + Q)}{Q^2} = \frac{Pp(P - Pq + Q)}{Q^2} = \frac{Pp(1 - Pq)}{Q^2}.$$

(b) For die A: $P = \frac{5}{6}$ and $Q = \frac{1}{6}$ giving $\chi_N(t) = 1/(6 - 5t)$.

For die B: $\Pr(X = 0) = \frac{3}{6}$, $\Pr(X = 1) = \frac{2}{6}$ and $\Pr(X = 2) = \frac{1}{6}$ giving $\Phi_X(t) = (3 + 2t + t^2)/6$.

The PGF for the game score S is thus

$$\Xi_S(t) = \frac{1}{6 - \frac{5}{6}(3 + 2t + t^2)} = \frac{6}{21 - 10t - 5t^2}.$$

We need to evaluate the first two derivatives of $\Xi_S(t)$ at $t = 1$, as follows:

$$\Xi'_S(t) = \frac{-6(-10 - 10t)}{(21 - 10t - 5t^2)^2} = \frac{60 + 60t}{(21 - 10t - 5t^2)^2}$$

$$\Rightarrow \quad E[S] = \Xi'_S(1) = \frac{120}{6^2} = \frac{10}{3} = 3.33,$$

$$\Xi''_S(t) = \frac{60}{(21 - 10t - 5t^2)^2} - \frac{2(60 + 60t)(-10 - 10t)}{(21 - 10t - 5t^2)^3}$$

$$\Rightarrow \quad \Xi''_S(1) = \frac{60}{36} - \frac{2(120)(-20)}{(6)^3} = \frac{215}{9}.$$

Substituting the calculated values gives $V[S]$ as

$$V[S] = \frac{215}{9} + \frac{10}{3} - \left(\frac{10}{3}\right)^2 = \frac{145}{9},$$

from which it follows that

$$\sigma_S = \sqrt{V[S]} = 4.01, \text{ i.e. greater than the mean.}$$

30.23 *A point P is chosen at random on the circle $x^2 + y^2 = 1$. The random variable X denotes the distance of P from $(1, 0)$. Find the mean and variance of X and the probability that X is greater than its mean.*

With O as the centre of the unit circle and Q as the point $(1, 0)$, let OP make an angle θ with the x-axis OQ. The random variable X then has the value $2\sin(\theta/2)$ with θ uniformly distributed on $(0, 2\pi)$, i.e.

$$f(x)\,dx = \frac{1}{2\pi}\,d\theta.$$

The mean of X is given straightforwardly by

$$\langle X \rangle = \int_0^2 X f(x)\,dx = \int_0^{2\pi} 2\sin\left(\frac{\theta}{2}\right)\frac{1}{2\pi}\,d\theta = \frac{1}{\pi}\left[-2\cos\frac{\theta}{2}\right]_0^{2\pi} = \frac{4}{\pi}.$$

For the variance we have

$$\sigma_X^2 = \langle X^2 \rangle - \langle X \rangle^2 = \int_0^{2\pi} 4\sin^2\left(\frac{\theta}{2}\right)\frac{1}{2\pi}\,d\theta - \frac{16}{\pi^2} = \frac{4}{2\pi}\frac{1}{2}2\pi - \frac{16}{\pi^2} = 2 - \frac{16}{\pi^2}.$$

When $X = \langle X \rangle = 4/\pi$, the angle $\theta = 2\sin^{-1}(2/\pi)$ and so

$$\Pr(X > \langle X \rangle) = \frac{2\pi - 4\sin^{-1}\dfrac{2}{\pi}}{2\pi} = 0.561.$$

30.25 *The number of errors needing correction on each page of a set of proofs follows a Poisson distribution of mean μ. The cost of the first correction on any page is α and that of each subsequent correction on the same page is β. Prove that the average cost of correcting a page is*

$$\alpha + \beta(\mu - 1) - (\alpha - \beta)e^{-\mu}.$$

Since the number of errors on a page is Poisson distributed, the probability of n errors on any particular page is

$$\Pr(n \text{ errors}) = p_n = e^{-\mu} \frac{\mu^n}{n!}.$$

The average cost per page, found by averaging the corresponding cost over all values of n, is

$$c = 0\, p_0 + \alpha p_1 + \sum_{n=2}^{\infty} [\alpha + (n-1)\beta]p_n$$

$$= \alpha\mu e^{-\mu} + (\alpha - \beta)\sum_{n=2}^{\infty} p_n + \beta \sum_{n=2}^{\infty} n p_n.$$

Now, $\sum_{n=0}^{\infty} p_n = 1$ and, for a Poisson distribution, $\sum_{n=0}^{\infty} n p_n = \mu$. These can be used to evaluate the above, once the $n = 0$ and $n = 1$ terms have been removed. Thus

$$c = \alpha\mu e^{-\mu} + (\alpha - \beta)(1 - e^{-\mu} - \mu e^{-\mu}) + \beta(\mu - 0 - \mu e^{-\mu})$$
$$= \alpha + \beta(\mu - 1) + e^{-\mu}(\alpha\mu - \alpha + \beta - \mu\alpha + \mu\beta - \mu\beta)$$
$$= \alpha + \beta(\mu - 1) + e^{-\mu}(\beta - \alpha),$$

as given in the question.

30.27 *Show that for large r the value at the maximum of the PDF for the gamma distribution of order r with parameter λ is approximately $\lambda/\sqrt{2\pi(r-1)}$.*

The gamma distribution takes the form

$$f(x) = \frac{\lambda}{\Gamma(r)} (\lambda x)^{r-1} e^{-\lambda x}$$

and its maximum will occur when $y(x) = x^{(r-1)}e^{-\lambda x}$ is maximal. This requires

$$0 = \frac{dy}{dx} = (r-1)x^{(r-2)}e^{-\lambda x} - \lambda x^{(r-1)}e^{-\lambda x} \quad \Rightarrow \quad \lambda x = r - 1.$$

509

The maximum value is thus

$$\gamma_{\max}(r) = \frac{\lambda}{\Gamma(r)} (r-1)^{(r-1)} e^{-(r-1)}.$$

Now, using Stirling's approximation,

$$\Gamma(n+1) = n! \sim \sqrt{2\pi n} \left(\frac{n}{e}\right)^n \quad \text{for large } n,$$

we obtain

$$\gamma_{\max}(r) \approx \frac{\lambda}{\sqrt{2\pi(r-1)}} \frac{e^{(r-1)}}{(r-1)^{(r-1)}} (r-1)^{(r-1)} e^{-(r-1)}$$

$$= \frac{\lambda}{\sqrt{2\pi(r-1)}}.$$

30.29 *The probability distribution for the number of eggs in a clutch is* $Po(\lambda)$, *and the probability that each egg will hatch is* p *(independently of the size of the clutch). Show by direct calculation that the probability distribution for the number of chicks that hatch is* $Po(\lambda p)$.

Clearly, to determine the probability that a clutch produces k chicks, we must consider clutches of size n, for all $n \geq k$, and for each such clutch find the probability that exactly k of the n chicks do hatch. We then average over all n, weighting the results according to the distribution of n.

The probability that k chicks hatch from a clutch of size n is ${}^nC_k p^k q^{n-k}$, where $q = 1 - p$. The probability that the clutch is of size n is $e^{-\lambda}\lambda^n/n!$. Consequently, the overall probability of k chicks hatching from a clutch is

$$\Pr(k \text{ chicks}) = \sum_{n=k}^{\infty} e^{-\lambda} \frac{\lambda^n}{n!} {}^nC_k p^k q^{n-k}$$

$$= e^{-\lambda} p^k \lambda^k \sum_{n=k}^{\infty} \frac{(\lambda q)^{n-k}}{n!} \frac{n!}{k!(n-k)!}, \quad \text{set } n - k = m,$$

$$= e^{-\lambda} \frac{(\lambda p)^k}{k!} \sum_{m=0}^{\infty} \frac{(\lambda q)^m}{m!}$$

$$= e^{-\lambda} \frac{(\lambda p)^k}{k!} e^{\lambda q}$$

$$= \frac{e^{-\lambda p}(\lambda p)^k}{k!},$$

since $q = 1 - p$. Thus $\Pr(k \text{ chicks})$ is distributed as a Poisson distribution with parameter $\mu = \lambda p$.

30.31 *Under EU legislation on harmonisation, all kippers are to weigh 0.2000 kg and vendors who sell underweight kippers must be fined by their government. The weight of a kipper is normally distributed with a mean of 0.2000 kg and a standard deviation of 0.0100 kg. They are packed in cartons of 100 and large quantities of them are sold.*

Every day a carton is to be selected at random from each vendor and tested according to one of the following schemes, which have been approved for the purpose.

(a) *The entire carton is weighed and the vendor is fined 2500 euros if the average weight of a kipper is less than 0.1975 kg.*

(b) *Twenty-five kippers are selected at random from the carton; the vendor is fined 100 euros if the average weight of a kipper is less than 0.1980 kg.*

(c) *Kippers are removed one at a time, at random, until one has been found that weighs more than 0.2000 kg; the vendor is fined $4n(n-1)$ euros, where n is the number of kippers removed.*

Which scheme should the Chancellor of the Exchequer be urging his government to adopt?

For these calculations we measure weights in grammes.

(a) For this scheme we have a normal distribution with mean $\mu = 200$ and s.d. $\sigma = 10$. The s.d. for a carton is $\sqrt{100}\,\sigma = 100$ and the mean weight is 20000. There is a penalty if the weight of a carton is less than 19750. This critical value represents a standard variable of

$$Z = \frac{19750 - 20000}{100} = -2.5.$$

The probability that $Z < -2.5 = 1 - \Phi(2.5) = 1 - 0.9938 = 0.0062$. Thus the average fine per carton tested on this scheme is $0.0062 \times 2500 = 15.5$ euros.

(b) For this scheme the general parameters are the same but the mean weight of the sample measured is 5000 and its s.d is $\sqrt{25}\,(10) = 50$. The Z-value at which a fine is imposed is

$$Z = \frac{(198 \times 25) - 5000}{50} = -1.$$

The probability that $Z < -1.0 = 1 - \Phi(1.0) = 1 - 0.8413 = 0.1587$. Thus the average fine per carton tested on this scheme is $0.1587 \times 100 = 15.9$ euros.

(c) This scheme is a series of Bernoulli trials in which the probability of success is $\frac{1}{2}$ (since half of all kippers weigh more than 200 and the distribution is normal). The probability that it will take n kippers to find one that passes the test is

$q^{n-1}p = (\frac{1}{2})^n$. The expected fine is therefore

$$f = \sum_{n=2}^{\infty} 4n(n-1)\left(\frac{1}{2}\right)^n = 4\frac{2(\frac{1}{4})}{(\frac{1}{2})^3} = 16 \text{ euros.}$$

The expression for the sum was found by twice differentiating the sum of the geometric series $\sum r^n$ with respect to r, as follows:

$$\sum_{n=0}^{\infty} r^n = \frac{1}{1-r} \quad \Rightarrow \quad \sum_{n=1}^{\infty} nr^{n-1} = \frac{1}{(1-r)^2}$$

$$\Rightarrow \quad \sum_{n=2}^{\infty} n(n-1)r^{n-2} = \frac{2}{(1-r)^3}$$

$$\Rightarrow \quad \sum_{n=2}^{\infty} n(n-1)r^n = \frac{2r^2}{(1-r)^3}.$$

There is, in fact, little to choose between the schemes on monetary grounds; no doubt political considerations, such as the current unemployment rate, will decide!

30.33 *A practical-class demonstrator sends his twelve students to the storeroom to collect apparatus for an experiment, but forgets to tell each which type of component to bring. There are three types, A, B and C, held in the stores (in large numbers) in the proportions 20%, 30% and 50%, respectively, and each student picks a component at random. In order to set up one experiment, one unit each of A and B and two units of C are needed. Let $\Pr(N)$ be the probability that at least N experiments can be setup.*

(a) *Evaluate $\Pr(3)$.*
(b) *Find an expression for $\Pr(N)$ in terms of k_1 and k_2, the numbers of components of types A and B, respectively, selected by the students. Show that $\Pr(2)$ can be written in the form*

$$\Pr(2) = (0.5)^{12} \sum_{i=2}^{6} {}^{12}C_i \, (0.4)^i \sum_{j=2}^{8-i} {}^{12-i}C_j \, (0.6)^j.$$

(c) *By considering the conditions under which no experiments can be set up, show that $\Pr(1) = 0.9145$.*

(a) To make three experiments possible the twelve components picked must be three each of A and B and six of C. The probability of this is given by the multinomial distribution as

$$\Pr(3) = \frac{(12)!}{3!\,3!\,6!}\,(0.2)^3(0.3)^3(0.5)^6 = 0.06237.$$

(b) Let the numbers of A, B and C selected be k_1, k_2 and k_3, respectively, and consider when *at least* N experiments can be set up. We have the obvious inequalities $k_1 \geq N$, $k_2 \geq N$ and $k_3 \geq 2N$. In addition $k_3 = 12 - k_1 - k_2$, implying that $k_2 \leq 12 - 2N - k_1$. Further, k_1 cannot be greater than $12 - 3N$ if at least N experiments are to be set up, as each requires three other components that are not of type A. These inequalities set the limits on the acceptable values of k_1 and k_2 (k_3 is not a third independent variable). Thus $\Pr(N)$ is given by

$$\sum_{k_1 \geq N}^{12-3N} \sum_{k_2 \geq N}^{12-2N-k_1} \frac{(12)!}{k_1!\,k_2!\,(12-k_1-k_2)!}\,(0.2)^{k_1}\,(0.3)^{k_2}\,(0.5)^{12-k_1-k_2}.$$

The answer to part (a) is a particular case of this with $N = 3$, when each summation reduces to a single term.

For $N = 2$ the expression becomes

$$\Pr(2) = \sum_{k_1 \geq 2}^{6} \sum_{k_2 \geq 2}^{8-k_1} \frac{(12)!}{k_1!\,k_2!\,(12-k_1-k_2)!}\,(0.2)^{k_1}\,(0.3)^{k_2}\,(0.5)^{12-k_1-k_2}$$

$$= (0.5)^{12} \sum_{i=2}^{6} \sum_{j=2}^{8-i} \frac{(12)!\,(0.2/0.5)^i}{i!\,(12-i)!}\,\frac{(12-i)!\,(0.3/0.5)^j}{j!\,(12-i-j)!}$$

$$= (0.5)^{12} \sum_{i=2}^{6} {}^{12}C_i\,(0.4)^i \sum_{j=2}^{8-i} {}^{12-i}C_j\,(0.6)^j.$$

(c) No experiment can be set up if any one of the following four events occurs: $A_1 = (k_1 = 0)$, $A_2 = (k_2 = 0)$, $A_3 = (k_3 = 0)$ and $A_4 = (k_3 = 1)$. The probability for the union of these four events is given by

$$\Pr(A_1 \cup A_2 \cup A_3 \cup A_4) = \sum_{i=1}^{4} \Pr(A_i) - \sum_{i,j} \Pr(A_i \cap A_j) + \cdots .$$

The probabilities $\Pr(A_i)$ are straightforward to calculate as follows:

$$\Pr(A_1) = (1 - 0.2)^{12}, \qquad \Pr(A_2) = (1 - 0.3)^{12},$$
$$\Pr(A_3) = (1 - 0.5)^{12}, \qquad \Pr(A_4) = {}^{12}C_1(1 - 0.5)^{12}(0.5).$$

The calculation of the probability for the intersection of two events is typified by

$$\Pr(A_1 \cap A_2) = [1 - (0.2 + 0.3)]^{12}$$
$$\text{and } \Pr(A_1 \cap A_4) = {}^{12}C_1[1 - (0.2 + 0.5)]^{11}(0.5)^1.$$

A few trial evaluations show that these are of order 10^{-4} and can be ignored by comparison with the larger terms in the first sum, which are (after rounding)

$$\sum_{i=1}^{4} \Pr(A_i) = (0.8)^{12} + (0.7)^{12} + (0.5)^{12} + 12(0.5)^{11}(0.5)$$

$$= 0.0687 + 0.0138 + 0.0002 + 0.0029 = 0.0856.$$

Since the probability of no experiments being possible is 0.0856, it follows that $\Pr(1) = 0.9144$.

30.35 *The continuous random variables X and Y have a joint PDF proportional to $xy(x-y)^2$ with $0 \leq x \leq 1$ and $0 \leq y \leq 1$. Find the marginal distributions for X and Y and show that they are negatively correlated with correlation coefficient $-\frac{2}{3}$.*

This PDF is clearly symmetric between x and y. We start by finding its normalisation constant c:

$$\int_0^1 \int_0^1 c(x^3 y - 2x^2 y^2 + xy^3) \, dx \, dy = c \left(\frac{1}{4}\frac{1}{2} - 2\frac{1}{3}\frac{1}{3} + \frac{1}{2}\frac{1}{4} \right) = \frac{c}{36}.$$

Thus, we must have that $c = 36$.

The marginal distribution for x is given by

$$f(x) = 36 \int_0^1 (x^3 y - 2x^2 y^2 + xy^3) \, dy$$

$$= 36(\tfrac{1}{2}x^3 - \tfrac{2}{3}x^2 + \tfrac{1}{4}x)$$

$$= 18x^3 - 24x^2 + 9x,$$

and the mean of x by

$$\mu_X = \bar{x} = \int_0^1 (18x^4 - 24x^3 + 9x^2) \, dx = \frac{18}{5} - \frac{24}{4} + \frac{9}{3} = \frac{3}{5}.$$

By symmetry, the marginal distribution and the mean for y are $18y^3 - 24y^2 + 9y$ and $\frac{3}{5}$, repectively.

To calculate the correlation coefficient we also need the variances of x and y and their covariance. The variances, obviously equal, are given by

$$\sigma_X^2 = \int_0^1 x^2(18x^3 - 24x^2 + 9x) \, dx - (\tfrac{3}{5})^2$$

$$= \frac{18}{6} - \frac{24}{5} + \frac{9}{4} - \frac{9}{25}$$

$$= \frac{900 - 1440 + 675 - 108}{300} = \frac{9}{100}.$$

The standard deviations σ_X and σ_Y are therefore both equal to 3/10.

The covariance is calculated next; it is given by

$$\begin{aligned}
\text{Cov}[X, Y] &= \langle XY \rangle - \mu_X \mu_Y \\
&= 36 \int_0^1 \int_0^1 (x^4 y^2 - 2x^3 y^3 + x^2 y^4)\, dx\, dy - \frac{3}{5}\frac{3}{5} \\
&= \frac{36}{5 \times 3} - \frac{72}{4 \times 4} + \frac{36}{3 \times 5} - \frac{9}{25} \\
&= \frac{12}{5} - \frac{9}{2} + \frac{12}{5} - \frac{9}{25} \\
&= \frac{120 - 225 + 120 - 18}{50} = -\frac{3}{50}.
\end{aligned}$$

Finally,

$$\text{Corr}[X, Y] = \frac{\text{Cov}[X, Y]}{\sigma_X \sigma_Y} = \frac{-\frac{3}{50}}{\frac{3}{10}\frac{3}{10}} = -\frac{2}{3}.$$

30.37 *Two continuous random variables X and Y have a joint probability distribution*

$$f(x, y) = A(x^2 + y^2),$$

where A is a constant and $0 \le x \le a,\ 0 \le y \le a$. Show that X and Y are negatively correlated with correlation coefficient $-15/73$. By sketching a rough contour map of $f(x, y)$ and marking off the regions of positive and negative correlation, convince yourself that this (perhaps counter-intuitive) result is plausible.

The calculations of the various parameters of the distribution are straightforward (see exercise 30.35). The parameter A is determined by the normalisation condition:

$$1 = \int_0^a \int_0^a A(x^2 + y^2)\, dx\, dy = A\left(\frac{a^4}{3} + \frac{a^4}{3}\right) \quad \Rightarrow \quad A = \frac{3}{2a^4}.$$

The two expectation values required are given by

$$\begin{aligned}
E[X] &= \int_0^a \int_0^a Ax(x^2 + y^2)\, dx\, dy \\
&= \frac{3}{2a^4}\left(\frac{a^5}{4 \times 1} + \frac{a^5}{2 \times 3}\right) = \frac{5a}{8}, \qquad (E[Y] = E[X]), \\
E[X^2] &= \int_0^a \int_0^a Ax^2(x^2 + y^2)\, dx\, dy \\
&= \frac{3}{2a^4}\left(\frac{a^6}{5 \times 1} + \frac{a^6}{3 \times 3}\right) = \frac{7a^2}{15}.
\end{aligned}$$

Hence the variance, calculated from the general result $V[X] = E[X^2] - (E[X])^2$, is

$$V[X] = \frac{7a^2}{15} - \left(\frac{5a}{8}\right)^2 = \frac{73}{960}a^2,$$

and the standard deviations are given by

$$\sigma_X = \sigma_Y = \sqrt{\frac{73}{960}}\, a.$$

To obtain the correlation coefficient we need also to calculate the following:

$$E[XY] = \int_0^a \int_0^a Axy(x^2 + y^2)\, dx\, dy$$
$$= \frac{3}{2a^4}\left(\frac{a^6}{4 \times 2} + \frac{a^6}{2 \times 4}\right) = \frac{3a^2}{8}.$$

Then the covariance, given by $\mathrm{Cov}[X, Y] = E[XY] - E[X]E[Y]$, is evaluated as

$$\mathrm{Cov}[X, Y] = \frac{3}{8}a^2 - \frac{5a}{8}\frac{5a}{8} = -\frac{a^2}{64}.$$

Combining this last result with the standard deviations calculated above, we then obtain

$$\mathrm{Corr}[X, Y] = \frac{-(a^2/64)}{\sqrt{\frac{73}{960}}\, a \sqrt{\frac{73}{960}}\, a} = -\frac{15}{73}.$$

As the means of both X and Y are $\frac{5}{8}a = 0.62a$, the areas of the square of side a for which $X - \mu_X$ and $Y - \mu_Y$ have the same sign (i.e. regions of positive correlation) are about $(0.62)^2 \approx 39\%$ and $(0.38)^2 \approx 14\%$ of the total area of the square. The regions of negative correlation occupy some 47% of the square.

However, $f(x, y) = A(x^2 + y^2)$ favours the regions where one or both of x and y are large and close to unity. Broadly speaking, this gives little weight to the region in which both X and Y are less than their means, and so, although it is the largest region in area, it contributes relatively little to the overall correlation. The two (equal area) regions of negative correlation together outweigh the smaller high probability region of positive correlation in the top right-hand corner of the square; the overall result is a net negative correlation coefficient.

30.39 Show that, as the number of trials n becomes large but $np_i = \lambda_i$, $i = 1, 2, \ldots, k - 1$, remains finite, the multinomial probability distribution,

$$M_n(x_1, x_2, \ldots, x_k) = \frac{n!}{x_1! x_2! \cdots x_k!} p_1^{x_1} p_2^{x_2} \cdots p_k^{x_k},$$

can be approximated by a multiple Poisson distribution with $k - 1$ factors:

$$M'_n(x_1, x_2, \ldots, x_{k-1}) = \prod_{i=1}^{k-1} \frac{e^{-\lambda_i} \lambda_i^{x_i}}{x_i!}.$$

(Write $\sum_i^{k-1} p_i = \delta$ and express all terms involving subscript k in terms of n and δ, either exactly or approximately. You will need to use $n! \approx n^\epsilon [(n - \epsilon)!]$ and $(1 - a/n)^n \approx e^{-a}$ for large n.)

(a) Verify that the terms of M'_n add up to unity when summed over all possible values of the random variables $x_1, x_2, \ldots, x_{k-1}$.

(b) If $k = 7$ and $\lambda_i = 9$ for all $i = 1, 2, \ldots, 6$, estimate, using the appropriate Gaussian approximation, the chance that at least three of $x_1, x_2, \ldots, x_6$ will be 15 or greater.

The probabilities p_i are not all independent, and $p_k = 1 - \sum' p_i$, where, for compactness and typographical clarity, we denote $\sum_{i=1}^{k-1}$ by $\sum'$. We further write $\sum' p_i$ as δ. In the same way, we denote $\sum' x_i$ by ϵ and can write $x_k = n - \epsilon$.

Now, as $n \to \infty$ with $p_i \to 0$, whilst the product np_i remains finite and equal to λ_i, we will have that $\delta \to 0$, $n\delta \to \sum' \lambda_i$ and $(n - \epsilon)/n \to 1$. Making these replacements in the factors that contain subscript k gives

$M_n(x_1, x_2, \ldots, x_k)$

$$= \frac{n!}{x_1! x_2! \cdots x_{k-1}! (n - \epsilon)!} p_1^{x_1} p_2^{x_2} \cdots p_{k-1}^{x_{k-1}} (1 - \delta)^{n-\epsilon}$$

$$\approx \frac{n^\epsilon (n - \epsilon)!}{x_1! x_2! \cdots x_{k-1}! (n - \epsilon)!} p_1^{x_1} p_2^{x_2} \cdots p_{k-1}^{x_{k-1}} \left(1 - \frac{n\delta}{n}\right)^{n-\epsilon}$$

$$= \frac{n^{x_1 + x_2 + \cdots + x_{k-1}}}{x_1! x_2! \cdots x_{k-1}!} p_1^{x_1} p_2^{x_2} \cdots p_{k-1}^{x_{k-1}} \left(1 - \frac{n\delta}{n}\right)^{n-\epsilon}$$

$$\to \frac{\lambda_1^{x_1} \lambda_2^{x_2} \cdots \lambda_{k-1}^{x_{k-1}}}{x_1! x_2! \cdots x_{k-1}!} e^{-(\lambda_1 + \lambda_2 + \cdots + \lambda_{k-1})}$$

$$= \prod_{i=1}^{k-1} \frac{e^{-\lambda_i} \lambda_i^{x_i}}{x_i!},$$

i.e. as $n \to \infty$ $M_n(x_1, x_2, \ldots, x_k)$ can be approximated by the direct product of $k - 1$ separate Poisson distributions.

(a) Since the modified expression $M'_n(x_1, x_2, \ldots, x_{k-1})$ consists of this multiple product of factors, the summation between 0 and ∞ over any particular variable, x_j say, can be carried out separately, with the factors not involving x_j treated as constant multipliers. A typical sum is

$$\sum_{x_j=0}^{\infty} \frac{e^{-\lambda_j} \lambda_j^{x_j}}{x_j!} = e^{-\lambda_j} e^{\lambda_j} = 1.$$

When all the summations have been carried out,

$$\sum_{\text{all } x_i} M'_n(x_1, x_2, \ldots, x_{k-1}) = (1)^{k-1} = 1.$$

(b) The Gaussian approximation to each Poisson distribution Po(9) is $N(9,9)$, for which the standard variable is given by

$$Z = \frac{X - 9}{\sqrt{9}}.$$

Thus the probability that one of the x_i will exceed 15 (after including a continuity correction) is

$$\Pr(x_i \geq 15) = \Pr\left(Z > \frac{14.5 - 9}{3}\right) = 1 - \Phi(1.833) = 1 - 0.966 = 0.0334.$$

That (any) three of them should do so has probability

$$^6C_3(0.0334)^3 = 20 \times 3.726 \; 10^{-5} = 7.5 \times 10^{-4}.$$

The probabilities that 4, 5 or 6 of the x_i will exceed 15 make negligible additions to this, which is already an approximation in any case.

31

Statistics

31.1 A group of students uses a pendulum experiment to measure g, the accelera- tion of free fall, and obtains the following values (in m s^{-2}): 9.80, 9.84, 9.72, 9.74, 9.87, 9.77, 9.28, 9.86, 9.81, 9.79, 9.82. What would you give as the best value and standard error for g as measured by the group?

We first note that the reading of 9.28 m s^{-2} is so far from the others that it is almost certainly in error and should not be used in the calculation. The mean of the ten remaining values is 9.802 and the standard deviation of the sample about its mean is 0.04643. After including Bessel's correction factor, the estimate of the population s.d. is $\sigma = 0.0489$, leading to a s.d. in the measured value of the mean of $0.0489/\sqrt{10} = 0.0155$. We therefore give the best value and standard error for g as 9.80 ± 0.02 m s^{-2}.

31.3 The following are the values obtained by a class of 14 students when measur- ing a physical quantity x: 53.8, 53.1, 56.9, 54.7, 58.2, 54.1, 56.4, 54.8, 57.3, 51.0, 55.1, 55.0, 54.2, 56.6.

   (a) *Display these results as a histogram and state what you would give as the best value for x.*
   (b) *Without calculation, estimate how much reliance could be placed upon your answer to (a).*
   (c) *Databooks give the value of x as 53.6 with negligible error. Are the data obtained by the students in conflict with this?*

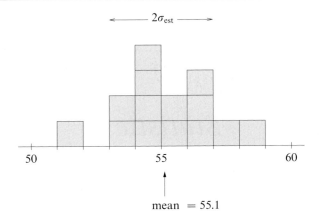

$$\text{mean} = 55.1$$

Figure 31.1   Histogram of the data in exercise 31.3.

(a) The histogram in figure 31.1 shows no reading that is an obvious mistake and there is no reason to suppose other than a Gaussian distribution. The best value for x is the arithmetic mean of the fourteen values given, i.e. 55.1.

(b) We note that eleven values, i.e. approximately two-thirds of the fourteen readings, lie within ± 2 bins of the mean. This estimates the s.d for the population as 2.0 and gives a standard error in the mean of $\approx 2.0/\sqrt{14} \approx 0.6$.

(c) Within the accuracy we are likely to achieve by estimating σ for the sample by eye, the value of Student's t is $(55.1 - 53.6)/0.6$, i.e. about 2.5. With fourteen readings there are 13 degrees of freedom. From standard tables for the Student's t-test, $C_{13}(2.5) \approx 0.985$. It is therefore likely at the $2 \times 0.015 = 3\%$ significance level that the data are in conflict with the accepted value.

[Numerical analysis of the data, rather than a visual estimate, gives the lower value 0.51 for the standard error in the mean and implies that there is a conflict between the data and the accepted value at the 1.0% significance level.]

31.5 *Measured quantities x and y are known to be connected by the formula*

$$y = \frac{ax}{x^2 + b},$$

where a and b are constants. Pairs of values obtained experimentally are

x:	2.0	3.0	4.0	5.0	6.0
y:	0.32	0.29	0.25	0.21	0.18.

Use these data to make best estimates of the values of y that would be obtained for (a) $x = 7.0$, and (b) $x = -3.5$. As measured by fractional error, which estimate is likely to be the more accurate?

In order to use this limited data to best advantage when estimating a and b graphically, the equation needs to be arranged in the linear form $v = mu + c$, since a straight-line graph is much the easiest form from which to extract parameters. The given equation can be arranged as

$$\frac{x}{y} = \frac{x^2}{a} + \frac{b}{a},$$

which is represented by a line with slope a^{-1} and intercept b/a when x^2 is used as the independent variable and x/y as the dependent one. The required tabulation is:

x	2.0	3.0	4.0	5.0	6.0
y	0.32	0.29	0.25	0.21	0.18
x^2	4.0	9.0	16.0	25.0	36.0
x/y	6.25	10.34	16.00	23.81	33.33

Plotting these data as a graph for $0 \le x^2 \le 40$ produces a straight line (within normal plotting accuracy). The line has a slope

$$\frac{1}{a} = \frac{28.1 - 2.7}{30.0 - 0.0} = 0.847 \quad \Rightarrow \quad a = 1.18.$$

The intercept is at $x/y = 2.7$, and, as this is equal to b/a, it follows that $b = 2.7 \times 1.18 = 3.2$. In fractional terms this is not likely to be very accurate as $b \ll x^2$ for all but two of the x-values used.

(a) For $x = 7.0$, the estimated value of y is

$$y = \frac{1.18 \times 7.0}{49.0 + 3.2} = 0.158.$$

(b) For $x = -3.5$, the estimated value of y is

$$y = \frac{1.18 \times (-3.5)}{12.25 + 3.2} = -0.267.$$

Although as a graphical extrapolation estimate (b) is further removed from the measured values, it is likely to be the more accurate because, using the fact that $y(-x) = -y(x)$, it is effectively obtained by (visual) *inter*polation amongst measured data rather than by *extra*polation from it.

> **31.7** *A population contains individuals of k types in equal proportions. A quantity X has mean μ_i amongst individuals of type i and variance σ^2, which has the same value for all types. In order to estimate the mean of X over the whole population, two schemes are considered; each involves a total sample size of nk. In the first the sample is drawn randomly from the whole population, whilst in the second (stratified sampling) n individuals are randomly selected from each of the k types.*
>
> *Show that in both cases the estimate has expectation*
>
> $$\mu = \frac{1}{k} \sum_{i=1}^{k} \mu_i,$$
>
> *but that the variance of the first scheme exceeds that of the second by an amount*
>
> $$\frac{1}{k^2 n} \sum_{i=1}^{k} (\mu_i - \mu)^2.$$

(i) For the first scheme the estimator $\hat{\mu}$ has expectation

$$\langle \hat{\mu} \rangle = \frac{1}{nk} \sum_{j=1}^{nk} \langle x_j \rangle,$$

where

$$\langle x_j \rangle = \frac{1}{k} \sum_{i=1}^{k} \mu_i \text{ for all } j,$$

since the k types are in equal proportions in the population. Thus,

$$\langle \hat{\mu} \rangle = \frac{1}{nk} \sum_{j=1}^{nk} \frac{1}{k} \sum_{i=1}^{k} \mu_i = \frac{1}{k} \sum_{i=1}^{k} \mu_i = \mu.$$

The variance of $\hat{\mu}$ is given by

$$
\begin{aligned}
V[\hat{\mu}] &= \frac{1}{n^2 k^2} \, nk \, V[x] \\
&= \frac{1}{nk} (\langle x^2 \rangle - \mu^2) \\
&= \frac{1}{nk} \left(\frac{1}{k} \sum_{i=1}^{k} \langle x_i^2 \rangle - \mu^2 \right),
\end{aligned}
$$

again since the k types are in equal proportions in the population.

Now we use the relationship $\sigma^2 = \langle x_i^2 \rangle - \mu_i^2$ to replace $\langle x_i^2 \rangle$ for each type, noting that σ^2 has the same value in each case. The expression for the variance becomes

$$V[\hat{\mu}] = \frac{1}{nk} \left[\frac{1}{k} \sum_{i=1}^{k} (\mu_i^2 + \sigma^2) - \mu^2 \right]$$

$$= \frac{\sigma^2 - \mu^2}{nk} + \frac{1}{nk^2} \sum_{i=1}^{k} (\mu_i - \mu + \mu)^2$$

$$= \frac{\sigma^2 - \mu^2}{nk} + \frac{1}{nk^2} \sum_{i=1}^{k} \left[(\mu_i - \mu)^2 + 2\mu(\mu_i - \mu) + \mu^2 \right]$$

$$= \frac{\sigma^2 - \mu^2}{nk} + \frac{1}{nk^2} \sum_{i=1}^{k} (\mu_i - \mu)^2 + 0 + \frac{k\mu^2}{nk^2}$$

$$= \frac{\sigma^2}{nk} + \frac{1}{nk^2} \sum_{i=1}^{k} (\mu_i - \mu)^2.$$

(ii) For the second scheme the calculations are more straightforward. The expectation value of the estimator $\hat{\mu} = (nk)^{-1} \sum_{i=1}^{k} \langle x_i \rangle$ is

$$\langle \hat{\mu} \rangle = \frac{1}{nk} \sum_{i=1}^{k} n\mu_i = \frac{1}{k} \sum_{i=1}^{k} \mu_i = \mu,$$

whilst the variance is given by

$$V[\hat{\mu}] = \frac{1}{n^2 k^2} \sum_{i=1}^{k} V[\langle x_i \rangle] = \frac{1}{n^2 k^2} \sum_{i=1}^{k} n\sigma_i^2 = \frac{1}{k^2} \frac{k\sigma^2}{n} = \frac{\sigma^2}{kn},$$

since $\sigma_i^2 = \sigma^2$ for all i.

Comparing the results from (i) and (ii), we see that the variance of the estimator in the first scheme is larger by

$$\frac{1}{nk^2} \sum_{i=1}^{k} (\mu_i - \mu)^2.$$

31.9 *Each of a series of experiments consists of a large, but unknown, number n ($\gg 1$) of trials, in each of which the probability of success p is the same, but also unknown. In the ith experiment, $i = 1, 2, \ldots, N$, the total number of successes is x_i ($\gg 1$). Determine the log-likelihood function.*

Using Stirling's approximation to $\ln(n - x)$, show that

$$\frac{d \ln(n - x)}{dn} \approx \frac{1}{2(n - x)} + \ln(n - x),$$

and hence evaluate $\partial(^nC_x)/\partial n$.

By finding the (coupled) equations determining the ML estimators $\hat{p}$ and $\hat{n}$, show that, to order n^{-1}, they must satisfy the simultaneous 'arithmetic' and 'geometric' mean constraints

$$\hat{n}\hat{p} = \frac{1}{N} \sum_{i=1}^{N} x_i \quad \text{and} \quad (1 - \hat{p})^N = \prod_{i=1}^{N} \left(1 - \frac{x_i}{\hat{n}}\right).$$

The likelihood function for these N Bernoulli trials is given by

$$L(\mathbf{x}; n, p) = \prod_{i=1}^{N} {}^nC_{x_i}\, p^{x_i}\, (1 - p)^{n - x_i}$$

and the corresponding log-likelihood function is

$$\ln L = \sum_{i=1}^{N} \ln {}^nC_{x_i} + \ln p \sum_{i=1}^{N} x_i + \ln(1 - p) \left[Nn - \sum_{i=1}^{N} x_i\right].$$

The binomial coefficient depends upon n and so we need to determine $\partial(^nC_x)/\partial n$. To do so, we first consider the derivative of $n!$. Stirling's approximation to $n!$ is

$$n! \sim \sqrt{2\pi n} \left(\frac{n}{e}\right)^n, \qquad \text{for large } n.$$

The derivative of n^n is found by setting $y = n^n$ and proceeding as follows:

$$\ln y = n \ln n \quad \Rightarrow \quad \frac{1}{y}\frac{dy}{dn} = \ln n + \frac{n}{n} \quad \Rightarrow \quad \frac{dy}{dn} = n^n(1 + \ln n).$$

It follows that

$$\frac{d(n!)}{dn} = \sqrt{2\pi} \left[\frac{1}{2\sqrt{n}} \left(\frac{n}{e}\right)^n + \frac{\sqrt{n}}{e^n} n^n(1 + \ln n) - \sqrt{n}\, n^n\, e^{-n}\right]$$

$$= \sqrt{2\pi} \left[\frac{1}{2\sqrt{n}} \left(\frac{n}{e}\right)^n + \frac{\sqrt{n}}{e^n} n^n \ln n\right]$$

$$= \sqrt{2\pi n} \left(\frac{n}{e}\right)^n \left[\frac{1}{2n} + \ln n\right] = n! \left(\frac{1}{2n} + \ln n\right).$$

An immediate consequence of this is

$$\frac{d(\ln n!)}{dn} = \frac{1}{n!}\frac{d(n!)}{dn} = \frac{1}{2n} + \ln n.$$

We now return to the log-likelihood function, the first term of which is

$$\sum_{i=1}^{N}\ln{}^{n}C_{x_i} = \sum_{i=1}^{N}[\ln n! - \ln x_i! - \ln(n - x_i)!],$$

with, for large n, a partial derivative with respect to n of

$$\sum_{i=1}^{N}\left[\frac{1}{2n} + \ln n - 0 - \frac{1}{2(n - x_i)} - \ln(n - x_i)\right]$$

$$= \sum_{i=1}^{N}\left[\ln\frac{n}{n - x_i} - \frac{x_i}{2n(n - x_i)}\right].$$

We are now in a position to find the partial derivatives of the log-likelihood function with respect to p and n and equate each of them to zero, thus yielding the equations $\hat{p}$ and $\hat{n}$ must satisfy.

Firstly, differentiating with respect to p gives

$$\frac{\partial(\ln L)}{\partial p} = \frac{1}{p}\sum_{i=1}^{N}x_i - \frac{1}{(1 - p)}\left[Nn - \sum_{i=1}^{N}x_i\right] = 0,$$

$$\frac{N\hat{n}}{1 - \hat{p}} = \left(\frac{1}{\hat{p}} + \frac{1}{1 - \hat{p}}\right)\sum_{i=1}^{N}x_i,$$

$$\frac{1}{\hat{p}}\sum_{i=1}^{N}x_i = N\hat{n} \quad\Rightarrow\quad \hat{n}\hat{p} = \frac{1}{N}\sum_{i=1}^{N}x_i.$$

Secondly, differentiation with respect to n yields

$$\frac{\partial(\ln L)}{\partial n} = \sum_{i=1}^{N}\left[\ln\frac{n}{n - x_i} - \frac{x_i}{2n(n - x_i)}\right] + N\ln(1 - p) = 0.$$

For large n (and, consequently, large x_i), the first term in the square brackets is of zero-order in n whilst the second is of order n^{-1}. Ignoring the second term and recalling that $\ln 1 = 0$, the equation is equivalent to

$$(1 - \hat{p})^{N}\prod_{i=1}^{N}\frac{\hat{n}}{\hat{n} - x_i} = 1 \quad\Rightarrow\quad (1 - \hat{p})^{N} = \prod_{i=1}^{N}\left(1 - \frac{x_i}{\hat{n}}\right).$$

> **31.11** *According to a particular theory, two dimensionless quantities X and Y have equal values. Nine measurements of X gave values of 22, 11, 19, 19, 14, 27, 8, 24 and 18, whilst seven measured values of Y were 11, 14, 17, 14, 19, 16 and 14. Assuming that the measurements of both quantities are Gaussian distributed with a common variance, are they consistent with the theory? An alternative theory predicts that $Y^2 = \pi^2 X$; are the data consistent with this proposal?*

On the hypothesis that $X = Y$ and both quantities have Gaussian distributions with a common variance, we need to calculate the value of t given by

$$t = \frac{\bar{w} - \omega}{\hat{\sigma}} \left(\frac{N_1 N_2}{N_1 + N_2} \right)^{1/2},$$

where $\bar{w} = \bar{x}_1 - \bar{x}_2$, $\omega = \mu_1 - \mu_2 = 0$ and

$$\hat{\sigma} = \left[\frac{N_1 s_1^2 + N_2 s_2^2}{N_1 + N_2 - 2} \right]^{1/2}.$$

The nine measurements of X have a mean of 18.0 and a value for s^2 of 33.33. The corresponding values for the seven measurements of Y are 15.0 and 5.71. Substituting these values gives

$$\hat{\sigma} = \left[\frac{9 \times 33.33 + 7 \times 5.71}{9 + 7 - 2} \right]^{1/2} = 4.93,$$

$$t = \frac{18.0 - 15.0 - 0}{4.93} \left(\frac{9 \times 7}{9 + 7} \right)^{1/2} = 1.21.$$

This variable follows a Student's t-distribution for $9 + 7 - 2 = 14$ degrees of freedom. Interpolation in standard tables gives $C_{14}(1.21) \approx 0.874$, showing that a larger value of t could be expected in about $2 \times (1 - 0.874) = 25\%$ of cases. Thus no inconsistency between the data and the first theory has been established.

For the second theory we are testing Y^2 against $\pi^2 X$; the former will not be Gaussian distributed and the two distributions will not have a common variance. Thus the best we can do is to compare the difference between the two expressions, evaluated with the mean values of X and Y, against the estimated error in that difference.

The difference in the expressions is $(15.0)^2 - 18.0\pi^2 = 47.3$. The error in the difference between the functions of Y and X is given approximately by

$$V(Y^2 - \pi^2 X) = (2Y)^2 V[Y] + (\pi^2)^2 V[X]$$

$$= (30.0)^2 \frac{5.71}{7 - 1} + (\pi^2)^2 \frac{33.33}{9 - 1}$$

$$= 1262 \quad \Rightarrow \quad \sigma \approx 35.5.$$

The difference is thus about $47.3/35.5 = 1.33$ standard deviations away from the theoretical value of 0. The distribution will not be truly Gaussian but, if it were, this figure would have a probability of being exceeded in magnitude some $2 \times (1 - 0.908) = 18\%$ of the time. Again no inconsistency between the data and theory has been established.

31.13 *The χ^2 distribution can be used to test for correlations between characteristics of sampled data. To illustrate this consider the following problem.*

During an investigation into possible links between mathematics and classical music, pupils at a school were asked whether they had preferences (a) between mathematics and english, and (b) between classical and pop music. The results are given below.

	Classical	None	Pop
Mathematics	23	13	14
None	17	17	36
English	30	10	40

By computing tables of expected numbers, based on the assumption that no correlations exist, and calculating the relevant values of χ^2, determine whether there is any evidence for

(a) *a link between academic and musical tastes, and*
(b) *a claim that pupils either had preferences in both areas or had no preference.*

You will need to consider the appropriate value for the number of degrees of freedom to use when applying the χ^2 test.

We first note that there were 200 pupils taking part in the survey. Denoting no academic preference between mathematics and english by NA and no musical preference by NM, we draw up an enhanced table of the actual numbers m_{XY} of preferences for the various combinations that also shows the overall probabilities p_X and p_Y of the three choices in each selection.

	C	NM	P	Total	p_X
M	23	13	14	50	0.25
NA	17	17	36	70	0.35
E	30	10	40	80	0.40
Total	70	40	90	200	
p_Y	0.35	0.20	0.45		

(a) If we now assume the (null) hypothesis that there are no correlations in the data and that any apparent correlations are the result of statistical fluctuations, then the expected number of pupils opting for the combination X and Y is $n_{XY} = 200 \times p_X \times p_Y$. A table of n_{XY} is as follows:

	C	NM	P	Total
M	17.5	10	22.5	50
NA	24.5	14	31.5	70
E	28	16	36	80
Total	70	40	90	200

Taking the standard deviation as the square root of the expected number of votes for each particular combination, the value of χ^2 is given by

$$\chi^2 = \sum_{\text{all XY combinations}} \left(\frac{n_i - m_i}{\sqrt{n_i}} \right)^2 = 12.3.$$

For an $n \times n$ correlation table (here $n = 3$), the $(n-1) \times (n-1)$ block of entries in the upper left-hand can be filled in arbitrarily. But, as the totals for each row and column are predetermined, the remaining $2n - 1$ entries are not arbitrary. Thus the number of degrees of freedom (d.o.f.) for such a table is $(n-1)^2$, here 4 d.o.f. From tables, a χ^2 of 12.3 for 4 d.o.f. makes the assumed hypothesis less than 2% likely, and so it is almost certain that a correlation between academic and musical tastes does exist.

(b) To investigate a claim that pupils either had preferences in both areas or had no preference, we must combine expressed preferences for classical or pop into one set labelled PM meaning 'expressed a musical preference'; similarly for academic subjects. The correlation table is now a 2×2 one and will have only one degree of freedom. The actual and expected ($n_{XY} = 200 p_X p_Y$) data tables are

	PM	NM	Total	p_X
PA	107	23	130	0.65
NA	53	17	70	0.35
Total	160	40	200	
p_Y	0.80	0.20		

	PM	NM	Total
PA	104	26	130
NA	56	14	70
Total	160	40	200

The value of χ^2 is

$$\chi^2 = \frac{(-3)^2}{104} + \frac{(3)^2}{26} + \frac{(3)^2}{56} + \frac{(-3)^2}{14} = 1.24.$$

This is close to the expected value (1) of χ^2 for 1 d.o.f. and is neither too big nor too small. Thus there is no evidence for the claim (or for any tampering with the data!).

31.15 *A particle detector consisting of a shielded scintillator is being tested by placing it near a particle source whose intensity can be controlled by the use of absorbers. It might register counts even in the absence of particles from the source because of the cosmic ray background.*

The number of counts n registered in a fixed time interval as a function of the source strength s is given as:

source strength s:	0	1	2	3	4	5	6
counts n:	6	11	20	42	44	62	61

At any given source strength, the number of counts is expected to be Poisson distributed with mean

$$n = a + bs,$$

where a and b are constants. Analyse the data for a fit to this relationship and obtain the best values for a and b together with their standard errors.

(a) *How well is the cosmic ray background determined?*

(b) *What is the value of the correlation coefficient between a and b? Is this consistent with what would happen if the cosmic ray background were imagined to be negligible?*

(c) *Do the data fit the expected relationship well? Is there any evidence that the reported data 'are too good a fit'?*

Because in this exercise the independent variable s takes only consecutive integer values, we will use it as a label i and denote the number of counts corresponding to $s = i$ by n_i. As the data are expected to be Poisson distributed, the best estimate of the variance of each reading is equal to the best estimate of the reading itself, namely the actual measured value. Thus each reading n_i has an error of $\sqrt{n_i}$, and the covariance matrix N takes the form $N = \text{diag}(n_0, n_1, \ldots, n_6)$, i.e. it is diagonal, but not a multiple of the unit matrix.

The expression for χ^2 is

$$\chi^2(a, b) = \sum_{i=0}^{6} \left(\frac{n_i - a - bi}{\sqrt{n_i}} \right)^2 \quad (*).$$

Minimisation with respect to a and b gives the simultaneous equations

$$0 = \frac{\partial \chi^2}{\partial a} = -2 \sum_{i=0}^{6} \frac{n_i - a - bi}{n_i},$$

$$0 = \frac{\partial \chi^2}{\partial b} = -2 \sum_{i=0}^{6} \frac{i(n_i - a - bi)}{n_i}.$$

As is shown more generally in textbooks on numerical computing (e.g. William H. Press *et al.*, *Numerical Recipes in C*, 2nd edn (Cambridge: Cambridge University Press, 1996), Sect. 15.2), these equations are most conveniently solved by defining the quantities

$$S \equiv \sum_{i=0}^{6} \frac{1}{n_i}, \quad S_x \equiv \sum_{i=0}^{6} \frac{i}{n_i}, \quad S_y \equiv \sum_{i=0}^{6} \frac{n_i}{n_i},$$

$$S_{xx} \equiv \sum_{i=0}^{6} \frac{i^2}{n_i}, \quad S_{xy} \equiv \sum_{i=0}^{6} \frac{in_i}{n_i}, \quad \Delta \equiv SS_{xx} - (S_x)^2.$$

With these definitions (which correspond to the quantities calculated and accessibly stored in most calculators programmed to perform least-squares fitting), the solutions for the best estimators of a and b are

$$\hat{a} = \frac{S_{xx}S_y - S_x S_{xy}}{\Delta},$$

$$\hat{b} = \frac{S_{xy}S - S_x S_y}{\Delta},$$

with variances and covariance given by

$$\sigma_a^2 = \frac{S_{xx}}{\Delta}, \quad \sigma_b^2 = \frac{S}{\Delta}, \quad \text{Cov}(a,b) = -\frac{S_x}{\Delta}.$$

The computed values of these quantities are: $S = 0.38664$; $S_x = 0.53225$; $S_y = 7$; $S_{xx} = 1.86221$; $S_{xy} = 21$; $\Delta = 0.43671$.

From these values, the best estimates of $\hat{a}$, $\hat{b}$ and the variances σ_a^2 and σ_b^2 are

$$\hat{a} = 4.2552, \quad \hat{b} = 10.061, \quad \sigma_a^2 = 4.264, \quad \sigma_b^2 = 0.8853.$$

The covariance is $\text{Cov}(a,b) = -1.2187$, giving estimates for a and b of

$$a = 4.3 \pm 2.1 \quad \text{and} \quad b = 10.06 \pm 0.94,$$

with a correlation coefficient $r_{ab} = -0.63$.

(a) The cosmic ray background must be present, since $n(0) \neq 0$, but its value of about 4 is uncertain to within a factor of 2.

(b) The correlation between a and b is negative and quite strong. This is as expected since, if the cosmic ray background represented by a were reduced towards zero, then b would have to be increased to compensate when fitting to the measured data for non-zero source strengths.

(c) A measure of the goodness-of-fit is the value of χ^2 achieved using the best-fit values for a and b. Direct resubstitution of the values found into (∗) gives $\chi^2 = 4.9$. If the weight of a particular reading is taken as the square root of the predicted (rather than the measured) value, then χ^2 rises slightly to 5.1. In either case the result is almost exactly that 'expected' for 5 d.o.f. – neither too good nor too bad.

There are five degrees of freedom because there are seven data points and two parameters have been chosen to give a best fit.

31.17 *The following are the values and standard errors of a physical quantity $f(\theta)$ measured at various values of θ (in which there is negligible error):*

θ	0	$\pi/6$	$\pi/4$	$\pi/3$
$f(\theta)$	3.72 ± 0.2	1.98 ± 0.1	-0.06 ± 0.1	-2.05 ± 0.1

θ	$\pi/2$	$2\pi/3$	$3\pi/4$	π
$f(\theta)$	-2.83 ± 0.2	1.15 ± 0.1	3.99 ± 0.2	9.71 ± 0.4

Theory suggests that f should be of the form $a_1 + a_2 \cos\theta + a_3 \cos 2\theta$. Show that the normal equations for the coefficients a_i are

$$481.3a_1 + 158.4a_2 - 43.8a_3 = 284.7,$$
$$158.4a_1 + 218.8a_2 + 62.1a_3 = -31.1,$$
$$-43.8a_1 + 62.1a_2 + 131.3a_3 = 368.4.$$

(a) *If you have matrix inversion routines available on a computer, determine the best values and variances for the coefficients a_i and the correlation between the coefficients a_1 and a_2.*

(b) *If you have only a calculator available, solve for the values using a Gauss–Seidel iteration and start from the approximate solution $a_1 = 2$, $a_2 = -2$, $a_3 = 4$.*

Assume that the measured data have uncorrelated errors. The quoted errors are not all equal and so the covariance matrix N, whilst being diagonal, will not be a multiple of the unit matrix; it will be

$$N = \text{diag}(0.04, 0.01, 0.01, 0.01, 0.04, 0.01, 0.04, 0.16).$$

Using as base functions the three functions $h_1(\theta) = 1$, $h_2(\theta) = \cos\theta$ and $h_3(\theta) = \cos 2\theta$, we calculate the elements of the 8×3 response matrix $R_{ij} = h_j(\theta_i)$. To save space we display its 3×8 transpose:

$$R^T = \begin{pmatrix} 1 & 1 & 1 & 1 & 1 & 1 & 1 & 1 \\ 1 & 0.866 & 0.707 & 0.500 & 0 & -0.500 & -0.707 & -1 \\ 1 & 0.500 & 0 & -0.500 & -1 & -0.500 & 0 & 1 \end{pmatrix}$$

Then

$$R^T N^{-1} = \begin{pmatrix} 25 & 100 & 100 & 100 & 25 & 100 & 25 & 6.25 \\ 25 & 86.6 & 70.7 & 50 & 0 & -50 & -17.7 & -6.25 \\ 25 & 50.0 & 0 & -50.0 & -25 & -50 & 0 & 6.25 \end{pmatrix}$$

and

$$R^{T}N^{-1}R = R^{T}N^{-1} \begin{pmatrix} 1 & 1 & 1 \\ 1 & 0.866 & 0.500 \\ 1 & 0.707 & 0 \\ 1 & 0.500 & -0.500 \\ 1 & 0 & -1 \\ 1 & -0.500 & -0.500 \\ 1 & -0.707 & 0 \\ 1 & -1 & 1 \end{pmatrix}$$

$$= \begin{pmatrix} 481.25 & 158.35 & -43.75 \\ 158.35 & 218.76 & 62.05 \\ -43.75 & 62.05 & 131.25 \end{pmatrix}.$$

From the measured values,

$$f = (3.72,\ 1.98,\ -0.06,\ -2.05,\ -2.83,\ 1.15,\ 3.99,\ 9.71)^{T},$$

we need to calculate $R^{T}N^{-1}f$, which is given by

$$\begin{pmatrix} 25 & 100 & 100 & 100 & 25 & 100 & 25 & 6.25 \\ 25 & 86.6 & 70.7 & 50 & 0 & -50 & -17.7 & -6.25 \\ 25 & 50.0 & 0 & -50 & -25 & -50 & 0 & 6.25 \end{pmatrix} \begin{pmatrix} 3.72 \\ 1.98 \\ -0.06 \\ -2.05 \\ -2.83 \\ 1.15 \\ 3.99 \\ 9.71 \end{pmatrix},$$

i.e. $(284.7,\ -31.08,\ 368.44)^{T}$.

The vector of LS estimators of a_i satisfies $R^{T}N^{-1}R\hat{a} = R^{T}N^{-1}f$. Substituting the forms calculated above into the two sides of the equality gives the set of equations stated in the question.

(a) Machine (or manual!) inversion gives

$$(R^{T}N^{-1}R)^{-1} = 10^{-3} \begin{pmatrix} 3.362 & -3.177 & 2.623 \\ -3.177 & 8.282 & -4.975 \\ 2.623 & -4.975 & 10.845 \end{pmatrix}.$$

From this (covariance matrix) we can calculate the standard errors on the a_i from the square roots of the terms on the leading diagonal as ±0.058, ±0.091 and ±0.104. We can further calculate the correlation coefficient r_{12} between a_1 and a_2 as

$$r_{12} = \frac{-3.177 \times 10^{-3}}{0.058 \times 0.091} = -0.60.$$

The best values for the a_i are given by the result of multiplying the column matrix $(284.7, -31.08, 368.44)^T$ by the above inverted matrix. This yields $(2.022, -2.944, 4.897)^T$ to give the best estimates of the a_i as

$$a_1 = 2.02 \pm 0.06, \quad a_2 = -2.99 \pm 0.09, \quad a_3 = 4.90 \pm 0.10.$$

(b) Denote the given set of equations by $\mathsf{A}\mathbf{a} = \mathbf{b}$ and start by dividing each equation by the quantity needed to make the diagonal elements of A each equal to unity; this produces $\mathsf{C}\mathbf{a} = \mathbf{d}$. Then, writing $\mathsf{C} = \mathsf{I} - \mathsf{F}$ yields the basis of the iteration scheme,

$$\mathbf{a}_{n+1} = \mathsf{F}\mathbf{a}_n + \mathbf{d}.$$

We use only the simplest form of Gauss–Seidel iteration (with no separation into upper and lower diagonal matrices).

The explicit form of $\mathsf{C}\mathbf{a} = \mathbf{d}$ is

$$\begin{pmatrix} 1 & 0.3290 & -0.0909 \\ 0.7239 & 1 & 0.2836 \\ -0.3333 & 0.4728 & 1 \end{pmatrix} \begin{pmatrix} a_1 \\ a_2 \\ a_3 \end{pmatrix} = \begin{pmatrix} 0.5916 \\ -0.1421 \\ 2.8072 \end{pmatrix}$$

and

$$\mathsf{F} = \begin{pmatrix} 0 & -0.3290 & 0.0909 \\ -0.7239 & 0 & -0.2836 \\ 0.3333 & -0.4728 & 0 \end{pmatrix}.$$

Starting with the approximate solution $a_1 = 2, a_2 = -2, a_3 = 4$ gives as the result of the first ten iterations

a_1	a_2	a_3
2.000	-2.000	4.000
1.613	-2.724	4.419
1.890	-2.563	4.633
1.856	-2.824	4.649
1.943	-2.804	4.761
1.947	-2.899	4.781
1.980	-2.907	4.827
1.987	-2.944	4.842
2.000	-2.953	4.861
2.005	-2.969	4.870
2.011	-2.975	4.879

This final set of values is in close agreement with that obtained by direct inversion; in fact, after eighteen iterations the values agree exactly to three significant figures. Of course, using this method makes it difficult to estimate the errors in the derived values.

31.19 *The F-distribution $h(F)$ for the ratio F of the variances of two samples of sizes N_1 and N_2 drawn from populations with a common variance is*

$$\left(\frac{n_1}{n_2}\right)^{n_1/2} \frac{F^{(n_1-2)/2}}{B\left(\frac{n_1}{2}, \frac{n_2}{2}\right)} \left(1 + \frac{n_1}{n_2}F\right)^{-(n_1+n_2)/2},$$

where, to save space, we have written $N_1 - 1$ as n_1 and $N_2 - 1$ as n_2.

Verify that the F-distribution $P(F)$ is symmetric between the two data samples, i.e. that it retains the same form but with N_1 and N_2 interchanged, if F is replaced by $F' = F^{-1}$. Symbolically, if $P'(F')$ is the distribution of F' and $P(F) = \eta(F, N_1, N_2)$, then $P'(F') = \eta(F', N_2, N_1)$.

We first write $F^{-1} = F'$ with $|dF| = |dF'|/F'^2$ and rewrite $h(F)$ as

$$\left(\frac{n_1}{n_2}\right)^{n_1/2} \frac{(F')^{-(n_1-2)/2}}{B\left(\frac{n_1}{2}, \frac{n_2}{2}\right)} \left(1 + \frac{n_1}{n_2 F'}\right)^{-(n_1+n_2)/2}$$

$$= \left(\frac{n_1}{n_2}\right)^{n_1/2} \frac{(F')^{-(n_1-2)/2}}{B\left(\frac{n_1}{2}, \frac{n_2}{2}\right)} (F')^{(n_1+n_2)/2} \left(\frac{n_2}{n_1}\right)^{(n_1+n_2)/2} \left(\frac{F' n_2}{n_1} + 1\right)^{-(n_1+n_2)/2}$$

$$= \left(\frac{n_2}{n_1}\right)^{n_2/2} \frac{(F')^{(n_2+2)/2}}{B\left(\frac{n_1}{2}, \frac{n_2}{2}\right)} \left(1 + \frac{n_2 F'}{n_1}\right)^{-(n_1+n_2)/2}.$$

Further,

$$h(F)\,|dF| = \left(\frac{n_2}{n_1}\right)^{n_2/2} \frac{(F')^{(n_2+2)/2}}{B\left(\frac{n_1}{2}, \frac{n_2}{2}\right)} \left(1 + \frac{n_2 F'}{n_1}\right)^{-(n_1+n_2)/2} \frac{1}{F'^2}\,|dF'|$$

$$= \left(\frac{n_2}{n_1}\right)^{n_2/2} \frac{(F')^{(n_2-2)/2}}{B\left(\frac{n_2}{2}, \frac{n_1}{2}\right)} \left(1 + \frac{n_2 F'}{n_1}\right)^{-(n_1+n_2)/2} |dF'|.$$

In the last step we have made use of the symmetry of the beta function $B(x, y)$ with respect to its arguments. To express the final result in the usual F-distribution form, we need to restore n_1 to $N_1 - 1$ and n_2 to $N_2 - 1$, but the symmetry between the data samples has already been demonstrated.